Ergebnisse der Mathematik und ihrer Grenzgebiete

3. Folge · Band 22

A Series of Modern Surveys in Mathematics

Gerd Faltings
Ching-Li Chai

Degeneration of Abelian Varieties

Springer-Verlag
Berlin Heidelberg New York
London Paris Tokyo
Hong Kong Barcelona

Gerd Faltings
Department of Mathematics
Princeton University
Princeton, NJ 08544, USA

Ching-Li Chai
Department of Mathematics
University of Pennsylvania
Philadelphia, PA 19104-6395, USA

Mathematics Subject Classification (1980): 14K10, 14D10

ISBN 3-540-52015-5 Springer-Verlag Berlin Heidelberg New York
ISBN 0-387-52015-5 Springer-Verlag New York Berlin Heidelberg

Library of Congress Cataloging-in-Publication Data
Faltings, Gerd. Degeneration of Abelian Varieties/Gerd Faltings, Ching-Li Chai. p. cm – (Ergebnisse der Mathematik und ihrer Grenzgebiete; 3. Folge, Bd. 22)
Includes bibliographical references.
ISBN 0-387-52015-5 (U.S.)
1. Abelian varieties. 2. Moduli theory. 3. Surfaces, Algebraic-Degenerations. I. Chai, Ching-Li. II. Title. III.Series. QA564.F35 1990 516.3'53–dc20 90-9451 CIP

Printed in the United States of America

2141/3140-543210 Printed on acid-free paper

Foreword

The topic of this book is the theory of degenerations of abelian varieties and its application to the construction of compactifications of moduli spaces of abelian varieties. These compactifications have applications to diophantine problems and, of course, are also interesting in their own right.

Degenerations of abelian varieties are given by maps $G \to S$ with S an irreducible scheme and G a group variety whose generic fibre is an abelian variety. One would like to classify such objects, which, however, is a hopeless task in this generality. But for more specialized families we can obtain more: The most important theorem about degenerations is the stable reduction theorem, which gives some evidence that for questions of compactification it suffices to study semi-abelian families; that is, we may assume that G is smooth and flat over S, with fibres which are connected extensions of abelian varieties by tori. A further assumption will be that the base S is normal, which makes such semi-abelian families extremely well behaved. In these circumstances, we give a rather complete classification in case S is the spectrum of a complete local ring, and for general S we can still say a good deal.

For a complete base $S = \operatorname{Spec}(R)$ (R a complete and normal local domain) the main result about degenerations says roughly that G is (in some sense) a quotient of a covering $\tilde{G}$ by a group of periods. Here $\tilde{G}$ is globally an extension of an abelian scheme A over R by a torus T/R, and the periods form a free abelian group in $\tilde{G}(K)$, K the fraction field of R. To make this more precise we should explain how to form such quotients:

The first example is the Tate-curve, where $\tilde{G} = G_m$ is the multiplicative group, and the periods are of the form $q^{\mathbb{Z}}$, $q \in R$ a non-unit different from zero. To form the quotient $G = \tilde{G}/q^{\mathbb{Z}}$ embed $\tilde{G}$ equivariantly into the projective line $\mathbb{P}^1_R$ over R, with coordinate T. Then blow up the closed subschemes defined by $\{q = 0, T = 0\}$ and $\{q = 0, T = \infty\}$. The resulting R-scheme is still equal to the old projective line over $R[1/q]$, while over $R/(q)$ we have obtained a chain of three copies of $\mathbb{P}^1$ glued together by identifying 0 in one copy to ∞ in the next. Blowing up the "end points" and going on we obtain an R-scheme $\tilde{P}$ (not of finite type) which over $R[1/q]$ (respectively over $R/(q)$) is equal to $\mathbb{P}^1$ (respectively an infinite chain of $\mathbb{P}^1$'s). $\tilde{G}$ still operates on $\tilde{P}$, and multiplication by q (over $R[1/q]$, where q defines a section of $\tilde{G}$) extends to an automorphism of $\tilde{P}$ which over $\{q = 0\}$ shifts the infinite chain of $\mathbb{P}^1$'s by one. It follows that we may form the quotient under q-multiplication of the formal scheme $\tilde{P}_{\text{for}}$ associated to $\tilde{P}$, which defines a formal scheme P_{for} over R. Over $\{q = 0\}$ it is a $\mathbb{P}^1$ with 0 and ∞ identified. It is easily checked that P_{for} is projective and thus algebraic, so we obtain a projective curve P over R. P has genus one and is smooth over $R[1/q]$, so defines an elliptic curve there. Moreover the smooth locus $G \subset P$ is a semi-abelian scheme which extends this elliptic curve. Thus we have given an explicit construction of degenerations, and it is known that for compactifying the moduli space of elliptic curves it suffices to add such degenerations.

Now the next question is how to generalize this to higher dimensions. First, we cannot use any set of periods as in the classical case but have to use some condition related to the existence of a polarization. Then we should again embed $\tilde{G}$ into a $\tilde{P}$, divide the formal completion $\tilde{P}_{\text{for}}$ of $\tilde{P}$ by the periods, show that the result P_{for} is projective and thus algebraic, and finally find $G \subset P$ which is the required degeneration. This sounds technically difficult (especially the construction of $\tilde{P}$), but it turns out that these problems have already been solved, by David Mumford, in 1974. He considered only the most degenerate case, where $\tilde{G}$ is a torus, but all the difficulties already show up there, and we had only to add some bookkeeping to tackle the general case. (As a bibliographical note, Morikawa generalized Tate's curve to higher dimensional abelian varieties in 1962, for the case of base dimension one.)

It remains to show that this construction gives all degenerations. Given such a $G \to S$ it is rather easy to define a $\tilde{G}$, but it is not at all obvious how to find the periods by which to divide. The new observation (which seems to be due to one of the authors) is that the periods are encoded in the coefficients of the theta functions, in analogy to the classical case. In the typical example, where $\tilde{G}$ is a torus, this means that we choose an ample line bundle $\mathcal{L}$ on G. This line bundle has a cubical structure (that is, it satisfies the theorem of the cube), and over the formal completion of G (which is equal to that of $\tilde{G}$, i.e. is a formal torus) this line bundle is trivial as a cubical bundle (this requirement makes the trivialization rather unique). So any algebraic section of $f \in \Gamma(G, \mathcal{L})$ defines a formal function on $G_{\text{for}} = \tilde{G}_{\text{for}}$, which can be expanded in a (convergent) Fourier-series $f = \sum a(\chi) \cdot \chi$ where χ runs over the characters of $\tilde{G}$. It turns out that the coefficients $a(\chi)$ are (multiplicatively) quadratic functions of χ, and that the associated bilinear form defines the periods. Once this is understood, the general case (where $\tilde{G}$ has an abelian part) requires some more notation but no new ideas. Of course, all these facts have a classical analogue over the complex numbers. We apply this to the study of degenerations (as we can construct them explicitly, we should be able to exhibit their properties). What remains to be done is a theory of degenerations over bases which are not normal, or where the fibres are no longer semi-abelian.

The next topic is the arithmetic compactification of the moduli space A_g, of principally polarized abelian varieties of dimension g. To do this one first has to construct this space itself. Over the complex numbers it is the quotient of the Siegel upper half-space $\mathbb{H}_g$ by the group $\mathrm{Sp}(2g, \mathbb{Z})$. Over the integers there are three approaches. The first two (due to D.Mumford) use geometric invariant theory or the theory of theta groups to define abelian varieties by explicitly given equations. This second approach leads to difficulties in characteristic two, due to the fact that it uses in an essential way symmetric ample line bundles. The third approach is due to M. Artin, using formal deformation theory and algebraization. This will be the method of choice in this book.

After constructing A_g we can start to compactify it. For simplicity we choose to ignore the difficulties arising from the non-trivial automorphisms of abelian varieties, which force our objects to be (separated) algebraic stacks (or orbifolds). Over the complex numbers there are two different ways to proceed, which lead to the minimal (= Satake- or Baily-Borel-) compactification A_g^*, or to the toroidal compactification $\overline{A}_g$. We have decided to use the term "minimal compactification" because this captures its properties very well, and because the list of names usually attached to it has become very long. It is the normal projective variety associated to the graded algebra of Siegel modular forms. As a set it is equal to the disjoint union of A_{g-r} for $0 \leq r \leq g$. The toroidal compactification (constructed in [AMRT]) $\overline{A}_g$ is a compact complex analytic space (in fact an algebraic space, or Moishezon space) which contains A_g as a dense open subset, whose complement is a divisor with normal crossings. It is not unique but depends on the choice of a "cone decomposition". Neither the minimal nor the toroidal compactification are known to represent any easily described functor (as happens for the compactification of the moduli space of curves), which makes it difficult to define them algebraically.

However after our study of degenerations the toroidal compactifications appear naturally as parameter spaces for the possible periods of degenerations. This enables us to construct them over the integers, following M. Artin's method: We first construct the completed local rings of those spaces, then algebraize and glue in the étale topology. We obtain algebraic spaces (more precisely algebraic stacks) which are proper and smooth over $\mathbb{Z}$. The universal abelian variety over A_g extends to a semi-abelian scheme over $\overline{A}_g$. Furthermore these compactifications are stratified with smooth strata, and their formal completions along any stratum can be described explicitly. Also, over such a completion the universal semi-abelian scheme is given by the Mumford construction, applied to explicitly given degeneration data. As a corollary of the existence of $\overline{A}_g$ we obtain, for example, that in any characteristic A_g is geometrically connected.

Once we have $\overline{A}_g$ we can start to construct the minimal compactification A_g^* $(g \geq 2)$ over the integers. We define it as the projective scheme associated to the graded algebra of integral modular forms. These forms are global sections of powers of a line bundle ω on $\overline{A}_g$ which is given by the invariant g-differentials on the universal semi-abelian scheme over $\overline{A}_g$. Our construction provides an arithmetic definition for them, and also the classical q-expansion becomes for us "evaluation on Mumford families".

However this definition of A_g^* does not reveal very much about its properties. To show that this algebra is finitely generated, and to describe A_g^* in more detail, we have to use certain ampleness properties of the line bundle ω. For example we have to show that some power of ω is generated by global sections, which amounts to finding some way to construct such sections. Classically one uses Poincaré and Eisenstein series, but of course these analytic devices cannot be used here. Instead we use thetaseries, whose algebraic definition comes from the relation between ω and theta nulls. This defines global sections of powers of ω over A_g, which extend to the compactification $\overline{A}_g$ by the q-expansion principle. That some power of ω is globally generated now follows from the fact that the theta divisor on an abelian variety cannot contain all torsion points. So the construction of A_g^* uses in an essential way the toroidal compactification $\overline{A}_g$ (classically one goes the opposite way). However, otherwise A_g^* has the "same" properties relative to $\mathbb{Z}$ as over the complex numbers. For example, it is stratified with strata A_{g-r} for $0 \leq r \leq g$. The A_g^* have applications in diophantine geometry, for instance, in defining the modular height of an abelian variety. We also have variants of $\overline{A}_g$ and A_g^* which incorporate level-n structures. These are only defined over $\mathbb{Z}[e^{2\pi i/n}, 1/n]$, and again the main properties of their generic fibres extend "by continuity".

These ideas apply to other Shimura varieties, if they are defined by moduli problems involving abelian varieties, polarizations and endomorphisms. An example is the case of Hilbert-Blumenthal varieties, where the toroidal compactification had been constructed previously by M. Rapoport. Another example is the case of non principal polarizations. However, in these cases there are usually some essential primes of bad reduction, where our methods do not apply, so that we only obtain good compactifications over $\mathbb{Z}[1/n]$ for some (well-determined) number n (n can be taken to be the product of the discriminant of the polarization and the discriminant of the endomorphism ring). As it is not clear in what generality one should pursue this question, and as our ideas usually either carry over immediately, or we are lead to very hard new problems which require new methods, we leave these generalizations to the reader.

Now let us explain the plan of this book. The first chapter gives references for certain well-known and (maybe) not so well-known facts about abelian and semi-abelian schemes. We mention questions of projectivity, the theory of toral parts of semi-abelian varieties, and the relation between theta nulls and differentials (here we improve a little bit on previous expositions). The reader might wish to skip this chapter, and only refer to it when needed. In chapter II we explain how degenerations induce periods. For simplicity we first treat the completely degenerate case (where $\tilde{G}$ is a torus), and then give the general argument. Improvements over previously published accounts include a treatment of polarizations of degree > 1 and in characteristic 2. In chapter III we give the converse, that is Mumford's construction. Here we follow Mumford's exposition closely. We also show that Mumford's construction gives all degenerations. As an application we compute the Kodaira-Spencer class of a degeneration, in terms of its degeneration data. This is used in the following chapter to prove "openness of versality". Furthermore for any semi-abelian degeneration (not necessarily over a complete base) we define a bilinear form $\underline{B}$ which gives periods up to units, and we compute this form for the Jacobians of semi-stable curves. In the next two chapters we construct the compactifications $\overline{A}_g$ (in chapter IV) and A_g^* (in V). We also discuss applications to diophantine geometry, the projectivity of $\overline{A}_g$ (in some cases), and certain extension theorems for semi-abelian schemes (which give some justification to the name "minimal compactification" for A_g^*).

Next (chapter VI) we study equivariant sheaves on A_g and $\overline{A}_g$. Over A_g these are given by direct images of constant sheaves on products of finitely many copies of the universal abelian variety, and to study extensions we compactify these products (using toroidal methods, of course). The sheaves in question can be either vector bundles, or crystals, or étale sheaves. The vector bundles are associated to representations of the algebraic group $\mathrm{GL}(g)$, and their cohomology groups (over $\overline{A}_g$) generalize modular forms. The crystalline or étale sheaves (both theories are entirely parallel) correspond to representations of the algebraic group $\mathrm{GSp}(2g)$. Connections between these theories are given by complex Hodge theory and the theory of Fontaine-Messing. In the construction of compactifications we also use the vector bundles of theta functions. However these are not associated to homogeneous bundles, and we do not study them in detail.

In the last chapter (VII) we explain how the algebra of Hecke operators acts on all these sheaves and their cohomologies. As there is still an enormous gap between what is conjectured and what is known, we mostly limit ourselves to giving the basic definitions. The only real result is the equation for Frobenius over the Hecke algebra, which previously had been obtained by P. Deligne using special points.

During our work on this book we received support from various sources, including the National Science Foundation, the Simon Guggenheim Memorial Foundation and the Institute for Advanced Study. We thank them for their support.

Many mathematicians, including Bosch, Elkik, Lang, Larsen, Lütkebohmert, Moret-Bailly, Rapoport, Raynaud and Taylor, have observerved inaccuracies in the manuscript and suggested improvements. We thank them all. We are especially indebted to M. Larsen, who edited the whole manuscript.

Princeton, New Jersey
August 1990

Gerd Faltings
Ching-Li Chai

Note

The reader is assumed to be familiar with basic algebraic geometry, up to the level of [EGA] I, II, III, or Hartshorne's text book. The method of descent is used freely, for which the reader can consult [SGA 1] exposé VI, VIII. Standard facts about abelian varieties can be found in [Mum 1], [L 1], and [Weil]. For abelian schemes, the reader can consult [GIT] chapter 6, and as exercise deduce basic properties of abelian schemes by base change to algebraically closed fields. The above constitute the prerequisite.

Chapters II to V form a logical sequence, and should be read in order. Mumford's paper [Mum 5] is essential for chapter III, and reprinted as an appendix. The reader will find it very helpful to know basic facts about torus embeddings (up to [KKMS] chapter I) when reading chapter IV. Knowledge about toroidal compactification over $\mathbb{C}$ (for instance [AMRT] or [Na 3]) is helpful for reading chapter IV but not logically necessary. Similarly knowledge about classical Siegel modular forms is desirable but not indispensable for reading chapter V. Results about Eichler integral in chapter VI can be read to some extent independently, if the reader knows something about toroidal compactification over $\mathbb{C}$ and is willing to accept results on formal cohomologies in chapter VI section 2. Chapter VII is totally independent of the rest of the book, and only requires familiarity with axiomatic properties of étale cohomology.

Table of Contents

Chapter I. Preliminaries

1. Abelian Schemes and Polarizations

1.1. Definition. An *abelian scheme* is a group scheme $\pi : A \to S$ which is smooth, proper with (geometrically) connected fibres. A basic fact is that an abelian scheme is actually a commutative group scheme.

1.2. Remark. (a) An abelian scheme is automatically of finite presentation, because properness implies quasi-compactness and smoothness implies local finiteness of presentation. Hence the technique of reduction to the noetherian case of EGA IV §8 can be applied.

(b) By the rigidity lemma of GIT prop. 6.1., every abelian scheme is commutative.

(c) The word "geometric" in the above definition can be removed, because a group scheme G over a field k is connected if $G \otimes_k \overline{k}$ is connected, where $\overline{k}$ = algebraic closure of k.

A fundamental fact about abelian schemes is the *theorem of cube*, which we formulate as follows:

1.3. Theorem. *Let $\pi : A \to S$ be an abelian scheme, $\mathcal{L}$ be an invertible sheaf on A. For any subset $I \subseteq \{1,2,3\}$, let $m_I : A \times_S A \times_S A \to A$ be the morphism "adding the i-th factor for all $i \in I$"; for example $m_\varnothing : A \to A$ is the zero morphism. Then the invertible sheaf*

$$\Theta(\mathcal{L}) = \otimes_{I \subseteq \{1,2,3\}} m_I^* \mathcal{L}^{\otimes (-1)^{\mathrm{Card}(I)}}$$

on $A \times_S A \times_S A$ is canonically *trivial, where I runs through all subsets of $\{1,2,3\}$.*

Mumford introduced the concept of G_m-*biextensions* in [Mum 3], which was then extensively analyzed and amplified in [SGA7 I] exp. VII, VIII by Grothendieck. Roughly speaking, a G_m-biextension of two commutative group schemes

G, H over S is a G_m-torsor over $G \times_S H$ together with two compatible partial group laws. The second and third sections of [Mum 3] give an excellent introduction to biextensions; see also the first few pages of [Br]. Exposés VII and VIII of [SGA7 I] contain a thorough treatment.

Breen formulated the notion of *cubical structures* in [Br]. Roughly speaking, an invertible sheaf $\mathcal{L}$ on a group scheme is cubical if it satisfies the theorem of cube. More precisely, a cubical structure on a G_m-torsor $\mathcal{L}$ over a commutative group scheme G is an isomorphism

$$\mathcal{O} \xrightarrow{\sim} \Theta(\mathcal{L}) \left(= \otimes_{I \subseteq \{1,2,3\}} m_I^* \mathcal{L}^{\otimes(-1)^{\mathrm{Card}(I)}}\right)$$

of G_m-torsors over G^3 which gives the G_m-torsor

$$\Lambda(\mathcal{L}) = \otimes_{I \subseteq \{1,2\}} m_I^* \mathcal{L}^{\otimes(-1)^{\mathrm{Card}(I)}}$$

over G^2 the structure of a symmetric G_m-biextension. Here we have used notations as in 1.1. See [Br] for detailed exposition. A major merit is that this notion somewhat rigidifies the notion of G_m-torsors, thereby making it easier to perform descent. Also the notions of theta groups and addition formulae for theta functions are naturally related to cubical structures.

1.4. Definition. Let X/S be an abelian scheme with unit-section $e : S \to X$.

(a) For any invertible sheaf $\mathcal{L}$, a *rigidification* of $\mathcal{L}$ is an isomorphism $\xi : \mathcal{O}_S \xrightarrow{\sim} e^*\mathcal{L}$.

(b) The relative Picard functor $\mathbf{Pic}(X/S)$ is canonically isomorphic to $\mathbf{Pic}_e(X/S) : T/S \rightsquigarrow$ {isomorphism classes of invertible sheaves $\mathcal{L}$ on $X \times_S T$ with rigidification along $e_T = e \times_S T$.}

(c) The subfunctor $\mathbf{Pic}^0(X/S)$ of $\mathbf{Pic}(X/S)$ is canonically isomorphic to $\mathbf{Pic}_e^0 : (X/S) : T/S \rightsquigarrow$ {isomorphism classes of invertible sheaves $\mathcal{L}$ on $X \times_S T$ with rigidification along e_T such that $\forall t \in T$, $\mathcal{L} \otimes \kappa(t)$ is algebraically equivalent to zero on X_t.}

1.5. Remark. (a) $\mathbf{Pic}(X/S)$ and $\mathbf{Pic}^0(X/S)$ are fppf sheaves over S. Let $f : X \to S$ be the structural morphism. By definition, $\mathbf{Pic}(X/S) = Rf_{*\mathrm{fppf}} G_m$ = the sheaf associated to the presheaf $T/S \mapsto \mathrm{Pic}(X \times_S T)$ w.r.t. the fppf topology. Since X has a section e, this is canonically isomorphic to $\mathbf{Pic}_e(X/S)$.

(b) $\mathbf{Pic}^0(X/S)$ is an open subfunctor of $\mathbf{Pic}(X/S)$.

(c) $\mathbf{Pic}^0(X/S)$ is formally smooth over S. In other words, given any nilpotent immersion $T \subset T'$, the restriction map $\mathbf{Pic}^0(X/S)(T') \to \mathbf{Pic}^0(X/S)(T')$ is surjective. However, $\mathbf{Pic}(X/S)$ is not smooth over S in general because the Picard number may jump.

Digressive Discussion About Representability of the Picard Functor of an Abelian Scheme A/S

Recall (EGA III 7.8.1) that $f : X \to S$ is called cohomologically flat in dimension zero if for every $S' \to S$ we have $f'_* \mathcal{O}_{X'} = \mathcal{O}_{S'} \otimes f_* \mathcal{O}_X$, where $X' = X \times_S S'$.

We have the following general result of Artin (c. [Ar 4]): Let $f : X \to S$ be proper, flat, finitely presented and cohomologically flat in dimension zero, where X and S are both algebraic spaces locally of finite type over a field or over an excellent Dedekind domain. Then the relative Picard functor $\mathbf{Pic}(X/S)$ is represented by an algebraic space locally of finite presentation over S, denoted by $\mathrm{Pic}(X/S)$. Denote by $\mathrm{Pic}^0(X/S)$ the open subspace corresponding to $\mathbf{Pic}^0(X/S)$.

Let A/S be an abelian scheme. For simplicity, we will denote by A^t the algebraic space $\mathrm{Pic}^0(A/S)$. A^t is proper and smooth over S, with a group structure, i.e. is an abelian algebraic space. The tautological invertible sheaf $\mathcal{P}_A$ on $A \times_S A^t$ corresponding to $\mathrm{id} : A^t \to A^t$ is called the Poincaré sheaf. $\mathcal{P}_A$ is rigidified both along $e \times_S A^t$ and $A \times_S e$, where e denotes the zero section of A and A^t. Moreover, the G_m-torsor associated to the Poincaré sheaf, still denoted by $\mathcal{P}_A$, is the canonical G_m-biextension of $A \times_S A^t$. (See [SGA7 I] exposé VII, VIII for details.)

The biduality theorem says that the natural map $A \to A^{tt} = (A^t)^t$ is an isomorphism, and $\mathcal{P}_A$ is (canonically isomorphic to) the Poincaré sheaf of A^t via the above isomorphism.

If A^t is a scheme, then it is an abelian scheme, called the dual abelian scheme of A/S. Whether A^t is always a scheme is a very technical question (It turns out that the answer is always "Yes".)

Let A/S be an abelian scheme. Here is a list of cases when A^t is known to be a scheme, with a published proof:

(a) $S = \mathrm{Spec}(R)$, where R is an artinian ring. The point is that the property of being a scheme depends only on the reduced structure. (Cf. [Kn])

(b) A is locally projective over S. This follows from the theory of Hilbert schemes.

(c) S is geometrically unibranch, e.g. S is normal. In this case, a theorem of Grothendieck asserts that A is globally projective over S. (Cf. [Ra 3] theorem XI 1.4.)

Notice that in all three cases above $\mathbf{Pic}(A/S)$ is also a scheme.

For the reader's convenience, we recall that an algebraic space X over a base scheme S is a sheaf in the big étale site over S, such that X is locally represented by a scheme. In other words, there is an S-scheme U and a morphism $U \to X$ such that for every scheme V and any morphism $V \to X$ the sheaf $V \times_X U$ is represented by a scheme W étale over V. For details see [Kn]. The definition of abelian scheme can be immediately generalized to abelian algebraic space. See the first few pages of [D 6] for a nice discussion.

In fact, Raynaud has proved that any abelian algebraic space over a scheme is automatically an abelian scheme. In particular, for any abelian scheme A/S, A^t is again a scheme. Raynaud's original proof uses non-flat descent, and was never published. At the end of this section, we shall sketch Deligne's proof, communicated to us by M. Raynaud.

Let $\mathcal{L}$ be an invertible sheaf on A. $\mathcal{L}$ defines a morphism

$$\lambda(\mathcal{L}) : A \to A^t \quad \text{given by} \quad a \mapsto T_a^*(\mathcal{L}) \otimes \mathcal{L}^{-1} \otimes a^*(\mathcal{L})^{-1} \otimes e^*\mathcal{L}$$

for any functorial point $a \in A(Z)$, for any scheme Z over S, where $T_a : A \to A$ is the morphism "translation by a". In other words, $\lambda(\mathcal{L}) \in A^t(A/S)$ corresponds to $\mu^*\mathcal{L} \otimes p_1^*\mathcal{L}^{\otimes -1} \otimes p_2^*\mathcal{L}^{\otimes -1} \otimes \varepsilon^*\mathcal{L}$ on $A \times_S A$ if we regard the second factor as parameter space, where $\mu, p_1, p_2, \varepsilon : A \times_S A \to A$ are the group law, the two projections, and the zero homomorphism respectively. $\lambda(\mathcal{L})$ is, in fact, a homomorphism w.r.t. the group structure on A and A^t, by the theorem of cube (Cf. [Mum 1] chap. 3, p. 91).

1.6. Definition. A *polarization* of $A \to S$ is a homomorphism $\lambda : A \to A^t$ such that for each geometric point $\overline{s}$ of S, $\lambda_{\overline{s}} = \lambda(\mathcal{L}_{\overline{s}})$ for some ample invertible sheaf $\mathcal{L}_{\overline{s}}$ on $A_{\overline{s}}$. Equivalently, locally on S_{fppf} (resp. $S_{\text{ét}}$, resp. S_{fpqc}) λ is defined by some ample invertible sheaf. On S_{fppf} (resp. $S_{\text{ét}}$, resp. S_{fpqc}), the invertible sheaves on A giving rise to λ form a principal homogeneous space under A^t (i.e. an A^t-torsor). Any polarization λ is symmetric: $\lambda^t = \lambda$ via the canonical isomorphism $A \xrightarrow{\sim} A^{tt}$.

If $\lambda : A \to A^t$ is a polarization, λ is finite and faithfully flat (i.e. is an isogeny). The pull-back of the Poincaré sheaf by $(\mathrm{id}_A, \lambda) : A \to A \times_S A^t$ is a symmetric, ample, invertible sheaf $\mathcal{L}$ such that $\lambda(\mathcal{L}) = 2\lambda$. We say that λ is a *principal polarization* if λ is an isomorphism.

If $\mathcal{L}$ is a relatively ample invertible sheaf on A, then $\pi_*(\mathcal{L})$ is a locally free sheaf on S. The rank of $\pi_*(\mathcal{L})$ is locally constant on S w.r.t. the Zariski topology S_{Zar}. Similarly $\deg(\lambda(\mathcal{L}))$ is locally constant on S w.r.t. S_{Zar}. In fact $\lambda(\mathcal{L}) : A \to A^t$ is a polarization and $\mathrm{rk}(\pi_*(\mathcal{L}))^2 = \deg(\lambda(\mathcal{L}))$; this is a consequence of the theorem of the cube. $\lambda(\mathcal{L})$ is an isomorphism iff $\pi_*(\mathcal{L})$ is an invertible sheaf, in which case $\lambda(\mathcal{L})$ is a principal polarization. The pull-back of the Poincaré sheaf by $(\mathrm{id}_A, \lambda(\mathcal{L})) : A \to A \times_S A^t$ is canonically isomorphic to $\mathcal{L} \otimes [-1]^*\mathcal{L}$, where $[n] : A \to A$ denotes the homomorphism "multiplication by n" $\forall n \in \mathbb{Z}$.

Another way to describe polarizations is via divisorial correspondences:

1.7. Definition. Let X_1, X_2 be schemes over a base scheme S. By definition, $\mathbf{Corr}_S(X_1, X_2)$ is the fppf sheaf cokernel of

$$\mathrm{can} : \mathbf{Pic}(X_1/S) \times \mathbf{Pic}(X_2/S) \to \mathbf{Pic}(X_1 \times_S X_2/S) ,$$

the canonical map defined by pull-backs. When X_1, X_2 have sections s_1, s_2, we can use the rigidified version of **Pic** and see easily that for any T/S, the sequence

$$\begin{aligned} 0 \to \mathbf{Pic}(X_1/S)(T) \times \mathbf{Pic}(X_2/S)(T) &\to \mathbf{Pic}(X_1 \times_S X_2/S)(T) \\ &\to \mathbf{Corr}_S(X_1, X_2)(T) \to 0 \end{aligned}$$

is exact. Moreover, we have canonical isomorphisms

$$\mathrm{Hom}_{s_1}(X_1, \mathbf{Pic}(X_2/S)) \xleftarrow{\sim} \mathbf{Corr}_S(X_1, X_2) \xrightarrow{\sim} \mathrm{Hom}_{s_2}(X_2, \mathbf{Pic}(X_1/S)) .$$

Here $\mathrm{Hom}_{s_1}(X_1, \mathbf{Pic}(X_2/S))$ denotes morphisms of the sheaf X_1 to the sheaf $\mathbf{Pic}(X_2/S)$ sending s_1 to the unit section and likewise for $\mathrm{Hom}_{s_2}(X_2, \mathbf{Pic}(X_1/S))$.

Applying to an abelian scheme A/S, we see that the fppf sheaf of polarizations, $\mathbf{Pol}(A/S)$, is a subsheaf of $\mathbf{Corr}_S(A, A)$. $\mathbf{Corr}_S(A, A)$ is canonically isomorphic to $\mathbf{Pic}_{e,e}(A \times_S A)$, and is a disjoint union of finite, unramified schemes over S.

Let $\phi : A/S \to B/S$ be an isogeny between abelian schemes, and let $\phi^t : B^t \to A^t$ denote the dual isogeny. Then *Cartier duality* says that $\ker(\phi)$ and $\ker(\phi^t)$ are canonically dual to each other. In other words, there is a canonical non-degenerate pairing $\ker(\phi) \times \ker(\phi^t) \to G_m$ giving rise to canonical isomorphisms $\ker \phi \xrightarrow{\sim} \mathbf{Hom}_{S-gp}(\ker \phi^t, G_m)$ and $\ker(\phi^t) \xrightarrow{\sim} \mathbf{Hom}_{S-gp}(\ker(\phi), G_m)$. For instance if $\lambda : A/S \to A^t/S$ is a polarization, then $\ker \lambda$ is a subgroup scheme of A/S self-dual under the Cartier pairing. Another example is $\phi = [n]_A =$ multiplication by n, which gives a non-degenerate pairing $A[n] \times A^t[n] \to G_m$, called the Weil pairing. When λ is a principal polarization, we get a canonical non-degenerate skew symmetric pairing $A[n] \times A[n] \to \mu_n$, also called the Weil pairing. (There is always a choice of sign in the pairing. We shall pin down this sign when we have to. A good reference on Cartier duality is [Oda] §1.)

1.8. Definition. Let $A \to S$ be an abelian scheme of relative dimension $g, 1 \leq n \in \mathbb{N}$. A *principal level n structure* is an isomorphism $\alpha : (\mathbb{Z}/n\mathbb{Z})^{2g} \xrightarrow{\sim} A[n]$, where $A[n] = \operatorname{Ker}[n]$ is the subgroup of n-torsion points.

Such an isomorphism exists only if n is invertible in $\mathcal{O}_S$. A lemma of Serre [Se 2] says that if $n \geq 3$, a principally polarized abelian scheme with level n structure has no non-trivial automorphism which preserves both the principal polarization and the level structure.

If there is a prime $p|n$ such that the residue characteristic of some point s of S is p, then there is no principal level n structure on $A \to S$, in the original sense described above. A variant of this is the following notion of *symplectic level n structure*: it is a symplectic isomorphism

$$\beta : (\mathbb{Z}/n\mathbb{Z})^g \times \mu_n^g \xrightarrow{\sim} A[n]$$

where $(\mathbb{Z}/n\mathbb{Z})^g \xrightarrow{\sim} \mu_n^g$ is endowed with the standard symplectic pairing and $A[n]$ with the Weil pairing. This definition incorporates the ordinary abelian varieties: if $S = \operatorname{Spec}(k)$ where k is a field of characteristic $p > 0$, then the existence of a symplectic level n structure on A implies that A is an ordinary abelian variety, and conversely if A is ordinary then such β exists if we pass to a suitable finite extension of k. (Recall that the p-rank of an abelian variety A over a field k of characteristic $p > 0$ is defined by $\operatorname{Card}(A[p](\bar{k})) = p^{p\text{-rk}(A)}$. The inequality $p\text{-rk}(A) \leq \dim(A)$ always holds, with equality if A is ordinary.)

We give Deligne's proof of Raynaud's theorem mentioned above:

1.9. Theorem (Raynaud). *Let S be a scheme, and A be an abelian algebraic space over S. Then*

(a) A is a scheme, hence an abelian scheme. Similarly $\mathbf{Pic}(A/S)$ *is a scheme.*

(b) Moreover, any finite set of points in A (or, $\mathbf{Pic}(A/S)$*) over an affine open subset of S is contained in an open affine subscheme of A.*

Reduction steps:

(1) We may assume that S *is affine* : $S = \mathrm{Spec}(R)$.

(2) Write $R = \cup R_i$, where each R_i is a $\mathbb{Z}$-algebra of finite type. Since A is of finite presentation over S, we are reduced to the case that *R is a $\mathbb{Z}$-algebra of finite type.*

(3) Since an algebraic space X is a scheme iff X_{red} is, we may assume that *S is reduced.*

Consider $\mathbf{Corr}_S(A)$ over S. It is an algebraic space (say, by [Ar 4] thm. 5.3.) which is unramified over S (Cf. [Ra 3] prop. III 4.1., p. 49). Hence $\mathbf{Corr}_S(A)$ is a scheme either by a result of Grothendieck (Cf. [Mur] p. 14), or Zariski's main theorem for algebraic spaces. $\mathbf{Corr}_S(A) \to S$ satisfies the valuative criterion for properness: this boils down to the well-known statement that over a discrete valuation ring V, a homomorphism between the generic fibers of 2 abelian schemes over V extends uniquely (see 2.7). Of course $\mathbf{Corr}_S(A)$ cannot be proper over S as it is not of finite type.

Let us first assume that S is irreducible and normal (in fact, geometrically unibranch is enough). We claim that A is (globally) projective over S.

Let $\nu \in S$ be the generic point of S. Let p_ν be a polarization of A_ν over $k(\nu)$, which certainly exists. It corresponds to a homomorphism $h_\nu : A_\nu \to A^t_\nu$. We want to show that $(a) h_\nu$ extends to a homomorphism $h : A \to A^t$, and $(b) h$ is a polarization of A.

(a) The question is local in the étale topology, so we may assume that $S = \mathrm{Spec}(R)$ with R a strictly local domain. For some non-zero element $f \in R$ we have over $\mathrm{Spec}(R_f)$ a map $h_f : A \to A^t$, which defines a section p_f of $\mathbf{Corr}_S(A)$ over $\mathrm{Spec}(R_f)$. Let $P \subset \mathbf{Corr}_S(A)$ denote its Zariski-closure. $P \to S$ is unramified and satisfies the valuative criterion for properness. If $U \subset P$ is an open subset which contains $p_f(\mathrm{Spec}(R_f))$ and which is of finite type over S, then $U = U_1 \amalg U_2$ where U_1 is isomorphic to a finite disjoint union of closed subschemes of S, and U_2 has empty special fibre. Also U is irreducible, so either U is isomorphic to S or U has empty special fibre. By the valuative criterion the first alternative must hold if U is big enough, and it follows that P is isomorphic to S. This proves (a).

(b) Let $\mathcal{L}$ be the pull-back to A of the Poincaré sheaf on $A \times A^t$ via (id_A, h). $\lambda(\mathcal{L}) = 2h$ by the theorem of cube. $\mathcal{L}_\nu$ is ample because h_ν is a polarization. Clearly h is an isogeny (i.e. is surjective) because h_ν is, and therefore $\mathcal{L}$ is non-degenerate. (By definition, an invertible sheaf $\mathcal{M}$ on an abelian variety over a field is non-degenerate iff $\mathrm{Ker}(\lambda(\mathcal{M}))$ is finite). Applying the Riemann-Roch theorem for abelian varieties (Cf. [Mum 1] p. 150) and semi-continuity for cohomology, we see that $H^p(A_s, \mathcal{L}) = 0$ and $h^0(A_s, \mathcal{L}) = (\mathcal{L}_\nu^{(g)})/g! > 0$ for every $s \in S$, and $p > 0$, where $g = \dim(A)$. This shows that $\mathcal{L}_s$ is ample for every $s \in S$ (Cf. [AV] p.60). Hence $\mathcal{L}$ is relatively ample and h is a polarization, and (b) is proved.

We have proved that for the normalization, $\tilde{T}$, of any irreducible component T of S, $A_{\tilde{T}}$ is a projective abelian scheme. By Grothendieck's theory of Hilbert schemes, $\mathbf{Pic}(A/S)_{\tilde{T}}$ is also a scheme which is a union of open and closed projective schemes. Clearly any finite set of points in either $A_{\tilde{T}}$ or $\mathbf{Pic}(A/S)_{\tilde{T}}$ is contained in an affine open subscheme. Now we can use Noetherian induction and assume that the desired properties hold over any proper subscheme of S. Because S is reduced, $A = A/S$ is obtained by gluing together its irreducible components, A_T, along their intersections. That is, A is the amalgamated sum of the A_T under their intersections. (On affine pieces amalgamated sum corresponds to fibred product of rings. Globalization is immediate.) Note that if I, J are ideals in a ring $B, B/I \cap J = B/I \times_{B/(I+J)} B/J$. It follows that if each A_T is a scheme, A itself is a scheme. Similarly, $\mathbf{Pic}(A/S)$ is a scheme if this is true for all $\mathbf{Pic}(A/S)_T$, and if the property about finite sets of points is preserved under gluing. So we may assume that $S = \operatorname{Spec}(R)$ is irreducible and affine. Let $\tilde{R} \supset R$ denote the normalization of R, and $I \subset R$ the conductor of $\tilde{R}/R$ (i.e. $I = \{x \in \tilde{R} | x\tilde{R} \subset R\}$), which is an $\tilde{R}$-ideal. Then R is the fibred product of $\tilde{R}$ and R/I over $\tilde{R}/I$, and as the assertion is known over $\tilde{R}$ and R/I (by induction) the argument about amalgamated sums works as before. □

1.10. Remarks. (a) The arguments above show that any abelian algebraic space over a noetherian normal scheme S is in fact a (globally) projective abelian scheme over S. Also any morphism between such schemes which is defined over an open dense subset of S extends to all of S.

(b) The arguments above show that for an abelian algebraic space A over an algebraic space S and any invertible $\mathcal{O}_A$-module $\mathcal{L}$, the locus of $s \in S$ such that $\mathcal{L}_s$ is ample is closed in S. Note that from the general base change theorem ([EGA] III 4.7.1.), it is also open.

2. Semi-Abelian Schemes and Tori

2.1. Definition. A *torus* T over a scheme S is a commutative group scheme over S such that locally on S_{fppf} (equivalently, $S_{\text{ét}}$ or S_{fpqc}) it is isomorphic to a product of finitely many copies of the multiplicative group scheme G_m. T is a *split torus* if it is S-isomorphic to a product of finitely many copies of the multiplicative group.

For a torus $T \to S, \underline{X}(T) = \mathbf{Hom}_S(T, G_m)$ will denote the character group of T. It is an $S_{\text{ét}}$ (or S_{fppf} resp. S_{fpqc}) sheaf of torsion-free $\mathbb{Z}$-modules of finite type. T is uniquely determined by $\underline{X}(T) : T = \mathbf{Hom}(\underline{X}(T), G_m)$. T is a split torus iff $\underline{X}(T)$ is constant.

A fundamental property of tori is that they are rigid: you cannot deform them. We describe this phenomenon as follows:

2.2. Theorem. *Let S be a scheme, $\mathcal{I} \subset \mathcal{O}_S$ be a sheaf of ideals such that $\mathcal{I}^2 = 0$. Let $S_0 = \mathbf{Spec}(\mathcal{O}_S/\mathcal{I})$ be the closed subscheme of S defined by $\mathcal{I}$. Let H_0 be a torus defined S_0, $G \to S$ be a commutative group scheme smooth over S, $G_0 = G \times_S S_0$. Then*

(i) H_0 can be uniquely lifted to a torus H over S. (Any two liftings are isomorphic by a unique isomorphism)

(ii) Any homomorphism $u_0 : H_0 \to G_0$ can be uniquely lifted to a homomorphism $u : H \to G$. If moreover u_0 is a closed embedding, so is u.

Proof. See [SGA 3] exp. IX §3.

2.3. Definition. A *semi-abelian scheme* is a smooth separated commutative group scheme $\pi : G \to S$ with (geometrically) connected fibers, such that each fibre G_s is an extension of an abelian variety A_s by a torus $T_s : 0 \to T_s \to G_s \to A_s \to 0$.

2.4. Remarks. In the above definition, A_s and T_s are uniquely determined by $G_s : T_s$ is the largest smooth connected affine subgroup scheme of G_s and A_s is the quotient of G_s by T_s. T_s is called the *toric part* of G_s.

For a semi-abelian scheme $G \to S$ and $s \in S$, let $X(s) = X(T_s)$ denote the character group of the toric part of G_s. Letting $\overline{s}$ be a geometric point lying over $s, X(\overline{s})$ is a free $\mathbb{Z}$-module of finite type. The Galois group $\mathrm{Gal}(\kappa(s)^{\mathrm{sep}}/\kappa(s))$ naturally acts on $X(\overline{s})$, and $X(s)$ is (or, can be interpreted as) just this Galois module.

It is easy to see (torsion-points!) that the function $s \mapsto \mathrm{rk}(X(s))$ is upper semi-continuous on S, i.e. the toric part may jump. We shall see later how the different $X(s)$ are related.

Let $\pi : G \to S$ be a commutative group scheme. We repeat the definition of cubical structure: for any subset $I \subseteq \{1,2,3\}$ we let $m_I : G \times_S G \times_S G \to G$ be the morphism "adding the i-th coordinate for all $i \in I$". Let $\mathcal{L}$ be a G_m-torsor on G, *rigidified* along the zero section. (As always we will switch freely between the two equivalent notions of G_m-torsors and invertible sheaves.) Let

$$\Theta(\mathcal{L}) = \otimes_{I \subseteq \{1,2,3\}} m_I^* \mathcal{L}^{\otimes(-1)^{\mathrm{Card}(I)}} .$$

Similarly, let

$$\Lambda(\mathcal{L}) = m^*\mathcal{L} \otimes p_1^*\mathcal{L}^{\otimes -1} \otimes p_2^*\mathcal{L}^{\otimes -1},$$

where $m, p_1, p_2 : G \times_S G \to G$ denote the group law and the two projections respectively. A *cubical structure* on $\mathcal{L}$ ($\mathcal{L}$ rigidified) is a trivialization $\Theta(\mathcal{L}) \xrightarrow{\sim} \mathbf{1}$ (= the trivial G_m-torsor) such that it induces on $\Lambda(\mathcal{L})$ a structure of symmetric biextension of $G \times_S G$ by G_m. For instance, the theorem of cube says that any rigidified invertible sheaf on an abelian scheme has a canonical cubical structure. In fact, any rigidified invertible sheaf on a smooth, commutative group scheme $G \to S$, with connected fibres over a *normal* base scheme S, such that the maximal fibres are semi-abelian varieties (e.g. a semi-abelian scheme over normal base) admits a unique cubical structure compatible with the rigidification. More precisely, the forgetful functor

$$\mathrm{CUB}(G, G_m) \to \mathrm{TORRIG}(G, G_m)$$

from the category of cubical G_m-torsors on G to the category of rigidified G_m-torsors on G is an equivalence of categories. For details, see [Br] prop. 2.4, p. 16.

A major merit of the notion of cubical structure is that it makes G_m-torsors more rigid, so that it is easier to do descent:

2.5. Proposition. *Let S be a strictly henselien local ring, $G \to S$ be a (global) extension of an abelian scheme A by a torus T: we have an exact sequence $0 \to T \to G \to A \to 0$. Let $\mathcal{L}$ be a cubical invertible sheaf (i.e. invertible sheaf with cubical structure) on G. Then $\mathcal{L}$ can be descended to a cubical invertible sheaf on A. The set of all possible descent data for $\mathcal{L}$ is a principal homogeneous space under $X(T)$ in a natural way.*

Proof. See [Br] prop. 3.10, p. 38.

We remark here that extensions of A by T are classified by $\mathrm{Hom}_{\mathbb{Z}}(X(T), A^t)$. We state the important semi-stable reduction theorem:

2.6. Theorem. *Let V be a discrete valuation ring, K be the fraction field of V, G_K be a semi-abelian variety over K. Then there exists a finite extension V' of V, with fraction field K', such that $G_{K'} = G_K \otimes_K K'$ extends to a semi-abelian scheme over* $\mathrm{Spec}(V')$.

When G_K is an abelian variety, the assertion means that the Néron model of $G_{K'}$ contains a semi-abelian open subgroup. In this case the proof is well-documented, see for instance [SGA 7 I] exp. IX, or [A-W]. The general case follows easily from this: We may assume that the toric part T_K of the generic fibre is split, and that the abelian part A_K of the generic fibre has a semi-abelian extension A over $\mathrm{Spec}\, V$. Now, if G is any smooth group scheme over V with connected fibres, any extension of G_K by G_m defines a rigidified line bundle $\mathcal{L}_K$ on G_K together with an isomorphism $\mu^*(\mathcal{L}_K) \approx \mathrm{pr}_1^*(\mathcal{L}_K) \otimes \mathrm{pr}_2^*(\mathcal{L}_K)$ over $G_K \times_K G_K$ (where $\mu : G \times G \to G$ is the group law). These extend uniquely to $\mathrm{Spec}\, V$, and so does the extension. Hence $0 \to T_K \to G_K \to A_K \to 0$ extends uniquely to $\mathrm{Spec}\, V$.

Remark. The argument above shows, in fact, that after finite extension of K, there exists an extension G of G_K such that the toral part of G_K extends to a closed subtorus of G.

We prove two useful results about semi-abelian schemes. The first one concerns extension of homomorphisms between semi-abelian schemes:

2.7. Proposition. *Let S be a noetherian normal scheme and G and H be two semi-abelian schemes over S. Suppose that over a dense open subscheme U of S there is a homomorphism $\phi_U : H_U \to G_U$. Then ϕ_U extends (uniquely) to a homomorphism $\phi : H \to G$ over S.*

Proof. We could just quote [Ra 3] IX cor. 1.4, p. 130, but we prefer to give a proof. We remark first that the uniqueness is obvious. For clarity we will proceed in steps.

(a) Reduction to the case $\dim(S) = 1$.

Assume that the proposition has been proved when $\dim(S) = 1$. Let Z be the scheme-theoretic closure of the graph of ϕ_U in $H \times_S G$. We would like to show that the first projection $p_1 : Z \to H$ is an isomorphism. The valuative criterion shows that p_1 is proper since we assumed the base dimension $= 1$ case. Clearly H is normal, since S is. Therefore, by Stein factorization, it suffices to show that $\forall s \in S$, $p_{1s} : Z_s \to H_s$ is quasi-finite. In fact, we will show that it is a homeomorphism. We may assume that S is local with closed point s, and that the residue field of S in s is algebraically closed. For any $z \in Z$ we can choose a morphism $f : T = \mathrm{Spec}(V) \to Z$, with V a discrete valuation ring, η the generic point of T, and t the closed point of T, such that the $f(\eta) \in \mathrm{graph}(\phi_U)$, $f(t) = z$, and f composed with $Z \to S$ induces an injection $\mathcal{O}_s \hookrightarrow V$. We also assume that the residue field in t is algebraically closed. By assumption there exists a morphism $\phi_T : G_T \to H_T$ which extends the pullback of ϕ_U over η. The induced homomorphism $G_t \to H_t$ is independent of the choices; that is, there exists a homomorphism $\phi_s : G_s \to H_s$ of (geometric) fibres over s, inducing ϕ_t, which is independent of the choice of z and T: Look at torsion points. Now z must be equal to $(1, \phi_{\bar{s}}) \cdot \mathrm{pr}_1(z)$. This finishes the reduction step.

From now on, $S = \mathrm{Spec}(V)$ where V is a discrete valuation ring, η the generic point of S, s the closed point of S, and $U = \{\eta\}$. We remark that, by descent, it suffices to prove the extendibility after (possibly ramified) finite base change of discrete valuation rings.

(b) Case G_η is an abelian variety.

Then G is the connected Néron model of G_η (i.e. G is the open subgroup scheme of the Néron model $\mathrm{Né}(G_\eta)$ of G_η each fibre of which is connected). This follows easily by looking at torsion points. Then $\phi_\eta : H_\eta \to G_\eta$ extends to $\phi : H \to G$ by the universal property of Néron model. We remark that this already finishes the proof of 2.5. if the generic fibre of G is abelian for general S, using the reduction step (a).

(c) Case G is a torus over S.

After an unramified base change, we may assume that G splits, i.e. $G \cong G_m^r/S$. Then we may assume that $r = 1$; i.e., $G \cong G_m$. The homomorphism ϕ_η is given by a regular function $f \in \Gamma(H_\eta, \mathcal{O}_H)$ with no zeros or poles on H_η. Looking upon f as a rational function on H, $\mathrm{div}(f)$ is a multiple of the closed fibre H_s. Since ϕ_η is a group homomorphism, $m^*(f) = \mathrm{pr}_1^*(f) \cdot \mathrm{pr}_2^*(f)$, where $m, \mathrm{pr}_1, \mathrm{pr}_2 : H \times_S H \to H$ denote the group law and the two projections respectively. Therefore, $m^* \mathrm{div}(f) = \mathrm{pr}_1^* \mathrm{div}(f) + \mathrm{pr}_2^* \mathrm{div}(f)$. Let π be a uniformizer of V and $\mathrm{div}(f) = a \cdot H_s = a \cdot \mathrm{div}(\pi)$ with $a \in \mathbb{Z}$. Since $m^* \mathrm{div}(\pi) = \mathrm{pr}_1^* \mathrm{div}(\pi) + \mathrm{pr}_2^* \mathrm{div}(\pi) =$ divisor of π in $H \times_S H =$ the closed fibre of $H \times_S H$, the equality $m^* \mathrm{div}(f) = \mathrm{pr}_1^* \mathrm{div}(f) + \mathrm{pr}_2^* \mathrm{div}(f)$ implies that $a = 0$. This means that f is a unit and therefore defines a morphism $\phi : H \to G_m$. This ϕ is a homomorphism because ϕ_η is.

(d) In general, G_η is an extension of an abelian variety B_η by a torus T_η : $0 \to T_\eta \to G_\eta \to B_\eta \to 0$. Let B be the connected Néron model of B_η. After possible base change, we may assume that B is semi-abelian. $G_\eta \to B_\eta$ extends to a homomorphism $\psi : G \to B$ by the universal property of Néron models. Similarly, the composition $f_\eta : H_\eta \to G_\eta \to B_\eta$ extends to a homomorphism $f : H \to B$.

A torsion point argument shows that ψ is flat and surjective. Let $T = \ker(\psi)$. Then T is a flat, separated, commutative group scheme over S whose generic fibre is a torus. Assume for a moment that condition (*) "T is a torus" holds. Then we can conclude as follows: Pulling back the extension $0 \to T \to G \to B \to 0$ by $f : H \to B$, we obtain an extension $0 \to T \to G' \to H \to 0$. Clearly G' is a semi-abelian scheme. The hypothesis means that over the generic point η, the extension $0 \to T_\eta \to G'_\eta \to H_\eta \to 0$ has a splitting $\xi_\eta : G'_\eta \to T_\eta$, and the desired conclusion is that ξ_η extends to a splitting of $0 \to T \to G' \to B \to 0$. By step (c), ξ_η extends to a homomorphism $\xi : G' \to T$, which must be a splitting of $0 \to T \to G' \to B \to 0$. Now, by the proof of 2.6, after passing to a finite extension of K, there exists an extension $\tilde{G}$ of G_K which satisfies (*), and the assertion of the proposition holds for maps into $\tilde{G}$. It follows that the identity map of G_K extends to a regular morphism $G \to \tilde{G}$, which must be an isogeny, hence faithfully flat. Its kernel is flat with trivial generic fibre, hence trivial, so $G = \tilde{G}$. This finishes the proof of Theorem 2.7. It also follows (using descent) that the closure in G of the toral part of G_K is a torus.

2.8. Remark. It follows that a semi-abelian scheme $G \to S$ all whose fibres are abelian varieties is proper over S, that is, that G is an abelian scheme: Use the valuative criterion to reduce to the case where S is the spectrum of a discrete valuation ring, and then use the fact that G is the Néron model of its generic fibre.

A variant of the above proposition concerns the toric part:

2.9. Proposition. *Let S be a noetherian normal scheme, U be a dense open subscheme of S, and G be semi-abelian scheme over S. If H_U is a closed subgroup of G_U which is a torus over U, then the closure of H_U in G is a torus $H \to S$ contained in G.*

Proof. A proof can be found in [Ra 3] IX 2.4, p. 137, but we also give a direct proof.

(a) Reduction step: By descent, we may assume that S is strictly henselian. Moreover, we may assume that U is the complement of the closed point s of S. The collection of n-torsion points is dense in H_U, as n runs over integers prime to $\mathrm{char}(\kappa(s))$. The action of $\mathrm{Gal}(\overline{\kappa}(s)/\kappa(s)$ on n-torsion points of G_U is unipotent, so the torus H_U splits over U.

(b) Denote by H the split torus over S extending H_U. By Prop. 2.4.1, the inclusion $H_U \hookrightarrow G_U$ uniquely extends to a homomorphism $h : H \to G$. We have to show that h is a closed immersion. The case when $S = \mathrm{Spec}(R)$ with R a discrete valuation ring can be reduced to i) G is a torus or ii) H_η is the maximal subtorus of G_η. Case i) is elementary, while ii) has been treated in the proof of 2.7. The general case when $\dim(S) > 1$ can be deduced from the case $\dim(S) = 1$ as in the proof of 2.7 (but much more easily), using the valuative criterion to show that h is proper.

We remark that if we drop the assumption that S is normal, Propositions 2.7 and 2.9 both become *false.* For an easy counter-example, let k be an algebraically closed field, C be the nodal curve over k obtained from $\mathbb{P}^1_k$ by identifying 0 and ∞. C has an infinite étale cover $\tilde{C}$ = an infinite chain of $\mathbb{P}^1$'s with Galois group $\mathbb{Z}$. Then we obtain a non-constant sheaf of free abelian groups $\underline{X}$ of rank 2 on C corresponding to the above covering. Let $U = C - \{$the (unique) singular point of $C\}$. By construction, $\underline{X}|_U$ is constant. Let $T = \mathbf{Hom}(\underline{X}, G_m)$ be the torus over C with character group $\underline{X}$. Since $\underline{X}|_U$ is constant, $T|_U \cong G_m^2/U$. Taking $S = C$, $H = G_m^2$, $G = T$, we get a counter-example to 2.7. The graph of $H \to G$ gives a counter-example to 2.9.

Now suppose $G \to S$ is a semi-abelian scheme, with an arbitrary base S which does not have to be normal. We want to study relations between the character groups $X(s)$ of the toric parts of the fibres G_s. If S is the spectrum of a discrete valuation ring V with generic point η and special point s, the generic torus T_η extends to a subtorus of G, whose special fibre is contained in T_s. This way we obtain a surjection $X(s) \to X(\eta)$. It follows easily that there exists an étale constructible $\underline{X}$ on S (= Spec V), whose fibre in $\bar{s}$ (respectively $\bar{\eta}$) is $X(s)$ (respectively $X(\eta)$), such that the map above is the specialization map associated to such a sheaf. For a general base scheme S, we might ask whether there exists such an $\underline{X}$, with geometric fibres $X(s)$, which under pullback to discrete valuation rings induces the previous one. The answer is yes:

2.10. Theorem. *Suppose $G \to S$ is a semi-abelian scheme. There exists a unique étale constructible sheaf $\underline{X} = \underline{X}(G)$ on S such that for any $s \in S$ the restriction of $\underline{X}$ to $\{s\}$ is equal to $X(s)$, and such that for any map* Spec$(V) \to S$ *with V a discrete valuation ring, the pullback of $\underline{X}$ to* Spec(V) *is the constructible sheaf constructed above. Furthermore formation of $\underline{X}$ is functorial and commutes with any base change, and for any torus T over S with character group $\underline{Y}$,* $\mathrm{Hom}_S(T, G)$ *is isomorphic to* $\mathrm{Hom}_S(\underline{X}, \underline{Y})$.

Proof. We may assume that S is of finite type over Spec$(\mathbb{Z})$, in particular that S is excellent. For any étale $S' \to S$ $\underline{X}(S')$ must be equal to the set of all families $\{x(s) | s \in S'\}$, where each $x(s)$ is a Gal$(\bar{s}/s)$-invariant element of $X(s)$, such that for any $Z = \mathrm{Spec}(V) \to S'$ as above, the values of x at the image of the generic and special points of Z define a section of $\underline{X}(G_Z)$. In fact this prescription defines an étale sheaf $\underline{X}$ on S, and we have to show that it does the job. It is also clear that it is functorial, and that any homomorphism $T \to G$ induces a map of sheaves $\underline{X} \to \underline{Y}$. We use the following reduction steps:

(a) We may assume that S is reduced.

(b) If $\underline{X} = \underline{X}(G)$ exists for $G \to S$, then for any morphism $S' \to S$, $\underline{X}(G_{S'})$ exists, and is equal to the pullback of $\underline{X}$ to S'.

(c) If $S = S_1 \cup S_2$ is the union of two closed subschemes such that $\underline{X}(G_{S_1})$ and $\underline{X}(G_{S_2})$ exist, then $\underline{X}(G)$ exists.

(d) Assume that S is irreducible, with normalization $\tilde{S}$, that $\underline{X}(G_{S'})$ exists for any proper closed subscheme $S' \subset S$, and that $\underline{X}(G_{\tilde{S}})$ exists. Then $\underline{X}(G)$ exists.

Only step (d) needs some explanation: It is a local problem. If $S = \mathrm{Spec}(R)$ is affine, $\tilde{S} = \mathrm{Spec}(\tilde{R})$, and $I \subset R$ denotes the conductor, then R is the fibred product of $\tilde{R} \to \tilde{R}/I \leftarrow R/I$. Now $\underline{X}(G)$ exists over $\tilde{S}$, $S' = \mathrm{Spec}(R/I)$ and $\tilde{S}' = \mathrm{Spec}(\tilde{R}/I)$. If $U = S - S' = \tilde{S} - \tilde{S}'$, and $j : U \to S$ and $\tilde{j} : U \to \tilde{S}$ denote the inclusions, $\underline{X}(G_{\tilde{S}})$ is obtained by gluing $\underline{X}(G_U)$ and $\underline{X}(G_{\tilde{S}'})$ along a map $\underline{X}(G_{\tilde{S}'}) \to \tilde{j}_*(\underline{X}(G_U))|\tilde{S}'$, which corresponds to a map $\underline{X}(G_{S'}) \to j_*(\underline{X}(G_U))|S'$. We use this to glue $\underline{X}(G_U)$ and $\underline{X}(G_{S'})$.

Now by noetherian induction we may assume that $\underline{X}(G_{S'})$ exists for any proper closed subscheme S' of S, and that S is irreducible and normal. By Proposition 2.9 the generic torus T_η extends to a subtorus $T \subset G$, with character-group $\underline{Y}$. By semi-continuity, there exists a non-trivial open subset $U \subset S$ such that for $s \in U$, T induces the toral part of the fibre $G_{s'}$. So obviously $\underline{X}(G_U) = \underline{Y}|U$ exists. If $S' = S - U$ (with reduced structure) and $i : S' \to S$ and $j : U \to S$ are the inclusions, we obtain a map $\underline{X}(G_{S'}) \to \underline{Y}|S' = i^* j_*(\underline{X}(G_U))$, which can be used to glue $\underline{X}(G_{S'})$ and $\underline{X}(G_U)$ to a global constructible sheaf $\underline{X}$. We have to check that its pull-back under every map $S'' = \mathrm{Spec}(V) \to S$ is what it should be. There is no problem if all of S'' maps either to U or to S'. Otherwise the pullback of $\underline{X}$ is obtained by the same gluing on $\mathrm{Spec}(V)$ as was used to construct $\underline{X}(G_{S''})$, and again we are done.

It remains to show that any morphism $\underline{X}(G) \to \underline{Y}$ defines a map $T \to G$. This follows from 2.9 if S is normal, and the rest is handled by Noetherian induction and gluing along amalgamated sums, as in the proof of Theorem 1.9.

2.11. Corollary. *Assume that the rank of $X(s)$ is locally constant on S. Then G is globally an extension of an abelian scheme by a torus, that is there exists an exact sequence $0 \to T \to G \to A \to 0$, T a torus, A abelian. Especially G is a torus if all its fibres are.*

Proof. Under these assumptions $\underline{X}(G)$ is locally constant, so corresponds to a torus T. The identity of $\underline{X}(G)$ defines a map $T \to G$ which must be a closed immersion (valuative criterion), and the quotient $A = G/T$ is abelian.

For any semi-abelian $G \to S$ we obtain a stratification of S by locally closed subsets $S_r = \{s \in S | \mathrm{rank}(X(s)) = r\}$. The closure of S_r is contained in the union of the $S_i, i \geq r$, and over each S_r G is globally the extension of an abelian scheme by a torus.

3. Deformation of Abelian Varieties

General discussions of deformation theory can be found in [Sch], [SGA 1 exposé III], [SGA 7 exposé VI]. The deformation theory of abelian varieties, due to Grothendieck and Mumford, works out nicely. A good exposition can be found in [Oort]. Let us summarize the results:

Let k be a field, and W denote either $W_\infty(k)$ (the ring of infinite Witt vectors over k) when $\mathrm{char}(k) = p > 0$, or k itself when $\mathrm{char}(k) = 0$. Let C_W be the category of artinian local W-algebras R such that the induced homomorphism $k = W/\mathfrak{m}_W \to R/\mathfrak{m}_R$ is an isomorphism. Let X_0 be an abelian variety over k, and $M = M_{X_0} : C_W \to \mathbf{Sets}$ be the deformation functor of X_0 given by:

$M(R) = \{$isomorphism classes of pairs (X, f), where X is an abelian scheme over R and f is an isomorphism $f : X \otimes_R k \xrightarrow{\sim} X_0\}$

Grothendieck showed that $M = M_{X_0}$ is formally smooth and that all obstructions vanish, and hence that M is pro-representable by a formally smooth algebra. Since $\dim H^1(X_0, \Omega^{1*}_{X_0}) = g^2$ it is pro-representented by Spf $W[[t_{i,j}]]$, where $0 \le i, j \le g$. The equi-characteristic tangent space is canonically isomorphic to $H^1(X_0, \Omega^{1*}_{X_0}) = H^1(X_0, \mathcal{O}_{X_0}) \otimes_k t_{X_0} = t_{X_0^t} \otimes_k t_{X_0} = \mathrm{Hom}(\Omega_{X_0^t}, t_{X_0})$. Here $\Omega^1_{X_0}$ denotes the sheaf of differentials on X_0 while t_{X_0}, $t_{X_0^t}$, and $\Omega_{X_0^t}$ denote the tangent space (at the origin) of X_0 and the tangent and cotangent spaces of the dual of X_0, respectively. An isomorphism is provided by the Kodaira-Spencer map (see Ch. III,9).

With notations as above, let (X_0, λ_0) be a principally polarized abelian variety over k. Let $P = P_{X_0,\lambda_0} : C_W \to \mathbf{Sets}$ be the deformation functor of (X_0, λ_0) given by:

$P(R) = \{$isomorphism classes of triples (X, λ, f), where (X, λ) is an principally polarized abelian scheme over R and f is an isomorphism $f : (X, \lambda) \otimes_R k \xrightarrow{\sim} (X_0, \lambda_0)\}$

Because the correspondence functor is unramified, P_{X_0,λ_0} is a subfunctor of M_{X_0}. Mumford proved that P is also formally smooth, and is pro-represented by Spf $W[[t_{i,j}]]/$(an ideal generated by $g(g-1)/2$ elements). The equi-characteristic tangent space is canonically isomorphic to $\mathrm{Hom}_{\mathrm{sym}}(\Omega_{X_0}, t_{X_0}) \subset \mathrm{Hom}(\Omega_{X_0}, t_{X_0})$ $= \mathrm{Hom}(\Omega_{X_0^t}, t_{X_0})$, where we identify Ω_{X_0} with $\Omega_{X_0^t}$ via λ_0, and $\mathrm{Hom}_{\mathrm{sym}}$ denotes symmetric homomorphisms.

The results above consist of computation of tangent spaces (easy) and vanishing of obstructions (harder). Better results are available. First if k has positive characteristic $p > 0$, a theorem of Serre and Tate asserts that the deforming an abelian variety is equivalent to deforming its associated Barsotti-Tate (= p-divisible) group. A good reference is section 1 of [Ka 3]. P. Cartier's covariant Dieudonné theory gives (as a special case) a translation of the equi-characteristic p ($p > 0$) deformation problem for a formal group into a problem in p-linear algebra. Via the Serre-Tate theorem, one can therefore control (part of) the char. p deformation of an abelian scheme coming from the local part of the Barsotti-Tate group. References for Cartier-Dieudonné theory are [Laz], [Haz]. The Grothendieck-Messing theory of crystals associated to Barsotti-Tate groups provides a deformation theory of Barsotti-Tate groups as well as abelian schemes, which is perhaps the easiest to use. References are [Gr 2], [Me], [Ma-Me]. Let us state this useful result in the case of abelian schemes:

Recall that an ideal I in a ring A is said to be equipped with a divided power structure if it possesses a family of maps $\gamma_n : I \to I$ $(n \geq 1)$, which satisfy suitable identities for maps of the form $(x \mapsto x^n/n!,\ n \geq 1)$. (A, I, γ) is said to be nilpotent if there is an N such that the ideal generated by all elements of the form $\gamma^{i_1}(x_1) \cdots \gamma^{i_k}(x_k)$ with $i_1 + \cdots + i_k \geq N$ is zero. These notions naturally globalize.

Let now S be a scheme with p locally nilpotent in $\mathcal{O}_S$, and $\mathcal{J}$ be a quasi-coherent ideal of $\mathcal{O}_S$ endowed with a locally nilpotent divided power structure γ. Let $S_0 = \operatorname{Spec}\mathcal{O}_S/\mathcal{J}$. Such pairs $(S_0 \subset S, \gamma)$ are objects of the locally nilpotent crystalline site of S_0. Given an abelian scheme X_0 over S_0 of dimension g, the Grothendieck-Messing theory provides us for every pair $(S_0 \subset S, \gamma)$ as above a locally free $\mathcal{O}_S$-module $D(X_0)_S$ of rank $2g$ which is functorial in $(S_0 \subset S, \gamma, X_0)$, such that for *any* abelian scheme X over S lifting X_0, $D(X_0)_S$ is canonically isomorphic to $H^1_{DR}(X, \mathcal{O}_{X/S})$. Thus we have locally free sheaves $D(X_0)_S$ and $D(X_0)_{S_0}$ over S depending only on X_0 with $D(X_0)_S \otimes_{\mathcal{O}_S} \mathcal{O}_{S_0} = D(X)_{S_0}$. On the other hand the Hodge filtration of de Rham cohomology gives a filtration $\Omega_{X_0} \subset D(X_0)_S$ with locally free quotient. Moreover for any abelian scheme X lifting X_0, we have a locally free $\mathcal{O}_S$-submodule Ω_X of $D(X_0)_S$ lifting Ω_{X_0}, with locally free quotient. Thus for fixed S_0, S, γ, X_0 as above, we have a functor F from the category of abelian schemes over S lifting X_0 to the category of locally free $\mathcal{O}_S$-submodules of $D(X_0)_S$ lifting Ω_{X_0} with locally free quotient. A main theorem of the Grothendieck-Messing theory asserts that F is an equivalence of categories. Similarly a principal polarization λ_0 defines non-degenerate symplectic forms on $D(X_0)_{S_0}$ and $D(X_0)_S$ such that Ω_{X_0} is isotropic, and λ_0 lifts to a principal polarization λ on an abelian scheme X which lifts X_0 if and only if Ω_X is isotropic. In the case $S_0 = \operatorname{Spec} k$, where k is a field, $S = \operatorname{Spec} k[\epsilon]/(\epsilon^2)$, we retrieve the results of Grothendieck and Mumford stated at the beginning of this section.

4. Review of Algebraic Stacks and Artin's Method

In this section we will give a rather sketchy review of algebraic stacks. Algebraic stacks are the scheme-theoretic analogue of what topologists call "orbifolds," although the notion of algebraic stacks seems to have appeared earlier. They generalize algebraic spaces and therefore also schemes. Because their definition is somewhat technical, many people seem to be afraid of and do not like them. But speaking from experience, the difficulty is purely psychological and can be overcome with time. In order to handle algebraic stacks, one has only to use descent systematically. The standard reference for algebraic stacks is [De-Mum]. It contains the basic definitions and results about algebraic stacks, but gives very few proofs. Unfortunately, it also seems to be the only available reference (with [Ar 6] as the arguable exception).

We begin with an informal discussion of fibered categories and stacks. A *fibered category* consists of a base category C and a category F over C (given by a functor $p : F \to C$) such that roughly speaking, the inverse image functor by f exists in F for every morphism f in C, and these pull-backs satisfy the usual relation "$(f \circ g)^* = g^* f^*$". For each $U \in \mathrm{Ob}\, C$, we denote by F_U (or $F(U)$ sometimes, abusing the notation) the fiber $p^{-1}(U)$. The category of fiber bundles (resp. vector bundles, resp. principal G-bundles for a Lie group G) over the category of all topological spaces (resp. manifolds) is an example of a fibered category. For the precise definition and more details, see [SGA1] exposé VI.

Let C a category with a Grothendieck topology. The official reference for Grothendieck topologies is [SGA4], but the first ten pages of Deligne's exposition [Arcata] in [SGA$4^{1/2}$] and section 0 of Artin's notes [Ar 1] are more accessible. A *stack* over C is a fibered category F over C such that for every covering morphism $U \to V$ in C, the canonical map $F_V \to \mathrm{Des}(U/V)$ is an equivalence of categories, where $\mathrm{Des}(U/V)$ denotes the category of descent data relative to $U \to V$. This formal definition basically says that we can glue both the objects and morphisms of F relative to the topology of C. The reference for the theory of stacks is Giraud's book [Gi], which unfortunately seems to intimidate beginners. The examples we gave for fibered categories are, in fact, all stacks. [DMOS] pp.220-226 summarizes the main definitions in [Gi].

We give some examples of stacks in algebraic geometry. Let Sch be the category of schemes, QCoh be the category of quasi-coherent sheaves on schemes, Coh be the category of coherent sheaves on schemes, VB be the category of locally free sheaves of finite rank on schemes. Then QCoh, Coh, VB are all stacks over Sch w.r.t. the Zariski topology on Sch. The theory of descent tells us that they are in fact all stacks over Sch w.r.t. the étale (resp. fppf, fpqc) topology on Sch.

Let F be a stack over C. If morphisms in F_U are all isomorphisms for every $U \in \mathrm{Ob}\, C$, we say that F is a *stack of groupoids*. Explicitly, this means that we are given a functor $p : F \to C$ and a Grothendieck topology on C, such that

a) For all $\phi : U \to V$ in C and $y \in \mathrm{Ob}(F_V)$, there is a morphism $f : x \to y$ in F with $p(f) = \phi$. We shall indulge an abuse of notation and write $x = f^* y$. Furthermore, we shall assume that for any ϕ and f, a choice of x has been made, and consider ϕ^* "as a functor from F_V to F_U". This amounts in effect to choosing a "splitting" of F.

b) Given morphisms $f : x \to z \leftarrow y : g$ in F with image $\alpha : U \to W \leftarrow V : \beta$. If $\gamma : U \to V$ is a morphism in C such that $\alpha = \beta\gamma$, then there is a unique morphism $h : x \to y$ in F such that $f = gh$ and $p(h) = \gamma$.

c) For any $V \in \mathrm{Ob}\, C$ and any objects x, y in F_V the functor from C/V to Sets which to any $\phi : U \to V$ in C assigns $\mathrm{Hom}_{C/V}(\phi^* x,\ \phi^* y)$ is a sheaf. This makes good sense, and is independent of the choice of ϕ^*, because of condition b).

d) If $\phi : U \to V$ is a covering in C, then any descent datum for objects in F relative to ϕ is effective.

The moduli stacks for proper smooth curves are good examples of stacks of groupoids. (In fact, they are algebraic stacks.) Let $g \geq 2$ (resp. $g = 1$) be an integer, and Sch be the category of schemes. Let M_g be the category defined as follows: $\operatorname{Ob} M_g$ = smooth proper morphisms of schemes $f : C \to S$ whose geometric fibers are all connected curves of genus g. (resp. endowed with a section e of f.) A morphism from $f_1 : C_1 \to S_1$ (resp. plus $e_1 : S_1 \to C_1$) to $f_2 : C_2 \to S_2$ (resp. plus $e_2 : S_2 \to C_2$) consists of a pair $(\tilde{h} : C_1 \to C_2,\ h : S_1 \to S_2)$ of morphisms of schemes such that $hf_1 = f_2\tilde{h}$ (resp. and $\tilde{h}e_1 = e_2h$), and $\tilde{h}$ induces an isomorphism $C_1 \xrightarrow{\sim} C_2 \times_{S_2} S_1$ Then the structural functor $M_g \to$ Sch is a stack of groupoids w.r.t. the étale (resp. fppf, fpqc, Zariski) topology on Sch. The effectivity of descent necessary for M_g to be a stack w.r.t. the flat topology follows from the fact that all curves are projective.

Let C be a category with Grothendieck topology. Stacks over C (resp. stacks of groupoids over C) form the objects of a 2-category. We shall give only the bare minimum of definitions for 2-categories. The notion of 2-category is due to J. Benabou and documented in his thesis, which is not widely available. The first ten pages or so of [Hakim] contain an excellent resumé, which is (more than) sufficient for our purposes.

4.1. Definition. A *2-category* $\mathfrak{C}$ consists of the following data:

i) a collection $\operatorname{Ob} \mathfrak{C}$, called the objects of $\mathfrak{C}$.

ii) for every pair (X, Y) of objects of $\mathfrak{C}$, a category $\operatorname{Hom}_{\mathfrak{C}}(X, Y)$, or simply $\operatorname{Hom}(X, Y)$, whose objects are called *1-morphisms* of $\mathfrak{C}$. Arrows (morphisms) in $\operatorname{Hom}_{\mathfrak{C}}(X, Y)$ are called *2-morphisms* of $\mathfrak{C}$.

iii) for every object X of $\mathfrak{C}$, a distinguished object $\mathrm{id}_X \in \operatorname{Ob}\operatorname{Hom}(X, X)$.

iv) for every triple (X, Y, Z) of objects of $\mathfrak{C}$, a functor ("composition")

$$\mu_{X,Y,Z} : \operatorname{Hom}(X, Y) \times \operatorname{Hom}(Y, Z) \to \operatorname{Hom}(X, Z),$$

These data should satisfy the following conditions:

a) for every triple (X, Y, Z) of objects of $\mathfrak{C}$, we have

$$\mu_{X,X,Y}(\mathrm{id}_X, -) = \mu_{X,Y,Y}(-, \mathrm{id}_Y) = \mathrm{id}_{\operatorname{Hom}(X,Y)}, \text{ the identity functor on } \operatorname{Hom}(X, Y).$$

b) ("associativity of composition") for every quadruple (X, Y, Z, W) of objects of $\mathfrak{C}$, we have

$$\mu_{X,Z,W} \circ (\mu_{X,Y,Z} \times \mathrm{id}_{\operatorname{Hom}(Z,W)}) = \mu_{X,Y,W} \circ (\mathrm{id}_{\operatorname{Hom}(X,Y)} \times \mu_{Y,Z,W}).$$

Examples of 2-categories:

1) The collection of all categories form a 2-category $\mathfrak{Cat}$. Its objects are categories. For any two categories X, Y in $\mathfrak{Cat}$, $\operatorname{Hom}(X, Y)$ is the category whose objects are functors from X to Y, whose morphisms are natural transformations of functors.

2) Let C be a category with a Grothendieck topology. The collection of all stacks form the objects of a 2-category $\mathfrak{St}/C$. For two stacks F_1, F_2, the category $\mathrm{Hom}_{\mathfrak{St}}(F_1, F_2)$ is defined by: $\mathrm{Ob}\,\mathrm{Hom}_{\mathfrak{St}}(F_1, F_2)$ = the collection of functors from F_1 to F_2, morphisms between objects are natural transformations of functors. The 2-category $\mathfrak{St}/C$ has a full subcategory $\mathfrak{Stgp}\,/C$ whose objects are stacks of groupoids.

3) Let C be a category with a Grothendieck topology, which is coarser than the canonical topology (i.e. For every $X \in \mathrm{Ob}\,C$, the representable functor h_X on C is a sheaf). To every $X \in \mathrm{Ob}\,C$ we associate a stack of groupoids $\tilde{X}$ with $\mathrm{Ob}\,\tilde{X} = \{\text{morphisms } U \to X \text{ in } C\}$ and a morphism from $(U \to X)$ to $(V \to X)$ is simply a morphism from U to V over X in C. Thus the only morphisms in $\tilde{X}_U$ are the $\mathrm{id}_{U\to X}$'s, for any $U \in \mathrm{Ob}\,C$. Hence in all aspects $\tilde{X}_U$ is just a set, and $U \rightsquigarrow \tilde{X}_U$ is nothing but the usual functor h_X on C. These $\tilde{X}$'s form a full subcategory $\tilde{C}$ of $\mathfrak{Stgp}\,/C$. For any $F \in \mathrm{Ob}(\mathfrak{Stgp}\,/C)$ and any $X \in \mathrm{Ob}\,C$, the category $\mathrm{Hom}_{\mathfrak{Stgp}\,/C}(\tilde{X}, F)$ is canonically equivalent to the category F_X (by analogy with Yoneda's lemma). Hence we can identify X with $\tilde{X}$ functorially, and simply write X instead of $\tilde{X}$.

One has the notion of fibred product in a 2-category $\mathfrak{C}$. The story begins with the notion of equivalence between two objects in $\mathfrak{C}$.

4.2. Definition. Two objects X, Y of a 2-category $\mathfrak{C}$ are *equivalent* if there are two 1-morphisms $u : X \to Y$, $v : Y \to X$ and two 2-isomorphisms $\alpha : v \circ u \to \mathrm{id}_X$, $\beta : u \circ v \to \mathrm{id}_Y$. Note that when $\mathfrak{C} = \mathfrak{Cat}$, this reduces to the familiar notion "equivalence of categories".

There is the notion "2-functor" between 2-categories. We shall not give the precise definition here. The collection of all 2-functors from $\mathfrak{C}$ to $\mathfrak{Cat}$ form a 2-category $\underline{\mathfrak{C}}$. There is an analogue of Yoneda embedding $\mathfrak{C} \subseteq \underline{\mathfrak{C}}$, given by $X \rightsquigarrow h_X = \mathrm{Hom}_{\mathfrak{C}}(-, X) \in \mathrm{Ob}\,\underline{\mathfrak{C}}$. Note we have an obvious notion of fibred product in $\mathfrak{Cat}$. Now given two 1-morphisms $u : X \to S$, $v : Y \to S$ in $\mathfrak{C}$, we can cook up $H_{u\times v} :=$ the fibred product of h_X and h_Y over $h_S \in \mathrm{Ob}\,\mathfrak{C}$. To be a bit more explicit, $H_{u\times v}$ assigns to any $T \in \mathrm{Ob}\,\mathfrak{C}$ the category $H_{u\times v}(T)$, whose objects are the triples (z, w, α) formed of two 1-morphisms $z : T \to X$, $w : T \to Y$ and a 2-isomorphism $\alpha : u \circ z \xrightarrow{\sim} v \circ w$. If $H_{u\times v}$ is equivalent in $\mathfrak{C}$ to h_Z for some $Z \in \mathrm{Ob}\,\mathfrak{C}$, then Z is uniquely determined up to equivalence, and will be said to be "the" fibred product of X and Y over S and denoted $X \times_S Y$.

When $\mathfrak{C} = \mathfrak{Stgp}\,/C$ for some category C with a Grothendieck topology, the fibred products can be described more explicitly: Let $u : X \to S$, $v : Y \to S$ be morphisms of stacks, and F be their fibred product. For $U \in \mathrm{Ob}\,C$, F_U consists of triples (x, y, α) where $x \in \mathrm{Ob}\,X$, $y \in \mathrm{Ob}\,Y$, and $\alpha : u(x) \xrightarrow{\sim} v(y)$ is an isomorphism in S_U. Morphisms and pull-backs in F are defined in the obvious way.

Now we come to the moduli stack of principally polarized abelian varieties and its variants. Let $g \geq 1$ be a natural number.

4.3. Definition. Let C be the category of schemes. The moduli stack of principally polarized abelian schemes $p : \mathsf{A}_g \to C$ is defined as follows: As a category, $\mathrm{Ob}(\mathsf{A}_g)$ = the collection of all principally polarized abelian schemes $(\mathsf{A}/S, \lambda)$ of relative dimension g, where S is some base scheme and λ is the principal polarization $\mathsf{A} \xrightarrow{\sim} \mathsf{A}^t$. A morphism from $(\mathsf{A}_1, \lambda_1)$ to $(\mathsf{A}_2, \lambda_2)$ is a homomorphism $\mu : \mathsf{A}_1 \to \mathsf{A}_2$ such that $\mu^*(\lambda_2) = \lambda_1$. Note that this condition implies that μ induces an isomorphism of abelian schemes $\mathsf{A}_1 \to \mathsf{A}_2 \times_{S_2} S_1$. Finally, $p : \mathsf{A}_g \to C$ is the forgetful functor $(\mathsf{A}/S, \lambda) \mapsto S$.

When C is given the étale (or fppf, fpqc, Zariski) topology, A_g becomes a stack over C.

A variant of A_g is the moduli stack of principally polarized abelian schemes with principal level-n structure $\mathsf{A}_{g,n}$.

4.4. Definition. $\mathsf{A}_{g,n}$ is the stack of groupoids defined by: $\mathrm{Ob}(A_{g,n})$ = the collection of all triples $(A/S, \lambda, \alpha)$, where $(A/S, \lambda)$ is a principally polarized abelian scheme of relative dimension g, and α is a symplectic isomorphism of finite, locally free group schemes $\alpha : A[n] \xrightarrow{\sim} (\mathbb{Z}/n\mathbb{Z})^{2g}$.

We can define other variants of A_g by varying the type of polarization, the type of level structure, or both.

4.5. Definition. Let C be a category with a Grothendieck topology. Let M_1, M_2 be stacks of groupoids over C. A 1-morphism $u : M_1 \to M_2$ is *relatively representable* if for any $X \in \mathrm{Ob}(C)$ and any 1-morphism $x : X \to M_2$, the fibred product $X \times_{M_2} M_1$ is represented by some $Y \in \mathrm{Ob}(C)$. Explicitly, this means the following:

(i) For any $f : Y \to X$ in C, the category M_f with $\mathrm{Ob}(M_f) = \{(y, \alpha) | y : Y \to M_1$ is a section of M_1 over Y, $\alpha : u(y) \xrightarrow{\sim} f^*(x)$ is an isomorphism$\}$ and the obvious notion of morphisms is equivalent to a category $\mathcal{M}_f$ all of whose morphisms are identities.

(ii) The functor $f \cdots \to \mathcal{M}_f$ is represented by some $v : Z \to X$ in C.

From now on, S denotes a fixed base scheme which is of finite type over a field or an excellent Dedekind domain, and C denotes the category of all schemes over the base scheme S endowed with the étale topology. The reason to put the restrictions on the base scheme S is that Artin's approximation theorem is available only in these cases. Let P be a property of morphisms of schemes which is stable under base change and étale local on the target. Let $u : M_1 \to M_2$ be a relatively representable 1-morphism of stacks of groupoids over C. Then *u is said to have property P* if for every 1-morphism $x : X \to M_2$ with $X \in \mathrm{Ob}(C)$, the morphism of schemes $Z = X \times_{M_2} M_1 \to X$ has property P.

A stack of groupoids M over C is said to be *limit preserving* if the following condition is satisfied: for any filtering inverse limit $T = \varprojlim T_i$ in C, the canonical map $\varinjlim M(T_i) \to M(T)$ is an equivalence of categories, where the left hand side is the obvious "2-category limit" in the 2-category $\mathfrak{Cat}$.

4.6. Definition. An *algebraic stack* is a limit preserving stack of groupoids M over C, such that

(i) The diagonal map $M \to M \times_S M$ is representable and locally of finite type. Explicitly, this means that for any $X, Y \in \mathrm{Ob}(C)$ and any pair of 1-morphisms $x : X \to M$, $y : Y \to M$, the fibred product $X \times_M Y$ is represented by some $Z \in \mathrm{Ob}(C)$. Furthermore, if X and Y are both locally of finite type over the base scheme S, we require that the fibred product Z is also locally of finite type over S.

(ii) There exist $X \in \mathrm{Ob}(C)$ locally of finite type over S and a 1-morphism $X \to M$ which is étale surjective. Explicitly, this means that for every $Y \in \mathrm{Ob}(C)$ and every 1-morphism $Y \to M$, the projection morphism of schemes $X \times_M Y \to Y$ is étale surjective.

4.7. Remarks. (i) Our definition of algebraic stacks differs from that of [D-M] in that we require them in addition to be limit preserving and satisfy the separation condition in 3.6.(i). According to the above definition, we can restrict the base category C to the category of all schemes locally of finite type over the base scheme S.

(ii) Artin introduced a more general definition of algebraic stacks in [Ar5]. He argued that it is natural to require $X \times_M Y$ to be an algebraic space locally of finite type over the base scheme S for any $X, Y \in \mathrm{Ob}(C)$ locally of finite type over S and any pair of 1-morphisms $x : X \to M$, $y : Y \to M$. Requiring $X \times_M Y$ to be a scheme would be unnatural. Moreover, he weakened condition (ii) in definition 4.6. by requiring $X \to M$ to be smooth surjective instead of étale surjective.

(iii) Our definition 3.6. for an algebraic stack M is nothing but an algebraic stack in the sense of Artin which admits an étale surjective covering. This is because any algebraic space unramified over a scheme is automatically a scheme. Note that the condition that M admits an étale surjective covering can also be formulated as a separation property: the diagonal morphism $M \to M \times_S M$ is unramified.

4.8. Definition. An algebraic stack M over the base scheme S is *separated over* S if the diagonal map $M \to M \times_S M$ (which is representable by 3.6. (i)) is finite. Note that this is equivalent to the apparently weaker condition that it is proper.

Let C be a category in which finite fibred products exist. Recall ([SGA3] exp. V) that a groupoid in C consists of the following data:

(i) Two objects X_0, X_1 in C and two morphisms s, t from X_1 to X_0. (s and t for source and target respectively.)

(ii) a morphism μ from $X_2 := X_1 \times_{X_0} X_1$ to X_1.

Let p_1, p_2 be the first and second projections from X_2 to X_1. These data satisfy the following conditions:

(a) The diagrams below are Cartesian.

$$\begin{array}{ccc} X_2 & \xrightarrow{\mu} & X_1 \\ {\scriptstyle p_1}\downarrow & & \downarrow{\scriptstyle t} \\ X_1 & \xrightarrow{t} & X_0 \end{array} \qquad \begin{array}{ccc} X_2 & \xrightarrow{\mu} & X_1 \\ {\scriptstyle p_2}\downarrow & & \downarrow{\scriptstyle s} \\ X_1 & \xrightarrow{s} & X_0 \end{array} \qquad \begin{array}{ccc} X_2 & \xrightarrow{p_2} & X_1 \\ {\scriptstyle p_1}\downarrow & & \downarrow{\scriptstyle t} \\ X_1 & \xrightarrow{s} & X_0 \end{array}$$

Note that the last one is Cartesian by the definition of X_2.
(b) The following diagram commutes. ("associativity")

$$\begin{array}{ccc} X_1 \times_{s,t} X_1 \times_{s,t} X_1 & \xrightarrow{\mu \times X_1} & X_1 \times_{s,t} X_1 \\ {\scriptstyle X_1 \times \mu}\downarrow & & \downarrow{\scriptstyle \mu} \\ X_1 \times_{s,t} X_1 & \xrightarrow{\mu} & X_1 \end{array}$$

(c) There exists a morphism $\epsilon : X_0 \to X_1$ such that $s \circ \epsilon = t \circ \epsilon = \mathrm{id}_{X_0}$. Clearly ϵ is uniquely determined if it exists.

When C is the category of all sets, the above definition reduces to the usual definition of groupoid. In general the conditions are equivalent to the fact that for any object Y of C the sets $\mathrm{Hom}_C(Y, X_0)$ and $\mathrm{Hom}_C(Y, X_1)$ are the objects (respectively morphisms) of a groupoid.

Let M be an algebraic stack over S. By definition, there exists a scheme X_0 locally of finite type over S and an étale surjective morphism $u : X_0 \to M$. Let X_1 be the the fibred product $X_0 \times_M X_0$, which is a scheme by the definition of algebraic stack. Recall that for any $T \in \mathrm{Ob}(C)$, the category $\mathbf{Hom}_C(T, X_1)^+$ whose objects are elements of $\mathrm{Hom}_C(T, X_1)$ with only identity morphisms is equivalent to the category C_T : $\mathrm{Ob}(C_T) = \{(f, g, \alpha) | f : T \to X_0,\ g : T \to X_0,\ \alpha : f^*(u) \xrightarrow{\sim} g^*(u)$ is an isomorphism in $M\}$. Moreover, the equivalence of categories between $\mathrm{Hom}_C(T, X_1)^+$ and C_T is functorial in T. This description defines source and target morphisms $s, t : X_1 \to X_0$ and composition morphism $\mu : X_2 (= X_1 \times_{X_0} X_1) \to X_1$, and shows the existence of $\epsilon : X_0 \to X_1$. The data (X_0, X_1, s, t, μ) defines a groupoid of schemes with s, t both étale, and X_0, X_1 are of finite type over the base scheme S. We shall call a groupoid of scheme (X_0, X_1, s, t, μ) *an étale groupoid* if s, t are both étale. It is said to be *locally of finite type over* S if X_0, X_1 are both locally of finite type over S. Summarizing, any étale covering $u : X \to M$ of an algebraic stack M gives rise to an étale groupoid locally of finite type over S. Intuitively, M is the quotient of the étale covering X_0 by the étale "equivalence relation" $R = X_1$.

In the above situation, if M is separated over S, then the morphism (s, t) : $X_1 \to X_0 \times_S X_0$ is finite. Conversely, if this morphism is finite for a covering X_0, then M is separated over S.

Convention. From now on we will deal only with locally noetherian and separated algebraic stacks. Therefore *whenever we speak of algebraic stacks, it is understood (by abuse of language) that they are locally noetherian and separated* *over the base scheme S*. We recall that S is assumed to be of finite type over either a field or an excellent Dedekind domain.

With the above convention in mind, we can now summarize our discussion about algebraic stacks in the following

4.9. Working Definition of (Separated) Algebraic Stacks. (A) Fix a base scheme S as before. An (S-separated) algebraic stack M is given by a pair of schemes U, R together with three S-morphisms $s, t : R \to U$ and $\mu : R \times_{U,t,s} R \to R$ such that

(i) s, t are both étale and the morphism $(s, t) : R \to U \times_S U$ is finite.

(ii) s, t, μ form a groupoid (in the category of S-schemes).

Intuitively (ii) means that $(s, t) : R \to U \times_S U$ is a categorical equivalence relation. We remark once again that M should be thought of as the quotient of U by the étale equivalence relation R. Therefore we will write $M = [U/R]$.

(B) Two quintuples $Q = (U, R, s, t, \mu)$, $\overline{Q} = (\overline{U}, \overline{R}, \overline{s}, \overline{t}, \overline{\mu})$ may define the same algebraic stack: If $U' \to U$ is an étale covering, then we get another quadruple (U', R', s', t', μ') by pull-back ($R' = R \times_{(U \times_S U)} (U' \times_S U')$, etc.) which defines the same algebraic stack. Thus Q and $\overline{Q}$ define the same algebraic stack if we can find a third quadruple $\tilde{Q}$ dominating both.

In case for some (and hence all) representative quintuples $Q = (U, R, s, t, \mu)$, the morphism $(s, t) : R \to U \times U$ is a closed embedding, $M = [U/R]$ is (represented by) a separated algebraic space (again simply called algebraic space) in the sense of Artin. This is simply the *definition* of algebraic space.

All notions and operations in algebraic geometry which are compatible with étale descent can be carried over to the context of algebraic stacks. For example, a quasi-coherent sheaf $\mathcal{F}$ on $M = [U/R]$ is given by a quasi-coherent sheaf $\mathcal{F}_U$ on U and an isomorphism $p_1^*\mathcal{F}_U \xrightarrow{\sim} p_2^*\mathcal{F}_U$ satisfying the transitivity condition. Similarly we can define relative schemes over algebraic stacks and abelian schemes over algebraic stacks. We can define fiber products of algebraic stacks. We can talk about flat, étale, smooth morphisms, etc.. The list is very long. In fact one can work out most part of EGA and SGA in the context of algebraic stacks.

As an easy example of an algebraic stack, let G be a finite étale group scheme over S, $U = S$, $R = G$, $s = t =$ the structural morphism, $\mu =$ group law. Consider the stack BG of all principal homogeneous G-schemes over S-schemes. The trivial principal homogeneous space G over S gives a map $S \to \mathrm{BG}$, which is easily seen to be an étale covering of BG. Moreover $S \times_{\mathrm{BG}} S$ is canonically isomorphic to G. Thus $\mathrm{BG} = [S/G]$, and the category $[S/G](T)$ of sections of $[S/G]$ over an S-scheme T consists of all G_T-torsors. We call $\mathrm{BG} = [S/G]$ the classifying stack of G. When $S = \operatorname{Spec} k = \operatorname{Spec} \overline{k}$ is a geometric point, we can identify G with an abstract finite group, still denoted G. In this case a k-valued point of $[S/G]$ is just a principal homogeneous G-set, and a quasi-coherent sheaf on $[S/G]$ is just a $k[G]$-module. The category $[S/G](k)$ is equivalent to the category $\mathcal{G}$ with one object and with $G(k)$ as morphisms, whose classifying space (i.e. geometric realization of its nerve) is the classifying space BG of G (in the topological sense).

4.10. Theorem. *Let $M = [U/R]$ be a noetherian algebraic stack. There is an algebraic space M and a morphism $M \to \mathsf{M}$ such that*

(i) for any algebraically closed field k, $\mathsf{M}(k) = M(k) = U(k)/R(k)$, and

(ii) any morphism from M to an algebraic space factors uniquely through M.

M is called the *coarse moduli space* of M and is obviously characterized by the second property alone. We will write U/R for M. The strict local rings of M can be described as follows: for a geometric point $\overline{x} : \operatorname{Spec} k \to M$, $\operatorname{Aut}(\overline{x}) = \overline{x} \times_M \overline{x}$ is a finite group acting on the strict local ring $\mathcal{O}_{\overline{x},M}$ of M. Then the strict local ring $\mathcal{O}_{\overline{x},\mathsf{M}}$ of the image of $\overline{x}$ in M is equal to $(\mathcal{O}_{\overline{x},M})^{\operatorname{Aut}(\overline{x})}$ = the $\operatorname{Aut}(\overline{x})$-invariants of $\mathcal{O}_{\overline{x},M}$. Note that if $\overline{x}$ factors through U, then $\mathcal{O}_{\overline{x},M} = \mathcal{O}_{\overline{x},U}$ and $\mathcal{O}_{\overline{x},\mathsf{M}}$ is the ring of invariants in $\mathcal{O}_{\overline{x},U}$ under the equivalence relation induced by R.

Thus for example even for a smooth M, its coarse moduli space M may have singularities, which are more or less of an artificial nature. The reason is that the map $U \to \mathsf{M}$ is no longer an étale covering if $\operatorname{Aut}(\overline{x}) \neq \{\mathrm{id}\}$, so M does not represent the quotient of U by R in the étale topology. This added complication is the price we must pay if we do not want to work with algebraic stacks.

The last property we mention is that M is proper over S iff its coarse moduli space M is.

4.11. To show that A_g is an algebraic stack we first have to show that there is an étale covering $U \to \mathsf{A}_g$ where U is an (affine) scheme. In fact we will be able to choose U to be of finite type over $\operatorname{Spec}\mathbb{Z}$, so that not only is A_g quasi-compact but indeed of finite type over $\operatorname{Spec}\mathbb{Z}$. There are at least two ways of doing it. The first way is to use the theory of Hilbert schemes, as in [Gr 2]. This theory together with the rigidity of abelian schemes tells us that there is a universal family $(A \to H,\ A \hookrightarrow \mathbb{P}^N_{/H},\ \lambda : A \xrightarrow{\sim} A^t)$ such that $A \to H$ is an abelian scheme of relative dimension g, λ is a principal polarization, $N = 4^g - 1$, $A \hookrightarrow \mathbb{P}^N_{/H}$ is a (closed) projective embedding such that the pull-back of $\mathcal{O}(1)$ is a rigidified symmetric H-ample invertible sheaf on A giving rise to 4λ (so that the Hilbert polynomial of $A \hookrightarrow \mathbb{P}^N_{/H}$ is $f(n) = 4^g \cdot n^g$). Moreover, H is quasi-projective and therefore of finite type over $\operatorname{Spec}(\mathbb{Z})$. The universal property tells us that PGL_{N+1} naturally acts on $A \to H$. Thus morally A_g "is" the quotient of H by the natural PGL_{N+1}-action. To get an actual étale covering of A_g, use an étale slice argument.

Another way is to use the method developed by M. Artin, c. [Ar 2], [Ar 3], [Ar 4], [Ar 6], which in some sense is more direct: Given any characteristic p geometric point $\overline{x} : \operatorname{Spec} k \to \mathsf{A}_g$, i.e., any g-dimensional principally polarized abelian variety A_k over an algebraically closed field k of characteristic $p > 0$, one first uses deformation theory (Cf. section 3) to get the universal deformation $A^\wedge$ of A_k. $A^\wedge$ is a formal principally polarized abelian variety over $\operatorname{Spf}(R^\wedge)$, where $R^\wedge$ is a formally smooth, complete, noetherian, local domain of dimension $1 + n = 1 + g \cdot (g+1)/2$, of mixed characteristic, with residue field k (hence is isomorphic to $W(k)[\![t_1, ..., t_n]\!]$). By Grothendieck's formal existence theorem ([EGA] III §5) $A^\wedge \to \mathrm{Spec}(R^\wedge)$ is uniquely algebraizable to a principally polarized abelian scheme $A \to \mathrm{Spec}(R^\wedge)$. We can approximate $A \to \mathrm{Spec}(R^\wedge)$ by another principally polarized abelian scheme $A' \to \mathrm{Spec}(R)$, where R is a (strictly) henselian local ring with formal completion isomorphic to $R^\wedge$. A standard limit argument ([EGA IV] §8) gives us a principally polarized abelian scheme $A^\dagger \to \mathrm{Spec}(R^\dagger)$ where $R^\dagger$ is a subring of R of finite type over $\mathbb{Z}$ such that $\mathrm{Spec}(R) \to \mathrm{Spec}(R^\dagger)$ is étale and $A^\dagger$ is still versal at $\overline{x} : \mathrm{Spec}\, k \to \mathrm{Spec}(R^\dagger)$. By the openness of versality (in this case it suffices to compute the Kodaira-Spencer map) we may assume that $A^\dagger \to \mathrm{Spec}(R^\dagger)$ is versal at every point of $\mathrm{Spec}(R^\dagger)$. Therefore we have found an étale morphism $\mathrm{Spec}(R^\dagger) \to \mathsf{A}_g$ covering a neighborhood of $\overline{x}$. This shows that we can find an étale covering of A_g. Since A_g is quasi-compact (which follows from the theory of Hilbert schemes), we can cover A_g by a finite number of affine schemes as above.

Having found an étale covering $U \to \mathsf{A}_g$, the equivalence relation R is just $U \times_{\mathsf{A}_g} U$ (what else can it be?). Let $(A_U \to U, \lambda_U)$ be the principally polarized abelian scheme corresponding to $U \to \mathsf{A}_g$ and $p_1^*(A_U, \lambda_U)$, $p_2^*(A_U, \lambda_U)$ be the pull-back of (A_U, λ_U) via the two projections $p_1, p_2 : U \times U \to U$. Clearly $U \times_{\mathsf{A}_g} U = \mathbf{Isom}_{U\times U}(p_1^*(A_U, \lambda_U), p_2^*(A_U, \lambda_U))$, a scheme finite over $U \times U$ which is clearly étale over U via either projection. The groupoid structure of R is just the composition of isomorphisms. Therefore, A_g is indeed an algebraic stack.

We introduce two useful coherent sheaves on A_g. The first one is a locally free sheaf Ω on A_g of rank g. The second is just $\Lambda^g(\Omega)$, denoted ω, an invertible sheaf. To define Ω, we will define for every $S \to \mathsf{A}_g$ a locally free rank g sheaf Ω_S on S which is compatible with pull-backs. This clearly suffices. A morphism $S \to \mathsf{A}_g$ corresponds to a principally polarized abelian scheme $(A \to S, \lambda)$, and we define Ω_S to be $e^*\Omega^1_{A/S}$, where $e : S \to A$ is the zero section. Clearly this construction is functorial in S.

We introduced (in 4.4) a variant of A_g by considering level structures. Let $1 \le n \in \mathbb{N}$ be an integer. Recall that $\mathsf{A}_{g,n}$ is the moduli stack which classifies principally polarized abelian schemes with level-n-structure. By exactly the same argument as before, $\mathsf{A}_{g,n}$ is an algebraic stack. If $n \ge 3$, Serre's lemma tells us that $\mathsf{A}_{g,n}$ is an algebraic space smooth over $\mathrm{Spec}(\mathbb{Z}[1/n])$. In fact, Mumford proved in [GIT] that $\mathsf{A}_{g,n}$ is actually a quasi-projective scheme over $\mathrm{Spec}(\mathbb{Z}[1/n])$.

5. Algebraic Theory of Theta Functions

In this section we give a basic review of the algebraic theory of theta functions. We will not even define the adelic algebraic theta functions since we do not need them. Readers are referred to [Mum 2] for a systematic development of the theory. All we will do is to explain the all-important notion of theta groups as conceived by Mumford. We note that the interpretation of theta groups and algebraic theta functions via cubical structures is due to Breen [Br 1], and the extension to the semi-abelian case has been documented by Moret-Bailly in [MB 1].

Let $A \to S$ be an abelian scheme, $\mathcal{L}$ be an invertible sheaf on A rigidified along the zero section. $\mathcal{L}$ defines an isomorphism $\lambda(\mathcal{L}) : A \to A^t$ given by the invertible sheaf $\Lambda(\mathcal{L}) = \mu^*\mathcal{L} \otimes p_1^*\mathcal{L}^{\otimes -1} \otimes p_2^*\mathcal{L}^{\otimes -1}$. Let $K(\mathcal{L}) = \ker \lambda(\mathcal{L})$. We will think of invertible sheaves and G_m-torsors interchangeably. By the definition of $K(\mathcal{L})$, $\Lambda(\mathcal{L})|A \times_S K(\mathcal{L})$ is a trivial biextension. This means that the G_m-torsor (attached to) $\mathcal{L}|K(\mathcal{L})$ has a canonical S-group scheme structure and acts on the G_m-torsor (attached to) $\mathcal{L}$. Denote this S-group scheme by $\mathcal{G}(\mathcal{L})$. It is easy to see that $\mathcal{G}(\mathcal{L})$ is a central extension of $\mathrm{K}(\mathcal{L})$ by G_m

$$0 \to G_m \to \mathcal{G}(\mathcal{L}) \to K(\mathcal{L}) \to 0,$$

and the commutator pairing

$$e(\mathcal{L}) : K(\mathcal{L}) \times_S K(\mathcal{L}) \to G_m$$

is equal to the Weil pairing associated to $\mathcal{L}$ if $K(\mathcal{L})$ is finite. Moreover, $\mathcal{G}(\mathcal{L})$ can be identified with the group $\{(x, \phi) | x \in K(\mathcal{L}),\ \phi : \mathcal{L} \xrightarrow{\sim} T_x^*\mathcal{L}\}$ (read as functorial points) because A is an abelian scheme. It will be called the theta group associated to $\mathcal{L}$.

We keep the above notations and assume moreover that $\mathcal{L}$ is an ample invertible sheaf. Then $K(\mathcal{L}) \to S$ is a finite flat subgroup scheme of $\pi : A \to S$. The center of the theta group $\mathcal{G}(\mathcal{L})$ is G_m, and the pairing $e(\mathcal{L}) : K(\mathcal{L}) \times_S K(\mathcal{L}) \to G_m$ is non-degenerate. This pairing determines $\mathcal{G}(\mathcal{L})$ up to a commutative extension of $K(\mathcal{L})$ by G_m. Also all automorphisms of $\mathcal{G}(\mathcal{L})$ which fix G_m and $K(\mathcal{L})$ are inner.

The representation $\pi_*\mathcal{L}$ of $\mathcal{G}(\mathcal{L})$ is a locally free sheaf of rank $(\operatorname{rk} \pi_*\mathcal{L}) = \deg(\lambda(\mathcal{L}))^{1/2}$, and $\mathcal{G}(\mathcal{L})$ acts on $\pi_*\mathcal{L}$ by functoriality and its center G_m acts via the identity representation. Tensoring by $\pi_*\mathcal{L}$ induces an equivalence of categories between $\mathcal{O}_S$-modules and $\mathcal{G}(\mathcal{L})$-representation on $\mathcal{O}_S$-modules where G_m acts by the identity. This fact characterizes $\pi_*\mathcal{L}$ up to tensor product with a line bundle.

Assume that $\mathcal{L}$ is relatively ample, symmetric, of degree d^2, and trivialized along the zero-section. In this case we can use the Grothendieck-Riemann-Roch theorem to determine the determinant of $\pi_*\mathcal{L}$ up to torsion of well defined order (this answers a question in [MB 1]). Let d denote the rank of $\pi_*\mathcal{L}$ (so $d^2 = \deg(\lambda(\mathcal{L}))$ which for simplicity we assume to be constant, $\Omega_{A/S}$ the sheaf of invariant differential forms on A, and $\omega_{A/S}$ its determinant.

5.1. Theorem. $8 \cdot d^3 \cdot \det(\pi_*\mathcal{L}) = -4 \cdot d^4 \cdot \omega_{A/S}$ *in* $\operatorname{Pic}(S)$.

Proof. Let us first assume that S is noetherian and quasi-projective. We denote by $K(S)$ the Grothendieck group of *vector bundles* on S. The determinant defines a homomorphism $\det : K(S) \to \operatorname{Pic}(S)$, and the direct image a transformation $\pi_* : K(A) \to K(S)$. We consider the function $f : \mathbb{Z} \to \operatorname{Pic}(S)$ defined by $f(n) = n^g \cdot \det(R\pi_*([\mathcal{L}^{\otimes n}])$, (where g is the relative dimension of A over S). By Serre duality we know that $f(-n) = -d \cdot n^{2g} \cdot [\omega_{A/S}] - f(n)$; hence the assertion (or rather, something even stronger) would follow from $f(-1) = f(1)$. We would know that $f(-1) = f(1)$ if for some integer D and every $n \geq 0$ and congruent to -1 modulo D, we had

$$(^*) \quad f(n) = n^{2g} \cdot f(1) \ .$$

This is because $f(\,)$ is a "numerical function"; that is, if Δ denotes the difference operator $\Delta(f)(n) = f(n) - f(n-1)$, then $\Delta^m(f) = 0$ for $m \gg 0$: To show this we may use the function $\tilde{f}(n) = \det(R\pi_*([\mathcal{L}^{\otimes n}])$ instead of f. If $\mathcal{L}$ is very ample we use it to embed A into some projective bundle $\mathbb{P}(\mathcal{E})$. Now $K(\mathbb{P}(\mathcal{E}))$ is generated over $K(S)$ by the powers of the tautological line bundle $\mathcal{O}(1)$. In particular, this applies to the direct image $i_*\mathcal{O}_A$ of $\mathcal{O}_A$, $i : A \hookrightarrow \mathbb{P}(\mathcal{E})$ the projective embedding, and the assertion follows easily. For a general ample $\mathcal{L}$ choose co-prime a and b such that $\mathcal{L}^{\otimes a}$ and $\mathcal{L}^{\otimes b}$ are very ample, and study the function (of two variables m and n) $\tilde{f}(a \cdot m + b \cdot n)$ by embedding into the product of the two projective bundles.

It follows that f is a linear combination (with coefficients in $\mathrm{Pic}(S)$) of binomial coefficients $(n+r, r) = \{n \cdot (n-1) \cdot \ldots \cdot (n-r+1)\}/r!$, $r = 0, 1, 2, \ldots$. However for such a function $f(-1) = f(1)$ if $f(n) = n^{2g} \cdot f(1)$ for any $n \equiv -1$ mod D: We may assume that the function takes values in a finitely generated abelian group, and then that this group is $\mathbb{Z}$ or $\mathbb{Z}/N \cdot \mathbb{Z}$. In the first case $f(n) = n^{2g} \cdot f(1)$ for all n, while in the second case one finds an $M \gg 0$ such that $f(n)$ depends only on n modulo M.

So it remains to show (*). We cannot do this for $\mathcal{L}$. We can, however, show it for $\mathcal{M} = \mathcal{L}^{\otimes 4}$ (exterior tensor product) over $B = A^4$. The point is that for any $n \geq 0$ there exists an isogeny $\alpha : B \to B$ such that $\alpha^*(\mathcal{M}) \approx \mathcal{M}^{\otimes n}$. The construction is suggested by ideas of J. Zarhin. α is given by a 4×4 matrix Z with $Z \cdot Z^t = n \cdot I$. To get such a Z write $n = a^2 + b^2 + c^2 + d^2$ as the sum of four squares, and consider multiplication by the quaternion $z = a + bi + cj + dk$ on the integral quaternions. Furthermore with this choice of Z the cokernel $\mathbb{Z}^4/Z(\mathbb{Z}^4)$ is of the form $L \oplus L$ for a finite abelian group L (since z lies in a quadratic subfield of the quaternions). It follows that for n prime to d the determinant of the bundle $\mathcal{O}_H$, of regular functions on the kernel H of α, is trivial: H is isomorphic to a direct sum of an even number of copies of $A[m]$, for divisors m of n. But as $A[m]$ is self-dual, the determinant of its algebra of functions has square one. This assertion of course also holds for the dual H^t.

Now assume that n is prime to the degree d^8 of $\mathcal{M}$. Then $K(\mathcal{M}^{\otimes n}) = K(\mathcal{M}) \oplus B[n]$ decomposes as an orthogonal direct sum, and so does its central extension $\mathcal{G}(\mathcal{M}^{\otimes n})$. Furthermore, $H = \mathrm{Ker}(\alpha)$ lifts to a subgroup $\mathcal{H}$ of $\mathcal{G}(\mathcal{M}^{\otimes n})$ with commutator $\mathcal{H}^c$ such that $\mathcal{H}^c/\mathcal{H} \approx \mathcal{G}(\mathcal{M})$, and $\pi_*(\mathcal{M}^{\otimes n})$ is the representation induced from $\pi_*(\mathcal{M})$ under the inclusion $\mathcal{H}^c \subset \mathcal{G}(\mathcal{M}^{\otimes n})$. If we restrict to the pre-image of $B[n]$ in $\mathcal{G}(\mathcal{M}^{\otimes n})$ we obtain the tensor product of $\pi_*(\mathcal{M})$ and some representation induced from a rank 1 representation of $\mathcal{H}^c$ on $\mathcal{O}_S$ whose n-th power extends to a representation of the whole group. The determinant of this induced representation is then, up to terms annihilated by n, equal to that of the affine algebra $\mathcal{O}_H$ of H; that is, it is trivial. Thus $n^{4g} \cdot \det(\pi_*(\mathcal{M}^{\otimes n})) = n^{8g} \cdot \det(\pi_*(\mathcal{M}))$.

Now the argument at the start implies that $8 \cdot d^3 \cdot \det(\pi_*(\mathcal{L})) = 2 \cdot \det(\pi_*(\mathcal{M})) = -4 \cdot d^4 \cdot [\omega_{A/S}]$, so that the theorem holds for quasi-projective S. For a more general base choose some $r \geq 3$ and consider the $\mathrm{PGL}(N)$-torsor S' over S $(N = r^g \cdot d - 1)$ which classifies bases of $\pi_*(\mathcal{L}^{\otimes r})$, up to scalars. It suffices to prove the equality for the induced $\mathrm{PGL}(N)$-equivariant line bundles on S'. However as $\mathrm{PGL}(N)$ has no non-trivial characters; a line bundle can have at most one $\mathrm{PGL}(N)$-action (see [GIT]), so we may forget about equivariance. However there exists a quasi-projective moduli space for tuples $(A, \mathcal{L}$, basis of $\pi_*(\mathcal{L}^{\otimes r}))$ (use Hilbert schemes, see [Gr 1]) over which the assertion holds, and by pullback to S' we are done. □

5.2. Remark. The result is sharp for principal polarizations $(d = 1)$. This follows from the 8-th roots of unity appearing in the transformation formulas for the classical theta series (see section 6). For general degree d one can probably eliminate the factor d^3. At least we may sharpen the argument (using equivariant K-groups) to show the following fact, which determines the $\mathcal{G}(\mathcal{L})$-module $\pi_*(\mathcal{L})$ (and thus its determinant) up to tensor product by a line bundle with trivial 8-th power:

Choose α as above for some $n \equiv -1 \bmod d$. α induces an automorphism of $K(\mathcal{M})$ which extends to an automorphism of $\mathcal{G}(\mathcal{M})$ equal to $-\mathrm{Id}$ on the center G_m. Via this automorphism $\mathcal{G}(\mathcal{M})$ operates on the dual of $\pi_*(\mathcal{M})$, and as $\mathcal{G}(\mathcal{M})$-module this dual is isomorphic to $\pi_*(\mathcal{M}) \otimes \omega_{A/S}^{\otimes 4}$.

Theorem 5.1. is important for the construction of modular forms. For example, assume that $d = 1$, that is, that $\mathcal{L}$ defines a principal polarization. If $s : S \to A$ is a section of f which is torsion $(n \cdot s = 0$ for some positive $n)$ $s^*(\mathcal{L})^{\otimes 2n}$ has the same class as $e^*(\mathcal{L})^{\otimes 2n}$ in $\mathrm{Pic}(S)$, i.e., is trivial. Now evaluation at s defines a map of line bundles $\pi_*\mathcal{L} \to s^*(\mathcal{L})$, whose $8 \cdot n$-th power defines a section of $\omega_{A/S}^{\otimes 4n}$ over S.

Theorem 5.1 was proved without explicit bound on torsion independently by Mumford (unpublished) and Moret-Bailly/Szpiro ([MB 1] Appendice 2), using Grothendieck-Riemann-Roch. Moret-Bailly first observed that one can use the classical theta transformation formula to obtain an explicit bound of torsion (to appear). The proof given here is new.

We also include for reference two classical results:

5.3. Proposition. *Suppose A is an abelian variety over an algebraically closed field k, $\mathcal{L}_1$ and $\mathcal{L}_2$ are ample line bundles on A. Then $\Gamma(A, \mathcal{L}_1 \otimes \mathcal{L}_2)$ is spanned by the images of $\Gamma(A, \mathcal{L}_1 \otimes \mathcal{M}) \otimes \Gamma(A, \mathcal{L}_2 \otimes \mathcal{M}^{-1}) \to \Gamma(A, \mathcal{L}_1 \otimes \mathcal{L}_2)$, $\mathcal{M}$ an element of* $\mathrm{Pic}^0(A) = A^t(k)$. *(Of course it suffices if $\mathcal{M}$ runs over a Zariski dense subset of $A^t(k)$).*

Proof. (Mumford) The space spanned by the images is stable under $\mathcal{G}(\mathcal{L}_1 \otimes \mathcal{L}_2)$.

5.4. Proposition. *Suppose A is an abelian variety over a field k, $\mathcal{L}$ an ample line bundle on A which defines a principal polarization, and $\Theta \subset A$ the zero-set of its unique global section. Then for any $n \geq 2$ Θ does not contain (in the sense of schemes) $A[n]$.*

Proof. The subspace of $\Gamma(A, [n]^*\mathcal{L})$ of sections which vanish on $A[n^2]$ is stable under $\mathcal{G}([n]^*\mathcal{L})$, thus zero as $[n]^*\mathcal{L}$ is generated by its global sections. Apply this to the pullback of the unique global section of $\mathcal{L}$.

We recall some more basic ideas about algebraic theta functions. However, except for statement (2) below, we shall not use them in the rest of the book:

(1) Clearly, evaluating elements of $\pi_*\mathcal{L}$ at the zero section gives a morphism $\tilde{\theta}_{\text{null}} : S \to \mathbb{V}(\pi_*\mathcal{L}) =$ the vector group associated to $\pi_*\mathcal{L}$. We *assume that* $\tilde{\theta}_{\text{null}}$ *induces a morphism* $\theta_{\text{null}}(A, \mathcal{L}) : S \to \mathbb{P}(\pi_*\mathcal{L})$. (This is always the case if $K(\mathcal{L})$ is "big".) We call $\theta_{\text{null}}(A, \mathcal{L})$ the theta null of $(A, \mathcal{L})$. This is the algebraic analogue (and a very good one) of the classical theta constants, because over $\mathbb{C}$ sections of ample invertible sheaves on abelian varieties are given by theta functions. Note that $\mathbb{P}(\pi_*\mathcal{L})$ is canonically determined by $\mathcal{G}(\mathcal{L})$.

(2) The addition formula in the theory of theta functions can be formulated algebraically in terms of the important isogeny

$$\phi : A \times_S A \to A, \ (x, y) \mapsto (x + y, x - y) :$$

If $\mathcal{L}$ is symmetric, then $\phi^*(pr_1^*\mathcal{L} \otimes pr_2^*\mathcal{L}) \cong pr_1^*\mathcal{L}^{\otimes 2} \otimes pr_2^*\mathcal{L}^{\otimes 2}$. Choose canonical bases for $\Gamma(pr_1^*\mathcal{L} \otimes pr_2^*\mathcal{L})$ and $\Gamma(pr_1^*\mathcal{L}^{\otimes 2} \otimes pr_2^*\mathcal{L}^{\otimes 2})$ using $\mathcal{G}(pr_1^*\mathcal{L} \otimes pr_2^*\mathcal{L})$ and $\mathcal{G}(pr_1^*\mathcal{L}^{\otimes 2} \otimes pr_2^*\mathcal{L}^{\otimes 2})$ respectively. The map

$$\phi^* : \Gamma(pr_1^*\mathcal{L} \otimes pr_2^*\mathcal{L}) \to \Gamma(pr_1^*\mathcal{L}^{\otimes 2} \otimes pr_2^*\mathcal{L}^{\otimes 2})$$

can then be expressed explicitly in terms of the canonical bases. This gives the addition formula.

(3) When $\mathcal{L}$ is actually very ample (which is the case when $K(\mathcal{L})$ is "big"), equations defining the projective embedding $A \hookrightarrow \mathbb{P}(\pi_*\mathcal{L})$ can be written down explicitly. The key point is a non-degeneracy condition on theta constants, which can be deduced from the addition formula. (Contrary to the elliptic curve case, *a lot* of equations are needed. This makes the equations harder to use than in the elliptic curve case.)

(4) More significantly, the theta null morphism gives the moduli, at least outside characteristic 2. More precisely, fixing the type of the polarization (of degree d), there is a finite flat covering M of $\mathsf{A}_g[1/d]$ étale over $\mathsf{A}_g[1/2d]$ and a theta structure on the pull-back of the universal principally polarized abelian scheme (with an ample invertible sheaf $\mathcal{L}$), such that the theta null morphism embeds M in $\mathbb{P}^N = \mathbb{P}(\pi_*\mathcal{L})$ as a locally closed subscheme. Moreover it is possible to give explicit equations (the number of which is really big) which cut out a closed subscheme Z of $\mathbb{P}^N$ containing M as an open subscheme. The universal family over M can also be described by equations.

The remarks above come from [Mum 2] I, II. One corollary of these ideas is that the explicit construction of the (very) ample line bundle $\mathcal{O}(1)$ on $M \subset \mathbb{P}^N$.

Theta constants are global sections of $\mathcal{O}(n)$. So far this method is *the* basis of constructing ample invertible sheaves on A_g.

We record in this connection a useful lemma of Zarhin, usually referred to as "Zarhin's trick" (Cf. [Za 4], also [MB 1] IX 1.1, p. 205.)

5.5. Lemma. *(Zarhin) If $(A \to S,\ \lambda : A \to A^t\ )$ is a polarized abelian scheme, then $(A \times_S A^t)^4$ has a principal polarization, canonically attached to (A, λ).*

6. Complex Uniformization of A_g

We can define the moduli stack of principally polarized abelian varieties over the category of complex analytic spaces and get a stack $\mathsf{A}_g^{\mathrm{an}}$, the analytic version of A_g. (We propose to call it a complex analytic stack, but it might be called an analytic orbifold.) Every one knows (or should know) what it is:

$$\begin{aligned} \mathsf{A}_g^{\mathrm{an}} &= [\mathbb{H}_g/\mathrm{Sp}(2g,\mathbb{Z})], \quad \text{where} \\ \mathbb{H}_g &= \{\Omega \in M_{g\times g}(\mathbb{C}) | \Omega = {}^t\Omega,\ \mathrm{Im}(\Omega) \gg 0\} \end{aligned}$$

is the Siegel upper half space of genus g, and

$$\mathrm{Sp}(2g,\mathbb{Z}) = \{\gamma \in \mathrm{GL}(2g,\mathbb{Z}) | \, {}^t\gamma J \gamma = J\}$$

with $J = \begin{pmatrix} 0 & I_g \\ -I_g & 0 \end{pmatrix}$; the elements of $\mathrm{Sp}(2g,\mathbb{Z})$ are of the form $\gamma = \begin{pmatrix} A & B \\ C & D \end{pmatrix}$ where A, B, C, D are integral $g \times g$ matrices with $A \cdot {}^tB = B \cdot {}^tA,\ C \cdot {}^tD = D \cdot {}^tC,\ A \cdot {}^tD - B \cdot {}^tC = I_g$. Thus $\mathbb{H}_g$ is an étale covering of $\mathsf{A}_g^{\mathrm{an}}$, $\mathrm{Sp}(2g,\mathbb{Z})$ acts on $\mathbb{H}_g$ via the classical formula $\gamma : \Omega \mapsto (A\Omega + B) \cdot (C\Omega + D)^{-1}$ for γ as above. The étale equivalence relation $p_1, p_2 : R \to \mathbb{H}_g \times \mathbb{H}_g$ is given by $R = \mathbb{H}_g \times \mathrm{Sp}(2g,\mathbb{Z})$, p_1 is the first projection and p_2 is the group action. The universal family of principally polarized abelian varieties A over $\mathbb{H}_g$ has $A_\Omega = \mathbb{C}^g/(\mathbb{Z}^g + \Omega \cdot \mathbb{Z}^g)$ as the fibre over $\Omega \in \mathbb{H}_g$ (think of $\mathbb{C}^g$ and $\mathbb{Z}^g$ as column vectors). The principal polarization on A_Ω is induced by the hermitian form $(u, v) \mapsto u \cdot \mathrm{Im}(\Omega)^{-1} \cdot \overline{v}$ for $u, v \in \mathbb{C}^g$.

Under the uniformization $\mathsf{A}_g^{\mathrm{an}} = [\mathbb{H}_g/\mathrm{Sp}(2g,\mathbb{Z})]$, the locally free sheaf Ω on $\mathsf{A}_g^{\mathrm{an}}$ is the trivial sheaf $(\mathcal{O}_{\mathbb{H}_g})^{\oplus g}$ with action of $\mathrm{Sp}(2g,\mathbb{Z})$ given by the 1-cocycle $(C\Omega + D)$. The invertible sheaf ω on $\mathsf{A}_g^{\mathrm{an}}$ is the trivial sheaf $\mathcal{O}_{\mathbb{H}_g}$ with action of $\mathrm{Sp}(2g,\mathbb{Z})$ given by the 1-cocycle $\det(C\Omega + D)$.

Recall that the classical Riemann theta function is the complex holomorphic function on $\mathbb{C}^g \times \mathbb{H}_g$ given by the theta series

$$\theta(\zeta, \Omega) = \sum\nolimits_{n \in \mathbb{Z}^g} e(1/2 \cdot {}^tn \cdot \Omega \cdot n + {}^tn \cdot \zeta)$$

where $e(\cdot) = \exp(2\pi\sqrt{-1}\cdot) : \mathbb{C} \to \mathbb{C}^*$ is the exponential function. $\theta(\zeta, \Omega)$ gives a non-zero section of a symmetric, ample, invertible, cubical sheaf $\mathcal{L}_\Omega$ on A_Ω, which defines the principal polarization. However $\mathcal{L}_\Omega$ does not descend to A. The pull-back of $\mathcal{L}_\Omega$ to $\mathbb{C}^g$ is cubically trivial. An easy calculation shows that

$$\theta(\zeta + m, \Omega) = \theta(\zeta, \Omega) \ \forall m \in \mathbb{Z}^g,$$
$$\theta(\zeta + m \cdot \Omega, \Omega) = e(-1/2 \cdot {}^t m \cdot \zeta - 1/2 \cdot {}^t m \cdot \Omega \cdot m) \cdot \theta(\zeta, \Omega) \ \forall m \in \mathbb{Z}^g.$$

Thus we see that if we write $E = e^g : \mathbb{C}^g \to (\mathbb{C}^*)^g$ for the homomorphism "exponentiating away half of the periods $\mathbb{Z}^g \subset \mathbb{Z}^g + \Omega \cdot \mathbb{Z}^g$," then

(i) $\theta(\zeta, \Omega)$ factorizes through E.

(ii) $A_\Omega = (\mathbb{C}^*)^g / E(\Omega \cdot \mathbb{Z}^g)$ and $\mathcal{L}_\Omega$ is defined by the 1-cocycle $e(-1/2 \cdot {}^t m \cdot \zeta - 1/2 \cdot {}^t m \cdot \Omega \cdot m)$ on $(\mathbb{C}^*)^g$. Note that this 1-cocycle has the form e(term linear in $\zeta +$ term constant in ζ)

(iii) as a function on $(\mathbb{C}^*)^g$, the theta series $\theta(\zeta, \Omega)$ assumes the form

$$\sum\nolimits_\chi \text{(quadratic function } a(\chi) \text{ in } \chi) \cdot \chi\ ,$$

where χ runs through all characters of the algebraic torus $(\mathbb{C}^*)^g$. Since $a(\chi) = e(1/2 \cdot {}^t n \cdot \Omega \cdot n)$ if χ is $E(\zeta) \mapsto E({}^t n \cdot \zeta)$, we see that *the quadratic function* $a(\cdot)$ *determines the (multiplicative) periods* $E(\Omega \cdot \mathbb{Z}^g)$! Pursuing analogue of the above in the algebraic context proves to be the key to understanding the degeneration of abelian varieties.

As remarked above the theta-series does not descend to A. However we have a covering $N_g \to A_g$ (of degree 2^{2g}) where N_g classifies principally polarized abelian varieties of dimension g, together with a symmetric theta divisor (or equivalently an ample symmetric sheaf which defines the polarization). N_g can be realized as the quotient of $\mathbb{H}_g \times (1/2 \cdot \mathbb{Z}/\mathbb{Z})^{2g}$ under an action of Γ which differs from the product action by a 1-cocycle $\rho : \Gamma \to (1/2 \cdot \mathbb{Z}/\mathbb{Z})^{2g}$. This 1-cocycle can be read off from the condition that for $(\Omega, a, b)(a, b \in (1/2 \cdot \mathbb{Z}/\mathbb{Z})^g)$ the zero set of $\theta(\zeta + a + b\Omega, \Omega)$ should descend to the universal symmetric theta divisor on A_Ω, that is ρ is determined by the behavior of $\theta(\zeta, \Omega)$ under transformations by Γ.

Another approach is to use the fact that the parity of $\theta(\zeta + a + b\Omega, \Omega)$ at the origin is given by the quadratic form e with $e(a, b) = (-1)^{4 \cdot a \cdot b}$. Under the action of $\gamma \in \Gamma$, e is changed to the sum of a constant and the $\rho(\gamma)$-translate of e. It follows that for a γ with entries A, B, C, D, $\rho(\gamma)$ is equal to the pair with entries $a = 1/2 \cdot \mathrm{diag}(B \cdot D^t)$ and $b = 1/2 \cdot \mathrm{diag}(A \cdot C^t)$. Now $\theta(\zeta + a + b\Omega, \Omega)$ becomes a global section of the universal symmetric line bundle on the pullback of A to N_g. So the corresponding Γ-equivariant line bundle on $\mathbb{H}_g \times (1/2 \cdot \mathbb{Z}/\mathbb{Z})^{2g}$ can be defined by the same factors of automorphy which appear in the transformation formulas for Θ. We remark that N_g has two connected components (corresponding to the sign of $e(a, b)$, or the parity at the origin of the theta divisor), which are quotients of $\mathbb{H}_g$ by subgroups of Γ.

The reader can also consult the first 6 pages of chapter VI §4 and chapter VII §1 for the relation with the group of symplectic similitudes. See also chapter V §3 for more information about theta series, and chapter IV §7 for theta level structures. Further references are [Fr], [I 3], [I 5], [Mum 7], [Mum 2].

Chapter II. Degeneration of Polarized Abelian Varieties

0. Introduction

The subject here is the degeneration of polarized abelian varieties over a noetherian normal domain complete w.r.t. some ideal. We begin by considering the following complex analogue. Let $S^* = (\Delta^*)^k$, $S = \Delta^k$, where Δ^* denotes the punctured unit disk and Δ the unit disk in $\mathbb{C}$. Let G^* be a family of abelian varieties of dimension g over S^*. In general, this family may degenerate over S. We shall assume that G^* can be extended to a semi-abelian family over S, i.e. $G = \{G_s\}_{s \in S}$ is a family of group varieties each fiber of which is an extension of an abelian variety by an algebraic torus. This is reasonable from the point of view of the semi-stable reduction theorem and the theory of toroidal embeddings (Cf. [AMRT], [Na 1].) To understand this degeneration, we consider the exponential map $\exp_S : \mathbf{Lie}(G/S) \to G$ and its restriction $\exp_{S^*} : \mathbf{Lie}(G^*/S) \to G^*$. $\mathcal{K} =$ kernel of $\exp_S$ is a sheaf of torsion free $\mathbb{Z}$-modules of finite rank extending the local system $\mathcal{K}^* = \mathcal{K}|_{S^*} = \{H_1(G_s, \mathbb{Z})\}_{s \in S^*}$. The fiber of $\mathcal{K}$ over $o \in S$, K_o, can be uniquely extended to a local system $\mathcal{I} \subset \mathcal{K}$ of rank $2g - r$, where $r =$ dimension of the torus part of G_o. The exponential map $\exp_S$ factors through $\mathcal{I}$, and G is canonically a quotient of $\tilde{G} = \mathbf{Lie}(G/S)/\mathcal{I} \ : \ G = \tilde{G}/(\mathcal{K}/\mathcal{I})$. Over S^*, $\mathcal{K}/\mathcal{I}$ is a local system of torsion free $\mathbb{Z}$-modules of rank r which we would like to think of as "periods" in $\tilde{G}$. These periods degenerate over S so that $\tilde{G}_o = G_o$.

Let us assume that we are given a family of polarizations of G^*. In other words we have an $\mathbb{R}$-bilinear alternating pairing

$$E \ : \ \mathbf{Lie}(G^*/S) \ \times \ \mathbf{Lie}(G^*/S) \to \mathbb{R}$$

which takes integer values on the lattice $\mathcal{K}^* \times \mathcal{K}^*$ such that the form $(z, w) \to E(\sqrt{-1}\, z, w)$ is symmetric and positive definite. If we pull back the family over S^* to the universal covering $\mathbb{H}^k$ of S^* and choose a standard symplectic basis of $\mathcal{K}^*$ w.r.t. E, we get a period map f from $\mathbb{H}^k$ to the Siegel upper half space. When the base is 1-dimensional, i.e. $k = 1$, this period map degenerates in the following fashion: Let z be the coordinate of $\mathbb{H}$. As z goes to infinity within a vertical strip, $f(z)$ can be given in matrix form w.r.t. suitably chosen basis of

$\mathcal{K}^*$ as

$$\begin{pmatrix} \omega(z) & \beta(z) \\ {}^t\beta(z) & \delta(z) \end{pmatrix} \qquad \omega : (g-r) \times (g-r) \qquad \delta : r \times r \qquad \beta : (g-r) \times r$$

such that $\omega(z)$ goes to some period matrix ω_0 of genus $g-r$, $\beta(z)$ stays bounded, and $\delta(z)$ goes to $\sqrt{-1}\,\infty$.

It is well known that the projective coordinates of abelian varieties are given by theta functions, which correspond to (algebraic) sections of ample line bundles. Hence analyzing degeneration of theta functions will certainly give all the relevant information about degeneration of polarized abelian varieties. This is really the key point of this chapter. Let us illustrate this in the "most degenerate case" $r = g$. The standard Riemann theta function is given by

$$\theta(z, \Omega) = \sum\nolimits_{n \in \mathbb{Z}^g} \exp(\pi\sqrt{-1}\,{}^t n\ \Omega\ n + 2\pi\sqrt{-1}\,{}^t n\ z)$$

where n and z are regarded as column vectors in $\mathbb{Z}^g$ and $\mathbb{C}^g$. Note that $\theta(z, \Omega)$ depends only on $\exp(2\pi\sqrt{-1}\,z)$ as a function of z. In other words, $\theta(z, \Omega)$ is really a function on $\mathbb{C}^g/2\pi\sqrt{-1}\,\mathbb{Z}^g \approx (\mathbb{C}^*)^g$. Letting $w_1, \ldots, w_n$ denote the coordinates on $(\mathbb{C}^*)^g$ and $q_{i,j} = \exp(\pi\sqrt{-1}\,\Omega_{i,j})$, we rewrite

$$\theta(z, \Omega) = \sum\nolimits_{n \in \mathbb{Z}^g} (\Pi_{1 \le i,j \le g}\, q_{i,j}^{n_i n_j}) \cdot (\Pi_{1 \le i \le g}\, w_i^n i)\,.$$

The point is that the coefficients of $(\Pi_{1 \le i \le g}\, w_i^n i)$ in the above expansion determine the periods and the polarization. In the case $r = g = 1$, Tate observed that the last series makes sense in the rigid analytic category and used it to construct the "Tate curve" as the quotient of G_m by a multiplicative period. His work was the genesis of works of Mumford and Raynaud to be discussed later.

Now we describe the goal of this chapter and the next. First some notations: Let R be a noetherian *normal* integral domain *complete* w.r.t. some ideal I, $S = \mathrm{Spec}R$, $S_0 = \mathrm{Spec}(R/I)$, η = generic point of S, K = quotient field of R. We make an additional hypothesis for technical reasons: *for any étale R/I -algebra $R_0^\dagger$, its unique lifting $R^\dagger$ to a formally étale R-algebra is normal.* The collection of all semi-abelian schemes over S whose generic fibre is abelian form (the objects of) a category DEG, the category of all semi-abelian degenerations over S. We would like to analyze such degenerations as follows:

1) Let G/S be a semi-abelian scheme $G \to S$ whose generic fiber G_η is abelian. Then we would like to assert that G arises as "the quotient" of a semi-abelian scheme $\tilde{G} \to S$ which is the extension of an abelian scheme A by a torus T over S, by a "polarizable period subgroup $\underline{Y}$ in $\tilde{G}(K)$" of rank $\dim(T/S)$. ($\underline{Y}$ will turn out to be an étale sheaf.) $\tilde{G}$ is related to G via formal completion: their I-adic completions are canonically isomorphic, and one should think of $\tilde{G}$ as an "infinite covering" of G.

2) Conversely if we have a semi-abelian scheme $\tilde{G} \to S$ and a "polarizable period subgroup $\underline{Y}$ in $\tilde{G}(K)$" as above, we would like to construct a "quotient"

$G \to S$, which is semi-abelian with abelian generic fibre and whose I-adic completion coincides with that of $\tilde{G}$.

3) The collection of all data $(\tilde{G} \to S, \underline{Y})$ as above form a category DD, the category of all semi-abelian degeneration data over S. We shall show that procedures 1) and 2) define two functors between DEG and DD, which are quasi-inverse to each other. Thus we have a fairly good understanding of semi-abelian degeneration of abelian varieties over complete normal bases.

Extending the above discussion, one would like to fit polarizations, invertible sheaves, ampleness, etc. into the picture of "quotient by period subgroups." This gives rise to various equivalences of category. The category of semi-abelian degenerations (resp. degeneration data) with additional structure x is denoted DEG_x (resp. DD_x.) As for subscripts, $_{\mathrm{pol}}$ stands for polarization, $_{\mathrm{ample}}$ for ample invertible sheaf, $_{\mathrm{IS}}$ for (general) invertible sheaf. If desired, one can also tack various numerical invariants onto the category names, e.g. $\dim(G/S)$, $\dim(T/S)$, degree of polarization, etc.. Formally we have equivalences of category

$$\mathrm{DEG} \approx \mathrm{DD}, \ \mathrm{DEG}_{\mathrm{pol}} \approx \mathrm{DD}_{\mathrm{pol}}, \ \mathrm{DEG}_{\mathrm{ample}} \approx \mathrm{DD}_{\mathrm{ample}}, \ \mathrm{DEG}_{\mathrm{IS}} \approx \mathrm{DD}_{\mathrm{IS}} .$$

In this chapter, we shall only carry out step 1) above, using the idea that periods are encoded in the coefficients of the partial Fourier expansion of theta functions. The rest will be done in the next chapter.

Historical Remarks. The case $r = g = 1$ and R a complete discrete valuation ring is due to Tate. 2) is due to Mumford, whose ingenious paper [Mum 5] treats the special case $r = g$ but contains all the key ideas. When R is a complete discrete valuation ring 1) was proved independently by Raynaud using techniques from rigid analytic geometry and by Mumford using algebraic theta functions for case res. char. $(R) \neq 2$.

1. Review of Raynaud Extensions

For now R is a noetherian ring complete w.r.t. some ideal I, $S = \mathrm{Spec}(R)$, $S_0 = \mathrm{Spec}(R/I)$. Let G/S be a semi-abelian scheme. We assume that $G_0 = G \times_S S_0$ is an extension of an abelian scheme by a torus over S_0. This is automatic if (R, I) is local. So we have an exact sequence $0 \to T_0 \to G_0 \to A_0 \to 0$, where T_0 is a torus and A_0 is an abelian scheme over S_0. Denote $\mathrm{Spec}(R/I^i)$ by S_i and $G_i = G \times_S S_i$, $S_{\mathrm{for}} = \lim S_i$, $G_{\mathrm{for}} = \lim G_i$. By [SGA3, IX Thm. 3.6], T_0 lifts uniquely to a multiplicative subgroup scheme T_i of G_i for every i. The quotient of G_i by T_i is an abelian scheme because it is smooth and trivially proper. In other words, we get an exact sequence $0 \to T_{\mathrm{for}} \to G_{\mathrm{for}} \to A_{\mathrm{for}} \to 0$, where T_{for} is a formal torus and A_{for} is a formal abelian scheme. We would like to know when the above sequence is algebraizable. When this is the case, the algebraized exact sequence $0 \to T \to \tilde{G} \to A \to 0$ is called the *Raynaud extension associated to* G (or G_{for}). The Raynaud extension attached to G is clearly functorial in G (under the algebraizability assumption).

A sufficient condition for algebraizability is that there exist a relatively ample invertible sheaf $\mathcal{L}$ on $G_{\rm for}$. For then $\mathcal{L}\otimes[-1]^*\mathcal{L}\otimes\mathcal{O}_{S_i}$ descends canonically to A_i, and $A_{\rm for}$ is algebraizable by [EGA III] 5.4.5. Thus $A_{\rm for}$ is the formal completion of a unique abelian scheme A. The torus T_0 is characterized by its dual $\underline{X}$, which is a locally constant étale sheaf of free abelian groups on S_0. It lifts uniquely to such a sheaf $\underline{X}$ on $S_{\rm for}$, and so the formal torus $T_{\rm for}$ is the formal completion (along S_0) of a unique torus T over S via $\underline{X}(T_{\rm for}) := \mathbf{Hom}(T_{\rm for}, G_m\ {}^\wedge) = \mathbf{Hom}(T, G_m) =: \underline{X}(T)$. We have $\mathrm{Ext}(A,T) = \mathrm{Hom}(\underline{X}(T), \mathbf{Ext}(A, G_m))$ and similarly $\mathrm{Ext}(A_{\rm for}, T_{\rm for}) = \mathrm{Hom}(\underline{X}(T), \mathbf{Ext}(A_{\rm for}, G_m\ {}^\wedge))$. Since $\mathbf{Ext}(A, G_m) = A^t$, $\mathbf{Ext}(A_{\rm for}, G_m\ {}^\wedge) = A^t_{\rm for}$, by [EGA III] §5 we conclude that $\mathrm{Ext}(A,T) = \mathrm{Ext}(A_{\rm for}, T_{\rm for})$. Hence the exact sequence $0 \to T_{\rm for} \to G_{\rm for} \to A_{\rm for} \to 0$ is uniquely algebraizable. Notice that the existence of an ample $\mathcal{L}$ is automatic when R is normal by a result of Raynaud [Ra 3] thm. XI 1.13.

We recall a result about descent of cubical G_m-torsors. Suppose we have an exact sequence

$$0 \longrightarrow T \stackrel{i}{\longrightarrow} G \stackrel{\pi}{\longrightarrow} H \longrightarrow 0$$

of smooth connected group schemes over a scheme S, where T is a torus. Then the category $\mathrm{CUB}(H, G_m)$ of cubical G_m-torsors on H is equivalent via π^* to the category of pairs $(\mathcal{L}, s)$ where $\mathcal{L} \in \mathrm{CUB}(G, G_m)$ and s is a trivialization of $i^*\mathcal{L} \in \mathrm{CUB}(T, G_m)$. Since the category $\mathrm{CUB}(T, G_m)$ is equivalent to the category $\mathrm{Ext}(T, G_m)$ of commutative extensions of T by G_m, we conclude that

1.1. Proposition. *(i) For any $\mathcal{L} \in \mathrm{CUB}(G, G_m)$, $\mathcal{L}\otimes[-1]*\mathcal{L}$ comes canonically from a cubical G_m-torsor on H.*

(ii) There is an étale surjective map $S' \to S$ such that $\mathcal{L}_{S'} = \mathcal{L}\times_S S'$ comes from some (non-unique) cubical G_m-torsor on $H_{S'} = H\times_S S'$, since any extension of tori is étale locally trivial.

(iii) If the base scheme S is normal (or more generally, geometrically unibranch) and T is split, any $\mathcal{L} \in \mathrm{CUB}(G, G_m)$ comes from some $\mathcal{M} \in \mathrm{CUB}(H, G_m)$, because $\mathrm{Ext}(T, G_m) = 0$ in this case.

A convenient reference is [MB 1] I.7. It also follows that $\mathcal{L}_{\rm for}$ is induced from a unique cubical $\tilde{\mathcal{L}}$ on $\tilde{G}$.

2. The Dual Variety

In this section R is a noetherian normal integral domain complete w.r.t. an ideal I, $S = \mathrm{Spec}R$, $S_0 = \mathrm{Spec}R/I$, η = generic point of S, K = fraction field of R. We may assume $I = \mathrm{rad}(I)$ if we like. Let G/S be a semi-abelian scheme whose generic fiber G_η is abelian. Assume given a rigidified invertible $\mathcal{O}_S$-module $\mathcal{L}$ such that $\mathcal{L}_\eta$ is ample. Then $\mathcal{L}$ has a unique cubical structure extending that of $\mathcal{L}_\eta$, because G has connected fibres. We remark that for any given cubical ample $\mathcal{L}_\eta$, if an extension $\mathcal{L}$ exists it is uniquely determined by $\mathcal{L}_\eta$, and it exists if either R is regular or $\mathcal{L}_\eta$ is symmetric (Cf. [MB 1] 3.2, 3.3).

We would like to construct a semi-abelian scheme G^t whose generic fibre is the dual abelian variety of G_η. This follows rapidly from the usual construction of the dual abelian variety as quotient of the original: The closure of the finite group scheme $K(\mathcal{L}_\eta)/\eta$ in G is a quasi-finite flat subgroup scheme $K(\mathcal{L})$ of G (c. [MB 1] IV thm. 2.4), and we take $G^t = G/K(\mathcal{L})$. The restriction of the biextension $\Lambda(\mathcal{L})$ to $K(\mathcal{L}) \times 1$ is canonically trivialized, and $\Lambda(\mathcal{L})$ descends to a biextension $\mathcal{P}$ on $G^t \times_S G$ which is the unique extension of the Poincaré sheaf on $G^t_\eta \times G_\eta$.

The torus part of $G^t \otimes R/I$ is split if this is so for G. Consider the Raynaud extensions associated to G and G^t, denoted $0 \to T \to \tilde{G} \to A \to 0$ and $0 \to T^t \to \tilde{G}^t \to A^t \to 0$ by abuse of notation. To justify this notation, we claim that A^t is the dual abelian scheme of A. In fact, we will show that the Poincaré sheaf $\mathcal{P} = \mathcal{P}_G$ on $G \times_S G^t$ constructed above can be descended to the Poincaré sheaf $\mathcal{P}_A$ on $A \times_S A^t$ after first passing to the formal completion. The formal completion $\mathcal{P}_{G,\text{for}} = \{\mathcal{P}_G \otimes R/I^n\}_{n \geq 0}$ of $\mathcal{P}_G$ is a biextension on $G_{\text{for}} \times G^t_{\text{for}}$ which can be uniquely descended to a biextension $\mathcal{P}_{A,\text{for}}$ on $A_{\text{for}} \times A^t_{\text{for}}$ by [SGA7 I] exposé VIII 3.5. Since A_{for} and A^t_{for} are both algebraic, so is $\mathcal{P}_{A,\text{for}}$ by [EGA] III §5. Denote by $\mathcal{P}_A$ the algebraization of $\mathcal{P}_{A,\text{for}}$. It remains to check that $\mathcal{P}_A$ actually defines a duality between A and A^t. It suffices to check this over η. Hence it suffices to check that the pairing $T_\ell(A) \times T_\ell(A^t) \to T_\ell(G_m)$ on Barsotti-Tate groups induced by $\mathcal{P}_A$ is a perfect pairing over η, for every prime ℓ. The pairing $T_\ell(\tilde{G}) \times T_\ell(\tilde{G}^t) \to T_\ell(G_m)$ induced by $\mathcal{P}_{G,\text{for}}$ factorizes through the previous pairing, because $\mathcal{P}_{G,\text{for}}$ is the pull-back of $\mathcal{P}_{A,\text{for}}$. On the other hand $T_\ell(\tilde{G}) \subset T_\ell(G)$, $T_\ell(\tilde{G}^t) \subset T_\ell(G^t)$ canonically, and the pairing $T_\ell(G) \times T_\ell(G^t) \to T_\ell(G_m)$ induced by $\mathcal{P}_G$ extends the previous pairing. Since G and G^t are abelian varieties over η, $T_\ell(G) \times T_\ell(G^t) \to T_\ell(G_m)$ is a perfect pairing over η. If we can show that the annihilator of $T_\ell(\tilde{G})_\eta$ in $T_\ell(G^t)_\eta$ is $T_\ell(T^t)\eta$ and the annihilator of $T_\ell(T)_\eta$ in $T_\ell(G^t)_\eta$ is $T_\ell(\tilde{G}^t)_\eta$, the claim will follow. But this follows from similar statements for pairings defined by $\Lambda(\mathcal{L})$. Since we may take $\mathcal{L}$ of the form $\mathcal{L}' \otimes [-1] * \mathcal{L}'$ for some ample cubical invertible sheaf $\mathcal{L}'$ so that $\mathcal{L}_{\text{for}}$ descends to an ample $\mathcal{M}_{\text{for}}$ on A_{for}, the result follows. More details about various facts used in the above argument can be found in [SGA7 I] exp. IX.

The two Raynaud extensions $0 \to T \to \tilde{G} \to A \to 0$, $0 \to T^t \to \tilde{G}^t \to A^t \to 0$ are coded by two homomorphisms $c : \underline{X}(T) \to A^t$, $c^t : \underline{X}(T^t) \to A$. We denote the two projections $\tilde{G} \to A$, $\tilde{G}^t \to A^t$ by π, π^t respectively and let $\underline{X} = \underline{X}(T)$, $\underline{Y} = \underline{X}(T^t)$. Pulling back the biextension $\mathcal{P}_A$ over $A \times A^t$ by $c^t \times c$, we get a G_m-biextension of the étale group scheme $\underline{Y} \times \underline{X}$ by G_m. If we are given a polarization $\lambda_\eta : G_\eta \to G^t_\eta$, it extends uniquely to a homomorphism $\lambda : G \to G^t$. The functoriality of Raynaud extensions gives us an induced homomorphism $\tilde{\lambda} : \tilde{G} \to \tilde{G}^t$. $\tilde{\lambda}$ gives rise to two homomorphisms $\phi : \underline{Y} \to \underline{X}$ and $\lambda_A : A \to A^t$ such that $c\phi = \lambda_A c^t$. ϕ and λ_A determines $\tilde{\lambda}$ uniquely because every homomorphism from an abelian scheme to a torus is trivial. Note that the two étale group schemes $\underline{X}$, $\underline{Y}$ have the same rank and ϕ is an injection with finite cokernel. Note that the degree of λ_G is $[\underline{X} : \phi(\underline{Y})]^2$-times the degree of λ_A.

3. The Setup

We start with the following data:

a) a noetherian normal integral domain R complete w.r.t. an ideal I, $I = \mathrm{rad}(I)$, $S = \operatorname{Spec} R$, $S_0 = \operatorname{Spec} R/I$, $\eta =$ generic point of S, $K =$ fraction field of R. *We assume that for any étale R/I-algebra $R_0^\dagger$ its unique lifting $R^\dagger$ to a formally étale I-adically complete R-algebra is normal.* This holds for example if R is regular, or if R is the I-adic completion of a normal excellent ring. We need this condition to do descent because sometimes certain conditions hold only after faithfully flat base change $R \to R^\dagger$ as above. We might ask when we have effective descent, that is, when every object over $R^\dagger$ whose two extensions to $R^{\dagger\dagger} = I$-adic completion of $R^\dagger \otimes_R R^\dagger$ are isomorphic (with the isomorphism satisfying some triple identity on $R^{\dagger\dagger\dagger}$), comes from R. This holds for morphisms between R-modules, but not for R-modules themselves (except if they are finitely generated). So in general we will be able to apply descent to morphisms, but not to objects.

b) a semi-abelian scheme G/S whose generic fiber G_η is abelian such that G_{S_0} is an extension of an abelian scheme $A_0 \to S_0$ by a torus $T_0 \to S_0$.

c) a rigidified invertible $\mathcal{O}_G$-module $\mathcal{L}$ such that $\mathcal{L}_\eta$ is ample.

Then $\mathcal{L}$ has a unique cubical structure compatible with the rigidification. We have the Raynaud extensions $0 \to T \to \tilde{G} \to A \to 0$, $0 \to T^t \to \tilde{G}^t \to A^t \to 0$, homomorphisms $c : \underline{X} = \underline{X}(T) \to A^t$, $c^t : \underline{Y} = \underline{X}(T^t) \to A$, $\lambda_\eta := \lambda(\mathcal{L}_\eta) : G_\eta \to G^t_\eta$, and $\lambda : G \to G^t$, $\phi : \underline{Y} \to \underline{X}$, $\lambda_A : A \to A^t$ attached to λ_η such that $c\phi = \lambda_A c^t$. We also have the biextension $(c^t \times c)^* \mathcal{P}_A$. We would like to analyze the above data, and decipher them in terms of "more explicit" ones. To proceed we shall make two auxiliary hypotheses; both hold étale locally on $\mathrm{Spf}(R)$:

(i) The formal cubical invertible sheaf $\mathcal{L}_{\mathrm{for}} \otimes \mathcal{O}_{T_{\mathrm{for}}}$ is trivial. We choose a trivialization of it, which is the same as choosing a descent data of $\mathcal{L}_{\mathrm{for}}$ w.r.t. π. Thus $\mathcal{L}_{\mathrm{for}}$ descends to a formal cubical sheaf $\mathcal{M}_{\mathrm{for}}$ on A_{for}. Clearly $\mathcal{M}_{\mathrm{for}}$ algebraizes to a cubical ample invertible sheaf $\mathcal{M}$ on A.

(ii) $\underline{X}$ and $\underline{Y}$ are constant, with values X respectively Y.

Several remarks about these two hypotheses are in order. First of all, we make them just to make later local computations easier. After the local computations, we will make global statements which can be verified étale locally on $\mathrm{Spf}(R)$. This observation leads to immediate generalization: the base S can be any noetherian normal formal scheme and not necessarily affine. Hypothesis (i) is true étale locally on $\mathrm{Spf}(R)$. When T is split it is automatic if $\mathrm{Spec}(R/I)$ is geometrically unibranch, in particular if (R, I) is local. Hypothesis (ii) holds after passing to a finite étale covering of $\mathrm{Spf}(R)$; especially, it holds locally in the étale topology.

Recall that when we imposed the two auxiliary hypotheses above, we made a choice: a descent data for $\mathcal{L}_{\mathrm{for}}$ with cubical structure w.r.t. π. The range of choices is given by the following statement: All trivializations of the cubical invertible sheaf $\mathcal{L}_{\text{for}} \otimes \mathcal{O}_{T_{\text{for}}}$ on T_{for}-form a torsor under $X = \underline{X}(T)$. This follows from "Rosenlicht's lemma", see [SGA7 I] exp. VIII 4.1.

We are now ready for an analysis of degeneration of polarized abelian varieties. The basic idea is to look at sections of $\mathcal{L}$ from the perspective of Raynaud extensions. This amounts to taking a "partial Fourier expansion" (or, Fourier-Jacobi expansion) of sections of $\mathcal{L}$, which corresponds to partial Fourier expansions of theta functions in the complex analytic case. Since theta functions determine projective embeddings of abelian varieties, we should be able to extract complete information about degeneration. This is indeed the case. The only tools we use are the addition formula of theta functions and representations of non-degenerate theta groups. Unfortunately the computations are quite messy as there are numerous isomorphisms in the picture, and the abelian part A is a great source of confusion. Thus we will first treat the simplest case when A is trivial, which already contains most of the key points.

4. The Completely Degenerate Case: A Trivial

In this section we keep the notation, data and auxiliary hypotheses of the previous section. In addition *we make one more assumption: The abelian part A is trivial.* This assumption greatly simplifies our notations and hypotheses. For then $\tilde{G} = T, \tilde{G}^t = T^t, c = c^t =$ trivial, and the only hypothesis we make is that the formal cubical invertible sheaf $\mathcal{L}_{\text{for}}$ on the formal torus G_{for} is trivial. We only have to choose a trivialization of the $\mathcal{L}_{\text{for}} \in \text{CUB}(G_{\text{for}}, G_m)$. Then $\mathcal{M}$ and $\mathcal{P}_A$ are trivial because they are rigidified. Since $c = c^t =$ trivial, there are natural trivializations of $c^{t*}\mathcal{M}$ and $(c \times c^t)^*\mathcal{P}_A$ coming from (pull-backs of) rigidifications of $\mathcal{M}$ and $\mathcal{P}_A$. We still have an injective homomorphism $\phi : Y \to X$ coming from the polarization of G_η defined by $\mathcal{L}_\eta$.

The formal torus $G_{\text{for}} = T_{\text{for}}$ acts on the trivial invertible sheaf $\mathcal{O}_{\text{for}} = \mathcal{O}_{G_{\text{for}}}$. Any element $f \in \Gamma(G_{\text{for}}, \mathcal{O}_{\text{for}})$ can be written uniquely as $f = \sum_{\mu \in X(T)} \sigma_\mu(f) \cdot \mu$, each $\sigma_\mu(f) \in R$ and $\forall n$, $\sigma_\mu(f) \equiv 0 \pmod{I^n}$ for all but finitely many μ. This is an analogue of classical Fourier expansion and can be seen easily by writing $f \bmod I^n$ as a finite sum of characters of $G \otimes R/I^n$ for each n, and taking the inverse limit as $n \to \infty$. Via the chosen trivialization of $\mathcal{L}_{\text{for}}$, any section $s \in \Gamma(G, \mathcal{L}) \hookrightarrow \Gamma(G_{\text{for}}, \mathcal{L}_{\text{for}})$ has a Fourier expansion $s = \sum_{\mu \in X} \sigma_\mu(s) \cdot \mu$, $\sigma_\mu(s) \in R\ \forall \mu \in X = X(T)$ and $\sigma_\mu(s) \to 0$ in the I-adic topology. More generally, any section $s \in \Gamma(G_\eta, \mathcal{L}_\eta) = \Gamma(G, \mathcal{L}) \otimes_A K \hookrightarrow \Gamma(G_{\text{for}}, \mathcal{L}_{\text{for}}) \otimes_A K$ admits an expansion $s = \sum_{\mu \in X} \sigma_\mu(s) \cdot \mu$, $\sigma_\mu(s) \in K\ \forall \mu \in X$ such that there is a non-zero element $y \in R$ with $y \cdot \sigma_\mu(s) \in R\ \forall \mu \in X$ and $y \cdot \sigma_\mu(s) \to 0$ in the I-adic topology. Remember that $\sigma_\mu(\cdot)$ depends on the cubical invertible sheaf $\mathcal{L}$ and the choice of trivialization of $\mathcal{L}_{\text{for}}$ we made.

Motivated by the classical Fourier expansion of theta functions, we would like to show that our Fourier expansion $\Gamma(G_\eta, \mathcal{L}_\eta) \to \Gamma(G_{\text{for}}, \mathcal{L}_{\text{for}}) \otimes_A K$ embodied by the collection $\{\sigma_\mu(\cdot)\}_{\mu \in X}$ of elements in $\text{Hom}_R(\Gamma(G, \mathcal{L}), R)$ carries a lot of information about degenerations. The following theorem is a precise formulation of the cubical relations between the $\sigma_\mu(\cdot)$'s.

4.1. Theorem. *The collection* $\{\sigma_\mu(\cdot)\}_{\mu\in X}$ *of elements in* $\mathrm{Hom}_R(\Gamma(G,\mathcal{L}),K)$ *satisfies the following properties:*

(i) $\sigma_\mu(\cdot)\neq 0\ \forall\mu\in X$.

(ii) There exist a function $a: Y\to K^*$ *and a bi-multiplicative function* $b: Y\times X\to K^*$ *such that* $\sigma_{\mu+\phi(y)}(\cdot)=a(y)\cdot b(y,\mu)\cdot\sigma_\mu(\cdot)\quad\forall y\in Y,\ \forall\mu\in X$. *(We shall say that* $a(\cdot)$ *is a* K^**-valued multiplicatively quadratic function on* Y*.)*

(iii) $a(0)=1$.

(iv) $a(y+z)=b(y,\phi(z))\cdot a(y)\cdot a(z)\quad\forall y,z\in Y$.

(v) $b(z,\phi(y))=b(y,\phi(z))\quad\forall y,z\in Y$.

(vi) $b(y,\phi(y))\in I$ *for every* $y\neq 0$ *in* Y.

(vii) $a(y)\in R$ *for almost all* y *in* Y. *More generally, for every* $n\geq 0$, *we have* $a(y)\in I^n$ *for all but finitely many* y *in* Y.

4.2. Remarks and Reductions. (a) The functions a,b are uniquely defined by the formula $\sigma_{\mu+\phi(y)}(\cdot)=a(y)\cdot b(y,\mu)\cdot\sigma_\mu(\cdot)$, once we have proved the important step (i) and that $\sigma_{\mu+\phi(y)}(\cdot)$ and $\sigma_\mu(\cdot)$ are proportional because $b(y,0)$ is required/defined to be $1\ \forall y\in Y$. That a,b satisfy (iii)–(v) follows immediately from the formula $\sigma_{\mu+\phi(y)}(\cdot)=a(y)\cdot b(y,\mu)\cdot\sigma_\mu(\cdot)$ and the requirement $b(y,0)=1\ \forall y\in Y$: (iii) is trivial. (iv) holds because $a(y+z)\cdot\sigma_0(\cdot)=\sigma_{\phi(y)+\phi(z)}(\cdot)=a(y)\cdot b(y,\phi(z))\cdot\sigma_{\phi(z)}(\cdot)=a(y)\cdot b(y,\phi(z))\cdot a(z)\cdot\sigma_0(\cdot)$. (v) holds because (iv) is symmetric in y and z. For the function $b(y,\mu)$, multiplicativity in y follows from multiplicativity in μ : $a(y+z)\cdot b(y+z,\mu)\cdot\sigma_\mu(\cdot)=\sigma_{\mu+\phi(y)+\phi(z)}(\cdot)=a(z)\cdot b(z,\mu+\phi(y))\cdot\sigma_{\mu+\phi(y)}(\cdot)=a(z)\cdot b(z,\mu+\phi(y))\cdot a(y)\cdot b(y,\mu)\cdot\sigma_\mu(\cdot)$, hence by (iv) $b(y,\phi(z))\cdot b(y+z,\mu)=b(z,\mu+\phi(y))\cdot b(y,\mu)$, and we deduce $b(y+z,\mu)=b(y,\mu)\cdot b(z,\mu)$ from (v) and the assumed multipicativity in μ. That $b(y,\mu+\nu)=b(y,\mu)\cdot b(z,\nu)\ \forall\mu,\nu\in X,\ \forall y\in Y$ still needs proof, which will follow easily from the addition formula.

(b) (vi), (vii) follows from (i)–(v) and the following observation: There is a non-zero element $c\in R$ such that $c\cdot a(y)\in R\ \forall y\in Y$ and $c\cdot a(y)\to 0$ in the I-adic topology. This is an immediate consequence of the fact that $\sum_{\mu\in X}\sigma_\mu(s)\cdot\mu$ converges I-adically $\forall s\in\Gamma(G,\mathcal{L})$. For every discrete valuation v of K with $v(R)\geq 0$ centered on $\mathrm{Spec}(R/I)$, $v(a(y))$ is an $\mathbb{R}$-valued quadratic polynomial function on Y bounded below such that $v(a(y))\to\infty$. Hence the associated quadratic form $v(b(\phi(y),y))$ is positive definite. But $\mathrm{rad}(I)=I$ and every prime ideal of R is the center of some discrete valuation v of K with $v(R)\geq 0$, hence (vi). Let $\{\mathcal{P}_1,\dots,\mathcal{P}_k\}$ be the set of height 1 prime ideals of R containing c, and $v_1,\dots,v_k$ the discrete valuations associated with $\mathcal{P}_1,\dots,\mathcal{P}_k$. Then $a(y)\in R$ for all but finitely many $y\in Y$ because $v_i(a(y))$ is quadratic in y and bounded below, $\forall 1\leq i\leq k$ and R is normal. Now that we know that $c\cdot a(y)\in I^{s(y)}$ for some $s: Y\to\mathbb{Z}_{\geq 0}$ with $s(y)\to\infty$ and $a(y)\in R$ for all but finitely many $y\in Y$, we deduce that $a(y)\in I^{n(y)}$ for some $n: Y\to\mathbb{Z}_{\geq 0}$ with $n(y)\to\infty$ by

[Z-S] ch. VIII §5 Cor. 1. So (vii) is proved. We remark that in fact (vii) is a consequence of (i)–(vi) and the normality of R by the argument in loc. cit..

(c) There exists an adic injection $R \to D$ from R into a complete discrete valuation ring D. Since our statement is about equalities and inequalities between elements of R, we are reduced to the case R is a complete discrete valuation ring.

(d) Now we may and will assume that R is a complete discrete valuation ring. The quickest proof of the theorem (whose general form will be formulated later) is to wave our hands and quote results of Raynaud announced in [Ra 4]. Raynaud's method uses rigid analytic geometry. Unfortunately a complete proof of his beautiful results is still in the "to appear" category. The proof given here uses the addition formula (no surprise), or more correctly a family of addition formulae.

Proof of Theorem. We may and will assume that R is a complete discrete valuation ring with maximal ideal $\mathfrak{m}$ as remarked above. Let $k = R/\mathfrak{m}$. Whenever necessary we will make further base change $R \to R'$ to larger complete discrete valuation rings. As remarked earlier it suffices to prove the theorem after the base change. Our major tool is the addition formula, which we will recall first and then translate into a statement about $\sigma_\mu(\cdot)$'s.

Denote by $\Phi : G \times_S G \to G \times_S G$ the fundamental isogeny used in the addition formula, defined by $(x, y) \to (x+y, x-y)$ for functorial points x, y. If $\mathcal{L}$ is a symmetric cubical invertible sheaf (i.e. $\mathcal{L} \cong [-1]^*\mathcal{L}$, either as G_m-torsors or with cubical structure), then $\Phi^*(p_1^*\mathcal{L} \otimes p_2^*\mathcal{L}) \cong p_1^*\mathcal{L}^{\otimes 2} \otimes p_2^*\mathcal{L}^{\otimes 2}$ where p_1, p_2 are the two projections. More generally, if $\mathcal{L}_1, \mathcal{L}_2$ are algebraically equivalent cubical invertible sheaves, then $\Phi^*(p_1^*\mathcal{L}_1 \otimes p_2^*\mathcal{L}_2) \cong p_1^*(\mathcal{L}_1 \otimes \mathcal{L}_2) \otimes p_2^*(\mathcal{L}_1 \otimes [-1]^*\mathcal{L}_2))$. Here the condition $\mathcal{L}_1, \mathcal{L}_2$ are algebraically equivalent means that their difference $\mathcal{N} = \mathcal{L}_1 \otimes \mathcal{L}_2^{\otimes -1}$ as G_m-torsor has the structure of commutative group scheme. Since G has connected fibres, this condition over S is equivalent to this condition over η, i.e. the usual notion of algebraic equivalence of invertible sheaves over abelian varieties. For a proof, notice that it suffices to check over the generic point η; this follows for example from [MB 1] II.1.1, p. 40. If $\mathcal{L}_1 = \mathcal{L}_2$, this follows directly from the theorem of the cube: Pull back from G_η^3 to G_η^2 via $(x, y)| \to (x, y, -y)$. So one is reduced to the case $\mathcal{L}_1 =$ trivial and $\mathcal{L}_2$ algebraically equivalent to the trivial invertible sheaf, which is immediate.

We need to recall some basic properties about the theta group $\mathcal{G}(\mathcal{L})$ and its representations. Everything we use is fairly well documented in [MB 1] V (see also [MB 1] IV for $K(\mathcal{L})$). Since R is henselian local, the quasi-finite group scheme $K(\mathcal{L})$ (= closure of $K(\mathcal{L}_\eta)$ in G by definition) has a unique open-and-closed subgroup scheme $K(\mathcal{L})^f$ finite over R such that $K(\mathcal{L})^f \otimes R/\mathfrak{m} = K(\mathcal{L}) \otimes R/\mathfrak{m}$. Let $\mathcal{G}(\mathcal{L})^f$ be the inverse image of $L \to G$ over $K(\mathcal{L})^f$, where L is the G_m-torsor corresponding to $\mathcal{L}$. $\mathcal{G}(\mathcal{L})^f$ has a natural structure as subgroup of the group scheme $\mathcal{G}(\mathcal{L}) =$ inverse image of $L \to G$ over $K(\mathcal{L})$. Both $\mathcal{G}(\mathcal{L})$ and $\mathcal{G}(\mathcal{L})^f$ act on $\mathcal{L}$. The trivialization of $\mathcal{L}_{\text{for}}$ induces a splitting of $0 \to G_m \to \mathcal{G}(\mathcal{L})^f \to K(\mathcal{L})^f \to 0$, through which $K(\mathcal{L})^f$ acts on $\mathcal{L}$ and hence on $\Gamma(G, \mathcal{L})$. Since $K(\mathcal{L})^f \otimes_R K$ is a maximal isotropic subgroup of $K(\mathcal{L}_\eta)$, $\Gamma(G, \mathcal{L})$ is a direct sum of $K(\mathcal{L})^f$-eigenspaces $\Gamma(G,\mathcal{L}) = \oplus_{\bar\mu}\Gamma(G,\mathcal{L})^{\bar\mu}$, where $\bar\mu$ runs through $X(K(\mathcal{L})^f) \cong X/\phi(Y)$, and each eigenspace has rank 1. (This follows from the representation theory of the theta group $\mathcal{G}(\mathcal{L})_\eta$.) The Fourier expansion of an element $s \in \Gamma(G,\mathcal{L})^{\bar\mu}$ has the form $s = \sum_{\nu\in\bar\mu} \sigma_\nu(s)\cdot\nu$. In other words $\sigma_\nu(\cdot)$ vanishes on $\Gamma(G,\mathcal{L})^{\bar\mu}$ if $\nu \notin \bar\mu$. Another way to formulate this is that any element s of $\Gamma(G,\mathcal{L})$ has the form $s = \sum_{\bar\mu\in X/\phi(Y)}\sum_{\nu\in\bar\mu}\sigma_\nu(s)\cdot\nu$ with each $\sum_{\nu\in\bar\mu}\sigma_\nu(s)\cdot\nu$ algebraic.

4.3. Lemma. *Let $f : H \to G$ be an isogeny of semi-abelian schemes over R such that G_η is abelian and $G_0 = G\otimes_R k$ is a split torus. Let $\mathcal{L}$ be a cubical invertible sheaf on G, and we fix a choice of trivialization of the formal cubical invertible sheaf $\mathcal{L}_{\mathrm{for}}$ on G_{for}. Via f_{for}, we have an induced trivialization of $(f^*\mathcal{L})_{\mathrm{for}}$. With respect to these compatible trivializations, we have $\sigma_\mu^{\mathcal{L}}(s) = \sigma_{f^*\mu}^{f^*\mathcal{L}}(f^*s) \quad \forall s \in \Gamma(G,\mathcal{L})$, $\forall\mu\in X$.*

The proof of this lemma is straight forward. Write $s = \sum_{\mu\in X}\sigma_\mu^{\mathcal{L}}(s)\cdot\mu$, hence $f^*s = \sum_{\mu\in X}\sigma_\mu^{\mathcal{L}}(s)\cdot f^*\mu$. As $f^* : X = X(G_{\mathrm{for}}) \to X(H_{\mathrm{for}}) := X'$ is injective and $f^*s = \sum_{\nu\in X'}\sigma_\mu^{f^*\mathcal{L}}(f^*s)\cdot\nu$ by definition, we get $\sigma_\mu^{\mathcal{L}}(s) = \sigma_{f^*\mu}^{f^*\mathcal{L}}(s)$.

Applying this lemma to our fundamental isogeny $\Phi : G\times_S G \to G\times_S G$, we deduce

Addition Formula. Let $\mathcal{L}_1, \mathcal{L}_2$ be algebraically equivalent ample cubical invertible sheaves on G, $\mathcal{F} = p_1^*\mathcal{L}_1 \otimes p_2^*\mathcal{L}_2$, $\mathcal{F}' = \Phi^*(p_1^*\mathcal{L}_1\otimes p_2^*\mathcal{L}_2) \cong p_1^*(\mathcal{L}_1\otimes\mathcal{L}_2)\otimes p_2^*(\mathcal{L}_1\otimes[-1]^*\mathcal{L}_2)$. Then $\sigma_{(\mu,\nu)}^{\mathcal{F}}(p_1^*s_1\otimes p_2^*s_2) = \sigma_{(\mu+\nu,\mu-\nu)}^{\mathcal{F}'}(\Phi^*(p_1^*s_1\otimes p_2^*s_2)) \quad \forall s_1,s_2\in\Gamma(G,\mathcal{L})$, $\forall\mu,\nu\in X$.

In the addition formula, the left hand side is a product: $\sigma_{(\mu,\nu)}^{\mathcal{F}}(p_1^*s_1\otimes p_2^*s_2) = \sigma_\mu^{\mathcal{L}_1}(s_1)\cdot\sigma_\nu^{\mathcal{L}_2}(s_2)$. We would like to reformulate it so that we get an equality both sides of which are products. For any pair $\bar\alpha,\bar\beta\in X/2\phi(Y)$, denote by $\Gamma(G\times G,\mathcal{F}')^{(\bar\alpha,\bar\beta)}$ the $(\bar\alpha,\bar\beta)$-eigenspace under $G_{\mathrm{for}}\times G_{\mathrm{for}}$-action. In the addition formula we decompose $\Phi^*(p_1^*s_1\otimes p_2^*s_2)$ into its $(\bar\alpha,\bar\beta)$-components. Since $\mathcal{F}'\cong p_1^*(\mathcal{L}_1\otimes\mathcal{L}_2)\otimes p_2^*(\mathcal{L}_1\otimes[-1]^*\mathcal{L}_2)$, $\Gamma(G\times G,\mathcal{F}')^{(\bar\alpha,\bar\beta)} \cong \Gamma(G,\mathcal{L}_1\otimes\mathcal{L}_2)^{\bar\alpha}\otimes\Gamma(G,\mathcal{L}_1\otimes[-1]^*\mathcal{L}_2)^{\bar\beta}$ under this somorphism. Note that both $\Gamma(G,\mathcal{L}_1\otimes\mathcal{L}_2)^{\bar\alpha}$ and $\Gamma(G,\mathcal{L}_1\otimes[-1]^*\mathcal{L}_2)^{\bar\beta}$ are free R-modules of rank one. Now we can reformulate the addition formula:

AF′. Let $\bar\mu,\bar\nu\in X/\phi(Y)$. $2\bar\mu, 2\bar\nu$ are well-defined elements of $X/2\phi(Y)$. Let $(\bar\alpha,\bar\beta)$ be a solution of $\bar\alpha+\bar\beta = 2\bar\mu$, $\bar\alpha-\bar\beta = 2\bar\nu$ in $X/2\phi(Y)$. (Solutions form a coset under $\phi(Y)/2\phi(Y)\subset X/2\phi(Y)$.) Let $\mathcal{L}_1,\mathcal{L}_2$ be as before, and s_1,s_2,s_3,s_4 be generators of the rank one free R-modules $\Gamma(G,\mathcal{L}_1)^{\bar\mu}$, $\Gamma(G,\mathcal{L}_2)^{\bar\nu}$, $\Gamma(G,\mathcal{L}_1\otimes\mathcal{L}_2)^{\bar\alpha}$ and $\Gamma(G,\mathcal{L}_1\otimes[-1]^*\mathcal{L}_2)^{\bar\beta}$. Then $\forall\mu\in\bar\mu$, $\forall\nu\in\bar\nu$ with $\mu+\nu\equiv\bar\alpha$, $\mu-\nu\equiv\bar\beta \pmod{2\phi(Y)}$, we have

$$\sigma_\mu^{\mathcal{L}_1}(s_1)\cdot\sigma_\nu^{\mathcal{L}_2}(s_2) = \kappa(\bar\alpha,\bar\beta)\cdot\sigma_{\mu+\nu}^{\mathcal{L}_1\otimes\mathcal{L}_2}(s_3)\cdot\sigma_{\mu-\nu}^{\mathcal{L}_1\otimes[-1]^*\mathcal{L}_2}(s_4)$$

for some $\kappa = \kappa(\bar{\alpha}, \bar{\beta}) \in R$ independent of μ, ν. Notice that κ depends on $\bar{\alpha}, \bar{\beta}$, and the choice of s_i's as well.

Proof of Theorem Continued. Clearly if we can show that all $\sigma_\mu(\cdot)$'s are non-zero, then $\sigma_\mu(\cdot)$ and $\sigma_{\mu+\phi(y)}(\cdot)$ are proportional $\forall y \in Y$. Recall that we asserted that $\sigma_\mu(\cdot) \neq 0$ $\forall \mu \in X$ for *every* cubical ample invertible sheaf $\mathcal{L}$. We prove this first for symmetric $\mathcal{L}$'s. Remark that after base change we may always assume that there exists a symmetric cubical invertible sheaf $\mathcal{L}'$ algebraically equivalent to a given $\mathcal{L}$ such that the symmetry is the identity on $G[2]^f = T[2]$.

Suppose that $\mathcal{L}$ is symmetric such that the symmetry is the identity on $G[2]^f$. After suitably adjusting the trivialization of $\mathcal{L}_{\text{for}}$, we may assume that the induced symmetry on $\mathcal{O}_{G_{\text{for}}}$ is $\sum x_\mu \cdot \mu \to \sum x_\mu \cdot \mu^{-1}$. We apply AF' to $\mathcal{L} = \mathcal{L}_1 = \mathcal{L}_2$, which now reads:

$$\sigma_\mu^{\mathcal{L}}(s_1) \cdot \sigma_\nu^{\mathcal{L}}(s_2) = \kappa(\overline{\mu+\nu}, \overline{\mu-\nu}) \cdot \sigma_{\mu+\nu}^{\mathcal{L}^2}(s_3) \cdot \sigma_{\mu-\nu}^{\mathcal{L}^2}(s_4)$$

where s_1, s_2 depend only on $\mu, \nu \,(\text{mod } \phi(Y))$ and $\kappa(\overline{\mu+\nu}, \overline{\mu-\nu})$, s_3, s_4 depend only on $\mu+\nu$, $\mu-\nu \,(\text{mod } 2\phi(Y))$. Consider first the case $\mu, \nu \in \phi(Y)$, and set $s = s_1 = s_2$. Since s_1 is a generator, $\exists \mu_0 \in \phi(Y)$ s.t. $\sigma_{\mu_0}^{\mathcal{L}}(s_1) \neq 0$. From $\sigma_{\mu_0}^{\mathcal{L}}(s) \cdot \sigma_{\mu_0}^{\mathcal{L}}(s) = \kappa(\bar{0}, \bar{0}) \cdot \sigma_{2\mu_0}^{\mathcal{L}^2}(s_3) \cdot \sigma_0^{\mathcal{L}^2}(s_4)$, where $s_3 = s_4 = s'$ say, we get $\kappa(\bar{0}, \bar{0}) \neq 0$, $\sigma_{2\mu_0}^{\mathcal{L}^2}(s') \neq 0$, $\sigma_0^{\mathcal{L}^2}(s') \neq 0$. Since $\sigma_0^{\mathcal{L}}(s) \cdot \sigma_0^{\mathcal{L}}(s) = \kappa(\bar{0}, \bar{0}) \cdot \sigma_0^{\mathcal{L}^2}(s') \cdot \sigma_0^{\mathcal{L}^2}(s')$, we get $\sigma_0^{\mathcal{L}}(s) \neq 0$. We claim that

$$(^*) \quad \sigma_\mu^{\mathcal{L}}(s) \neq 0 \quad \forall \mu \in \phi(Y) \quad \text{for this symmetric } \mathcal{L}.$$

Assuming (*) for the time being, let us show that $\sigma_\mu^{\mathcal{L}}(s) \neq 0$ $\forall \mu \in X$ for a general $\mathcal{L}$ (not necessarily symmetric) follows from this and AF$'$ quite easily. Thus let $\mathcal{L}_1$ be an arbitrary ample cubical invertible sheaf on G, and $\mathcal{L}_2$ be a symmetric cubical invertible sheaf on G algebraically equivalent to $\mathcal{L}_1$ such that the symmetry is the identity on $T[2]$. By assumption (*) holds for $\mathcal{L}_2$. AF$'$ now reads:

$$\sigma_\mu^{\mathcal{L}_1}(s_1) \cdot \sigma_\nu^{\mathcal{L}_2}(s_2) = \kappa(\overline{\mu+\nu}, \overline{\mu-\nu}) \cdot \sigma_{\mu+\nu}^{\mathcal{L}_1 \otimes \mathcal{L}_2}(s_3) \cdot \sigma_{\mu-\nu}^{\mathcal{L}_1 \otimes [-1]^* \mathcal{L}_2}(s_4)$$

where s_1, s_2 depends only on $\mu, \nu \,(\text{mod } \phi(Y))$ and $\kappa(\overline{\mu+\nu}, \overline{\mu-\nu})$, s_3, s_4 depends only on $\mu+\nu, \mu-\nu \,(\text{mod } 2\phi(Y))$. Given any $\mu \in X$, $\exists \mu_0 \equiv \mu \,(\text{mod } \phi(Y))$ s.t. $\sigma_{\mu_0}^{\mathcal{L}_1} \neq 0$. AF$'$ gives $\kappa(\overline{\mu_0+\nu}, \overline{\mu_0-\nu}) \neq 0$, $\sigma_{\mu_0+\nu}^{\mathcal{L}_1 \otimes \mathcal{L}_2} \neq 0$, $\sigma_{\mu_0-\nu}^{\mathcal{L}_1 \otimes [-1]^* \mathcal{L}_2} \neq 0$, $\forall \nu \in \phi(Y)$. Write $\mu = \mu_0 + \nu$, we have in particular $\kappa(\bar{\mu}, \bar{\mu}) \neq 0$, $\sigma_\mu^{\mathcal{L}_1 \otimes \mathcal{L}_2} \neq 0$, $\sigma_\mu^{\mathcal{L}_1 \otimes [-1]^* \mathcal{L}_2} \neq 0$. Hence $\sigma_\mu^{\mathcal{L}_1} \neq 0$ by AF$'$ again.

We still have to prove the claim (*). Let $Z = \{\mu \in \phi(Y) | \sigma_\mu^{\mathcal{L}} \neq 0\}$. We show first that Z is a subgroup of $\phi(Y)$. We have already seen that $0 \in Z$. If $\mu, \nu \in Z$, AF$'$ gives $\sigma_{2\mu}^{\mathcal{L}^2} \neq 0$, $\sigma_{2\nu}^{\mathcal{L}^2} \neq 0$. Applying AF$'$ again to $\mu+\nu$, $\mu-\nu$, we get $\sigma_{\mu+\nu}^{\mathcal{L}}(s_1) \cdot \sigma_{\mu-\nu}^{\mathcal{L}}(s_2) = \kappa(\bar{0}, \bar{0}) \cdot \sigma_{2\mu}^{\mathcal{L}^2}(s_3) \cdot \sigma_{2\nu}^{\mathcal{L}^2}(s_4)$ with $s_1 = s_2, s_3 = s_4$. As $\kappa(\bar{0}, \bar{0}) \neq 0$, we get $\sigma_{\mu+\nu}^{\mathcal{L}} \neq 0$, $\sigma_{\mu-\nu}^{\mathcal{L}} \neq 0$. So $\mu+\nu \in Z$, $\mu-\nu \in Z$, and Z is a subgroup. If $Z \neq \phi(Y)$, there is a proper subgroup W of $\phi(Y)$

containing Z which is of finite index in $\phi(Y)$. We must show that this leads to a contradiction. W corresponds to a finite flat subgroup H of T containing $K(\mathcal{L})^f$. The trivialization of $\mathcal{L}_{\text{for}}$ induces a section of $\mathcal{G}(\mathcal{L})^f \to K(\mathcal{L})^f$, hence an action of $K(\mathcal{L})^f$ on $\mathcal{L}$. $\mathcal{L}$ descends to a cubical ample invertible sheaf $\underline{\mathcal{L}}$ on the quotient semi-abelian scheme $\mathcal{G} = G/K(\mathcal{L})^f$. Since $K(\mathcal{L})^f_\eta$ is a maximal isotropic subgroup of $K(\mathcal{L})\eta$, $\underline{\mathcal{L}}$ induces a principal polarization on $\mathcal{G}$. The existence of W means that any $K(\mathcal{L})^f$ -invariant section $s \in \Gamma(G, \mathcal{L})$ has a Fourier expansion $s = \sum_{\mu \in W} \sigma_\mu^{\mathcal{L}}(s) \cdot \mu$. Passing to $\mathcal{G}$, we are reduced to the case $\phi(Y) = X$. Translating $\mathcal{L}$ by elements of $G(R)$, we conclude that for any $\mathcal{L}'$ algebraically equivalent to $\mathcal{L}$ and any section s' of $\mathcal{L}'_\eta$, the Fourier expansion of s' involves only those $\mu \in W$, provided we choose a suitable trivialization of $\mathcal{L}_{\text{for}}$. But we know (I, Prop. 5.3) that if we make suitable base change $R \to R', \Gamma(G, \mathcal{L}_\eta^{\otimes n})$ is generated by elements of the form $T_{x_1}^*(s) \otimes \ldots \otimes T_{x_n}^*(s)$ for $n \geq 1$, where $x_1, \ldots, x_n \in G(R)$ with $x_1 + \ldots + x_n = 0$. Thus every element of $\Gamma(G, \mathcal{L}^{\otimes n})$ has a Fourier expansion involving only $\mu \in W$. Hence elements of $\Gamma(G, \mathcal{L}^{\otimes n})$ factor through H. But as $\mathcal{L}$ is ample on G over R, this is a contradiction. So we have proved the difficult part (i) of the theorem. As we remarked before, the functions $a : Y \to K^*$, $b : Y \times X \to K^*$ are then well-defined.

It remains only to prove the multiplicativity of b in the second variable. For any $\mu, \nu \in X$, $y \in Y$, we have by definition
$\sigma^{\mathcal{L}}_{\mu+\phi(y)}(\cdot) = a(y) \cdot b(y, \mu) \cdot \sigma^{\mathcal{L}}_\mu(\cdot)$, $\sigma^{\mathcal{L}}_{\nu+\phi(y)}(\cdot) = a(y) \cdot b(y, \nu) \cdot \sigma^{\mathcal{L}}_\nu(\cdot)$ and
$\sigma^{\mathcal{L}}_{\mu+\nu+\phi(y)}(\cdot) = a(y) \cdot b(y, \mu + \nu) \cdot \sigma^{\mathcal{L}}_{\mu+\nu}(\cdot)$. AF' gives
$\sigma^{\mathcal{L}}_{\mu+\phi(y)}(s_1) \cdot \sigma^{\mathcal{L}}_{\nu+\phi(y)}(s_1) = \kappa(\overline{\mu + \nu}, \overline{\mu - \nu}) \cdot \sigma^{\mathcal{L}^2}_{\mu+\nu+2\phi(y)}(s_3) \cdot \sigma^{\mathcal{L} \otimes [-1]^* \mathcal{L}}_{\mu-\nu}(s_4)$ and
$\sigma^{\mathcal{L}}_\mu(s_1) \cdot \sigma^{\mathcal{L}}_\nu(s_1) = \kappa(\overline{\mu + \nu}, \overline{\mu - \nu}) \cdot \sigma^{\mathcal{L}^2}_{\mu+\nu}(s_3) \cdot \sigma^{\mathcal{L} \otimes [-1]^* \mathcal{L}}_{\mu-\nu}(s_4)$.
Hence we get $a^{\mathcal{L}}(y)^2 \cdot b^{\mathcal{L}}(y, \mu) \cdot b^{\mathcal{L}}(y, \nu) = a^{\mathcal{L}^2}(2y) \cdot b^{\mathcal{L}^2}(2y, \mu + \nu)$. Similarly we have $a^{\mathcal{L}}(y)^2 \cdot b^{\mathcal{L}}(y, \mu+\nu) = a^{\mathcal{L}^2}(2y) \cdot b^{\mathcal{L}^2}(2y, \mu+\nu)$, say by substituting $\mu+\nu$ for μ and 0 for ν. Together these two equations give $b^{\mathcal{L}}(y, \mu+\nu) = b^{\mathcal{L}}(y, \mu) \cdot b^{\mathcal{L}}(y, \nu)$. This finishes the proof of the theorem.

5. The General Case: Abelian Part Non-Trivial

In the previous section we simplify matters by assuming that the abelian part is trivial. After choosing a trivialization of $\mathcal{L}_{\text{for}}$ we obtain two functions $a : Y \to K^*$, $b : X \times Y \to K^*$. They satisfy the following properties: $a(0) = 1$, $a(y+z) = b(\phi(y, z)) \cdot a(y) \cdot a(z)$ $\forall y, z \in Y$. Their effects are: b defines a homomorphism $\imath : Y \to \tilde{G}^*(K)$ and $a(\cdot)$ defines a cubical structure on $Y \to \tilde{G}^*(K)$ (as defined in chapter 1) which gives rise to $\mathcal{L}$ via Mumford's construction to be described later. In this chapter we will see that results of the previous section can be naturally generalized to our present setting by essentially the same method. The complication introduced by the presence of the abelian part is basically notational, if we are careful enough to keep track of innumerable canonical isomorphisms. The new ingredient, if any, is the non-degenerate theta group coming from the abelian part A.

We keep the notation and the two auxilliary hypotheses of section 3. Again, choose a descent data for $\mathcal{L}_{\text{for}}$ w.r.t. π. We would like to define a "partial Fourier expansion" of global sections of $\mathcal{L}$. Considering $0 \to T_{\text{for}} \to G_{\text{for}} \to A_{\text{for}} \to 0$ as direct limit of $0 \to T_i \to G_i \to A_i \to 0$ and the action of T_{for} on G_{for} as the direct limit of the action of T_i on G_i, we see that $\pi_* \mathcal{O}_{G_{\text{for}}}$ is the I-adic topological direct sum of T_{for}-eigenspaces $\mathcal{O}^{\chi}_{G_{\text{for}}}$, where $\mathcal{O}^{\chi}_{G_{\text{for}}} = \{f \in \mathcal{O}_{G_{\text{for}}} | f(tx) = \chi(t)f(x)\ \forall t \in T_{\text{for}},\ \forall x \in G_{\text{for}}\}$ and χ runs over X = character group of T. Similarly, T_{for} acts on $\mathcal{L} = \pi^* \mathcal{M}$, hence $\pi_* \mathcal{L}_{\text{for}}$ is the I-adic topological direct sum of T_{for}-eigenspaces $\mathcal{M}^{\chi}_{\text{for}}$, where $\mathcal{M}^{\chi}_{\text{for}} = \{f \in \mathcal{L}_{\text{for}} | f(tx) = \chi(t)f(x)\ \forall t \in T_{\text{for}},\ \forall x \in G_{\text{for}}\}$ and χ runs over X. Thus each $s \in \Gamma(G, \mathcal{L})$ can be uniquely written $s = \sum_{\chi \in X} s_\chi$ where each s_χ is a section of $\mathcal{M}^{\chi}_{\text{for}}$ and the sum converges in the I-adic topology. $\Gamma(G_{\text{for}}, \mathcal{M}^{\chi}_{\text{for}})$ is naturally identified with the space of global sections $\Gamma(A, \mathcal{M}_\chi)$ of an ample cubical invertible sheaf $\mathcal{M}_\chi$ as follows: If we change the action of T_{for} on $\mathcal{L}_{\text{for}}$ by χ, then $\mathcal{L}_{\text{for}}$ descends to an ample cubical invertible sheaf $\mathcal{M}_{\chi,\text{for}}$ on A_{for}. Equivalently, $\mathcal{M}_{\chi,\text{for}}$ is the subsheaf of $\pi_{\text{for}*} \mathcal{L}_{\text{for}}$ consisting of χ-eigenfunctions. $\mathcal{M}_{\chi,\text{for}}$ is related to $\mathcal{M}^{\chi}_{\text{for}}$ by: $\mathcal{M}^{\chi}_{\text{for}} = \pi^{\cdot}_{\text{for}} \mathcal{M}_{\chi,\text{for}}$, where $\pi^{\cdot}_{\text{for}}$ is the sheaf-theoretic pull-back. $\mathcal{M}_{\chi,\text{for}}$ algebraizes to an ample cubical invertible sheaf $\mathcal{M}_\chi$ on A by Grothendieck's existence theorem, hence $\Gamma(G_{\text{for}}, \mathcal{M}^{\chi}_{\text{for}}) = \Gamma(A, \mathcal{M}_\chi)$. The partial Fourier expansion for $s \in \Gamma(G, \mathcal{L})$ can be written $s = \sum_{\chi \in X} \sigma_\chi(s)$, where σ_χ is an R-linear homomorphism from $\Gamma(G, \mathcal{L})$ to $\Gamma(A, \mathcal{M}_\chi)$. σ_χ depends on $\mathcal{L}$ and the chosen action of T_{for} on $\mathcal{L}_{\text{for}}$. We will write $\sigma^{\mathcal{L}}_\chi$ to specify the dependence on $\mathcal{L}$ when necessary. The $\sigma^{\mathcal{L}}_\chi$'s extend to $\Gamma(G, \mathcal{L}_\eta) \to \Gamma(A, \mathcal{M}_{\chi,\eta})$ and give a partial Fourier expansion of elements of $\Gamma(G, \mathcal{L}_\eta)$ in the obvious sense.

To proceed further it is necessary to examine here some unpleasant details about various invertible sheaves on the abelian scheme A, so that we are not confused later on. For each $\chi \in X$ denote by $\mathcal{O}_\chi$ the subsheaf of $\pi_* \mathcal{O}_G$ consisting of χ-eigenfunctions. Each $\mathcal{O}_\chi$ is an invertible sheaf on A with natural cubical structure, and $\pi_* \mathcal{O}_G = \oplus_{\chi \in X} \mathcal{O}_\chi$. $\mathcal{O}^\chi := \pi^{\cdot}_{\text{for}} \mathcal{O}_\chi$ is the subsheaf of χ-eigenfunctions in $\mathcal{O}_G$, and $\mathcal{O}_G$ is the direct sum of all $\mathcal{O}^\chi$'s. In §2 we defined the two classifying homomorphisms $c : X \to A^t$, $c^t : Y \to A$ by push-out. Unfortunately, the push-out of $1 \to T \to \tilde{G} \to A \to 1$ by $\chi : T \to G_m$ is an G_m-torsor over A whose associated invertible invertible $\mathcal{O}_A$-module is $\mathcal{O}_{-\chi}$, as can be checked by easy local computation, keeping in mind the duality between spaces and functions on them. For this reason *we shall change the sign of the classifying homomorphisms c and c^t from now on*, so that $c(\chi) \in A^t(R)$ is $\mathcal{O}_\chi$, and we still have $c \circ \phi = \lambda_A \circ c^t$. Since the polarization attached to $\mathcal{M}$ is λ_A, we have $c^t(y)^* \mathcal{M} \cong \mathcal{M}_{\phi(y)} \cong \mathcal{M} \otimes \mathcal{O}_{\phi(y)}$ for each y in Y. (Without the change of sign, we would have $c^t(y)^* \mathcal{M} \cong \mathcal{M}_{-\phi(y)}$, which entails a lot of minus signs in formulas to appear later. Later on whenever a principal homogeneous spaces of a torus over an abelian scheme occur, we shall explicitly define the sign of the classifying homomorphism if there is any possible confusion.)

As in §4, we would like to relate $\sigma_\chi : \Gamma(G,\mathcal{L}) \to \Gamma(A,\mathcal{M}_\chi)$ with $\sigma_{\chi+\phi(y)} : \Gamma(G,\mathcal{L}) \to \Gamma(A,\mathcal{M}_{\chi+\phi(y)})$. Since $T^*_{c^t(y)}\mathcal{M}_\chi \cong \mathcal{M}_{\chi+\phi(y)} \otimes \mathcal{M}_\chi(c^t(y)) \otimes \mathcal{M}_{\chi+\phi(y)}(0)^{-1} \cong \mathcal{M}_{\chi+\phi(y)} \otimes \mathcal{M}_\chi(c^t(y))$ via the rigidification of $\mathcal{M}_{\chi+\phi(y)}$, we obtain by composition a map

$$T^*_{c^t(y)} \circ \sigma_\chi : \Gamma(G,\mathcal{L}) \to \Gamma(A,\mathcal{M}_{\chi+\phi(y)}) \otimes_R \mathcal{M}_\chi(c^t(y)).$$

Notice that $\mathcal{M}_\chi(c^t(y)) \cong \mathcal{M}(c^t(y))\otimes\mathcal{O}_\chi(c^t(y))\otimes\mathcal{M}_\chi(0)\otimes\mathcal{M}(0)^{-1}\otimes\mathcal{O}_\chi(0)^{-1} \cong \mathcal{M}(c^t(y)) \otimes \mathcal{O}_\chi(c^t(y))$ via the rigidifications (of $\mathcal{M}_\chi$, $\mathcal{M}$ and $\mathcal{O}_\chi$). For each element $m \in \mathcal{M}_\chi(c^t(y))^{-1}_\eta = \mathcal{M}_\chi(c^t(y))^{-1} \otimes_R K$, we obtain by "multiplication with m" a map

$$m \cdot T^*_{c^t(y)} \circ \sigma_\chi : \Gamma(G,\mathcal{L}) \to \Gamma(A,\mathcal{M}_{\chi+\phi(y)}) \otimes_R K.$$

The main theorem of this chapter asserts that $\forall y \in Y$, $\forall \chi \in X$, there is a unique $m(y,\chi) \in \mathcal{M}_\chi(c^t(y))^{-1}_\eta$ such that $m \cdot T^*_{c^t(y)} \circ \sigma_\chi = \sigma_{\chi+\phi(y)}$ (as maps from $\Gamma(G,\mathcal{L})$ to $\Gamma(A,\mathcal{M}_{\chi+\phi(y)}) \otimes_R K$). This element $m(y,\chi)$ will be presented as a product of an element $\psi(y) \in \mathcal{M}(c^t(y))^{-1}_\eta$ and an element $\tau(y,\chi) \in \mathcal{O}_\chi(c^t(y))^{-1}_\eta \cong \mathcal{P}_A(c^t(y),c(\chi))^{-1}_\eta$, where $\mathcal{P}_A$ is the Poincaré sheaf over $A \times_S A^t$. It will be seen that $\psi(y)$ is multiplicatively quadratic in $y \in Y$ w.r.t. the biextension structure of $\mathcal{P}_A^{-1}$. In other words, $\psi(\cdot)$ defines a cubic trivialization of $(c^t)^*\mathcal{M}_\eta^{-1}$, while $\tau(\cdot,\cdot)$ defines a trivialization of the biextension $(c^t \times c)^*\mathcal{P}_\eta^{-1}$. When the abelian part A is trivial, then $\psi(y), \nu(y,\chi)$ reduces to $a(y)$, $b(y,\chi)$ as in §4.

Now we formulate the main theorem of this chapter.

5.1. Theorem. The collection of R-module homomorphisms $\sigma_\chi : \Gamma(G,\mathcal{L}) \to \Gamma(A,\mathcal{M}_\chi)$, $\chi \in X$, have the following properties:

(i) $\sigma_\chi \neq 0 \quad \forall \chi \in X$.

(ii) There exists a cubic trivialization $\psi : 1_Y \to (c^t)^*\mathcal{M}_\eta^{-1}$ and a trivialization of biextension $\tau : 1_{Y\times X} \to (c^t \times c)^*\mathcal{P}_\eta^{-1}$ such that $\sigma_{\chi+\phi(y)}(s) = \psi(y)\cdot\tau(y,\chi)\cdot T^*_{c^t(y)} \circ \sigma_\chi(s)$ $\forall s \in \Gamma(G,\mathcal{L})$, $\forall y \in Y$, $\forall \chi \in X$. In particular, $\psi(0) = 1$, $\tau(0,\chi) = 1$, $\tau(y,0) = 1$ $\forall y \in Y$, $\forall \chi \in X$, where "1" denote the obvious sections given by the rigidification.

(iii) $\tau \circ (\mathrm{id}_Y \times \phi) : 1_{Y\times Y} \to (c^t \times \lambda_A \circ c^t)^*\mathcal{P}_\eta^{-1} = (c^t \times c^t)^*\Lambda(\mathcal{M})_\eta^{-1}$ is a symmetric trivialization of $(c^t \times c^t)^*\Lambda(\mathcal{M})_\eta^{-1}$ with respect to the canonical structure of symmetric biextension on $\Lambda(\mathcal{M})^{-1}$. (symbolically we shall write $\tau(y,\phi(z)) = \tau(z,\phi(y))$ with the relevant canonical isomorphisms and symmetry understood.)

(iv) ψ and τ are compatible with respect to ϕ, in the sense that $\Lambda(\psi) = \tau\circ(\mathrm{id}_Y\times\phi)$ as trivializations of the biextension $(c^t\times c^t)^*\Lambda(\mathcal{M})_\eta^{-1}$. (Symbolically, we shall write $\psi(y+z) = \psi(y)\cdot\psi(z)\cdot\tau(y,\phi(z))$ $\forall y,z \in Y$, with all the canonical isomorphisms understood.)

(v) $\forall y \in Y$, $\tau(y,\phi(y))$ extends to a section of the invertible sheaf $\Lambda(\mathcal{M})^{-1}(c^t(y),\ c^t(y))$ over S, and is congruent to 0 modulo I if $y \neq 0$.

(vi) For all but finitely many y in Y, $\psi(y)$ extends to a section of the invertible sheaf $\mathcal{M}^{-1}(c^t(y))$ over S, and is congruent to 0 modulo I. More generally, for every $n \geq 0$, $\psi(y)$ is congruent to 0 modulo I^n for all but finitely many y in Y.

5.2. Remarks and Reduction. (a) As in theorem 4.1, the important step is to show that $\sigma_\chi \neq 0\ \forall \chi \in X$ and that the two homomorphisms $T^*_{c^t(y)} \circ \sigma_\chi : \Gamma(G, \mathcal{L}) \to \Gamma(A, \mathcal{M}_{\chi+\phi(y)}) \otimes_R \mathcal{M}_\chi(c^t(y))$ and $\sigma_{\chi+\phi(y)} : \Gamma(G, \mathcal{L}) \to \Gamma(A, \mathcal{M}_{\chi+\phi(y)})$ are proportional in the sense that $\exists m(y,\chi) \in \mathcal{M}_\chi(c^t(y))_\eta^{-1}$ such that $\sigma_{\chi+\phi(y)} = m(y,\chi) \cdot T^*_{c^t(y)} \circ \sigma_\chi$. From this one deduces immediately that ψ and τ are uniquely defined by the equation $\sigma_{\chi+\phi(y)} = \psi(y) \cdot \tau(y,\chi) \cdot T^*_{c^t(y)} \circ \sigma_\chi$, and (iii) (iv) are satisfied. As before, the multiplicativity in y of $\tau(\chi, y)$ follows from that in χ. Parts (v) and (vi) hold by the same argument as in theorem 4.1. Also, (vi) is a formal consequence of (i)–(v) and the normality of R.

(b) Again as in theorem 4.1 we may assume that R is a complete discrete valuation ring. The only point is that the quantities involved are defined over R and functorial with respect to base change, hence we can test equalities over complete discrete valuation rings.

Invariant Formulation. We have made a choice of descent data of $\mathcal{L}_{\text{for}}$ w.r.t. π and obtain $\mathcal{M}$ by descending $\mathcal{L}_{\text{for}}$ and algebraizing. Let us try to understand what ψ and τ mean intrinsically. First of all τ is independent of the choice, as two such choices differ by a character of T which amounts to shifting the indices of the $\mathcal{M}_\chi$'s by this character. For each y in Y, $\{\tau(y,\chi)\}_{\chi \in X}$ defines a homomorphism $c^t(y)^*(\oplus_{\mu \in X} \mathcal{O}_\mu) \to K$, hence a point in $\tilde{G}(K)$ over $c^t(y) \in A(R)$. Hence τ defines a homomorphism $\imath : Y \to \tilde{G}(K)$ lifting c^t. In terms of $\imath$, ψ defines a cubical trivialization of $\imath^* \tilde{\mathcal{L}}_\eta^{-1}$, which is independent of the choice of descent data: Again two different choices just produce a shift in the indexing of $\mathcal{M}_\chi$'s. The resulting cubical trivialization of $\imath^* \tilde{\mathcal{L}}_\eta^{-1}$, will again be denoted by ψ by abuse of notation. ψ is compatible with τ with respect to ϕ in the sense that $\Lambda(\psi) = (\pi \times \pi)^* \tau \circ (\mathrm{id}_Y \times \phi)$ as trivializations of the symmetric biextension $(c^t \times c^t)^* \Lambda(\mathcal{M})_\eta^{-1}$. $\forall y \in Y$, $\forall \mu \in X$, "multiplication by $\psi(y) \cdot \tau(y,\mu)$" defines a map $T^*_{c^t(y)}(\mathcal{M} \otimes \mathcal{O}_\mu) \otimes_R K \to \mathcal{M} \otimes \mathcal{O}_{\mu+\phi(y)} \otimes_R K$. Putting them together as μ varies over X, we obtain for each y in Y a map $T^*_{c^t(y)} \tilde{\mathcal{L}}_\eta \to \tilde{\mathcal{L}}_\eta$. This gives an action of Y on $\tilde{\mathcal{L}}_\eta$ w.r.t. $\imath : Y \to \tilde{G}(K)$. (Conversely, it can be easily checked that any action of Y on $\tilde{\mathcal{L}}_\eta$ w.r.t. $\imath : Y \to \tilde{G}(K)$ such that the commutator of the action of Y and T is given by the character $\phi(y)$ of Y, $\forall y \in Y$, is of this form.) The positivity conditions (v), (vi) can be reformulated as follows:

(v)′ $\forall y \in Y$, $(\pi \times \pi)^* \tau(y, \phi(y))$ extends to a section of the invertible sheaf $\Lambda(\tilde{\mathcal{L}})^{-1}(\imath(y), \imath(y))$ over S, and is congruent to 0 modulo I if $y \neq 0$.

(vi)″ For every integer $n \geq 0$, $\psi(y)$ extends to a section of the invertible sheaf $\tilde{\mathcal{L}}^{-1}(\imath(y))$ over S, and is congruent to 0 modulo I^n for all but finitely many y in Y.

Remark about trivializing $(c^t)^*\mathcal{M}^{-1}$ and $(c^t \times c)^*\mathcal{P}_A^{-1}$: Locally in Zariski topology, one may trivialize the cubical torsor $(c^t)^*\mathcal{M}^{-1}$ and the biextension $(c^t \times c)^*\mathcal{P}_A^{-1}$, the latter means that one can lift c^t to $\tilde{G}$ Zariski locally (under our two auxiliary hypotheses, especially the second one). But one may not be able to trivialize both in a compatible way even étale locally, although this holds over η. The problem is really that one may not be able to find a trivialization of $(c^t\times c)^*\mathcal{P}_A^{-1}$ étale locally whose pull-back by $\mathrm{id}_Y\times\phi$ is a symmetric trivialization of $(c^t \times c^t)^*\Lambda(\mathcal{M})^{-1}$.

Remark about Globalization. Remember that we have made two hypotheses about the triviality of cubical sheaves and constancy of sheaves of characters of tori; both hold locally in the étale topology. We also assumed a condition on R which says roughly that "normality of $\mathrm{Spf}(R)$ is preserved when passing to étale coverings". This condition is not used in theorem 5.1: we need it only to localize in the étale topology. Moreover we need not restrict ourselves to affine formal schemes $\mathrm{Spf}(R)$. The appropriate globalization is in the setting of (relative) schemes over a (noetherian normal in our case) formal base scheme S_{for} w.r.t the étale topology of S, such that the affine coordinate rings satisfies the condition above. In particular S_{for} need not be affine. Recall also that we assumed that the torus part of $G \times_S S_0$ is split. In this more general setting we can eliminate this hypothesis because it is always true étale locally on S_0: the invariant formulation descends! (see theorem 6.2 below.) We shall do this in the end of this chapter.

From now on we assume that R is a complete discrete valuation ring, and we will freely enlarge R by base change whenever necessary or convenient, sometimes without saying so explicitly. Our basic tool is again the addition formula, which will not surprise the reader. Since the rank of $\Gamma(A,\mathcal{M})$ is not necessarily one, our situation is a bit more complicated than before. The trick to overcome this minor difficulty is to introduce for every $\chi \in X$ a non-degenerate theta group $\mathcal{G}(\mathcal{L})^f_{\bar{\chi}}$ $(\bar{\chi} \in X/\phi(Y))$ acting on $\Gamma(A,\mathcal{M}_\chi)$ and an R-submodule $\Gamma(G,\mathcal{L})^{\bar{\chi}}$ of $\Gamma(G,\mathcal{L})$, such that σ_χ is an equivariant homomorphism between representations of $\mathcal{G}(\mathcal{L})^f_{\bar{\chi}}$ for every χ in X. Since $T^*_{c^t(y)} \circ \sigma_\chi : \Gamma(G,\mathcal{L}) \to \Gamma(A,\mathcal{M}_{\chi+\phi(y)}) \otimes_R \mathcal{M}_\chi(c^t(y))$ is also such an equivariant map (just as $\sigma_{\chi+\phi(y)}$) as will become clear when we give the relevant definitions, $\sigma_{\chi+\phi(y)} : \Gamma(G,\mathcal{L}) \to \Gamma(A,\mathcal{M}_{\chi+\phi(y)})$ and $T^*_{c^t(y)} \circ \sigma_\chi$ are proportional (in the sense that they differ by a factor $m(y,\chi) \in \mathcal{M}(c^t(y))_\eta^{-1}$) by a fundamental result about representations of non-degenerate theta groups. Here we should insert the remark that $\Gamma(G,\mathcal{L})\otimes K$ decomposes into a direct sum of subspaces $\Gamma(G,\mathcal{L})^{\bar{\mu}} \otimes K$, $\bar{\mu} \in X/\phi(Y)$, and σ_χ is zero on $\Gamma(G,\mathcal{L})^{\bar{\mu}}$ if $\bar{\chi} \neq \bar{\mu}$. Using these non-degenerate theta groups makes our present situation analogous to the previous one when the abelian part is trivial, and we will prove the theorem by very similar methods.

Let $K(\mathcal{L})$ be the scheme theoretic closure of $K(\mathcal{L})_\eta$ in G, which is a quasi-finite flat group scheme over the complete discrete valuation ring R. The inverse image $\mathcal{G}(\mathcal{L})$ of $K(\mathcal{L})$ in the G_m-torsor L defined by $\mathcal{L}$ has a natural group structure as a central extension of $K(\mathcal{L})$ by G_m extending $\mathcal{G}(\mathcal{L}_\eta)$. Denote by $K(\mathcal{L})^f$ the "finite part" of $K(\mathcal{L})$, so that $K(\mathcal{L}) = K(\mathcal{L})^f \amalg$ (an open subscheme with empty special fibre). Let $K(\mathcal{L})^m := K(\mathcal{L})^f \cap T$ be the "multiplicative part" of $K(\mathcal{L})$. There is an alternating pairing $e_S^{\mathcal{L}_\eta} : K(\mathcal{L}) \times K(\mathcal{L}) \to G_m$ extending $e^{\mathcal{L}_\eta} : K(\mathcal{L}_\eta) \times K(\mathcal{L}_\eta) \to G_{m,\eta}$, which is simply the commutator pairing of $\mathcal{G}(\mathcal{L})$. $K(\mathcal{L})^m$ and $K(\mathcal{L})^f$ are annihilator of each other with respect to $e_S^{\mathcal{L}_\eta}$, and $e_S^{\mathcal{L}_\eta}$ induces an alternating perfect pairing on $K(\mathcal{L})^f/K(\mathcal{L})^m = K(\mathcal{M})$ which is nothing else but $e^{\mathcal{M}}$. The reference for all the assertions above is [MB] IV 2.4.

Let $\mathcal{G}(\mathcal{L})^f$ be the inverse image of $K(\mathcal{L})^f$ in $\mathcal{G}(\mathcal{L})$, $\mathcal{G}(\mathcal{L})^m$ the inverse image of $K(\mathcal{L})^m$ in $\mathcal{G}(\mathcal{L})$. We have an exact sequence

$$0 \to \mathcal{G}(\mathcal{L})^m \to \mathcal{G}(\mathcal{L})^f \to K(\mathcal{M}) \to 0.$$

Via the chosen action of T_{for} on $\mathcal{L}_{\text{for}}$ (equivalently, a descent data for $\mathcal{L}_{\text{for}}$ w.r.t. π), we get a splitting of the central extension $0 \to G_m \to \mathcal{G}(\mathcal{L})^m \to K(\mathcal{L})^m \to 0$. Hence we have an algebraic action of $K(\mathcal{L})^m$ on $\Gamma(G,\mathcal{L})$. Let $d_t = \text{rk}(K(\mathcal{L})^m)$ and $d_a = \text{rk}(\Gamma(A,\mathcal{M}))$, so that $\Gamma(G,\mathcal{L})$ has rank $d = d_a d_t$. Basic results about representations of non-degenerate theta groups tell us that each $K(\mathcal{L})^m$-eigenspace $\Gamma(G,\mathcal{L})^{\bar{\chi}}$ has rank d_a, where $\bar{\chi} \in X/\phi(Y) =$ character group of $K(\mathcal{L})^m$. Every $\bar{\chi} \in X/\phi(Y)$ gives rise to a homomorphism $h_\chi : \mathcal{G}(\mathcal{L})^m \to G_m$ which is the identity on $G_m \subset \mathcal{G}(\mathcal{L})^m$ and is $\bar{\chi}$ on $K(\mathcal{L})^m \subset \mathcal{G}(\mathcal{L})^m$ via the splitting. The push-out of $0 \to \mathcal{G}(\mathcal{L})^m \to \mathcal{G}(\mathcal{L})^f \to K(\mathcal{M}) \to 0$ by h_χ is a central extension $0 \to G_m \to \mathcal{G}(\mathcal{L})^f_{\bar{\chi}} \to K(\mathcal{M}) \to 0$ naturally isomorphic to $0 \to G_m \to \mathcal{G}(\mathcal{M}_\chi) \to K(\mathcal{M}_\chi) \to 0$. $\Gamma(G,\mathcal{L})^{\bar{\chi}}$ is a $\mathcal{G}(\mathcal{L})^f$-invariant submodule of $\Gamma(G,\mathcal{L})$, and $\mathcal{G}(\mathcal{L})^f_{\bar{\chi}}$ acts on $\Gamma(G,\mathcal{L})^{\bar{\chi}}$ by the definition of push-out. It is clear that $\sigma_\chi : \Gamma(G,\mathcal{L})^{\bar{\chi}} \to \Gamma(A,\mathcal{M}_\chi)$ is a $\mathcal{G}(\mathcal{L})^f_{\bar{\chi}}$-equivariant homomorphism between free R-modules of rank d_a. Hence $\Gamma(G,\mathcal{L})^{\bar{\chi}}$ and $\Gamma(A,\mathcal{M}_\chi)$ are both irreducible $\mathcal{G}(\mathcal{L})^f_{\bar{\chi}}$-modules, and σ_χ is either zero or "the" intertwining operator between these two irreducible representations, well-defined up to non-zero multiple. Note that σ_χ is zero on $\Gamma(G,\mathcal{L})^{\bar{\mu}}$ if $\bar{\chi} \neq \bar{\mu}$ in $X/\phi(Y)$.

The addition formula describes the behavior of ample cubical invertible sheaves and their sections under pull-back by the fundamental isogeny $\Phi : G \times_S G \to G \times_S G$, $\Phi : (x,y) \to (x+y, x-y)$. Thus we first study behavior under a general isogeny:

5.3. Lemma. *Let $f : H \to G$ be an isogeny of semi-abelian schemes over R such that G_η is abelian and the torus part of $G_0 = G \otimes_R k$ is a split torus. f induces an isogeny from $0 \to T_{H,\text{for}} \to H_{\text{for}} \to A_{H,\text{for}} \to 0$ to $0 \to T_{G,\text{for}} \to G_{\text{for}} \to A_{G,\text{for}} \to 0$, where T_H, T_G denote the torus part and A_H, A_G denote the abelian part of H_{for}, G_{for} respectively. Let $\mathcal{L}$ be a cubical invertible sheaf on G, and fix a choice of trivialization of the formal cubical invertible sheaf $\mathcal{L}_{\text{for}} \otimes \mathcal{O}_{T_{G,\text{for}}}$ on $T_{G,\text{for}}$. Via f_{for}, we have an induced trivialization of $(f^*\mathcal{L})_{\text{for}} \otimes \mathcal{O}_{T_{H,\text{for}}}$. With respect to these compatible trivializations, we have $f^*(\sigma_\chi^{\mathcal{L}}(s)) = \sigma_{f^*\chi}^{f^*\mathcal{L}}(f^*s)$ $\forall s \in \Gamma(G,\mathcal{L}), \forall \chi \in X$. Here $f^*\chi \in X(T_H)$ denotes the image of χ*

under $f_T^* : X(T_G) \to X(T_H)$, *and* $f^*(\sigma_\chi^{\mathcal{L}}(s))$ *denotes the image of* $\sigma_\chi^{\mathcal{L}}(s)$ *under* $f_A^* : \Gamma(A_G, \mathcal{M}_\chi) \to \Gamma(A_H, \mathcal{M}_{f^*\chi})$, *where* f_A *is the isogeny between the abelian parts induced by* f.

The proof of this lemma is straight forward. Write $s = \sum_{\chi\in X} \sigma_\chi^{\mathcal{L}}(s)$, hence $f^*s = \sum_{\chi\in X} f_A^*(\sigma_\chi^{\mathcal{L}}(s))$, where f_A is the isogeny between the abelian parts induced by f. As $f^* : X = X(G_{\text{for}}) \to X(H_{\text{for}}) := X'$ is injective and $f^*s = \sum_{\nu\in X'} \sigma_\nu^{f^*\mathcal{L}}(f^*s)$ by definition, we get $f^*(\sigma_\chi^{\mathcal{L}}(s)) = \sigma_{f^*\chi}^{f^*\mathcal{L}}(s)$.

Applying this lemma to our fundamental isogeny $\Phi : G \times_S G \to G \times_S G$, we deduce

Addition Formula. Let $\mathcal{L}_1, \mathcal{L}_2$ be algebraically equivalent ample cubical invertible sheaves on G, $\mathcal{F} = p_1^*\mathcal{L}_1 \otimes p_2^*\mathcal{L}_2$, $\mathcal{F}' = \Phi^*(p_1^*\mathcal{L}_1 \otimes p_2^*\mathcal{L}_2) \cong p_1^*(\mathcal{L}_1 \otimes \mathcal{L}_2) \otimes p_2^*(\mathcal{L}_1 \otimes [-1]^*\mathcal{L}_2)$. Then $\Phi_A^*(\sigma_{(\chi,\nu)}^{\mathcal{F}}(p_1^*s_1 \otimes p_2^*s_2)) = \sigma_{(\chi+\nu,\chi-\nu)}^{\mathcal{F}'}(\Phi^*(p_1^*s_1 \otimes p_2^*s_2))$ $\forall s_1, s_2 \in \Gamma(G, \mathcal{L})$, $\forall \mu, \nu \in X$, where Φ_A is the fundamental isogeny for the abelian part.

Since $\sigma_{(\chi,\nu)}^{\mathcal{F}} = \sigma_\chi^{\mathcal{L}_1} \otimes \sigma_\nu^{\mathcal{L}_2}$ and $\sigma_{(\chi+\nu,\chi-\nu)}^{\mathcal{F}'} = \sigma_{\chi+\nu}^{\mathcal{L}_1\otimes\mathcal{L}_2} \otimes \sigma_{\chi-\nu}^{\mathcal{L}_1\otimes[-1]^*\mathcal{L}_2}$, we can rewrite the addition formula

$$(\mathrm{AF}')\quad \Phi_A^* \circ (\sigma_\chi^{\mathcal{L}_1} \otimes \sigma_\nu^{\mathcal{L}_2}) = (\sigma_{\chi+\nu}^{\mathcal{L}_1\otimes\mathcal{L}_2} \otimes \sigma_{\chi-\nu}^{\mathcal{L}_1\otimes[-1]^*\mathcal{L}_2}) \circ \Phi^*.$$

Here

$$\begin{aligned}
&\sigma_\chi^{\mathcal{L}_1} \otimes \sigma_\nu^{\mathcal{L}_2} : \Gamma(G, \mathcal{L}_1) \otimes \Gamma(G, \mathcal{L}_2) \to \Gamma(A, \mathcal{M}_{1,\chi}) \otimes \Gamma(G, \mathcal{M}_{2,\nu}),\\
&\sigma_{\chi+\nu}^{\mathcal{L}_1\otimes\mathcal{L}_2} \otimes \sigma_{\chi-\nu}^{\mathcal{L}_1\otimes[-1]^*\mathcal{L}_2} : \Gamma(G, \mathcal{L}_1 \otimes \mathcal{L}_2) \otimes \Gamma(G, \mathcal{L}_1 \otimes [-1]^*\mathcal{L}_2)\\
&\qquad \to \Gamma(A, (\mathcal{M}_1 \otimes \mathcal{M}_2)_{\chi+\nu}) \otimes \Gamma(A, (\mathcal{M}_1 \otimes [-1]^*\mathcal{M}_2)_{\chi-\nu}),\\
&\Phi_A^* : \Gamma(A, \mathcal{M}_{1,\chi}) \otimes \Gamma(A, \mathcal{M}_{2,\nu})\\
&\qquad \to \Gamma(A, (\mathcal{M}_1 \otimes \mathcal{M}_2)_{\chi+\nu}) \otimes \Gamma(A, (\mathcal{M}_1 \otimes [-1]^*\mathcal{M}_2)_{\chi-\nu}),\\
&\Phi^* : \Gamma(G, \mathcal{L}_1) \otimes \Gamma(G, \mathcal{L}_2) \to \Gamma(G, \mathcal{L}_1 \otimes \mathcal{L}_2) \otimes (G, \mathcal{L}_1 \otimes [-1]^*\mathcal{L}_2).
\end{aligned}$$

5.4. Remark. Here is how we use (AF′): The left hand side of (AF′) is non-zero iff $\sigma_\chi^{\mathcal{L}_1} \neq 0$ and $\sigma_\nu^{\mathcal{L}_2} \neq 0$, in which case it is injective. This follows from the representation theory of non-degenerate theta groups. For the same reason, the right hand side of (AF′) is non-zero iff it is injective iff $\sigma_{\chi+\nu}^{\mathcal{L}_1\otimes\mathcal{L}_2} \neq 0$, $\sigma_{\chi-\nu}^{\mathcal{L}_1\otimes[-1]^*\mathcal{L}_2} \neq 0$, and $P_{\overline{\chi+\nu},\overline{\chi-\nu}} \circ \Phi^* \neq 0$, where $P_{\overline{\chi+\nu},\overline{\chi-\nu}}$ is the projection to the $(\overline{\chi+\nu}, \overline{\chi-\nu})$-eigenspace under $K(\mathcal{F}')$-action, with $(\overline{\chi+\nu}, \overline{\chi-\nu}) \in X/2\phi(Y) \times X/2\phi(Y) =$ character group of $K(\mathcal{F}')$.

With the addition formula at our disposal, we can finish the proof of theorem 5.1 just as before. The strategy is to prove non-vanishing of $\sigma_\chi^{\mathcal{L}}$ first for symmetric $\mathcal{L}$'s with the additional property that the induced symmetry on $\mathcal{L}_{\text{for}}$ maps $\mathcal{M}_{\text{for}}^{\chi}$ to $\mathcal{M}_{\text{for}}^{-\chi}$. We remark that given an arbitrary ample cubical invertible sheaf on G, after base change there always exists a symmetric $\mathcal{L}$ algebraically equivalent to the given one with the above property on symmetry. Assume for the time being that we have such an $\mathcal{L}$. Because $\Gamma(G,\mathcal{L})^{\bar{\chi}} \neq 0\ \forall \bar{\chi} \in X/\phi(Y)$, we know that $\forall \bar{\chi} \in X$, $\exists \chi \in \bar{\chi}$ s.t. $\sigma_\chi^{\mathcal{L}} \neq 0$. In particular $\sigma_{\phi(y_0)}^{\mathcal{L}} \neq 0$ for some $y_0 \in Y$. (AF′) tells us that $P_{\bar{0},\bar{0}} \circ \Phi* \neq 0$, $\sigma_{\phi(2y_0)}^{\mathcal{L}^2} \neq 0$, $\sigma_0^{\mathcal{L}^2} \neq 0$. Consider the subset $Z := \{\chi \in \phi(Y) | \sigma_\chi^{\mathcal{L}} \neq 0\}$ of Y. We show that Z is a subgroup of $\phi(Y)$: We have seen that $0 \in Z$. If $\chi, \nu \in Z$, (AF′) gives $\sigma_{2\chi}^{\mathcal{L}^2} \neq 0$, $\sigma_{2\nu}^{\mathcal{L}^2} \neq 0$, another application of (AF′) gives $\sigma_{\chi+\nu}^{\mathcal{L}} \neq 0$, $\sigma_{\chi-\nu}^{\mathcal{L}} \neq 0$, hence $\chi+\nu \in Z$, $\chi-\nu \in Z$. We claim that $Z = \phi(Y)$. If this is not true, then there is a proper subgroup W of $\phi(Y)$ containing Z and of finite index in $\phi(Y)$. We must show that this leads to a contradiction. W corresponds to a finite flat subgroup H of T containing $K(\mathcal{L})^m$. The trivialization of $\mathcal{L}_{\text{for}} \otimes \mathcal{O}_{T_{\text{for}}}$ induces a section of $\mathcal{G}(\mathcal{L})^m \to K(\mathcal{L})^m$, hence an action of $K(\mathcal{L})^m$ on $\mathcal{L}$. $\mathcal{L}$ descends to a cubical ample invertible sheaf $\underline{\mathcal{L}}$ on the quotient semi-abelian scheme $\mathcal{G} = G/K(\mathcal{L})^m$. The existence of W means that any $K(\mathcal{L})^m$-invariant section $s \in \Gamma(G,\mathcal{L})$ has a Fourier expansion $s = \sum_{\chi \in W} \sigma_\chi^{\mathcal{L}}(s)$. Passing to $\mathcal{G}$, we are reduced to the case $\phi(Y) = X$. Translating $\mathcal{L}$ by elements of $G(R)$, we conclude that for any $\mathcal{L}'$ algebraically equivalent to $\mathcal{L}$ and any section s' of $\mathcal{L}'_\eta$, the Fourier expansion of s' involves only those $\chi \in W$, provided we choose suitable trivialization of $\mathcal{L}_{\text{for}}$. But we know (I, Prop. 5.3) that if we make suitable base change $R \to R'$, $\Gamma(G,\mathcal{L}_\eta^{\otimes n})$ is generated by elements of the form $T_{x_1}^*(s) \otimes \cdots \otimes T_{x_n}^*(s)$ for $n \geq 1$, where $x_1, \ldots, x_n \in G(R)$ with $x_1 + \cdots + x_n = 0$. Thus every element of $\Gamma(G,\mathcal{L}^{\otimes n})$ has a Fourier expansion involving only those $\chi \in W$. Hence elements of $\Gamma(G,\mathcal{L}^{\otimes n})$ factor through H. But $\mathcal{L}$ is ample on G over R; this is a contradiction. So we have proved the claim that $\sigma_\chi^{\mathcal{L}} \neq 0\ \forall \chi \in \phi(Y)$.

Now let $\mathcal{L}_1$ be an arbitrary given ample cubical invertible sheaf on G. From above we know that there exist another ample invertible sheaf $\mathcal{L}_2$ on G algebraically equivalent to $\mathcal{L}_1$ with the additional property that $\sigma_\nu^{\mathcal{L}_2} \neq 0\ \forall \nu \in \phi(Y)$. Given any $\chi \in X$, $\exists \xi \equiv \chi \,(\text{mod}\ \phi(Y))$ s.t. $\sigma_\xi^{\mathcal{L}_1} \neq 0$. (AF′) gives
$P_{\overline{\zeta+\nu},\overline{\zeta-\nu}} \circ \Phi^* \neq 0$, $\sigma_{\zeta+\nu}^{\mathcal{L}_1 \otimes \mathcal{L}_2} \neq 0$, $\sigma_{\zeta-\nu}^{\mathcal{L}_1 \otimes [-1]^* \mathcal{L}_2} \neq 0$, $\forall \nu \in \phi(Y)$. Write $\chi = \zeta + \nu$, we have in particular $P_{\bar{\chi},\bar{\chi}} \circ \Phi^* \neq 0$, $\sigma_\chi^{\mathcal{L}_1 \otimes \mathcal{L}_2} \neq 0$, $\sigma_\chi^{\mathcal{L}_1 \otimes [-1]^* \mathcal{L}_2} \neq 0$. Hence $\sigma_\chi^{\mathcal{L}_1} \neq 0$ by (AF′) again.

So far we have proved (i). As remarked before $\sigma_{\chi+\phi(y)}$ and $T_{c^t(y)}^* \circ \sigma_\chi$ are proportional, because both are non-zero equivariant homomorphisms between irreducible representations of $K(\mathcal{L})_{\bar{\chi}}^f$. Hence the trivializations $\psi : 1_Y \to (c^t)^* \mathcal{M}_\eta^{-1}$ and $\tau : 1_{Y \times X} \to (c^t \times c)^* \mathcal{P}_\eta^{-1}$ are welldefined by the equation $\sigma_{\chi+\phi(y)}(s) = \psi(y) \cdot \tau(y,\chi) \cdot T_{c^t(y)}^* \sigma_\chi(s)$ and the requirement that $\psi(0) = 1$, $\tau(\chi, 0) = 1$, where "1" denotes the relevant rigidifications.

It remains only to prove the multiplicativity of τ in the first variable. For any $\chi, \nu \in X$, $y \in Y$, we have by definition
$$\sigma_{\chi+\phi(y)}(\cdot) = \psi(y) \cdot \tau(y,\chi) \cdot T_{c^t(y)}^* \circ \sigma_\chi(\cdot),$$
$$\sigma_{\chi+\nu+\phi(y)}(\cdot) = \psi(y) \cdot \tau(y,\chi+\nu) \cdot T_{c^t(y)}^* \circ \sigma_{\chi+\nu}(\cdot).$$

(AF′) gives

$$\Phi_A^* \circ (\sigma_{\chi+\phi(y)}^{\mathcal{L}} \otimes \sigma_{\nu+\phi(y)}^{\mathcal{L}}) = (\sigma_{\chi+\nu+2\phi(y)}^{\mathcal{L}\otimes\mathcal{L}} \otimes \sigma_{\chi-\nu}^{\mathcal{L}\otimes[-1]^*\mathcal{L}}) \circ \Phi^* \quad \text{and}$$
$$\Phi_A^* \circ (\sigma_\chi^{\mathcal{L}} \otimes \sigma_\nu^{\mathcal{L}}) = (\sigma_{\chi+\nu}^{\mathcal{L}\otimes\mathcal{L}} \otimes \sigma_{\chi-\nu}^{\mathcal{L}\otimes[-1]^*\mathcal{L}}) \circ \Phi^*.$$

Since translation by $(y,y) \in G \times G$ goes under Φ to translation by $(2y,0)$, we conclude that

$$\psi^{\mathcal{L}}(y)^2 \cdot \tau^{\mathcal{L}}(y,\chi) \cdot b^{\mathcal{L}}(y,\nu) = \psi^{\mathcal{L}^2}(2y) \cdot \tau^{\mathcal{L}^2}(2y,\chi+\nu).$$

Similarly we have

$$\psi^{\mathcal{L}}(y)^2 \cdot \tau^{\mathcal{L}}(y,\chi+\nu) = \psi^{\mathcal{L}^2}(2y) \cdot \tau^{\mathcal{L}^2}(2y,\chi+\nu),$$

say by substituting $\chi+\nu$ for χ and 0 for ν. Together these two equations give $\tau^{\mathcal{L}}(y,\chi+\nu) = \tau^{\mathcal{L}}(y,\chi) \cdot \tau^{\mathcal{L}}(y,\nu)$. This finishes the proof of the theorem.

6. Complements

In the process above we have obtained further information about G, which we now make explicit. We also give the descent argument needed to remove the two simplifying hypotheses used above.

We first assume that $(c^t \times c)^* \mathcal{P}_A$ is trivialized, that $\underline{X}$ and $\underline{Y}$ are constant, and that $\tilde{\mathcal{L}}$ descends to a line bundle $\mathcal{M}$ on A. We then can describe the space of global sections $\Gamma(G_\eta, \mathcal{L}_\eta) \subset \Gamma(G_{\text{for}}, \mathcal{L}_{\text{for}}) \otimes K$, as follows: On $\tilde{G}$ a global section $s \in \Gamma(G, \mathcal{L})$ has a Fourier expansion $s = \sum \sigma_\chi(s)$, where $\sigma_\chi(s) \in \Gamma(A, \mathcal{M}_\chi)$ and $\sigma_{\chi+\phi(y)}(\cdot) = \psi(y) \cdot \tau(y,\chi) \cdot T_{c^t(y)}^* \circ \sigma_\chi(\cdot)$. It follows that $\Gamma(G_\eta, \mathcal{L}_\eta)$ injects into the space of families $s_\chi \in \Gamma(A_\eta, \mathcal{M}_{\chi,\eta})$ which satisfy $s_{\chi+\phi(y)} = \psi(y) \cdot \tau(y,\chi) \cdot T_{c^t(y)}^* s_\chi$. These families form a space of dimension $(X : \phi(Y)) \cdot \dim(\Gamma(A_\eta, \mathcal{M}_\eta))$ over K, which is also the dimension of $\Gamma(G_\eta, \mathcal{L}_\eta)$. Hence we have equality; that is, all possible Fourier expansions must occur.

Next we study the dependence of $\psi(\cdot)$ and $\tau(\cdot,\cdot)$ on $\mathcal{L}$. We assume given two ample line bundles $\mathcal{L}_1$ and $\mathcal{L}_2$ on G, such that the associated bundles $\tilde{\mathcal{L}}_1$ and $\tilde{\mathcal{L}}_2$ descend to $\mathcal{M}_1$ resp. $\mathcal{M}_2$ on A. If $\mathcal{L} = \mathcal{L}_1 \otimes \mathcal{L}_2$, then the associated $\tilde{\mathcal{L}} = \tilde{\mathcal{L}}_1 \otimes \tilde{\mathcal{L}}_2$ descends to $\mathcal{M} = \mathcal{M}_1 \otimes \mathcal{M}_2$. Our construction gives cubical trivializations ψ, ψ_1 and ψ_2 on Y, and bimultiplicative trivializations τ, τ_1 and τ_2 on $Y \times X$. We assume that $\tau_1 = \tau_2$. Later we shall see that this is always true, but for the moment note that it holds at least for $\mathcal{L}_2 = [-1]^*(\mathcal{L}_1)$.

6.1. Lemma. *Under these hypotheses we have $\psi = \psi_1 \cdot \psi_2$, $\tau = \tau_1 = \tau_2$. In invariant terms $\imath = \imath_1 = \imath_2$, and the Y-action on $\tilde{\mathcal{L}}_\eta \approx \tilde{\mathcal{L}}_{1,\eta} \otimes \tilde{\mathcal{L}}_{2,\eta}$ is the product of the actions on $\tilde{\mathcal{L}}_{1,\eta}$ and $\tilde{\mathcal{L}}_{2,\eta}$.*

To prove this, note first that $\phi = \phi_1 + \phi_2$. Now choose non-zero sections $s_1 \in \Gamma(G, \mathcal{L}_1)$ and $s_2 \in \Gamma(G, \mathcal{L}_2)$, and let $s = s_1 \otimes s_2 \in \Gamma(G, \mathcal{L})$. Then $\sigma_\chi^{\mathcal{L}}(s) = \sum_{\alpha+\beta=\chi} \sigma_\alpha^{\mathcal{L}_1}(s_1) \otimes \sigma_\beta^{\mathcal{L}_2}(s_2)$. We look what happens to both sides if we replace χ by $\chi + \phi(y)$: In the sum we can change α to $\alpha + \phi_1(y)$ and β to $\beta + \phi_2(y)$, which changes each summand by a factor $\psi_1(y) \cdot \psi_2(y) \cdot \tau_1(y, \alpha) \cdot \tau_2(y, \beta) = \psi_1(y) \cdot \psi_2(y) \cdot \tau_1(y, \chi)$ since $\tau_1 = \tau_2$. Similarly on the left we obtain a factor $\psi(y) \cdot \tau(y, \chi)$, and if $\sigma_\chi(s) \neq 0$ it follows that these two factors must be equal. It first follows that $\tau(y,)$ and $\tau_1(y,)$ coincide on the subgroup $\tilde{X}$ of X generated by the differences $\{\chi_1 - \chi_2\}$, for elements $\chi_{1/2} \in X$ with $\sigma_\chi(s_1 \otimes s_2) \neq 0$ for some choice of s_1 and s_2. This subgroup contains $\phi(Y)$, and we conclude that $\psi = \psi_1 \cdot \psi_2$.

Next we have remarked that τ does not depend on the choice of $\mathcal{M}$ (with $\pi^*(\mathcal{M}) \approx \tilde{\mathcal{L}}$), as two such choices differ by a character of T, which amounts to shifting the indices of the $\mathcal{M}_\chi$ by this character. The same holds if we replace $\mathcal{L}$ by a translate $T_g^*(\mathcal{L})$, for an element $g \in G(R)$, as $\tilde{\mathcal{L}}$ and $\mathcal{M}$ are replaced by their corresponding translates. As before we may assume R to be a complete discrete valuation ring with algebraically closed residue field. Now consider the set of pairs $(g, h) \in G(R) \times G(R)$ such that $\lambda_1(g) + \lambda_2(h) = 0$ (in $G^t(R)$). For any such pair $T_g^*(\mathcal{L}_1) \otimes T_h^*(\mathcal{L}_2) \approx \mathcal{L}_1 \otimes \mathcal{L}_2 \approx \mathcal{L}$, and for suitable choices of (g, h)'s we have $\tilde{X} = X$ ($\tilde{X}$ here is the sum of the previous $\tilde{X}$'s corresponding to the (g, h)'s): This follows from the fact that the pairs (g, h) as above are Zariski-dense in the connected component of the identity of the kernel of $\lambda_1 + \lambda_2 : G_\eta \times G_\eta \to G_\eta$, and that $\Gamma(G_\eta, \mathcal{L}_\eta)$ is spanned by the images of $\Gamma(G_\eta, T_g^*(\mathcal{L}_1)) \otimes \Gamma(G_\eta, T_h^*(\mathcal{L}_2))$ for elements (g, h) in the identity-component of that kernel [Mum 4], or I, Prop.5.3.

We can now conclude that $\tau = \tau_1$, and the lemma has been proved.

For any $n \geq 1$ we derive a description of $\Gamma(G_\eta, \mathcal{L}_\eta^{\otimes n})$ in terms of Fourier-expansions. We apply this as follows: Suppose that for two pairs $(G_1, \mathcal{L}_1)$ and $(G_2, \mathcal{L}_2)$ we are given an isomorphism $(\tilde{G}_1, \tilde{\mathcal{L}}_1) \approx (\tilde{G}_2, \tilde{\mathcal{L}}_2)$ which respects the $\imath$'s and the action of Y on $\tilde{L}_{1,\eta}$ and $\tilde{L}_{2,\eta}$. This induces an isomorphism between $(G_{1,\text{for}}, \mathcal{L}_{1,\text{for}})$ and $(G_{2,\text{for}}, \mathcal{L}_{2,\text{for}})$ which for each $n \geq 1$ maps $\Gamma(G_{1,\eta}, \mathcal{L}_{1,\eta}^{\otimes n}) \subset \Gamma(G_{1,\text{for}}, \mathcal{L}_{1,\text{for}}^{\otimes n}) \otimes K$ into $\Gamma(G_{2,\eta}, \mathcal{L}_{2,\eta}^{\otimes n}) \subset \Gamma(G_{2,\text{for}}, \mathcal{L}_{2,\text{for}}^{\otimes n}) \otimes K$. As the $\mathcal{L}_i$ are ample it follows that we get an induced isomorphism $(G_{1,\eta}, \mathcal{L}_{1,\eta}) \approx (G_{2,\eta}, \mathcal{L}_{2,\eta})$ which extends to an isomorphism $(G_1, \mathcal{L}_1) \approx (G_2, \mathcal{L}_2)$ inducing the isomorphism $(\tilde{G}_1, \tilde{\mathcal{L}}_1) \approx (\tilde{G}_2, \tilde{\mathcal{L}}_2)$ we started with. It follows that our data determine $(G, \mathcal{L})$ up to isomorphism.

Finally in the general case ($(c^t \times c)^* \mathcal{P}_A$ not trivial, $\underline{X}$ and $\underline{Y}$ not constant, or $\tilde{\mathcal{L}}$ not trivial on T) we have to solve a descent problem: There exists an affine étale covering of S_0 or S_{for} such that these conditions hold after base-change. We then glue and obtain the following result:

6.2. Theorem. *The construction above associates to any pair $(G, \mathcal{L})$ (G semi-abelian, G_η abelian, $\mathcal{L}_\eta$ ample) the following data, which determine $(G, \mathcal{L})$ up to unique isomorphism:*

(1) an abelian scheme A/S, a torus T/S and an extension $\tilde{G}/S$ of A by T. This extension determines and is determined by a homomorphism $c : \underline{X} \to A^t$ via (negative of) push-out, where $\underline{X}$ = character group of T, and A^t is the dual abelian scheme of A.

(2) an étale sheaf of free abelian groups $\underline{Y}$ of rank $r = \dim(T/S)$.

(3) a homomorphism $c^t : \underline{Y} \to A$. c^t determines an extension $\tilde{G}^t$ of A^t/S by a torus T^t/S with character group $\underline{Y}$.

(4) an injective homomorphism $\phi : \underline{Y} \to \underline{X}$ with finite cokernel.

(5) a homomorphism $\imath : \underline{Y} \to \tilde{G} \otimes K$ lying over c^t. $\imath$ determines and is determined by a trivialization $\tau^{-1} : 1_{\underline{Y}\times\underline{X}} \xrightarrow{\simeq} (c^t \times c)^*\mathcal{P}_{A,\eta}$ of the biextension $(c^t \times c)^*\mathcal{P}_{A,\eta}$ over the étale group scheme $\underline{X} \times \underline{Y}$. Let $\tau : 1_{\underline{Y}\times\underline{X}} \xrightarrow{\simeq} (c^t \times c)^*\mathcal{P}_{A,\eta}^{-1}$ be the inverse of τ.

(6) a cubical ample invertible sheaf $\tilde{\mathcal{L}}$ on $\tilde{G}$ which induces a polarization $\lambda_A : A \to A^t$ of A/S such that $\lambda_A c^t = c\phi$. In other words $\tilde{\mathcal{L}}$ induces a homomorphism from $\tilde{G}$ to $\tilde{G}^t$.

(7) an action of $\underline{Y}$ on $\tilde{\mathcal{L}}_\eta$ over $\imath$ compatible with ϕ. This action determines and is determined by a trivialization $\psi : 1_{\underline{Y}} \xrightarrow{\simeq} \imath^*\tilde{\mathcal{L}}_\eta^{-1}$ compatible with $\tau \circ (\mathrm{id}_{\underline{Y}} \times \phi) : 1_{\underline{Y}\times\underline{Y}} \xrightarrow{\simeq} (c^t \times c)^*\mathcal{P}_{A,\eta}^{-1}$. Compatibility makes sense because the biextension $\Lambda(\mathcal{L})$ uniquely descends to a biextension of $A \times_S A$ which by (6) is just $(\mathrm{id}_A \times \lambda_A)^*\mathcal{P}_A$.

The trivialization ψ in (7) is required to satisfy the following positivity condition: If $\underline{Y}$ is constant (with value Y) then for all but finitely many y in Y, $\psi(y)$ extends to a section of the invertible sheaf $\imath^*\tilde{\mathcal{L}}_y^{-1}$ on S, and is congruent to 0 modulo I. In general this holds after passing to a finite étale covering of S which trivializes $\underline{Y}$. It follows that for every non-zero element y in Y, $\tau(y, \phi(y))$ extends to a section of the invertible sheaf $(c^t \times c\phi)^*\mathcal{P}_A^{-1}|_{(y,y)}$ on S and is congruent to zero modulo I.

As indicated in 5.2, theorem 6.2 can be formulated in the setting of relative semi-abelian scheme G over a formal scheme S (w.r.t. the étale topology), such that the affine coordinate rings satisfy the strong normality condition (as given at the beginning of §3).

6.3. Remark. Assume for simplicity that $\underline{X}$ and $\underline{Y}$ are constant. The trivializations in (5) and (7) determine invertible fractional ideals (Cartier divisors) $I_{y,\mu}$ (for $(y,\mu) \in Y \times X$) and I_y (for $y \in Y$) on S, which are bilinear in (y,μ) respectively quadratic in y. I_y depends on the way we descend $\tilde{\mathcal{L}}$ to $\mathcal{M}$, but the $I_{y,\mu}$ are canonical. That is, we obtain a bilinear form $B : Y \times X \to \mathrm{Div}(S)$ which a priori depends on $\mathcal{L}$, but we shall see later that is does not. $B(y, \phi(y))$ is positive for any non-zero $y \in Y$, and B measures the denominators of $\imath : Y \to \tilde{G}(K)$. As the form B is often easier to describe than the whole set of data above, it will be very important in what follows.

Chapter III. Mumford's Construction

0. Introduction

In this chapter R denotes a noetherian normal domain complete w.r.t. an ideal I, and we always assume $\mathrm{rad}(I) = I$ for convenience. S always denote $\mathrm{Spec}(R)$, $\eta =$ generic point of S, $S_0 = \mathrm{Spec}(R/I)$, $K =$ quotient field of R. As before we *assume that for any étale R/I-algebra $R_0^\dagger$ its unique lifting $R^\dagger$ to a formally étale I-adically complete R-algebra is normal.* This holds for example if R is regular, or if R is the I-adic completion of a normal excellent ring. In fact everything could be done over a formal scheme which is not necessarily affine whose coordinate rings satisfy the above strong normality condition, but we leave it to the reader to take care of this additional generality. (The reference for relative schemes is [Hak].) We have seen in chapter II that starting from a semi-abelian scheme G/S and an invertible sheaf $\mathcal{L}$ on G ample over G_η, we obtain the following data: an extension $\tilde{G}/S$ of an abelian scheme A/S by a torus T/S, a homomorphism $\imath : \underline{Y} \to \tilde{G} \otimes K$ where $\underline{Y}$ is an étale sheaf whose fibres are free abelian groups of rank equal to the dimension of T, an ample cubical invertible sheaf $\tilde{\mathcal{L}}$ on $\tilde{G}$, and an action of $\underline{Y}$ on $\tilde{\mathcal{L}}_\eta$ satisfying some positivity condition. Data of the above type will be referred to as "$\mathrm{DD}_{\mathrm{ample}}$", which stands for "degeneration data with ample line bundle". Following this line we use "DD" for data consisting of just $\tilde{G}$, $\underline{Y}$, and $\imath : \underline{Y} \to \tilde{G} \otimes K$ which can be extended to a $\mathrm{DD}_{\mathrm{ample}}$. There are other variants which we will not try to describe here. To be honest, so far we have not proved that $\imath : \underline{Y} \to \tilde{G} \otimes K$ is independent of $\mathcal{L}$, but we will later.

In this chapter we will describe an ingenious construction due to D. Mumford which produces a G when given a DD $\imath : \underline{Y} \to \tilde{G} \otimes K$, and a $(G, \mathcal{L})$ when given a $\mathrm{DD}_{\mathrm{ample}}$. Mumford's construction defines several related functors (defined for suitable category of $\mathrm{DD}_{\mathrm{ample}}$'s, DD's or their variants, all of which we will denote temporarily by M. It turns out that they are all equivalences of categories, and their quasi-inverses F are described in chapter 2. Ideally we should have proved directly in chapter 2 that F is a functor. Unfortunately our method seems to work well only for isogenies but not for arbitrary homomorphisms $G_{1,\eta} \to G_{2,\eta}$, awkward even for proving independence of $\imath$ from $\mathcal{L}$. What we will do instead is to prove that M is fully faithful; then it is clear that it is an equivalence of categories and that F is its quasi-inverse.

Mumford's construction is beautifully presented in his 1972 Compositio paper (the second one in that issue), quoted as [Mum 5] here. There only the case $\tilde{G} = a$ split torus is treated, thereby saving the reader all the confusions stemming from the abelian part. But it contains all the essential ideas and generalizes in a natural way provided one is careful about necessary bookkeeping, so as not to be confused by the numerous isomorphisms. It is simply impossible to improve upon [Mum 5]. Therefore we will only give a quick sketch of it in §1 and let readers enjoy Mumford's paper for themselves. [Mum 5] is an essential prerequisite for this chapter and will be freely invoked whenever necessary or convenient. We should remark that in [Mum 5] the base is supposed to be excellent, which however is not necessary: In the notation of [Mum 5] excellency is only used once, to show that P is irreducible (Prop. 4.3 in [Mum 5]), which also follows by basechange to complete discrete valuation rings (which are excellent).

1. Review of the Basic Ideas

In this section we will try to give the broad outline of the Mumford's construction. No details are given here. Even some crucial definitions are skipped in order to focus only on the key ideas. The input of the construction is a $\mathrm{DD}_{\mathrm{ample}}$, i.e. data consisting of

(i) an extension $0 \to T \to \tilde{G} \to A \to 0$ where A/S an abelian scheme and T/S is a torus over S. Let π be the map from $\tilde{G}$ to A above, and $\underline{X}$ the character group of T (which is an étale sheaf like $\underline{Y}$).

(ii) a homomorphism $\imath : \underline{Y} \to \tilde{G} \otimes K$, where $\underline{Y}$ is an étale sheaf on S whose fibres are free abelian groups of rank $r = \dim(T/S)$.

(iii) a cubical invertible sheaf $\tilde{\mathcal{L}}$ of the form $\tilde{\mathcal{L}} \approx \pi^*\mathcal{M}$ on $\tilde{G}$, for some cubical ample invertible sheaf $\mathcal{M}$ on A.

(iv) an action of $\underline{Y}$ on $\tilde{\mathcal{L}}_\eta$ over $\imath$ satisfying suitable positivity conditions as specified in chapter 2.

In the case A is trivial and $\underline{X}$ and $\underline{Y}$ are constant (with values X resp. Y), the above data just means:

(a) a bimultiplicative pairing $b : Y \times X \to K^*$,

(b) a homomorphism $\phi : Y \to X$,

(c) a function $a : Y \to K^*$ such that $a(y+z) = a(y) \cdot a(z) \cdot b(y, \phi(z))$ subject to the following positivity condition: $b(y, \phi(y)) \in I$ for every non-zero element y in Y. Also for all but finitely many y in Y $a(y)$ lies in R and converges to 0 in the I-adic topology. Notice that $\imath$ is necessarily injective.

Given the above data, the goal is to construct a "quotient" $(G, \mathcal{L})$ of $(\tilde{G}, \tilde{\mathcal{L}})$ by Y in a suitable sense, so that G is semi-abelian and $\mathcal{L}$ is ample over the generic fibre G_η, which is an abelian variety. The method is to use the Grothendieck existence theorem (c. [EGA III] §5) to construct a compactification P of G together with an ample invertible sheaf $\mathcal{L}$ on P extending the given invertible sheaf $\mathcal{L}$ on G. Thus one should construct a suitable proper formal scheme P_{for} together with a formal invertible sheaf $\mathcal{L}_{\text{for}}$ on P_{for}. $(P_{\text{for}}, \mathcal{L}_{\text{for}})$ will come from $(\tilde{P}_{\text{for}}, \tilde{\mathcal{L}}_{\text{for}})$ as quotient by Y in the formal category, where $\tilde{P}_{\text{for}}$ is a formal scheme and $\tilde{\mathcal{L}}_{\text{for}}$ is a formal invertible sheaf on $\tilde{P}_{\text{for}}$. In turn $(\tilde{P}_{\text{for}}, \tilde{\mathcal{L}}_{\text{for}})$ is the formal completion of $(\tilde{P}, \tilde{\mathcal{L}})$ where $\tilde{P}$ is a scheme containing $\tilde{G}$ as a dense open subscheme and $\tilde{\mathcal{L}}$ is an ample invertible sheaf on $\tilde{P}$ extending the given invertible sheaf $\tilde{\mathcal{L}}$ on $\tilde{G}$. $\tilde{P}$ should be regarded as a partial compactification of $\tilde{G}$. The construction of $\tilde{P}$ is one very ingenious part of [Mum 5]. Instead of giving an explicit $\tilde{P}$, Mumford axiomatized the properties of good pairs $(\tilde{P}, \tilde{\mathcal{L}})$, which he baptized "relatively complete models". Any relatively complete model works, and this gives one great flexibility when working out examples. The conditions involved in the definition of relatively complete models are roughly that

(1) $\tilde{P}$ is integral and locally of finite type over S and contains $\tilde{G}$ as an open dense subscheme. It turns out that $\tilde{P}_\eta = \tilde{G}_\eta$.

(2) There exists a Y-action on $\tilde{P}$ extending the translation action of Y on G_η.

(3) There exists a Y-action and a T-action on $\tilde{\mathcal{L}}$. These two actions do not commute, but satisfy a certain commutation relation.

(4) Although $\tilde{P}$ is not of finite type over S, there exists an open $\tilde{G}$-invariant subscheme $U \subset \tilde{P}$ of finite type over S whose Y-translates cover $\tilde{P}$.

(5) $\tilde{\mathcal{L}}$ is ample over S.

(6) $\tilde{P}$ is required to satisfy certain completeness condition in terms of valuations of the function field of $\tilde{G}$. This is a bit tricky and will not be spelled out here.

Relatively complete models can be shown to exist explicitly by writing down some big graded algebra and taking Proj. When R is a complete discrete valuation ring one can use the technique of torus embedding to construct relatively complete models. This technique also works in other cases, notably in some moduli problems. Given any relatively complete model $(\tilde{P}, \tilde{\mathcal{L}})$, its formal completion $(\tilde{P}_{\text{for}}, \tilde{\mathcal{L}}_{\text{for}})$ inherits a Y-action. The completeness condition (6) implies that each irreducible component of $\tilde{P} \times_S S_0$ satisfies the valuative criterion for properness, hence is proper because it can be shown to be quasi-compact. The commutation relation implies that Y acts freely on $\tilde{P}_{\text{for}}$. Moreover if $\overline{U}_0$ denotes the closure of $U \times_S S_0$ in $\tilde{P} \times_S S_0$ where U is as in (4), $\overline{U}_0$ intersects only a finite number of Y-translates of itself. The Y-quotient $(P_{\text{for}}, \mathcal{L}_{\text{for}})$ of $(\tilde{P}_{\text{for}}, \tilde{\mathcal{L}}_{\text{for}})$ is a proper formal scheme with an ample invertible sheaf, hence algebraizes to a projective scheme P by Grothendieck's existence theorem. To find the open subscheme G inside P, it suffices to find its complement B, which is a closed subscheme of P. Naturally one takes the complement $\tilde{B}$ of the union of Y-translates of $\tilde{G}$ in $\tilde{P}$ (with reduced structure). Passing first to formal completion and then to the Y-quotient, one gets a closed formal subscheme B_{for} of P_{for}. By Grothendieck's existence theorem again B_{for} algebraizes to a closed subscheme of B of P, whose complement G is the quotient "$\tilde{G}/Y$" we sought. The fact that $\tilde{P}_\eta = \tilde{G}_\eta$ implies that $P_\eta = G_\eta$ if one works through the above construction. Similarly G is smooth over S because $\tilde{G}$ is.

To show that G has a natural group structure one proves more generally that the above construction is functorial in $(\tilde{G}, \imath : Y \to \tilde{G}(K))$. The functoriality is reduced by the usual graph consideration to constructing some closed subscheme of $P_1 \times_S P_2$ which is the closure of the graph of the desired homomorphism. This subscheme is constructed by a method similar to those described above. Once one has the group structure, G_η is an abelian variety by definition. There is also a description of torsion points of G in terms of $(\tilde{G}, \imath : Y \to \tilde{G}(K))$, from which one deduces that G is semi-abelian. This finishes our rough description of Mumford's construction.

2. Semi-Abelian Degeneration Data

This section consists of an assembly of definitions. They are more or less implicit in the previous chapter, and we would like to make them explicit. Remember the objects we analyzed in that chapter are pairs $(G, \mathcal{L})$, where G is a semi-abelian scheme over S such that G_η is an abelian variety, $G_0 = G \times_S S_0$ is an extension of an abelian scheme A_0/S_0 by a torus T_0/S_0, and $\mathcal{L}$ is a invertible sheaf on G rigidified along the zero-section (hence endowed with a canonical cubical structure) such that $\mathcal{L}_\eta$ is ample over G_η. The category of all such pairs $(G, \mathcal{L})$ will be denoted by $\mathrm{DEG}_{\mathrm{ample}}$, where DEG is abbreviation for degeneration. The morphisms are group homomorphisms of the G's which respect the $\mathcal{L}$'s. Forgetting about the ample invertible sheaf, the category of all semi-abelian schemes G over S as above will be denoted DEG. Note that by a result of Raynaud, any G in DEG has an invertible sheaf $\mathcal{L}$ with $\mathcal{L}_\eta$ ample over G_η. There is a natural functor $\mathrm{DEG}_{\mathrm{ample}} \to \mathrm{DEG}$, which is far from being fully faithful. In fact any morphism f from $(G_1, \mathcal{L}_1)$ to $(G_2, \mathcal{L}_2)$ must be an isogeny because $f^*\mathcal{L}_2 = \mathcal{L}_1$ and $\mathcal{L}_1$ and $\mathcal{L}_2$ are ample, while any homomorphism g from G_1 to G_2 is a morphism in DEG. We can specify the dimension of G (usually denoted by g) and the dimension of the torus part of G_0 (usually denoted by r) if we desire to do so. The reader may label the corresponding subcategories by introducing subscripts into the notation, like $\mathrm{DEG}_{\mathrm{ample},g,r}$ or $\mathrm{DEG}_{g,r}$, but we refrain from doing so.

We can define related categories by considering polarizations or invertible sheaves which are not necessarily ample. For instance $\mathrm{DEG}_{\mathrm{pol}}$ is the category consisting of pairs (G, λ_η) where G is in DEG and $\lambda_\eta : G_\eta \to G_\eta^t$ is a polarization of G_η. Homomorphisms should respect polarizations. A morphism f from $(G_1, \lambda_{1,\eta})$ to $(G_2, \lambda_{2,\eta})$ is an isogeny f from G_1 to G_2 pulling $\lambda_{2,\eta}$ back to $\lambda_{1,\eta}$. If we require the degree of the polarization λ_η to be a fixed integer $d^2 > 0$, we obtain a full subcategory which is a groupoid, i.e. all morphisms are isomorphisms. A possible further refinement requires that the homomorphism (λ_t, λ_a)

induced by λ between the Raynaud extensions $0 \to T \to \tilde{G} \to A \to 0$, $0 \to T^t \to G^t \to A^t \to 0$ is such that $\deg(\lambda_t) = d_t$, $\deg(\lambda_a) = d_a^2$ with $d_t d_a = d$. (It is understood that all the integers involved are positive.) The category $\mathrm{DEG_{IS}}$ consists of pairs $(G, \mathcal{L})$ where G is an object of DEG and $\mathcal{L}$ is an invertible sheaf on G rigidified along the zero section, hence endowed with a natural cubical structure. A morphism f from $(G_1, \mathcal{L}_1)$ to $(G_2, \mathcal{L}_2)$ is a homomorphism f from G_1 to G_2 such that $f^*\mathcal{L}_2 = \mathcal{L}_1$.

In chapter 2 we started with a $(G, \mathcal{L})$ in $\mathrm{DEG_{ample}}$ and ended up with the following data:

(1) an abelian scheme A/S, a torus T/S, and an extension $\tilde{G}/S$ of A by T. This extension determines and is determined by a homomorphism $c : \underline{X} \to A^t$ via (negative of) push-out, where $\underline{X}$ = character group of T, and A^t is the dual abelian scheme of A.

(2) an étale sheaf of free abelian groups $\underline{Y}$ of rank $r = \dim(T/S)$.

(3) a homomorphism $c^t : \underline{Y} \to A$. c^t determines an extension $\tilde{G}^t$ of A^t/S by a torus T^t/S with character group $\underline{Y}$ via (negative of) push-out.

(4) an injective homomorphism $\phi : \underline{Y} \to \underline{X}$ with finite cokernel.

(5) a homomorphism $\imath : \underline{Y} \to \tilde{G} \otimes K$ lying over c^t. $\imath$ determines and is determined by a trivialization $\tau : 1_{\underline{Y}\times\underline{X}} \xrightarrow{\sim} (c^t \times c)^*\mathcal{P}_{A,\eta}^{-1}$ of the biextension $(c^t \times c)^*\mathcal{P}_{A,\eta}^{-1}$ over the étale group scheme $\underline{X} \times \underline{Y}$.

(6) a cubical ample invertible sheaf $\tilde{\mathcal{L}}$ on $\tilde{G}$ which induces a polarization $\lambda_A : A \to A^t$ of A/S such that $\lambda_A c^t = c\phi$. In other words $\tilde{\mathcal{L}}$ induces a homomorphism from $\tilde{G}$ to $\tilde{G}^t$.

(7) an action of $\underline{Y}$ on $\tilde{\mathcal{L}}_\eta$ over $\imath$ compatible with ϕ. This action determines and is determined by a trivialization $\psi : 1_{\underline{Y}} \xrightarrow{\sim} \imath^*\tilde{\mathcal{L}}_\eta^{-1}$ compatible with $\tau \circ (\mathrm{id}_{\underline{Y}} \times \phi) : 1_{\underline{Y}\times\underline{Y}} \xrightarrow{\sim} (c^t \times c\phi)^*\mathcal{P}_{A,\eta}^{-1}$. Compatibility makes sense because the biextension $\Lambda(\tilde{\mathcal{L}})$ uniquely descends to a biextension of $A \times_S A$ which is just $(\mathrm{id}_A \times \lambda_A)^*\mathcal{P}_A$ by (6).

The trivialization ψ in (7) is required to satisfy the following positivity condition: If $\underline{Y}$ is constant (with value Y) then for all but finitely many y in Y, $\psi(y)$ extends to a section of the invertible sheaf $\imath^*\tilde{\mathcal{L}}_y^{-1}$ on S, and is congruent to 0 modulo I. In general this holds after passing to a finite étale covering of S which trivializes $\underline{Y}$. It follows that for every non-zero element y in Y, $\tau(y, \phi(y))$ extends to a section of the invertible sheaf $(c^t \times c\phi)^*\mathcal{P}_A^{-1}|_{(y,y)}$ on S and is congruent to zero modulo I.

By definition, $\mathrm{DD_{ample}}$ is the category whose objects consist of data (1)–(7) as above satisfying the *positivity condition.* Morphisms are defined in the obvious way; a morphism from $(\tilde{G}_1, \underline{X}_1, \underline{Y}_1, c_1, c_1^t, \imath_1, \phi_1, \tau_1, \lambda_1, \tilde{\mathcal{L}}_1, \psi_1)$ to $(\tilde{G}_2, \underline{X}_2, \underline{Y}_2, c_2, c_2^t, \imath_2, \phi_2, \tau_2, \lambda_2, \tilde{\mathcal{L}}_2, \psi_2)$ is a pair $(f : (G_1, \tilde{\mathcal{L}}_1) \to (G_2, \tilde{\mathcal{L}}_2)$, $g : \underline{Y}_1 \to \underline{Y}_2)$ compatible with the rest of the data involved.

We define another category $\mathrm{DD_{pol}}$ as follows: An object of $\mathrm{DD_{pol}}$ consists of data (1)–(6) listed below satisfying a symmetry condition and a positivity condition. The list is:

(1) an abelian scheme A/S, a torus T/S, and an extension $\tilde{G}/S$ of A by T. This extension determines and is determined by a homomorphism $c : \underline{X} \to$

$A^t(S)$ via (negative of) push-out, where as usual $\underline{X}$ = character group of T and A^t is the dual abelian scheme of A.

(2) an étale sheaf $\underline{Y}$ of free abelian groups of rank $r = \dim(T/S)$.

(3) a homomorphism $c^t : \underline{Y} \to A(S)$. c^t determines an extension $\tilde{G}^t$ of A^t/S by a torus T^t/S with character group $\underline{Y}$ via (negative of) push-out.

(4) a polarization $\lambda_A : A \to A^t$ over S of A/S.

(5) an injective homomorphism $\phi : \underline{Y} \to \underline{X}$ with finite cokernel such that $\lambda_A c^t = c\phi$.

(6) a homomorphism $\imath : \underline{Y} \to \tilde{G} \otimes K$ lying over c^t. $\imath$ determines and is determined by a trivialization $\tau : 1_{\underline{Y}\times\underline{X}} \xrightarrow{\sim} (c^t \times c)^*\mathcal{P}_{A,\eta}^{-1}$ of the biextension $(c^t \times c)^*\mathcal{P}_{A,\eta}^{-1}$ over $\underline{Y} \times \underline{X}$.

Let $s : A \times_S A \to A \times_S A$ be the flipping, and $\tilde{s} : (c^t \times c\phi)^*\mathcal{P}_A^{-1} \to (c^t \times c\phi)^*\mathcal{P}_A^{-1}$ be the canonical symmetry over s via $(c^t \times c\phi)^*\mathcal{P}_A^{-1} \cong (c^t \times c^t)^*(\mathrm{id}_A \times \lambda_A)^*\mathcal{P}_A^{-1}$ and the natural structure of symmetric biextension on $(\mathrm{id}_A \times \lambda_A)^*\mathcal{P}_A$. τ is required to satisfy the symmetry condition: $\tau(y_1, \phi(y_2)) = \tilde{s}(\tau(y_2, \phi(y_1)))$ for all y_1, y_2 in Y. Equivalently, $\tau \circ (\mathrm{id}_Y \times \phi) : 1_{\underline{Y}\times\underline{X}} \xrightarrow{\sim} (c^t \times c\phi)^*\mathcal{P}_{A,\eta}^{-1}$ is a symmetric trivialization w.r.t. $\tilde{s}$.

The *positivity condition* alluded to is that, after trivializing $\underline{Y}$, for every non-zero element y in Y, $\tau(y, \phi(y))$ extends to a section of the invertible sheaf $(c^t \times c\phi)^*\mathcal{P}_A^{-1}|_{(y,y)}$ on S and is congruent to zero modulo I. We have seen this before. A morphism from $(\tilde{G}_1, \underline{X}_1, \underline{Y}_1, c_1, c_1^t, \imath_1, \tau_1, \phi_1, \lambda_1)$ to $(\tilde{G}_2, \underline{X}_2, \underline{Y}_2, c_2, c_2^t, \imath_2, \tau_2, \phi_2, \lambda_2)$ is a pair $(f : G_1 \to G_2,\ g : \underline{Y}_1 \to \underline{Y}_2)$ compatible with the rest of the data involved.

We remark that any object $(\tilde{G}, \underline{X}, \underline{Y}, c, c^t, \imath, \lambda_A, \phi, \tau)$ of $\mathrm{DD}_{\mathrm{pol}}$ can be canonically extended to an object $(\tilde{G}, \underline{X}, \underline{Y}, c, c^t, \imath', \lambda_A, \phi, \tau, \tilde{\mathcal{L}}, \psi)$ of $\mathrm{DD}_{\mathrm{ample}}$: Let the line $\mathcal{M}$ denote the pull-back of the Poincaré-bundle on $A \times A^t$ under $(1, \lambda_A)$, and $\tilde{\mathcal{L}} = \pi^*(\mathcal{M})$, which admits (over η) a canonical $\underline{Y}$-action. However this object induces twice the original polarization.

We define DD as the category whose objects are pairs $(\tilde{G}, \imath : \underline{Y} \to \tilde{G}(K))$ which can be extended to objects of $\mathrm{DD}_{\mathrm{pol}}$ or equivalently which can be extended to objects of $\mathrm{DD}_{\mathrm{ample}}$. Here $\tilde{G}/S$ denotes an extension of an abelian scheme A/S by a torus T/S and $\imath : \underline{Y} \to \tilde{G} \otimes K$ stands for a homomorphism from an étale sheaf $\underline{Y}$ of free abelian group of rank $r = \dim(T/S)$ to $\tilde{G}(K)$ as before. A morphism from $(\tilde{G}_1, \imath_1 : \underline{Y}_1 \to \tilde{G}_1 \otimes K)$ to $(\tilde{G}_2, \imath_2 : \underline{Y}_2 \to \tilde{G}_2 \otimes K)$ is a pair $(f : G_1 \to G_2,\ g : \underline{Y}_1 \to \underline{Y}_2)$ such that $f \circ \imath_1 = \imath_2 \circ g$.

Finally we define $\mathrm{DD}_{\mathrm{IS}}$ as the category whose objects are 11-tuples $(\tilde{G}, \underline{X}, \underline{Y}, c, c^t, \imath, f_A, f_Y, \tau, \tilde{\mathcal{F}}, \zeta)$ where

(1) $\tilde{G}/S$ is an extension of an abelian scheme A/S by a torus T/S. This extension determines and is determined by a homomorphism $c : \underline{X} \to A^t$ via (negative of) push-out, where $\underline{X}$ = character group of T and A^t is the dual abelian scheme of A.

(2) $\underline{Y}$ is an étale sheaf on S of free abelian groups of rank $r = \dim(T/S)$.

(3) c^t is a homomorphism $c^t : \underline{Y} \to A$ over S. c^t determines an extension $\tilde{G}^t$ of A^t/S by a torus T^t/S with character group $\underline{Y}$ via (negative of) push-out.

(4) $\imath : \underline{Y} \to \tilde{G} \otimes K$ is a homomorphism from $\underline{Y}$ to $\tilde{G} \otimes K$ which lies over c^t. $\imath$ determines and is determined by a trivialization $\tau : 1_{\underline{Y} \times \underline{X}} \xrightarrow{\sim} (c^t \times c)^* \mathcal{P}_{A,\eta}^{-1}$ of the biextension $(c^t \times c)^* \mathcal{P}_{A,\eta}^{-1}$ over $\underline{Y} \times \underline{X}$.

(5) a cubical invertible sheaf $\tilde{\mathcal{F}}$ on $\tilde{G}$ which induces a homomorphism $f_A : A \to A^t$ of abelian schemes over S.

(6) $f_Y : \underline{Y} \to \underline{X}$ is a homomorphism such that $f_A c^t = c f_Y$. In other words f_Y induces a homomorphism from $\tilde{G}$ to $\tilde{G}^t$.

(7) an action of $\underline{Y}$ on $\tilde{\mathcal{F}}_\eta$ over $\imath$ compatible with f_Y. This action determines and is determined by a trivialization $\zeta : 1_{\underline{Y}} \xrightarrow{\sim} \imath^* \tilde{\mathcal{F}}_\eta^{-1}$ compatible with $\tau \circ (\mathrm{id}_{\underline{Y}} \times f_A) : \tau : 1_{\underline{Y} \times \underline{Y}} \xrightarrow{\sim} (c^t \times c f_Y)^* \mathcal{P}_{A,\eta}^{-1}$. Compatibility makes sense because the biextension $\Lambda(\mathcal{L})$ uniquely descends to a biextension of $A \times_S A$ which is just $(\mathrm{id}_A \times f_A)^* \mathcal{P}_A$ by (5).

The condition for an 11-tuple $(\tilde{G}, \underline{X}, \underline{Y}, c, c^t, \imath, f_A, f_Y, \tau, \tilde{\mathcal{F}}, \zeta)$ to be an object of $\mathrm{DD}_{\mathrm{IS}}$ is that $(\tilde{G}, \underline{X}, \underline{Y}, c, c^t, \imath)$ defines an object of DD, and that the denominators of ζ satisfy a certain finiteness condition, namely that (if Y is split) there exist an $n \geq 1$ such that for $y \in Y$ the product of $I^n_{y,\phi(y)}$ and the invertible fractional ideal I_y defined by ζ is integral. Equivalently there should exist $\tilde{\mathcal{L}}_1$ and $\tilde{\mathcal{L}}_2$ on $\tilde{G}$ extending our data to an object of $\mathrm{DD}_{\mathrm{ample}}$ such that $\tilde{\mathcal{F}} \approx \tilde{\mathcal{L}}_1 \otimes \tilde{\mathcal{L}}_2^{-1}$. Informally, the condition is that the period homomorphism $\imath$ admits a polarization, although we do not specify a particular one, and that the denominators in ζ are no worse than those in $\imath$.

We shall need to consider a full subcategory of $\mathrm{DD}_{\mathrm{ample}}$ whose objects will be called split objects, defined as follows: an object $(\tilde{G}, \underline{X}, \underline{Y}, c, c^t, \imath', \lambda_A, \phi, \tau, \tilde{\mathcal{L}}, \psi)$ is split iff the sheaves $\underline{X}, \underline{Y}$ are constant and $\tilde{\mathcal{L}}$ can be descended to a cubical invertible sheaf $\mathcal{M}$ on A (i.e. the restriction of $\tilde{\mathcal{L}}$ to T is a trivial cubical sheaf). Every object of $\mathrm{DD}_{\mathrm{ample}}$ becomes split after passing to a suitable étale covering of $\mathrm{Spf}(R)$. The Mumford's construction will first be done explicitly on a full subcategory $\mathrm{DD}^*_{\mathrm{ample}}$ whose objects are split objects satisfying an additional condition (*), to be found in §3. Then we shall extend Mumford's construction to general objects (not necessarily split) by a gluing process in the étale topology, to be done in §6. One has to be careful during the gluing process, since in general one cannot glue objects in the étale topology of $\mathrm{Spf}(R)$.

We remark here that the major goal of this chapter is to show that the category $\mathrm{DD}_{\mathrm{ample}}$ (resp. $\mathrm{DD}_{\mathrm{pol}}$, DD, $\mathrm{DD}_{\mathrm{IS}}$) is equivalent to the category $\mathrm{DEG}_{\mathrm{ample}}$ (resp. $\mathrm{DEG}_{\mathrm{pol}}$, DEG, $\mathrm{DEG}_{\mathrm{IS}}$), i.e. "DD = DEG". These equivalences of categories preserve the dimension of the semi-abelian schemes and their torus parts, as well as the degrees of polarizations.

3. Relatively Complete Models

Strictly speaking, Mumford's construction is a map from a certain subcollection of $\mathrm{Ob}(\mathrm{DD}_{\mathrm{ample}})$ to $\mathrm{Ob}(\mathrm{DEG}_{\mathrm{ample}})$. We will show later that this construction induces a functor from DD to DEG which turns out to be an equivalence of categories. From this it follows that Mumford's construction induces an equivalence of categories from $\mathrm{DD}_{\mathrm{pol}}$ to $\mathrm{DEG}_{\mathrm{pol}}$ which preserves the numerical invariants d_t, d_a, g, r. We first treat the "split case", where we assume that all tori are split, and that "$\tilde{\mathcal{L}}$ descends to $\mathcal{M}$". Then we will show that Mumford's construction "glues together in the étale topology" and defines a functor from $\mathrm{DD}_{\mathrm{IS}}$ to $\mathrm{DEG}_{\mathrm{IS}}$, which is again an equivalence of categories. Recall that the main result of chapter II (theorem 6.2) was the construction of a functor F from $\mathrm{DEG}_{\mathrm{ample}}$ to $\mathrm{DD}_{\mathrm{ample}}$. The functoriality is easily checked because every morphism in $\mathrm{DEG}_{\mathrm{ample}}$ induces an isogeny of semi-abelian schemes. It is true that there is a functor F from DEG to DD whose restriction on $\mathrm{DEG}_{\mathrm{ample}}$ is F, but this does not seem to be an immediate consequence, because there are morphisms in DEG which do not come from isogenies. We remark that the partial Fourier expansion gives the link between F and Mumford's construction, which loosely speaking are "inverse" to each other. The unpleasant fact that we do not know the existence of relatively complete models for all objects of $\mathrm{DEG}_{\mathrm{ample}}$ (we are jumping ahead now) makes it a little more complicated to show that F is an equivalence of categories.

We start with a split object $(\tilde{G}, X, Y, c, c^t, \imath, \lambda, \phi, \tau, \tilde{\mathcal{L}}, \psi, \mathcal{M})$ in $\mathrm{DD}_{\mathrm{ample}}$, where $\tilde{\mathcal{L}} = \pi^*\mathcal{M}$ and π is the canonical homomorphism from $\tilde{G}$ to A. Thus ψ can be regarded as a trivialization of the cubical invertible sheaf $\imath^*\mathcal{M}_\eta^{-1}$ over the constant group Y_η, which extends to a section of $\imath^*\mathcal{M}^{-1}$ over Y_S and is congruent to zero modulo I. Our goal is to construct an object $(G, \mathcal{L}, \mathcal{M}_{\mathrm{for}})$ of $\mathrm{DEG}_{\mathrm{ample}}$. One of Mumford's insights was to axiomatize the desirable conditions of partial compactifications of $(\tilde{G}, \tilde{\mathcal{L}})$ on which Y acts to facilitate "the quotient by Y" construction. Such partial compactifications are called "relatively complete models". Here is the precise definition:

3.1. Definition. A relatively complete model of a split object $(\tilde{G}, X, Y, c, c^t, \imath, \lambda, \phi, \tau, \tilde{\mathcal{L}}, \psi, \mathcal{M})$ in $\mathrm{DD}_{\mathrm{ample}}$ consists of the following 5 pieces of data:

(a) an integral scheme $\tilde{P}$ locally of finite type over A ($\Leftrightarrow$ locally of finite type over S) containing $\tilde{G}$ as a dense open subset. We denote again by $\pi : \tilde{P} \to A$ the structural morphism.

(b) an invertible sheaf on $\tilde{P}$ extending the invertible sheaf $\pi^*\mathcal{M}$ on $\tilde{G} \subset \tilde{P}$, which we denote again by $\tilde{\mathcal{L}}$, by abuse of notation.

(c) an action of $\tilde{G}$ on $\tilde{P}$ extending the translation action of $\tilde{G}$ on itself. We denote this action by $T_g : \tilde{P} \to \tilde{P}$ for any functorial point g of $\tilde{G}$.

(d) an action of Y on $(\tilde{P}, \tilde{\mathcal{L}})$ extending the action of Y on $(\tilde{G}_\eta, \pi^*\mathcal{M}_\eta)$ defined by $\imath$ and ψ. This action will be denoted by $S_y : \tilde{P} \to \tilde{P}$ and $\tilde{S}_y : S_y^*\tilde{\mathcal{L}} \to \tilde{\mathcal{L}}$ for any y in Y. Note that this assumption implies that $\imath(y)$ extends to a section of $\tilde{P}$ over S for every y in Y.

(e) an action of $\tilde{G}$ on the invertible sheaf $\mathcal{N} = \tilde{\mathcal{L}} \otimes \pi^*\mathcal{M}^{-1}$ on $\tilde{P}$ extending the translation action of $\tilde{G}$ on its structural sheaf, where $\pi : \tilde{P} \to A$ is the structural morphism in (a). This action will be denoted by $\tilde{T}_g : T_g^*\mathcal{N} \to \mathcal{N}$ for any functorial point g of $\tilde{G}$.

The above data are required to satisfy the following three conditions:

(1) There exists an open $\tilde{G}$-invariant subscheme $U \subset \tilde{P}$ of finite type over S such that $\tilde{P} = \bigcup_{y\in Y} S_y(U)$.

(2) $\tilde{\mathcal{L}}$ is ample on $\tilde{P}$ in the sense that the complements of the zero-sets of global sections of $\tilde{\mathcal{L}}^n$, $n \geq 1$, define a basis of the (Zariski) topology of $\tilde{P}$.

(3) (completeness condition) Let $\sum(\tilde{G})$ be the set of all valuations v of the rational function field $K(\tilde{G})$ of $\tilde{G}$ such that $v(R) \geq 0$, i.e. the underlying set of the Zariski's Riemann surface of $K(\tilde{G})/R$. For any v in $\sum(\tilde{G})$, denote by x_v the center of v on A (which exists because A is proper over S). Recall that $\pi_*\mathcal{O}_{\tilde{G}} = \oplus_{\mu\in X}\mathcal{O}_\mu$ in the notations of chapter 2. We denote by $I_{y,\mu}$ the locally principal fractional ideal of R defined by $\tau(y,\mu)$ w.r.t. the integral structure given by $(c^t \times c)^*\mathcal{P}_A^{-1}|_{(y,\mu)}$. Thus $(y,\mu) \to I_{y,\mu}$ is a bimultiplicative map from $Y \times X$ to the group of invertible ideals of R. Under these notations, the completeness condition is: $\forall v \in \sum(\tilde{G})$, v has a center on $\tilde{P} \Leftrightarrow \forall \mu \in X,\ \exists y \in Y$ such that $v(I_{y,\mu} \cdot (\mathcal{O}_\mu)_{x_v}) \geq 0$.

This finishes the definition of relatively complete models. Any single relatively complete model can be used to carry out Mumford's construction, as we will see shortly. Logically we should show first that relatively complete models exist for any split object $(\tilde{G}, X, Y, c, c^t, \imath, \lambda, \phi, \tau, \tilde{\mathcal{L}}, \psi, \mathcal{M})$ of $\mathrm{DD}_{\mathrm{ample}}$. The natural way is to write one down explicitly as Proj of "terms of partial Fourier expansion". Unfortunately this works only when the given degeneration data $(\tilde{G}, X, Y, c, c^t, \imath, \lambda, \phi, \tau, \tilde{\mathcal{L}}, \psi, \mathcal{M})$ satisfies some additional conditions to be specified later.

We need some more notation. For every y in Y, let I_y be the locally principal fractional ideal of R defined by $\psi(y)$ w.r.t. the integral structure given by $c^{t*}\mathcal{M}^{-1}|_y$. For almost all y, I_y is actually an ideal of R instead of a fractional ideal, because of the positivity assumption. It is easy to see from the compatibility of ψ and τ that that, as invertible ideals, $I_{y+z} = I_y \cdot I_z \cdot I_{z,\phi(y)}$ for all $y, z \in Y$. Next we introduce two graded quasi-coherent $\mathcal{O}_A$-algebras $\mathcal{S}_1$ and $\mathcal{S}_2$. By definition, $\mathcal{S}_1 = \oplus_{n\geq 0}\pi_*(\mathcal{O}_{\tilde{G}})\theta^n = \sum_{n\geq 0}(\oplus_{\mu\in X}\mathcal{O}_\mu)\theta^n$, where θ is an indeterminate of degree 1 giving rise to the grading and $\pi : \tilde{G} \to A$ is the canonical morphism. Similarly we define $\mathcal{S}_2 = \sum_{n\geq 0}(\oplus_{\mu\in X}\mathcal{O}_\mu \otimes \mathcal{M}^n)\theta^n$, where θ is again an indeterminate of degree 1. Note that the Y-action on $\pi^*\mathcal{M} \otimes_R K$ defined by τ and ψ gives rise to a Y-action on $\mathcal{S}_2 \otimes_A K$, which we denote by $\tilde{S}_y : S_y^*(\mathcal{S}_2 \otimes_A K) \to \mathcal{S}_2 \otimes_A K$. On degree 1 components we have $\tilde{S}_y : S_y^*(\mathcal{O}_\mu \otimes \mathcal{M}) \to \mathcal{O}_{\mu+\phi(y)} \otimes \mathcal{M}$. Following [Mum 5], we define a *star* Σ in X to be a finite subset of X generating X (as abelian group) such that $0 \in \Sigma$ and $-\Sigma = \Sigma$. Letting Σ be a star, we define two graded quasi-coherent $\mathcal{O}_A$-algebras which are sub-algebras of $\mathcal{S}_1 \otimes_R K$ and $\mathcal{S}_2 \otimes_R K$ respectively: $\mathcal{R}_{1,\Sigma} = \mathcal{O}_A[(I_y \cdot I_{\alpha,y} \cdot \mathcal{O}_{\alpha+\phi(y)})\theta]_{y\in Y,\alpha\in\Sigma}$ and $\mathcal{R}_{2,\Sigma} = \mathcal{O}_A[(I_y \cdot I_{\alpha,y} \cdot \mathcal{O}_{\alpha+\phi(y)} \otimes \mathcal{M})\theta]_{y\in Y,\alpha\in\Sigma} = \mathcal{O}_A[\tilde{S}_y(S_y^*(\mathcal{O}_\alpha \otimes \mathcal{M}))\theta]_{y\in Y,\alpha\in\Sigma}$. Clearly we have $\mathrm{Proj}(\mathcal{R}_{1,\Sigma}) = \mathrm{Proj}(\mathcal{R}_{2,\Sigma})$, which we denote by $\tilde{P}_{(\phi,\psi),\Sigma}$. Regarding $\tilde{P}_{(\phi,\psi),\Sigma}$ as $\mathrm{Proj}(\mathcal{R}_{1,\Sigma})$ endows it with a invertible sheaf $\mathcal{N}$ which is relatively ample over A, while its structure as $\mathrm{Proj}(\mathcal{R}_{2,\Sigma})$ endows it with an invertible sheaf $\tilde{\mathcal{L}} = \mathcal{N} \otimes \pi^*\mathcal{M}$, where $\pi : \tilde{P} \to A$ is the structural morphism.

$\tilde{\mathcal{L}}$ is ample over S because $\mathcal{M}$ is ample on A over S. The pair $(\tilde{P}_{(\phi,\psi),\Sigma}, \tilde{\mathcal{L}})$ inherits a natural Y-action from $\mathcal{S}_2 \otimes_R K$.

$\tilde{P}_{(\phi,\psi),\Sigma}$ is our candidate for a relatively complete model. Unfortunately it sometimes fails, and an additional condition is needed to ensure that it works. The condition is:

$$(^*) \quad I_y \cdot I_{y,\alpha} \subset R \quad \forall y \in Y,\ \forall \alpha \in \Sigma\ .$$

Since writing down the $\tilde{P}_{(\phi,\psi),\Sigma}$'s is the only known general way for proving the existence of relatively complete models, we should briefly discuss when condition (*) is satisfied. It is certainly not always satisfied for a given object $(\tilde{G}, X, Y, c, c^t, \lambda, \phi, \tau, \tilde{\mathcal{L}}, \psi, \mathcal{M})$ in $\mathrm{DD}_{\mathrm{ample}}$. But if we replace ϕ by $(2n+1)\phi$ (resp. $2n\phi$), ψ by $\psi^{n+1} \otimes [-1]^*\psi^n$ (resp. $\psi^n \otimes [-1]^*\psi^n$), $\mathcal{M}$ by $\mathcal{M}^{n+1} \otimes [-1]^*\mathcal{M}^n$ (resp. $\mathcal{M}^n \otimes [-1]^*\mathcal{M}^n$) hence $\tilde{\mathcal{L}}$ by $\tilde{\mathcal{L}}^{n+1} \otimes [-1]^*\tilde{\mathcal{L}}^n$ (resp. $\tilde{\mathcal{L}}^n \otimes [-1]^*\tilde{\mathcal{L}}^n$), while keeping the other data unchanged, the resulting object in $\mathrm{DD}_{\mathrm{ample}}$ will satisfy the condition (*) if n is sufficiently big. This follows from

3.2. Lemma. *Let R be a noetherian normal domain with fraction field K. Let $a : Y \to K^*$ be a multiplicatively quadratic function, $b : X \times Y \to K^*$ be a bimultiplicative function and $\phi : Y \to X$ be an injective homomorphism between free abelian groups of equal rank $r < \infty$, such that*

$$a(y+z) = a(y) \ \cdot\ a(z) \ \cdot\ b(y, \phi(z)) \quad \forall y, z \in Y.$$

Suppose moreover that $a(y) \in R$ for all but finitely many y's in Y. Then for every α in X, there exists $n_0 > 0$ such that

$$a(y) \cdot b(y, 2n\phi(y)) \cdot b(y, \alpha) \in R\ \forall y \in Y, \text{ and}$$
$$b(y, 2n\phi(y)) \cdot b(y, \alpha) \in R\ \forall y \in Y$$

for every $n \geq n_0$.

Proof of Lemma. Since X, Y are finitely generated, there are finitely many height 1 prime ideals $\{\mathcal{P}_1, \ldots, \mathcal{P}_k\}$ such that $a(y)$, $b(\mu, y) \in R_{\mathcal{P}}\ \forall y \in Y$, $\forall \mu \in X$ if $\mathcal{P} \neq \mathcal{P}_i$ for $1 \leq i \leq k$. Since $R = \cap R_{\mathcal{P}}$ where $\mathcal{P}$ runs over all prime ideals of height one in R, we are reduced to the case that R is a discrete valuation ring. Let v be the associated discrete valuation. $v \circ a$ is a real-valued quadratic function whose associated bilinear form $v \circ b(y, \phi(z))$ is positive semi-definite. The lemma now follows from a standard compactness argument.

3.3. Proposition. *Assume that the condition (*) is satisfied, then $(\tilde{P}_{(\phi,\psi),\Sigma}, \tilde{\mathcal{L}})$ is a relatively complete model.*

Proof. (We will not give complete details because the proof consists of paraphrasing [Mum 5] in our present setting.) $\tilde{P}_{(\phi,\psi),\Sigma}$ is covered by relatively affine open subschemes $U_{\alpha,y}$'s, where α runs over Σ and y over Y and (by definition)

$$U_{\alpha,y} = \mathbf{Spec}_A\, \mathcal{O}_A[\cdots, I_y^{-1} \cdot I_{y,\alpha}^{-1} \cdot I_z \cdot I_{z,\beta} \cdot \mathcal{O}_{\beta-\alpha+\phi(z-y)}\ , \cdots]_{\beta \in \Sigma,\ z \in Y}.$$

It is easy to see that S_z carries $U_{\alpha,y}$ to $U_{\alpha,y+z}$ for every z in Y, so that there are only finitely many Y-orbits in this collection of relatively affine open subschemes. Using condition (*), one sees that $U_{0,0} = \tilde{G}$ because Σ is a star in X. As in [Mum 5], each $U_{\alpha,y}$ is of finite type over A and hence also of finite type over S. Only the completeness condition remains to be checked: if v is a valuation on $K(\tilde{G})$ with $v(R) \geq 0$ on, then v has a center on $\tilde{P}_{(\phi,\psi),\Sigma} \Leftrightarrow \forall \mu \in X,\ \exists y \in Y$ such that $v(I_{y,\mu} \cdot (\mathcal{O}_\mu)_{x_v}) \geq 0$.

As in [Mum 5], we verify easily that v has a center on $\tilde{P}_{(\phi,\psi),\Sigma} \Leftrightarrow \min_{\alpha\in\Sigma, y\in Y}$ $v(I_y \cdot I_{y,\alpha} \cdot \mathcal{O}_{\alpha+\phi(y),x_v})$ exists and

$$\begin{aligned}
&\forall \mu \in X,\ \exists y \in Y \text{ such that } v(I_{y,\mu} \cdot (\mathcal{O}_\mu)_{x_v}) \geq 0\\
&\Leftrightarrow \text{ (i) } \forall z \in Y,\ \exists y \in Y \text{ such that } v(I_{y,\phi(z)} \cdot (\mathcal{O}_{\phi(z)})_{x_v}) \geq 0\\
&\Leftrightarrow \text{ (ii) } \forall z \in Y,\ \exists \mu \in X \text{ such that } v(I_{z,\mu} \cdot (\mathcal{O}_{\phi(z)})_{x_v}) \geq 0\\
&\Leftrightarrow \text{ (iii) } \forall z \in Y,\ \exists n \geq 1 \text{ such that } v(I^n_{z,\phi(z)} \cdot (\mathcal{O}_{\phi(z)})_{x_v}) \geq 0.
\end{aligned}$$

The implication "$\Rightarrow$" of the completeness condition can be easily verified exactly as in [Mum 5]. A minor nuisance is that the proof of "$\Leftarrow$" in [Mum 5] p. 247 is not completely correct: the inequality proved in the last seven lines of [Mum 5] p. 247 does not imply that the minimum in question exists. But this is easy enough to correct. Choose a basis $\{y_1, \dots, y_r\}$ of Y. Assume (iii) holds. Then there is an integer $k > 1$ such that $v(I^k_{y_i,\phi(y_i)} \cdot (\mathcal{O}_{\phi(y_i)})_{x_v}) \geq 0$ and $v(I^k_{y_i,\phi(y_i)} \cdot (\mathcal{O}_{\phi(-y_i)})_{x_v}) \geq 0$ for $1 \leq i \leq r$. Clearly the desired conclusion that $\min_{\alpha\in\Sigma, y\in Y}$ $v(I_y \cdot I_{y,\alpha} \cdot \mathcal{O}_{\alpha+\phi(y),x_v})$ exists is implied by

(iv) $\forall \mu \in X$, $\exists$ a finite subset Q of Y such that $v(I_y \cdot I_{y,\mu} \cdot (\mathcal{O}_{\mu+\phi(y)})_{x_v}) \geq v((\mathcal{O}_\mu)_{x_v})$ $\forall y \in Y - Q$.

By our choice of k, (iv) will hold if we can prove

(v) for all but finitely many elements $\mathbf{m} = (m_1, \dots, m_r)$ of $\mathbb{Z}^r$, we have

$$I_{\Sigma m_i y_i} \cdot I_{\Sigma m_i y_i, \alpha} \cdot \prod_{1\leq i\leq r} I^{-k\cdot m_i}_{y_i,\phi(y_i)} \subset R.$$

Since R is a normal domain, it suffices to check this inclusion for prime ideals of height 1 in R. Thus we may assume that R is a discrete valuation ring. The statement (v) now translates into an inequality involving $\mathbb{R}$-valued quadratic functions, which is implied by

(vi) Let $f : \mathbb{R}^r \to \mathbb{R}$ be a quadratic function bounded below with $f(0) = 0$ whose associated bilinear form $Q(x,y) := f(x+y) - f(x) - f(y)$ is positive semi-definite. Then for any elements $a_1, \dots, a_j \in \mathbb{R}^r$, $f(x) - Q(x,a_1) - \cdots - Q(x,a_j) \geq 0$ for all $x \in \mathbb{R}^r$ outside of some compact subset. The proof of (vi) is left to the reader as an exercise in linear algebra. This finishes the (sketched) proof of the proposition.

3.4. Remark. The relatively complete model $\tilde{P}$ is functorial once we have specified the star Σ in X. We are interested in what happens if we replace the line bundle $\mathcal{M}$ on A by a different descended form of $\tilde{\mathcal{L}}$, while leaving unchanged the other data: The difference between the two $\mathcal{M}$'s is measured by a character of T, that is an object of X. If this object happens to lie in $\phi(Y)$, that is of the form $\phi(y)$, one easily sees that the new relatively complete model $\tilde{P}$ can be obtained from the old one by replacing the inclusion $\tilde{G} \subset \tilde{P}$ by its transform under S_y, and retaining the other data except for $\mathcal{M}$. We shall need this remark later, to deal with the difficulty that in general the line bundle $\tilde{\mathcal{L}}$ cannot be descended globally to an $\mathcal{M}$.

4. The Construction of the Quotient

Suppose now that we are given a relatively complete model $(\tilde{P}, \tilde{\mathcal{L}})$ of a split object $(\tilde{G}, X, Y, c, c^t, \lambda, \phi, \tau, \tilde{\mathcal{L}}, \psi, \mathcal{M})$ in $\mathrm{DD}_{\mathrm{ample}}$ and we wish to construct a "quotient" of $(\tilde{G}, \tilde{\mathcal{L}})$ by Y. We first prove a proposition corresponding to (3.1) of [Mum 5], which says roughly that $\tilde{P}$ is not much bigger than $\tilde{G}$:

4.1. Proposition. *For any $y \in Y$, $\mathcal{O}_{\phi(y)} \subset \pi_*\mathcal{O}_{\tilde{P}} \otimes R[I_{y,\phi(y)}^{-1}]$, i.e. $\mathcal{O}_{\phi(y)}$ defines regular functions on $\tilde{P}$ over* $\mathrm{Spec}(R[I_{y,\phi(y)}^{-1}])$.

4.2. Corollary. $\tilde{G}_\eta = \tilde{P}_\eta$.

Proof of 4.2. Over η, $\mathcal{O}_{\phi(y)}$ defines regular functions on $\tilde{P}_\eta$ for all y in Y, so $\mathcal{O}_\mu$ defines regular functions on the normalization of $\tilde{P}$ over η for every μ in X. Thus the normalization of $\tilde{P}$ and $\tilde{G}$ are equal over η, so a fortiori we have $\tilde{G}_\eta = \tilde{P}_\eta$.

Proof of 4.1. This is a consequence of the assumption that $\tilde{G}$ acts on the sheaf of $\mathcal{O}_{\tilde{P}}$-modules $\mathcal{N} = \pi^*\mathcal{M}^{-1} \otimes \tilde{\mathcal{L}}$ extending the translation action on $\mathcal{O}_{\tilde{G}}$: First, $\imath(y)$ extends to an element of $\tilde{G}(R[I_{y,\phi(y)}^{-1}])$. This is because for every μ in X, $I_{y,\phi(y)+\mu}^n \in R$ for $n \gg 0$ by the usual trick of reducing to prime ideals of height one. Over $\mathrm{Spec}(R[I_{y,\phi(y)}^{-1}])$, we have two isomorphisms:

(a) $\mathcal{N} \xrightarrow{\sim} \mathcal{N} \otimes \pi^*\mathcal{O}_{\phi(y)}$ deduced from $\tilde{S}_y : \tilde{\mathcal{L}} \xrightarrow{\sim} \tilde{\mathcal{L}}$, and

(b) $\tilde{T}_{i(y)} : \tilde{T}_{i(y)}\mathcal{N} \xrightarrow{\sim} \mathcal{N}$ from the $\tilde{G}$-action on $\mathcal{N}$.

Hence we have an isomorphism $\zeta : \mathcal{O}_{\tilde{P}} \xrightarrow{\sim} \mathcal{O}_{\tilde{P}} \otimes \pi^*\mathcal{O}_{\phi(y)}$ extending the obvious isomorphism when both sides are restricted to $\tilde{G}$.

Consider the decomposition of $\mathcal{O}_{\tilde{P}}$ w.r.t. the action of T. $\mathcal{O}_{\tilde{P}}$ is the direct sum of subsheaves of μ-eigenfunctions $\mathcal{O}_{\tilde{P}}^\mu$, where μ runs over X. Taking the direct image under $\pi : \tilde{P} \to A$, we see that $\pi_*\mathcal{O}_{\tilde{P}}$ is the direct sum of its subsheaves of μ-eigenfunctions $(\pi_*\mathcal{O}_{\tilde{P}})_\mu$ with μ running over X. Since $\tilde{G} \subset \tilde{P}$, $(\pi_*\mathcal{O}_{\tilde{P}})_\mu$ is naturally a subsheaf of $\mathcal{O}_\mu$ = μ-eigenfunctions of $\pi_*\mathcal{O}_{\tilde{G}}$. In particular, when $\mu = 0$ we have $(\pi_*\mathcal{O}_{\tilde{P}})_0 = \mathcal{O}_A$ because $\mathcal{O}_A$ is contained in $\pi_*\mathcal{O}_P$ via $\pi : \tilde{P} \to A$.

The isomorphism $\zeta : \mathcal{O}_{\tilde{P}} \xrightarrow{\sim} \mathcal{O}_{\tilde{P}} \otimes \pi^*\mathcal{O}_{\phi(y)}$ over $\mathrm{Spec}(R[I^{-1}_{y,\phi(y)}])$ gives rise to isomorphisms $(\pi_*\mathcal{O}_{\tilde{P}})_{\mu+\phi(y)} \xrightarrow{\sim} (\pi_*\mathcal{O}_{\tilde{P}})_{\mu} \otimes \mathcal{O}_{\phi(y)}$ over $\mathrm{Spec}(R[I^{-1}_{y,\phi(y)}])$ because ζ extends the obvious isomorphism over $\tilde{G}$. Taking $\mu = 0$, we see that $(\pi_*\mathcal{O}_{\tilde{P}})_{\phi(y)} \otimes R[I^{-1}_{y,\phi(y)}] \xrightarrow{\sim} \mathcal{O}_{\phi(y)} \otimes R[I^{-1}_{y,\phi(y)}]$ is nothing but the canonical inclusion of $(\pi_*\mathcal{O}_{\tilde{P}})_{\phi(y)}$ in $\mathcal{O}_{\phi(y)}$. In other words, $(\pi_*\mathcal{O}_{\tilde{P}})_{\phi(y)} \otimes R[I^{-1}_{y,\phi(y)}] = \mathcal{O}_{\phi(y)} \otimes R[I^{-1}_{y,\phi(y)}]$. □

4.3. Proposition. *Every irreducible component of* $\tilde{P}_0 := \tilde{P} \otimes R/I$ *is proper over* $A_0 := A \otimes R/I$*, or equivalently, over* R/I*.*

The proof of 4.3. is identical with that of [Mum 5], 3.3. One deduces from 4.1 and the completeness condition that $\pi_0 : \tilde{P}_0 \to A_0$ satisfies the valuative criterion for properness for all valuations. Lemma 3.4. of [Mum 5] then implies that π_0 is proper.

4.4. Corollary. *The closure* $\overline{U}_0$ *of* U_0 *in* $\tilde{P}_0$ *is proper over* S_0*, where* $U_0 := U \times_S S_0$ *with* U *as in condition (1) of the definition of relatively complete models.*

4.5. Proposition. *There is a finite subset* $Q \subset Y$ *such that* $S_y(\overline{U}_0) \cap S_z(\overline{U}_0) = \emptyset$ *if* $y - z \notin Q$*.*

Proof. First we remark that the actions of Y and T on $(\tilde{P}, \tilde{\mathcal{L}})$ do not commute, but satisfy a commutation relation: for a functorial point $a \in T(S')$ for some S-scheme S' and $y \in Y$, there are two isomorphisms $\tilde{T}_a \circ T_a^*(\tilde{S}_y)$, $\tilde{S}_y \circ \tilde{S}_y(\tilde{T}_a) : T_a^* \circ S_y^*(\mathcal{M} \otimes \mathcal{O}_\mu) = S_y^* \circ T_a^*(\mathcal{M} \otimes \mathcal{O}_\mu) \to \mathcal{M} \otimes \mathcal{O}_{\mu+\phi(y)}$, and we have $\tilde{T}_a \circ T_a^*(\tilde{S}_y) = \phi(y)(a) \cdot \tilde{S}_y \circ \tilde{S}_y(\tilde{T}_a)$. This identity can be easily checked over the generic point η, so it holds on $(\tilde{P}, \tilde{\mathcal{L}})$.

The trick to prove 4.5. is to examine the characters for the T-action on $\tilde{\mathcal{L}}$ on the (irreducible components of the) fixed-point subscheme F of $\tilde{P}_0$. Borel's fixed point theorem guarantees that F intersects every irreducible component of $\tilde{P}_0$ by 4.3. $\overline{U}_0 \cap F$ has only a finite number of connected components $C_1, \ldots, C_n$. T acts on $(C_i, \tilde{\mathcal{L}} \otimes \mathcal{O}_{C_i})$ via a character μ_i. By the commutation relation above, T acts on the restriction of $\tilde{\mathcal{L}}$ to the connected components $S_y(C_1), \ldots, S_y(C_n)$ of $S_y(\overline{U}_0)$ via characters $\mu_1 + \phi(y), \ldots, \mu_n + \phi(y)$. It is clear that $Q := \phi^{-1}\{\mu_i - \mu_j \mid 1 \le i, j \le n\}$ has the property required in 4.5.

4.6. Corollary. Y *acts freely on* $\tilde{P}_0$*.*

4.7. Proposition. $\tilde{P}_0$ *is connected.*

The proof of 4.7. is a direct translation of [Mum 5] 3.8.

4.8. Proposition/Definition. *For every* $n \ge 1$*, let* $\tilde{P}_n := \tilde{P} \times_S S_n$*. There exists a scheme* P_n *projective over* $S_n = \mathrm{Spec}\, R/I^n$ *and an étale surjective morphism*

$\pi_n : \tilde{P}_n \longrightarrow P_n$ such that P_n is the quotient of $\tilde{P}_n$ as fpqc-sheaf and the ample invertible sheaf $\tilde{\mathcal{L}} \otimes R/I^n$ on $\tilde{P}_n$ descends to an ample invertible sheaf $\mathcal{L}_n$ on P_n. The schemes P_n fit together as n vary to form a formal scheme P_{for}, and the ample invertible sheaves $\mathcal{L}_n$ also fit together to give a formal ample invertible sheaf $\mathcal{L}_{\text{for}}$ on P_{for}. The pair $(P_{\text{for}}, \mathcal{L}_{\text{for}})$ algebraizes to $(P, \mathcal{L})$ where P is a projective scheme over S and $\mathcal{L}$ is an ample invertible sheaf on P.

Proof. This is a direct consequence of previous results and Grothendieck's formal existence theorem; see also [Mum 5] 3.10..

4.9. Definition. We will pick out an open subscheme G inside P, which is to be regarded as "the" quotient of $\tilde{G}$ by $\imath(Y)$. Moreover, $(G, \mathcal{L})$ with $\mathcal{L} = \mathcal{L}|_G$ is to be regarded as "the" quotient of $(\tilde{G}, \tilde{\mathcal{L}})$ by $\imath(Y)$. Equivalently, we define the complement Z of G in P. We do this in several steps:

(1) $\cup_{y \in Y} S_y(\tilde{G}) \subset \tilde{P}$ is an open subscheme of $\tilde{P}$. $G_n := (\cup_{y \in Y} S_y(\tilde{G}) \times_S S_n)/\imath(Y) \subset P_n$ is an open subscheme of P_n. $G_{\text{for}} := \varinjlim G_n \subset P_{\text{for}}$ is an open formal subscheme of P_{for}. G_{for} is canonically isomorphic to the formal completion $\tilde{G}_{\text{for}}$ of $\tilde{G}$ since G_n is canonically isomorphic to $\tilde{G}_n := \tilde{G} \times_S S_n$.

(2) $\tilde{C} := (\tilde{P} - \cup_{y \in Y} S_y(\tilde{G}))_{\text{red}}$ is a closed reduced subscheme of $\tilde{P}$. $C_n := (\tilde{C} \times_S S_n)/\imath(Y)$ is a closed subscheme of P_n whose underlying topological space coincides with $P_n - G_n$. $C_{\text{for}} := \varinjlim C_n$ is a closed formal subscheme of P_{for} whose underlying topological space coincides with $P_{\text{for}} - G_{\text{for}}$.

(3) C_{for} algebraizes (uniquely) to a closed subscheme C of P. Define G to be the open subscheme $P - C$ of P. Clearly the formal completion of G is canonically isomorphic to G_{for} as defined in (1), hence canonically isomorphic to $\tilde{G}_{\text{for}}$. This G is the desired open subscheme of P which should be regarded as "the quotient of $\tilde{G}$ by $\imath(Y)$". We will prove in the next section that G is a semi-abelian scheme whose group law is compatible with that of $\tilde{G}$ via the canonical isomorphism between G_{for} and $\tilde{G}_{\text{for}}$.

Let us record some basic properties of G:

4.10. Proposition. *G is smooth over S.*

Proof. See [Mum 5] 4.2.

4.11. Proposition. *P is irreducible.*

Proof. See [Mum 5] 4.3.

4.12. Proposition. *$G_\eta = P_\eta$. In particular, G_η is proper over η.*

Proof. Since $\tilde{G}_\eta = \tilde{P}_\eta$, $\mathcal{O}_{\tilde{C}}$ is killed by a non-zero element x of R. Therefore $\mathcal{O}_C$ is also killed by x. (Cf. [Mum 5] 4.9.) □

5. Functoriality of Mumford's Construction

We keep the previous notation. In particular $(\tilde{P}, \tilde{\mathcal{L}})$ is a relatively complete model of a split object $(\tilde{G}, X, Y, c, c^t, \lambda, \phi, \tau, \tilde{\mathcal{L}}, \psi, \mathcal{M})$ in $\mathrm{DD}_{\mathrm{ample}}$, and P, G are constructed from $(\tilde{P}, \tilde{\mathcal{L}})$ as in §4. So far we have described G as a scheme, but the group law has yet to be defined. By general nonsense it suffices to prove the functoriality of Mumford's construction, from which the group law of G follows immediately.

The key to the proof of functoriality is the notion of "integrable semi-abelian scheme":

5.1. Definition. (c.f. [Mum 5] 4.4.) Let $\tilde{G}_1 \subset \tilde{G}$ be a semi-abelian subscheme of $\tilde{G}$ which is an extension of an abelian scheme A_1 by a (necessarily split) torus T_1 over S. Let Y_1 be the subgroup $\imath^{-1}(\tilde{G}_1(K))$ of Y. $\tilde{G}$ is said to be *integrable* if $\mathrm{rank}(Y_1) = \dim(T_1/S)$.

5.2. Remark. A morphism of Raynaud extensions from $(O \to T_1 \to \tilde{G}_1 \to A_1 \to 0)$ to $(0 \to T \to \tilde{G} \to A \to 0)$ corresponds to a pair of homomorphisms $(f_T^t : X \to X_1,\ f_A^t : A^t \to A_1^t)$ such that $f_A^t \circ c = c_1 \circ f_T^t$, where c and c_1 give the two Raynaud extensions via (negative of) push-out. In the situation in 5.1 above, but without the integrability assumption $\mathrm{rank}(Y_1) = \dim(T_1/S)$, define $\phi_1 := f_T^t \circ \phi$, $j_A : A_1 \to A$, $j_Y : Y_1 \to Y$ the inclusions, $\lambda_1 = j_A^t \circ \lambda \circ j_A$, $c_1^t := \pi_1 \circ \imath \circ j : Y_1 \to A_1$ where $\pi_1 : \tilde{G}_1 \to A_1$ is the canonical map. The trivialization τ of $(c^t \times c)^* \mathcal{P}_{A,\eta}^{-1}$ induces a trivialization τ_1 of $(c_1^t \times c_1)^* \mathcal{P}_{A_1,\eta}^{-1}$, which corresponds to the homomorphism $\imath_1 : Y_1 \to G_1(K)$ induced by $\imath$. Since τ satisfies the positivity condition, so does τ_1: for every non-zero element y in Y^1, $\tau_1(y, \phi(y))$ extends to a section of $(c_1^t \times c_1)^* \mathcal{P}_{A_1,\eta}|_{(y,\phi(y))}$ over S which is congruent to zero modulo I. In particular, ϕ_1 is injective and $\mathrm{rank}(Y_1) \leq \dim(T_1/S)$, which holds whenever $(\tilde{G}, X, Y, c, c^t)$ is an object of DD. If the equality holds, i.e. $\tilde{G}_1$ is integrable, then $(\tilde{G}_1, X_1, Y_1, c_1, c_1^t, \lambda_1, \phi_1, \tau_1)$ (resp. $(\tilde{G}_1, X_1, Y_1, c_1, c_1^t, \lambda_1, \phi_1, \tau_1, \tilde{\mathcal{L}}|_{\tilde{G}_1}, \psi|_{Y_1}, \mathcal{M}|_{Y_1})$) is an object of $\mathrm{DD}_{\mathrm{pol}}$ (resp. $\mathrm{DD}_{\mathrm{ample}}$) provided $(\tilde{G}, X, Y, c, c^t, \lambda, \phi, \tau)$ (resp. $(\tilde{G}, X, Y, c, c^t, \lambda, \phi, \tau, \tilde{\mathcal{L}}, \psi, \mathcal{M})$) is an object of $\mathrm{DD}_{\mathrm{pol}}$ (resp. $\mathrm{DD}_{\mathrm{ample}}$).

5.3. We return to the situation that $(\tilde{P}, \tilde{\mathcal{L}})$ is a relatively complete model of an object $(\tilde{G}, X, Y, c, c^t, \phi, \tau, \tilde{\mathcal{L}}, \psi, \mathcal{M})$ in $\mathrm{DD}_{\mathrm{ample}}$, and $\tilde{G}_1 \subset \tilde{G}$ is an integrable semi-abelian subscheme of $\tilde{G}$. We would like to produce a closed subscheme W of P such that the intersection of W with G is an open subscheme G_1 of W whose formal completion along I is canonically isomorphic to $\tilde{G}_1$. It will be shown that G_1 is the quotient of $\tilde{G}_1$ by Y_1 via Mumford's construction. We proceed in several steps:

(a) Let $\tilde{W}_1$ be the scheme-theoretic closure of $\tilde{G}_1$ in $\tilde{P}$. $\tilde{W}_1$ is clearly Y_1-invariant, where the action of Y_1 is induced by that of Y. $\tilde{W}_1$ is reduced because $\tilde{G}_1$ is.

(b) Let $\tilde{W}_{1,\text{for}}$ be the I-adic completion of $\tilde{W}_1$, which is also Y_1-invariant. Define a formal scheme $\tilde{W}_{\text{for}}$ by

$$\tilde{W}_{\text{for}} := \cup_{\, y\in Y/Y_1}\, S_y(\tilde{W}_{1,\text{for}}) \text{ as a subscheme of } \tilde{P}_{\text{for}}.$$

$\tilde{W}_{\text{for}}$ is well-defined because, as we will see in a moment, the right hand side is a locally finite union.

(c) Let $W_{\text{for}} := \tilde{W}_{\text{for}}/Y$, which is a closed subscheme of P_{for}.

(d) W_{for} algebraizes to a closed subscheme W of P. Define G_1 by $G_1 := W \cap G$.

In (a)–(d) above, the assertion in (b) that the union is locally finite is the only non-trivial step where the integrability of $\tilde{G}_1$ is used. It follows from

5.4. Proposition. *Let $\tilde{G}_1$ be a integrable semi-abelian subscheme of $\tilde{G}$ and Y_1 be the subgroup $\imath^{-1}(\tilde{G}_1(K))$ of Y. Let $\tilde{W}_1$ be the scheme theoretic closure of $\tilde{G}_1$ in $\tilde{P}$. Then there is a finite subset $Q \subset Y$ such that*

$$(\tilde{W}_1 \times_S S_0) \cap S_y(U_0) = \emptyset \quad \textit{for every } y \notin Q + Y_1.$$

Proof. (cf. [Mum 5] 4.5) The subtorus T_1 of T corresponds to a quotient free abelian group X_1 of X. So we have an exact sequence $0 \to X' \to X \to X_1 \to 0$, where X' is the group of characters of T which are identically 1 on T_1. Since $\tau_1(z,\chi) = 1$ for any χ in X' and any z in Y_1, we have $\phi^{-1}(X') \cap Y_1 = \{0\}$. Therefore $Y_1 + \phi^{-1}(X')$ has finite index in Y. Let $Y' := \{y \in Y | ny \in Y_1$ for some integer $n \geq 1\}$, we have $Y = Y' \oplus \phi^{-1}(X')$. Consider the invertible fractional ideals $I_{z,\chi}$ for $\chi \in X'$ and $z \in Y$, which depend only on z modulo Y'.

For any non-zero element y of Y, choose an integer $n_y \geq 1$ such that $I^{n_y}_{y,\phi(y)} \cdot \mathcal{O}_{\phi(y)}$ is contained in $(\pi|_U)_*(\mathcal{O}_U)$ and is congruent to zero modulo I. This is possible by 4.1. because U is of finite type. By an argument similar to [Mum 5] 1.3, which uses our old trick of reducing to the case that R is a discrete valuation ring, there is a finite subset $\{y_1, \dots, y_n\}$ of $\phi^{-1}(X')$ and a finite subset Q' of Y such that

$$\forall z \in Y \text{ with } z \notin Q' + Y',\ \exists i\ 1 \leq i \leq n \text{ such that } I_{z,\phi(y_i)} \subset (I_{y_i,\phi(y)})^{n_{y_i}}.$$

Consider $\mathcal{O}_\mu|_{S_z(U)}$ regarded as (the direct image of) a sheaf of rational functions on $S_z(U)$. It corresponds to the sheaf $I_{\mu,z} \cdot \mathcal{O}_\mu|_U$ on U via the isomorphism $S_z : U \to S_z(U)$. From above we see that if $z \notin Q' + Y'$, then the sheaf of rational functions $I_{z,\phi(y_i)} \cdot \mathcal{O}_{\phi(y_i)}|_U$ on U consists of regular functions that are congruent to zero modulo I. Hence the sheaf of regular functions $\mathcal{O}_{\phi(y_i)}|_{S_z(U)}$ on $S_z(U)$ consists of regular functions that are congruent to zero modulo I. On the other hand, $\mathcal{O}_{\phi(y_i)} \otimes \mathcal{O}_{\tilde{W}_1}$ contains the constant functional 1. This proves the proposition 5.4. if we choose Q to be a set of representatives of $Q' + Y'/Y_1$.

The precise form of functoriality of Mumford's construction can now be stated:

5.5. Theorem. *(cf. [Mum 5] 4.6) Let* $(\tilde{G}_1, X, \imath_1 : Y_1 \to \tilde{G}_1(K), c_1, c_1^t, \lambda_1, \phi_1, \tau_1, \psi_1)$ *and* $(\tilde{G}_2, X, \imath_2 : Y_2 \to \tilde{G}_2(K), c_2, c_2^t, \lambda_2, \phi_2, \tau_2, \psi_2)$ *be two objects of* $\mathrm{DD}_{\mathrm{ample}}$ *with relatively complete models* $(\tilde{P}_1, \tilde{\mathcal{L}}_1)$, $(\tilde{P}_1, \tilde{\mathcal{L}}_1)$, *and* G_1, G_2 *be the two schemes constructed following the procedure in §4. Then for any given pair consisting of a homomorphism* $h : Y_1 \to Y_2$ *and an* S*-homomorphism* $\tilde{\alpha} : \tilde{G}_1 \to \tilde{G}_2$ *such that* $\imath_2 \circ h = \alpha \circ \imath_1$, *there is a unique* S*-homomorphism* $\alpha : G_1 \to G_2$ *such that under the canonical isomorphisms of the* I*-adic completions of* G_i *and* $\tilde{G}_i$, α *and* $\tilde{\alpha}$ *are formally identical.*

Proof. Easy translation of the proof of [Mum 5] 4.6, hence omitted. Notice that the uniqueness is trivial. The existence part follows from 5.3.

5.6. Corollary. *The scheme* G *depends only on* $(\tilde{G}, \imath : Y \to \tilde{G}(K))$ *as an object of* DD, *and is independent of the choice of* ϕ, ψ, *and the relatively complete model* $(\tilde{P}, \tilde{\mathcal{L}})$.

Proof. Apply 5.5. to $(\tilde{G}, X, \imath : Y \to \tilde{G}(K), c, c^t, \lambda_1, \phi_1, \tau, \psi_1, \tilde{P}_1, \tilde{\mathcal{L}}_1)$, $(\tilde{G}_2, X, \imath : Y \to \tilde{G}(K), c, c^t, \phi_2, \lambda_2, \tau, \psi_2, \tilde{P}_2, \tilde{\mathcal{L}}_2)$, and the identities $\mathrm{id}_{\tilde{G}} : \tilde{G} \to \tilde{G}$, $\mathrm{id}_Y : Y \to Y$.

5.7. Corollary. G *is a group scheme over* S.

Proof. Apply 5.5. to $(\tilde{G} \times_S \tilde{G},\ X \times X,\ \imath \times \imath : Y \times Y \to \tilde{G} \times_S \tilde{G}(K),\ c \times c,\ c^t \times c^t,\ \phi \times \phi,\ \tau \times \tau,\ \psi \times \psi,\ \tilde{P} \times_S \tilde{P},\ p_1^* \tilde{\mathcal{L}} \otimes p_2^* \tilde{\mathcal{L}})$, $(\tilde{G}, X, \imath : Y \to \tilde{G}(K), c, c^t, \phi, \tau, \psi, \tilde{P}, \tilde{\mathcal{L}})$, and the multiplication maps $\mu_{\tilde{G}} : \tilde{G} \times_S \tilde{G} \to \tilde{G}$, $+Y : Y \times Y \to Y$. This yields a morphism $\mu_G : G \times_S G \to G$. Apply 5.5. to the inverses $[-1]_{\tilde{G}} : \tilde{G} \to \tilde{G}$, $-1 : Y \to Y$, we get a morphism $[-1]_G : G \to G$. μ_G and $[-1]_G$ satisfies the conditions required for a group law by of the uniqueness in 5.5.

5.8. Corollary. G_η *is an abelian variety.*

To prove that G has connected fibres and is indeed a semi-abelian scheme over S, we need a description of torsion points of G. Let $G^* := \cup_{y \in Y} S_y(\tilde{G}) \subset \tilde{P}$. For any $y \in Y$ and any non-zero integer n, let $\sigma_y : S \to G^*$ be the unique S-section of G^* such that $\sigma_y(\eta) = y$. Define $Z_y^{(n)}$ by the fibre product

$$\begin{array}{ccc} Z_y^{(n)} & \longrightarrow & G^* \\ \big\downarrow & & \big\downarrow {\scriptstyle [n]} \qquad = \quad \text{mult. by } n. \\ \sigma_y : S & \longrightarrow & G^* \end{array}$$

It is easy to see that for any $z \in Y$, S_z induces an isomorphism from $Z_y^{(n)}$ to $Z_{y+nz}^{(n)}$ for all y. Thus the disjoint union $\amalg_{y \in Y/nY} Z_y^{(n)}$ can be given the structure of a commutative group scheme over S in a canonical way.

5.9. Theorem. $G[n] :=$ *Kernel of* $[n]_G$ *is canonically isomorphic to the* S*-group scheme* $\amalg_{y \in Y/nY} Z_y^{(n)}$ *defined above.*

Proof. Entirely similar to [Mum 5] 4.10, hence omitted.

5.10. Proposition. *For any $s \in S$ and fixed y and n, we have*

$$\begin{aligned} & Z_y^{(n)} \times_S \operatorname{Spec} \kappa(s) \neq \emptyset \\ \Leftrightarrow & \exists z \in Y,\ \exists u \in S_z(\tilde{G})(s) \text{ such that } nu = \sigma_z(s). \\ \Leftrightarrow & \exists z \in Y \text{ such that } y - nz \in \tilde{G}(\mathcal{O}_{S,s}). \\ \Leftrightarrow & \exists z \in Y \text{ such that } I_{nz,\phi(y-nz)} \cdot \mathcal{O}_{S,s} = \mathcal{O}_{S,s}. \\ \Leftrightarrow & \ y \in Y_s + nY, \text{ where } Y_s = \{z \in Y | I_{z,\phi(z)} \cdot \mathcal{O}_{S,s} = \mathcal{O}_{S,s}\} = \\ & = \{z \in Y | z \in \tilde{G}(\mathcal{O}_{S,s})\}. \end{aligned}$$

Proof. The first two equivalences follows directly from the definitions. For the third equivalence, we show that $\forall w \in Y$,

$$w \in \tilde{G}(\mathcal{O}_{S,s}) \Leftrightarrow I_{w,\phi(w)} \cdot \mathcal{O}_{S,s} = \mathcal{O}_{S,s}.$$

The implication "$\Rightarrow$" is easy. Now we claim that if $I_{w,\phi(w)} \cdot \mathcal{O}_{S,s} = \mathcal{O}_{S,s}$, then $I_{w,\mu} \cdot \mathcal{O}_{S,s} = \mathcal{O}_{S,s}$ $\forall \mu \in X$. As usual we are reduced to the case that R is a discrete valuation ring and s is the closed point of $S = \operatorname{Spec}(R)$. Taking valuations, we see that the claim follows from the fact that for a positive semi-definite quadratic form Q on a finite dimensional real vector space V, a vector v lies in the radical of Q iff $Q(v,v) = 0$. This proves the claim, and the implication "$\Leftarrow$" follows easily. The fourth and last equivalence is just a reformulation of the third one. We remark that Y_s is a subgroup of Y.

5.11. Corollary. *For a non-zero integer n, let $\tilde{G}[n]$ be the kernel of $[n]_{\tilde{G}}$ and $G[n]$ be the kernel of $[n]_G$. For any $s \in S$, there is a natural exact sequence*

$$0 \to \tilde{G}[n] \times_S \operatorname{Spec} \kappa(s) \to G[n] \times_S \operatorname{Spec} \kappa(s) \to (1/n)Y_s/Y_s \to 0.$$

Taking the limit, we obtain an exact sequence

$$\begin{aligned} 0 \to (\textit{torsion of } \tilde{G} \times_S \operatorname{Spec} \kappa(s)) \to (\textit{torsion of } G \times_S \operatorname{Spec} \kappa(s)) \to \\ \to (\mathbb{Q}/\mathbb{Z}) \otimes_{\mathbb{Z}} Y_s \to 0. \end{aligned}$$

5.12. Corollary. *The geometric fibres of G over S are all connected with trivial unipotent radical, i.e. G is a semi-abelian scheme over S.*

Proof. This follows from the fact that if a commutative algebraic group H over an algebraically closed field has the property that the p-primary torsion of H is p-divisible (in the sense of group schemes) and scheme-theoretically dense in H for every prime number p, then H is connected with trivial unipotent radical (hence is an extension of an abelian variety by a torus).

We summarize the results proved so far: Mumford's construction defines a functor $M : \mathrm{DD}' \to \mathrm{DEG}$, where DD' is the full subcategory of DD consisting of split objects which have relatively complete models. This functor preserves the dimension of groups and their torus parts.

6. Behavior of $\otimes$, Pullback, Dual Under Mumford's Construction

The goal of this section is to show that Mumford's construction induces functors

$$M_{\mathrm{IS}} : \mathrm{DD}_{\mathrm{IS}} \to \mathrm{DEG}_{\mathrm{IS}}$$
$$M_{\mathrm{pol}} : \mathrm{DD}_{\mathrm{pol}} \to \mathrm{DEG}_{\mathrm{pol}}$$
$$M_{\mathrm{ample}} : \mathrm{DD}_{\mathrm{ample}} \to \mathrm{DEG}_{\mathrm{ample}}$$

in addition to $M : \mathrm{DD} \to \mathrm{DEG}$, which are compatible with one another in the obvious sense. For this purpose it is necessary to represent, for the quotient G of $\tilde{G}$ by $\imath : Y \to \tilde{G}(K)$, the dual abelian variety of G_η also as a quotient. We remark that in the case when the abelian part is trivial and $\underline{X}, \underline{Y}$ are constant, this has been done in [Rap] pp.305–319. In the next section we will show that these four functors are all equivalences of categories. So in a sense we have completely analyzed semi-abelian degenerations of abelian varieties over complete noetherian domains satisfying the strong normality condition w.r.t. étale coverings. Of course, the (more) interesting cases when the special fibres are not connected or the degenerations are not semi-stable remain as a great challenge.

Let $\mathrm{DD}^*_{\mathrm{ample}}$ denote the full subcategory of $\mathrm{DD}_{\mathrm{ample}}$ whose objects satisfy condition (*) of §3. So far Mumford's construction works on split objects in $\mathrm{DD}^*_{\mathrm{ample}}$. Our first step is to extend it to all (not necessarily split) objects of $\mathrm{DD}^*_{\mathrm{ample}}$. Ideally this should follow from étale descent, but this poses problems because we only deal with I-adically complete base schemes: As the semi-abelian schemes G are not proper over the base, we only can descend G's together with their compactification P. However descent always works for finite étale coverings, which allows us to assume that all tori involved are split. Let us show how to deal with an object $(\tilde{G}, X, Y, c, c^t, \imath, \lambda_A, \phi, \tau, \tilde{\mathcal{L}}, \psi, \mathcal{M})$ of $\mathrm{DD}_{\mathrm{ample}}$ which satisfies condition (*): Extending to a split object comes down to defining a T-action on $\tilde{\mathcal{L}}$. This can be done locally in the Zariski topology on S_{for} or S_0, and the obstruction to doing it globally is given by an X-torsor. Passing to a finite étale covering we may assume that the structure group reduces to $\phi(Y) \subset X$. It follows that we can find an open covering of S_{for} by affines $\mathrm{Spf}(R_i)$ such that over R_i $\tilde{\mathcal{L}}$ descends to $\mathcal{M}_i$, and such that over R_{ij} (= the affine ring of the intersection $\mathrm{Spf}(R_i) \cap \mathrm{Spf}(R_j)$) $\mathcal{M}_i$ and $\mathcal{M}_j$ differ by an element of Y. The construction in section 5 now gives relatively complete models $\tilde{G} \subset \tilde{P}_i$ over R_i such that over R_{ij}, $\tilde{P}_i \approx \tilde{P}_j$ canonically, with the two inclusions of $\tilde{G}$ differing by an S_y. It follows that the associated formal schemes $\mathcal{P}_i$ glue to a formal scheme $\tilde{\mathcal{P}}$ over $S^\wedge$, that we may form the Y-quotient $\mathcal{P}$ which is the formal completion of a proper P over R, and that P contains an open subscheme G which is semi-abelian: The fact that the different inclusions of $\tilde{G}$ into $\tilde{P}_i$ differ by Y does not matter as we divide by the Y-action anyway.

This settles the case of objects of $\mathrm{DD}_{\mathrm{ample}}$ satisfying (*).

Our next task is to construct a functor $M_{\mathrm{IS}} : \mathrm{DD}_{\mathrm{IS}} \to \mathrm{DEG}_{\mathrm{IS}}$. Note that $\mathrm{DD}_{\mathrm{IS}}$ is a Picard category fibred over DD in the obvious sense. In down-to-earth terms, this simply means that for any two object $(\tilde{G}, X, Y, c, c^t, \imath, f_{A,1}, f_{Y,1}, \tau_1, \tilde{\mathcal{F}}_1, \zeta_1), (\tilde{G}, X, Y, c, c^t, \imath, f_{A,2}, f_{Y,2}, \tau_2, \tilde{\mathcal{F}}_2, \zeta_2)$ of $\mathrm{DD}_{\mathrm{IS}}$ lying over an object $(\tilde{G}, X, Y, c, c^t, \imath)$ of DD, we can produce a third one $(\tilde{G}, X, Y, c, c^t, \imath, f_{A,1} + f_{A,2}, f_{Y,1} + f_{Y,2}, \tau_1 \otimes \tau_2, \tilde{\mathcal{F}}_1 \otimes \tilde{\mathcal{F}}_2, \zeta_1 \otimes \zeta_2)$, denoted by $(\tilde{G}, X, Y, c, c^t, \imath, f_{A,1}, f_{Y,1}, \tau_1, \tilde{\mathcal{F}}_1, \zeta_1) \otimes (\tilde{G}, X, Y, c, c^t, \imath, f_{A,2}, f_{Y,2}, \tau_2, \tilde{\mathcal{F}}_2, \zeta_2)$. This tensor operation satisfies the usual properties required for a (strictly commutative) Picard category. In particular there is an neutral element w.r.t. tensor product in each fibre, the tensor inverse of $(\tilde{G}, X, Y, c, c^t, \imath, f_A, f_Y, \tau, \tilde{\mathcal{F}}, \zeta)$ is denoted by $(\tilde{G}, X, Y, c, c^t, \imath, f_A, f_Y, \tau, \tilde{\mathcal{F}}, \zeta)^{\otimes -1}$; more generally the n-th tensor power of $(\tilde{G}, X, Y, c, c^t, \imath, f_A, f_Y, \tau, \tilde{\mathcal{F}}, \zeta)$ is denoted by $(\tilde{G}, X, Y, c, c^t, \imath, f_A, f_Y, \tau, \tilde{\mathcal{F}}, \zeta)^{\otimes n}$.

We start with an object $(\tilde{G}, X, Y, c, c^t, \imath, f_A, f_Y, \tau, \tilde{\mathcal{F}}, \zeta)$ of $\mathrm{DD}_{\mathrm{IS}}$. To produce an object in $\mathrm{DEG}_{\mathrm{IS}}$, we proceed as follows:
Exactly as in §3 where we discussed the existence of relatively complete models, there are objects $(\tilde{G}, X, Y, c, c^t, \imath, \lambda_1, \phi_1, \tau, \tilde{\mathcal{L}}_1, \psi_1)$, $(\tilde{G}, X, Y, c, c^t, \imath, \lambda_2, \phi_2, \tau, \tilde{\mathcal{L}}_2, \psi_2)$ in $\mathrm{DD}_{\mathrm{ample}}$ over $(\tilde{G}, X, Y, c, c^t, \imath)$ which satisfy the condition (*) such that $(\tilde{G}, X, Y, c, c^t, \imath, f_A, f_Y, \tau, \tilde{\mathcal{F}}, \zeta) = (\tilde{G}, X, Y, c, c^t, \imath, \lambda_1, \phi_1, \tau, \tilde{\mathcal{L}}_1, \psi_1) \otimes (\tilde{G}, X, Y, c, c^t, \imath, \lambda_2, \phi_2, \tau, \tilde{\mathcal{L}}_2, \psi_2)^{\otimes -1}$. We apply the Mumford construction to both these objects to obtain pairs $(G, \mathcal{L}_1)$ and $(G, \mathcal{L}_2)$: The underlying G is the same, as this holds locally and we have étale descent for morphisms. We claim that the pair $(G, \mathcal{L}_1 \otimes \mathcal{L}_2^{-1})$ is independent of all choices. This holds because if the Mumford construction on $\mathrm{DD}^*_{\mathrm{ample}}$ preserves tensor products, which follows easily from the usual trick of diagonal embeddings. One also sees easily that we have obtained a functor. It also follows that the Mumford construction on $\mathrm{DD}_{\mathrm{IS}}$ can be described using relatively complete models; that is, we can find a relatively complete model $\tilde{P}$ for $(G, \imath)$ such that $\tilde{\mathcal{F}}$ extends to a Y-equivariant bundle on $\tilde{P}$ (satisfying some axioms), and $\mathcal{F}$ is obtained by dividing out Y: Choose for $\tilde{P}$ the closure of $\tilde{G}$ in the product $\tilde{P}_1 \times_S \tilde{P}_2$ of relatively complete models for $\tilde{\mathcal{L}}_1$ and $\tilde{\mathcal{L}}_2$.

All in all we have defined a tensor functor

$$M_{\mathrm{IS}} : \mathrm{DD}_{\mathrm{IS}} \to \mathrm{DEG}_{\mathrm{IS}} ,$$

and one checks easily that it induces

$$M_{\mathrm{ample}} : \mathrm{DD}_{\mathrm{ample}} \to \mathrm{DEG}_{\mathrm{ample}}.$$

It follows that we also obtain (as any object of DD admits some ample line bundle)

$$M : \mathrm{DD} \to \mathrm{DEG}.$$

Suppose now that $(\tilde{G}, \imath : \underline{Y} \to \tilde{G} \otimes K)$ is an object of DD and $G/S = M((\tilde{G}, \imath : \underline{Y} \to \tilde{G} \otimes K))$, i.e. G is the quotient of $\tilde{G}$ by $\underline{Y}$. We would like to construct a semi-abelian scheme G^t/S via Mumford's construction, whose generic fibre is the dual abelian variety of G_η. To prove this we need to construct an invertible sheaf $\mathcal{P}$ on $G \times_S G^t$ such that $\mathcal{P}_\eta$ is the Poincaré sheaf on $G_\eta \times G^t_\eta$. The homomorphism $\imath : \underline{Y} \to \tilde{G} \otimes K$ corresponds to a trivialization $\tau : 1_{\underline{Y} \times \underline{X}} \xrightarrow{\sim} (c^t \times c)^* \mathcal{P}^{-1}_{A,\eta}$ in the previous notations. The homomorphism $c^t : \underline{Y} \to A$ determines (and is determined by) an extension $\tilde{G}^t$ of the abelian scheme A^t/S by a torus T^t/S with character group $\underline{Y}$. Let $\sigma : A^t \times_S A \to A \times_S A^t$ be the isomorphism flipping the two factors A and A^t, $\rho : \underline{Y} \times \underline{X} \to \underline{X} \times \underline{Y}$ the isomorphism flipping the two factors $\underline{X}$ and $\underline{Y}$. We emphasis that *the Poincaré sheaf lives on* $A \times A^t$ in our notation, as order becomes important now. From τ we obtain a trivialization $\sigma^* \circ \tau \circ \rho^{-1} : 1_{\underline{X} \times \underline{Y}} \xrightarrow{\sim} (c \times c^t)^* (\sigma^* \mathcal{P}_A)^{-1}_\eta$ of the biextension $(c \times c^t)^* (\sigma^* \mathcal{P}_A)^{-1}_\eta$ lying over $c \times c^t : \underline{X} \times \underline{Y} \to A^t \times_S A$. Let $\tau^t = \sigma^* \circ \tau \circ \rho^{-1}$, $\mathcal{P}^t_A = \sigma^* \mathcal{P}_A =$ Poincaré sheaf for A^t on $A \times_S A^t$. We have by simple flipping obtained a trivialization $\tau^t : 1_{\underline{X} \times \underline{Y}} \xrightarrow{\sim} (c \times c^t)^* \mathcal{P}^{-1}_{A^t,\eta}$ of the biextension $(c \times c^t)^* \mathcal{P}^{-1}_{A^t,\eta}$, which corresponds to a homomorphism $\imath^t : \underline{X} \to G^t \otimes K$. It is not too difficult to show that $(\tilde{G}, \imath^t : \underline{X} \to \tilde{G}^t \otimes K)$ is an object of DD. Applying the functor M we obtain a semi-abelian scheme $G^t/S = M((\tilde{G}^t, \imath^t : \underline{X} \to \tilde{G}^t \otimes K))$ whose generic fibre G^t_η will be shown to be the dual abelian variety of G_η.

Now we want an object of $\mathrm{DD_{IS}}$ lying over the object $(\tilde{G} \times_S \tilde{G}^t,\ \imath \times \imath^t : \underline{Y} \times \underline{X} \to \tilde{G} \times_S \tilde{G}^t \otimes K)$ of DD. We need several things:

(1) a homomorphism from $\underline{Y} \times \underline{X}$ to $\underline{X} \times \underline{Y}$. We take this to be the flipping ρ.

(2) a homomorphism from $A \times_S A^t$ to its dual $A^t \times_S A$. We take this to be the flipping ρ^{-1}.

(3) an invertible sheaf on $A \times_S A^t$. We take this to be the Poincaré sheaf $\mathcal{P}_A$.

(4) a trivialization of the cubical invertible sheaf $(c^t \times c)^* \mathcal{P}^{-1}_A$. We take this to be τ. This makes sense because any biextension has a canonical cubical structure. Notice that $\imath \times \imath^t$ corresponds to the trivialization $\Psi = \pi^*_{13} \circ \tau \circ \pi_{13} \otimes \pi^*_{24} \circ \tau^t \circ \pi_{24}$ of the biextension $(c^t \times c \times c \times c^t)^* (\pi^*_{13} \mathcal{P}_A \otimes \pi^*_{24} \mathcal{P}_A)^{-1}_\eta$, and τ is compatible with Ψ as can be easily checked using the biextension structure of the Poincaré sheaf.

These data together define an object $\mathcal{P}_{(\tilde{G},i)}$ of $\mathrm{DD_{IS}}$ lying over the object $(\tilde{G} \times_S \tilde{G}^t,\ \imath \times \imath^t : \underline{Y} \times \underline{X} \to \tilde{G} \times_S \tilde{G}^t \otimes K)$ of DD. Applying the functor M_{IS} we obtain an invertible sheaf $\mathcal{P}$ on $G \times_S G^t/S$. We want to show that $\mathcal{P}_\eta$ is (canonically isomorphic to) the Poincaré sheaf on $G_\eta \times_\eta G^t_\eta$. We need to know that the $\chi(\mathcal{P}_\eta) = \pm 1$ (it is actually equal to $(-1)^{\dim(G/S)}$). As $\mathcal{P}$ is birigidified, so is $\mathcal{P}_\eta$; hence $\mathcal{P}_\eta$ defines an algebraic correspondence. Then $\chi(\mathcal{P}_\eta) = \pm 1$ implies that G^t_η is isomorphic to the dual abelian variety of G_η and $\mathcal{P}_\eta$ is isomorphic to the Poincaré sheaf on $G_\eta \times_\eta G^t_\eta$.

To prove $\chi(\mathcal{P}_\eta) = \pm 1$, we will prove a more general result which computes $|\chi(\mathcal{F}_\eta)|$ where $(G, \mathcal{F}) = M_{\mathrm{IS}}((\tilde{G}, \underline{X}, \underline{Y}, c, c^t, \imath, f_A, f_Y, \tau, \tilde{\mathcal{F}}, \zeta)$ for an object $(\tilde{G}, \underline{X}, \underline{Y}, c, c^t, \imath, f_A, f_Y, \tau, \tilde{\mathcal{F}}, \zeta)$ of $\mathrm{DD}_{\mathrm{IS}}$.

6.1. Theorem. *Notations as above. Then* $\chi(\mathcal{F}_\eta)^2 = \chi(\mathcal{N}_\eta)^2 \cdot \deg(f_Y)^2 = \deg(f_A) \cdot \deg(f_Y)^2$.

Proof. The second equality is immediate from the Riemann-Roch theorem for abelian varieties (c.f. [Mum 1] p.150). We would like to reduce the theorem to the special case when $(\tilde{G}, \underline{X}, \underline{Y}, c, c^t, \imath, f_A, f_Y, \tau, \tilde{\mathcal{F}}, \zeta)$ is a split object of $\mathrm{DD}_{\mathrm{ample}}$ which admits a relatively complete model: By étale localization we may assume that $\underline{X}$ and $\underline{Y}$ are constant, and that $\tilde{\mathcal{F}}$ descends to a line bundle $\mathcal{N}$ on A. There is a split object $(\tilde{G}, X, Y, c, c^t, \imath, \lambda_A, \phi, \tau, \tilde{\mathcal{L}}, \psi, \mathcal{M})$ of $\mathrm{DD}_{\mathrm{ample}}$ such that $(\tilde{G}, X, Y, c, c^t, \imath, f_A, f_Y, \tau, \tilde{\mathcal{F}}, \zeta, \mathcal{N}) \otimes (\tilde{G}, X, Y, c, c^t, \imath, \lambda_A, \phi, \tau, \tilde{\mathcal{L}}, \psi, \mathcal{M})^{\otimes n}$ lies in $\mathrm{DD}^*_{\mathrm{ample}}$ for $n \gg 0$. Let $(G, \mathcal{L}) = M_{\mathrm{IS}}((\tilde{G}, X, Y, c, c^t, \imath, \lambda_A, \phi, \tau, \tilde{\mathcal{L}}, \psi, \mathcal{M})^{\otimes n})$. Consider $\chi(\mathcal{F}_\eta \otimes \mathcal{L}_\eta^{\otimes n})^2$ and $\chi(\mathcal{N}_\eta \otimes \mathcal{M}_\eta^{\otimes n})^2 \cdot \deg(f_Y + n\phi)^2$. Both are polynomials in n by the Riemann-Roch theorem for abelian varieties. Thus if equality holds for $n \gg 0$, it holds for all n, especially when $n = 0$. This completes the reduction step. The case when $(\tilde{G}, X, Y, c, c^t, \imath, f_A, f_Y, \tau, \tilde{\mathcal{F}}, \zeta, \mathcal{N})$ is a split object of $\mathrm{DD}_{\mathrm{ample}}$ admitting a relatively complete model is a consequence of the following description of the global sections of $\mathcal{F}_\eta$ in terms of partial Fourier expansion.

6.2. Theorem. *Let* $(\tilde{G}, X, Y, c, c^t, \imath, \lambda_A, \phi, \tau, \tilde{\mathcal{L}}, \psi, \mathcal{M})$ *a split object of* $\mathrm{DD}_{\mathrm{ample}}$ *which has a relatively complete model* $(\tilde{P}, \tilde{\mathcal{L}})$, $(G, \mathcal{L}) = M_{\mathrm{ample}}(\tilde{G}, X, Y, c, c^t, \imath, \lambda_A, \phi, \tau, \tilde{\mathcal{L}}, \psi)$. *Then via the partial Fourier expansion,* $\Gamma(G_\eta, \mathcal{L}_\eta)$ *is identified with the subspace of* $\Gamma(G_{\mathrm{for}}, \mathcal{L}_{\mathrm{for}}) \otimes_R K = \Gamma(\tilde{G}_{\mathrm{for}}, \tilde{\mathcal{L}}_{\mathrm{for}}) \otimes_R K$ *spanned by the infinite sums* $\{\sum_{\mu \in X/Y} \sum_{y \in Y} \tilde{S}_y(S^*_{-y}\theta_\mu) | \theta_\mu \in \Gamma(A_\eta, \mathcal{M}_\eta \otimes \mathcal{O}_\mu)\}$.

Proof of 6.2. Recall that the partial Fourier expansion of a section $s \in \Gamma(G, \mathcal{L})$ is simply the decomposition of the image of s in $\Gamma(G_{\mathrm{for}}, \mathcal{L}_{\mathrm{for}}) = \Gamma(\tilde{G}_{\mathrm{for}}, \tilde{\mathcal{L}}_{\mathrm{for}})$ into an I-adically convergent sum of eigenvectors w.r.t. the T_{for}-action on $\mathcal{L}_{\mathrm{for}} \approx \tilde{\mathcal{L}}_{\mathrm{for}}$. Since we have the relatively complete model $(\tilde{P}, \tilde{\mathcal{L}})$, we can say a little more: $\exists a \in R$, $a \neq 0$, such that $a \cdot s \in \Gamma(P, \mathcal{L})$, since $P_\eta = G_\eta$. Because $(P_{\mathrm{for}}, \mathcal{L}_{\mathrm{for}})$ is the Y-quotient of $(\tilde{P}_{\mathrm{for}}, \tilde{\mathcal{L}}_{\mathrm{for}})$, $a \cdot s \in \Gamma(\tilde{P}_{\mathrm{for}}, \tilde{\mathcal{L}}_{\mathrm{for}})^Y \subset \Gamma(\tilde{G}_{\mathrm{for}}, \tilde{\mathcal{L}}_{\mathrm{for}})$, and an analysis of Fourier decompositions gives that $a \cdot s$ is an infinite sum as above.

To prove the reverse inclusion $\Gamma(\tilde{G}^*_{\mathrm{for}}, \tilde{\mathcal{L}}_{\mathrm{for}})^Y \subset \Gamma(G_\eta, \mathcal{L}_\eta)$, it suffices to show that $\forall \theta_\mu \in \Gamma(A_\eta, \mathcal{M}_\eta \otimes \mathcal{O}_\mu)$, $\mu \in X$, $\exists a \in R$, $a \neq 0$, such that $a \cdot \sum_{y \in Y} \tilde{S}_y(S^*_{-y}\theta_\mu) \in \Gamma(\tilde{P}_{\mathrm{for}}, \tilde{\mathcal{L}}_{\mathrm{for}})^Y$. The only point is to show that this infinite sum converges I-adically, using the fact that there is a G-invariant open subscheme $U \subset \tilde{P}$ of finite type over S such that $\cup_{y \in Y} S_y(U) = \tilde{P}$. The convergence is implied by the following statement, whose proof falls into the familiar pattern and hence is left to the reader:

$$\forall n > 0,\ \exists \text{ a finite subset } Q \subset Y \text{ such that } I_y \cdot I_{y,\mu} \subset I^n \ \forall y \notin Q\ .$$

This completes the proof of theorem 6.1. and theorem 6.2. $\square$

6.3. Corollary. $\chi(\mathcal{L}_\eta, 0) = \dim_K \Gamma(G_\eta, \mathcal{L}_\eta) = \dim_K \Gamma(A_\eta, \mathcal{M}_\eta) \cdot \deg(\phi) = \chi(\mathcal{M}_\eta) \cdot \deg(\phi)$.

Recall that in chapter II we started with an $(G, \mathcal{L}) \in \mathrm{Ob}(\mathrm{DEG}_{\mathrm{ample}})$ and obtained using partial Fourier expansion an object of $\mathrm{DD}_{\mathrm{ample}}$. This defines a functor F_{ample} from $\mathrm{DEG}_{\mathrm{ample}}$ to $\mathrm{DD}_{\mathrm{ample}}$. While in this chapter we used Mumford's construction to obtain an object of $\mathrm{DEG}_{\mathrm{ample}}$ from an object of $\mathrm{DD}_{\mathrm{ample}}$. The proof of 6.2 (and the usual reductions to the split case, and the existence of a relative complete model) show that the composition $F_{\mathrm{ample}} \cdot M_{\mathrm{ample}}$ is canonically isomorphic to the identity. As F_{ample} detects isomorphisms this also holds for the composition $M_{\mathrm{ample}} \cdot F_{\mathrm{ample}}$, so these two processes are inverse to each other, and we have as a byproduct:

6.4. Proposition. *The functor* $M_{\mathrm{ample}} : \mathrm{DD}_{\mathrm{ample}} \to \mathrm{DEG}_{\mathrm{ample}}$ *induces an equivalence of categories*

$$M_{\mathrm{ample}} : \mathrm{DD}_{\mathrm{ample}} \to \mathrm{DEG}_{\mathrm{ample}}$$

whose quasi-inverse is a functor

$$F_{\mathrm{ample}} : \mathrm{DEG}_{\mathrm{ample}} \to \mathrm{DD}_{\mathrm{ample}}$$

given by the process of chapter II.

We can understand the category $\mathrm{DD}_{\mathrm{pol}}$ better now. Given an object $(\tilde{G}, \underline{X}, \underline{Y}, c, c^t, \lambda, \phi, \tau)$ of $\mathrm{DD}_{\mathrm{pol}}$, we get an object $(\tilde{G}, \imath)$ of DD from the triple $(\tilde{G}, \underline{X}, \underline{Y}, c, c^t, \tau)$, from which we get $(\tilde{G}^t, \imath^t) \in \mathrm{Ob}(\mathrm{DD})$ and $\mathcal{P}_{(\tilde{G},i)} \in \mathrm{Ob}(\mathrm{DD}_{\mathrm{IS}})$ as before. It is now very easy to see that $(\tilde{G}, \underline{X}, \underline{Y}, c, c^t, \lambda, \phi, \tau)$ determines and is determined by a morphism $\lambda_{\mathrm{pol}} : (\tilde{G}, \imath) \to (\tilde{G}^t, \imath^t)$ in DD. Hence $(\tilde{G}, \underline{X}, \underline{Y}, c, c^t, \lambda, \phi, \tau)$ of $\mathrm{DD}_{\mathrm{pol}}$ determines a homomorphism $\lambda_G : G \to G^t$ which defines a polarization on G_η, as it is not hard to see that the pull-back of $\mathcal{P}_{(\tilde{G},i)}$ by $\mathrm{id} \times \lambda_{\mathrm{pol}}$ is in fact an object of $\mathrm{DD}_{\mathrm{ample}}$. Hence we get a functor

$$M_{\mathrm{pol}} : \mathrm{DD}_{\mathrm{pol}} \to \mathrm{DEG}_{\mathrm{pol}} .$$

Similarly, an object $(\tilde{G}, \underline{X}, \underline{Y}, c, c^t, \imath, f_A, f_Y, \tau, \tilde{\mathcal{F}}, \zeta)$ of $\mathrm{DD}_{\mathrm{IS}}$ defines a morphism from $f_{\mathrm{IS}} : (\tilde{G}, \imath) \to (\tilde{G}^t, \imath^t)$ in DD such that the pull back of $\mathcal{P}_{(\tilde{G},i)}$ by $\mathrm{id} \times f_{\mathrm{IS}}$ is simply $\Lambda((\tilde{G}, \underline{X}, \underline{Y}, c, c^t, \imath, f_A, f_Y, \tau, \tilde{\mathcal{F}}, \zeta))$. Applying the functor M, we obtain a homomorphism $f = M(f_{\mathrm{IS}})$ such that f_η is nothing else but the homomorphism $\lambda(\mathcal{F}_\eta)$ associated to the invertible sheaf $\mathcal{F}_\eta$, where $(G, \mathcal{F}) = M_{\mathrm{IS}}((\tilde{G}, \underline{X}, \underline{Y}, c, c^t, \imath, f_A, f_Y, \tau, \tilde{\mathcal{F}}, \zeta))$. By now we have several functors:

$$M : \mathrm{DD} \to \mathrm{DEG},$$
$$M_{\mathrm{IS}} : \mathrm{DD}_{\mathrm{IS}} \to \mathrm{DEG}_{\mathrm{IS}},$$
$$M_{\mathrm{ample}} : \mathrm{DD}_{\mathrm{ample}} \to \mathrm{DEG}_{\mathrm{ample}},$$
$$M_{\mathrm{pol}} : \mathrm{DD}_{\mathrm{pol}} \to \mathrm{DEG}_{\mathrm{pol}}.$$

In the next section we will see that they are all equivalences of categories. Note that we have already seen that

$$M_{\text{ample}} : \mathrm{DD}_{\text{ample}} \to \mathrm{DEG}_{\text{ample}}$$

is an equivalence of categories.

7. Full Faithfulness of Mumford's Construction

7.1. Theorem. *Let $(\tilde{G}_i, \imath_i : \underline{Y}_i \to \tilde{G}_i \otimes K)$ be two objects in* DD, $i = 1, 2$. *Let $G_i = M(G_i, \imath_i)$ be the semi-abelian quotient of $\tilde{G}_i$ by $\imath_i(\underline{Y}_i)$ via Mumford's construction. Suppose $f_\eta : G_{1,\eta} \to G_{2,\eta}$ is a homomorphism between the abelian varieties $G_{i,\eta}$, $i = 1, 2$. Then there is a uniquely determined pair of homomorphisms $(\tilde{f} : \tilde{G}_1 \to \tilde{G}_2,\ h : \underline{Y}_1 \to \underline{Y}_2)$ defining a morphism in* DD*'s such that $M(\tilde{f}, h) = f_\eta$.*

Proof. From functoriality of Raynaud extensions we immediately obtain a homomorphism $\tilde{f} : \tilde{G}_1 \to \tilde{G}_2$. Since $\underline{Y}_i$ is the character group of the torus part T_i^t of $\tilde{G}_i^t$, the dual $f_\eta^t : G_{2,\eta}^t \to G_{1,\eta}^t$ of f_η induces $\tilde{f}^t : \tilde{G}_2^t \to \tilde{G}_1^t$ and hence a homomorphism $h : \underline{Y}_1 \to \underline{Y}_2$. Thus we already have our candidates. We have to show that they define a homomorphism in DD which induces f. For this we may assume that f is an isomorphism (consider $G_1 \times G_2$ and its endomorphism $\mathrm{id} + j_2 \circ f \circ pr_1$, where j_2 is the second inclusion and pr_1 the first projection). Choose an ample line bundle $\mathcal{L}_2$ on G_2 and let $\mathcal{L}_1 = f^*(\mathcal{L}_2)$. One checks easily that on the $\mathrm{DD}_{\text{ample}}$ side, the degeneration data we get from $f^*(\mathcal{L}_2)$ is the pullback of the degeneration data obtained from $\mathcal{L}_2$ by the map induced by f. Thus f defines a morphism $(G_1, \mathcal{L}_1) \to (G_2, \mathcal{L}_2)$ in $\mathrm{DEG}_{\text{ample}}$, which is induced from a morphism in $\mathrm{DD}_{\text{ample}}$, and the underlying morphism in DD is independent of the choice of the ample invertible sheaf. □

7.2. Corollary. *The functors*

$$\begin{aligned} M &: \mathrm{DD} \to \mathrm{DEG}, \\ M_{\mathrm{IS}} &: \mathrm{DD}_{\mathrm{IS}} \to \mathrm{DEG}_{\mathrm{IS}}, \\ M_{\text{ample}} &: \mathrm{DD}_{\text{ample}} \to \mathrm{DEG}_{\text{ample}}, \\ M_{\text{pol}} &: \mathrm{DD}_{\text{pol}} \to \mathrm{DEG}_{\text{pol}} \end{aligned}$$

are all equivalences of categories, whose quasi-inverses F, F_{IS}, F_{ample} and F_{pol} are given by the construction in chapter 2. These functors preserve the dimensions of G and T, and moreover, the degree of $\lambda_G : G \to G^t$ associated to an object of DD_{pol} is equal to $[\underline{X} : \phi(\underline{Y})]^2 \cdot \deg(\lambda_A)$. In particular, the period homomorphism $\imath : \underline{Y} \to \tilde{G}(K)$ obtained from $(G, \mathcal{L})$ in chapter 2 depends only on the semi-abelian scheme G, not on the choice of an ample line bundle $\mathcal{L}$.

We have also obtained information about torsion points. Suppose G is an object of DEG, with dual G^t, associated to $(\tilde{G}, \imath : \underline{Y} \to \tilde{G} \otimes K)$.

7.3. Corollary. *For any integer n the n-torsion $G_\eta[n]$ fits into an exact sequence (of finite group schemes over K) $0 \to \tilde{G}_\eta[n] \to G_\eta[n] \to \underline{Y}/n\underline{Y} \to 0$, which determines an element of $\mathrm{Ext}_{\mathbb{Z}/n\mathbb{Z}}(\underline{Y}/n\underline{Y}, \tilde{G}[n]) = \mathrm{Ext}(\underline{Y}, \tilde{G}[n])$. This element is the image of $\imath$ under the connecting homomorphism $\mathrm{Hom}(\underline{Y}, \tilde{G}) \to \mathrm{Ext}(\underline{Y}, \tilde{G}[n])$ associated to the exact sequence $0 \to \tilde{G}[n] \to \tilde{G} \to \tilde{G} \to 0$. Especially the extension above determines modulo n the pairing $\underline{B} : \underline{Y} \times \underline{X} \to \mathrm{Div}(S)$ which sends (y, μ) to the divisor of $I_{y,\mu}$.*

7.4. Corollary. *Consider the two extensions $0 \to \tilde{G}_\eta[n] \to G_\eta[n] \to \underline{Y}/n\underline{Y} \to 0$ and $0 \to \tilde{G}^t_\eta[n] \to G^t_\eta[n] \to \underline{X}/n\underline{X} \to 0$. Under the perfect duality $G[n] \times G^t[n] \to G_m$, $\tilde{G}_\eta[n]$ is perpendicular to $T^t_\eta[n] \subset \tilde{G}^t_\eta[n]$, and the induced pairings between $T^t_\eta[n]$ and $\underline{Y}/n\underline{Y}$ and between $\tilde{G}_\eta[n]/T_\eta[n] = A[n]$ and $\tilde{G}^t_\eta[n]/T^t_\eta[n] = A^t[n]$ are the canonical dualities.*

This is checked easily following the constructions.

7.5. Corollary. *Let $\mathfrak{p} \subset R$ denote a prime ideal, corresponding to a point $s \in S$. Assume for simplicity that $\underline{X}$ and $\underline{Y}$ are constant (the general case follows by descent). Then the character group of the torus part in the fibre of G^t_s over s is the quotient of Y by the subgroup Y_s of elements $y \in Y$ for which $I_{y,\phi(y)}$ is not contained in $\mathfrak{p}$. The character group of the torus part of the fibre of G over s is the quotient of s by the subgroup of elements for which some non-trivial multiple lies in $\phi(Y_s)$.*

This follows from the determination of torsion points in section 5, and the fact that the toric part of $G^t_s[n]$ is the perpendicular of $G_s[n]$.

8. Examples

Let us consider the special case where R is a complete discrete valuation ring and I its maximal ideal. Again consider $G \in$ DEG associated to $(\tilde{G}, \imath : \underline{Y} \to \tilde{G} \otimes K) \in$ DD, and assume that $\underline{Y}$ is constant (hence $\underline{X}$ is also constant). Under these hypotheses G is the "rigid analytic quotient of $\tilde{G}$ by Y", as we shall show next. First choose a relatively complete model $\tilde{P}$. We claim that $\tilde{P}(K) = \tilde{P}(R)$, that is, that every K-valued point p of $\tilde{P}$ extends to R. Indeed, $p : \mathrm{Spec}(K) \to \tilde{P}$ will induce a valuation which satisfies the assumptions of the completeness condition for the relatively complete model $\tilde{P}$, except that p does not dominate $\tilde{G}$. To remedy this, choose a valuation of $K(\tilde{P}) = K(\tilde{G})$ centered at p. Composing with an extension of the valuation of K we obtain a new valuation which satisfies the assumptions in the completeness condition for

$\tilde{P}$. It follows that the composed valuation has a center on $\tilde{P}$, and the assertion follows.

8.1. Proposition. *Under these assumptions* $G(K) = \tilde{G}(K)/\imath(Y)$.

Proof. $G(K) = P(K) = P(R) = P_{\text{for}}(R) = \tilde{P}_{\text{for}}(R)/Y = \tilde{P}(R)/Y = \tilde{P}(K)/Y = \tilde{G}(K)/\imath(Y)$.

If $\mathcal{G}$ denotes the Néron-model of G_η over R, G is the connected component of the identity of $\mathcal{G}$. As an application we can compute the group of connected components of the special fibre $\mathcal{G}_s$, that is, the quotient $\mathcal{G}_s/G_s$: Extend R such that its residue field k becomes separably closed. Then $\mathcal{G}(R)$ is dense in $\mathcal{G}$, and the group of connected components is $\mathcal{G}(R)/G(R) = G(K)/G(R) = \tilde{G}(K)/G(R) \cdot \imath(Y) = \tilde{G}(K)/\tilde{G}(R) \cdot \imath(Y)$. Now $\tilde{G}(K)/\tilde{G}(R) \approx T(K)/T(R) \approx \mathrm{Hom}(X, K^*/R^*) \approx X^* = \mathrm{Hom}(X, \mathbb{Z})$ (under the valuation $v : K^* \to \mathbb{Z}$), and the map $Y \to X^*$ induced by $\imath$ is given by the bilinear pairing $B : Y \times X \to \mathbb{Z}$ which sends a pair (y, μ) to the valuation of the ideal $I_{y,\mu}$. Hence:

8.2. Corollary. *$\mathcal{G}_s/G_s$ is isomorphic to the quotient of X^* by the image of Y under the mapping defined by the pairing above. Especially the order of this group is equal to the determinant of the pairing B.*

Another application is the study of the Jacobians of stable curves. Let $S = \mathrm{Spec}(R)$ denote a base as before (satisfying the running assumptions of this chapter, but not necessarily 1-dimensional), and consider a stable curve $C \to S$ in the sense of Deligne-Mumford. That is, C is proper and flat over S, the geometric fibres are reduced and connected and have only simple double points as singularities, and the fibres have no non-trivial vector fields. We also assume that the generic fibre C_η is smooth. Then it is known that the relative Picard functor $\mathbf{Pic}(X/S)$ is representable, and that $\mathbf{Pic}^0(X/S)$ is semi-abelian. We assume that $C_0 = C \times_S S_0$ is the union of smooth connected curves $\{C_v | v \in \mathcal{V}\}$ which intersect transversally along sections $e : S_0 \to C_0$ (self-intersection is allowed) indexed by $\mathcal{E}$. Then $\mathcal{V}$ and $\mathcal{E}$ form the vertices respectively edges of a connected graph $\mathcal{G}$, where any $e \in \mathcal{E}$ connects the two (not necessarily distinct) vertices $v \in \mathcal{V}$ which correspond to the two components C_v intersecting along e. We orient the edges arbitrarily and form $X = H_1(\mathcal{G}, \mathbb{Z}) = \mathrm{Kernel}(\mathbb{Z}^{\mathcal{E}} \to \mathbb{Z}^{\mathcal{V}})$. It is easily seen that $\mathbf{Pic}^0(C_0/S_0)$ is an extension of an abelian variety (the product of all $\mathbf{Pic}^0(C_v)$) by the split torus with character group X. It follows that $\mathbf{Pic}^0(C/S)$ defines an object in $\mathrm{DEG}_{\mathrm{pol}}$, of degree one (so $X = \phi(Y)$), which is given by data in $\mathrm{DD}_{\mathrm{pol}}$. It is difficult to describe these explicitly, but at least we can compute the symmetric pairing $B : X \times X \to \mathrm{Div}(S)$. This is done more or less by general principles. In the fully degenerate case more is known by Manin-Drinfel'd [M-D].

We first describe a natural candidate for the pairing, and then show that it is the right choice. For the natural candidate we observe that for each $e \in \mathcal{E}$ there exists a Cartier-divisor $D_e \subset S$ which describes the singularity of C along

e : D_e can be defined using Fitting ideals for $\Omega_{C/S}$. Another way is to observe that locally in the étale topology C looks like the spectrum of $R[u,v]/(u \cdot v - f_e)$, for some $f_e \in R$ which is a local equation for D_e. Now for each $e \in \mathcal{E}$ the projection on the e-component defines a linear form l_e on $\mathbb{Z}^{\mathcal{E}}$, and we consider the bilinear-form on $X \times X$ given by $(\mu, \nu) \to \sum l_e(\mu) \cdot l_e(\nu) \cdot D_e$. Note that either l_e vanishes on X (if removing e disconnects $\mathcal{G}$) or else l_e is surjective.

8.3. Theorem. *$\underline{B}$ is equal to the bilinear form described above.*

Proof. This behaves well under base change. Locally in the étale topology C is the pull-back of a versal object, so it suffices to consider such an object. Then the base is regular and the D_e have equations f_e which form part of a regular system of parameters. Especially they are pairwise distinct and irreducible. The fibre of C over the generic point of D_e has e as its only double point, and it follows that the character group of the toric part of $\mathbf{Pic}^0(C/S)$ at this point is the quotient of X by the kernel of l_e. The bilinear form in question takes linear combinations of the D_e as values, and the coefficient of D_e is a quadratic function on X modulo the kernel of l_e. It follows that the form is given by $\mu \mapsto \sum n_e \cdot l_e(\mu)^2 \cdot D_e$, with positive integral coefficients n_e. We have to show that $n_e = 1$, provided l_e does not vanish identically on X (otherwise the value of n_e is irrelevant.)

To do this we now change base to a discrete valuation ring R (so $\mathrm{Div}(S) = \mathbb{Z}$) such that all D_e get valuation one. Then the total space of C is regular. We have to compare the two positive definite quadratic forms $\sum l_e^2$ and $\sum n_e \cdot l_e^2$ on X, and it suffices to show that they have the same determinant (as one is $\geq$ the other). The determinant for $\sum n_e \cdot l_e^2$ is equal to the number of connected components of the Néron-model of $\mathrm{Pic}^0(C_\eta/K)$ (8.2 above), which (by [Ra 5]) is equal to the determinant of the intersection pairing on the space of divisors supported on the special fibre, modulo the full fibre. This space can be identified with the dual M^* of the subgroup $M := \mathrm{Im}(\mathbb{Z}^{\mathcal{E}} \to \mathbb{Z}^{\mathcal{V}}) \subset \mathbb{Z}^{\mathcal{V}}$, as $H_0(\mathcal{G}, \mathbb{Z}) = \mathbb{Z}$. Now $\mathbb{Z}^{\mathcal{E}}$ admits the positive definite unimodular quadratic form $\sum l_e^2$, and by adjointness we get an injection of M^* into $\mathbb{Z}^{\mathcal{E}}$ which identifies M^* with the perpendicular of X. One checks that the intersection form on M^* is equal to the restriction to M^* of the form on $\mathbb{Z}^{\mathcal{E}}$, and one knows from general considerations that the restrictions of this form to X and to M^* both have the same determinant. The last part can be simplified by reducing to the case of a single double-point, using the bilinear form $\underline{B}$ of section 10. □

9. Kodaira-Spencer Class

The aim of this section is to compute the Kodaira-Spencer class of an abelian scheme $G_V \to V$ obtained from Mumford's construction, where V is the largest open subscheme of S over which G is abelian. This will be accomplished by studying various Kodaira-Spencer maps, and relating them to vector extensions.

Let us first recall the basic definition of the Kodaira-Spencer class. A good reference is [Ka 1], esp. sec. 3. Let $X \to S \to \Sigma$ be morphism of schemes such that $f : X \to S$ is proper smooth and $g : S \to \Sigma$ is smooth. Then we have the exact sequence of sheaves of relative 1-forms

$$0 \to f^*\Omega^1_{S/\Sigma} \to \Omega^1_{X/\Sigma} \to \Omega^1_{X/S} \to 0 .$$

This exact sequence gives rise to the Kodaira-Spencer class: It corresponds to a global extension class

$$\rho \in \mathrm{Ext}^1_{\mathcal{O}_X}(\Omega^1_{X/S}, f^*\Omega^1_{S/\Sigma}) \xrightarrow{\sim} H^1(X, \mathbf{Der}(X/S) \otimes_{\mathcal{O}_X} f^*\Omega^1_{S/\Sigma}).$$

The local-to-global spectral sequence gives a homomorphism

$$\begin{aligned}
&\mathrm{Ext}^1_{\mathcal{O}_X}(\Omega^1_{X/S}, f^*\Omega^1_{S/\Sigma}) \xrightarrow{\sim} H^1(X, \mathbf{Der}(X/S) \otimes_{\mathcal{O}_X} f^*\Omega^1_{S/\Sigma}) \\
&\to H^0(S, R^1f_*(\mathbf{Der}(X/S) \otimes_{\mathcal{O}_X} f^*\Omega^1_{S/\Sigma})) \xleftarrow{\sim} H^0(S, R^1f_*\mathbf{Der}(X/S) \otimes_{\mathcal{O}_S} \Omega^1_{S/\Sigma}) \\
&= \mathrm{Hom}_{\mathcal{O}_S}(\mathbf{Der}(S/\Sigma), R^1f_*\mathbf{Der}(X/S)).
\end{aligned}$$

The image of ρ under the above homomorphism is the Kodaira-Spencer homomorphism, still denoted by ρ. The reader can find a Čech-cocycle representation of ρ in loc. cit., which is the form usually given for the Kodaira-Spencer class in the context of deformation theory.

When $X = A$ is an abelian scheme over S, the Kodaira-Spencer mapping $\rho \in \mathrm{Hom}_{\mathcal{O}_S}(\mathbf{Der}(S/\Sigma), R^1f_*\mathbf{Der}(A/S)) \cong \mathrm{Hom}_{\mathcal{O}_S}(\Omega_{A^t} \otimes \Omega_A, \Omega^1_{S/\Sigma})$, where $\Omega_A = e^*\Omega^1_{A/S}$(resp. $\Omega_{A^t} = e^*\Omega^1_{A^t/S}$) is the sheaf of translation invariant relative 1-forms on A (resp. A^t). Another way to define it is to consider the first de Rham cohomology $\mathcal{H} = \mathcal{H}(A) = \mathcal{H}^1_{\mathrm{DR}}(A/S) = R^1f_*(\Omega^{\cdot}_{A/S})$: This is a vector bundle on S with an integrable connection (the Gauss-Manin connection) $\nabla : \mathcal{H} \to \mathcal{H} \otimes \Omega^1_{S/\Sigma}$. Furthermore it has a subbundle $F^1(\mathcal{H}) \approx f_*(\Omega^1_{A/S}) \approx \Omega_A$ such that $\mathcal{H}/F^1(\mathcal{H}) \approx R^1f_*(\mathcal{O}_A) \approx \Omega^*_{A^t}$, and the map $F^1(\mathcal{H}) \to \mathcal{H} \otimes \Omega^1_{S/T} \to (\mathcal{H}/F^1(\mathcal{H})) \otimes \Omega^1_{S/\Sigma}$ induced by ∇ is another incarnation of ρ. We also note that for this construction we need not assume that $S \to \Sigma$ is smooth or even of finite type, so we might choose for Σ the final object $\mathrm{Spec}(\mathbb{Z})$ and obtain a Kodaira-Spencer map with values in Ω_S, from which all other such maps are derived by composing with $\Omega_S \to \Omega_{S/\Sigma}$. This will also work for the constructions which follow. Any line bundle $\mathcal{L}$ on A has a Chern class $c(\mathcal{L}) \in \mathcal{H}^2_{\mathrm{DR}}(A/S) = \Lambda^2\mathcal{H}$, which lies in F^1 of the induced filtration and is ∇-parallel (as it is the image of the first Chern class in $\mathcal{H}^2_{\mathrm{DR}}(A/\Sigma)$). Its image in $F^1/F^2 = R^1f^*(\Omega^1_{A/S})$ can be described as follows: If $\mathcal{L}$ is given by some 1-cocycle $f_{ij} \in \Gamma(U_i \cap U_j, \mathcal{O}^*_A)$, for some open covering $\{U_i\}$ of A, this image is represented by the 1-cocycle $d\log(f_{ij}) \in \Omega^1_{A/S}$. It follows that if $\phi = \phi(\mathcal{L}) : A \to A^t$ is the map defined by $\mathcal{L}$, (which sends $a \in A$ to $T^*_a(\mathcal{L}) \otimes \mathcal{L}^{-1}$), then this image is equal to the class in $\Omega_A \otimes R^1 f_*(\mathcal{O}_A)$ induced by the differential $d\phi : \Omega_A^* \to \Omega_{A^t}^* = R^1 f_*(\mathcal{O}_A)$: An infinitesimal change ∂a of the argument replaces $f_{ij}(a)$ by $f_{ij}(a+\partial a) = f_{ij} \cdot (1+\langle \partial a, d\log(f_{ij})\rangle)$; hence $d\lambda_{\mathcal{L}}$ maps ∂a to the cohomology class represented by the 1-cocycle $\langle \partial a, d\log(f_{ij})\rangle$, etc. We apply this to the Poincaré bundle $\mathcal{P}$ on $A \times A^t$: As it is trivial on the factors its class $c(\mathcal{P})$ lies in $\mathcal{H}(A) \otimes \mathcal{H}(A^t) \subset \mathcal{H}^2_{\mathrm{DR}}(A \times A^t)$, and it induces a perfect duality between $\mathcal{H}(A)$ and $\mathcal{H}(A^t)$ which is parallel, and which induces the canonical pairings between $F^1(\mathcal{H}(A)) = \Omega_A$ and $(\mathcal{H}(A^t)/F^1) = \Omega_A^*$, and between $\mathcal{H}(A)/F^1 = \Omega_{A^t}^*$ and $F^1(\mathcal{H}(A^t)) = \Omega_{A^t}$. We derive that the Kodaira-Spencer classes for $\rho_A : \Omega_A \otimes \Omega_{A^t} \to \Omega_S$ and $\rho_{A^t} : \Omega_{A^t} \otimes \Omega_A \to \Omega_S$ differ only by the flipping isomorphism $\Omega_A \otimes \Omega_{A^t} \approx \Omega_{A^t} \otimes \Omega_A$: We obtain one minus-sign from passing to the dual, and the other from the fact that the two multiplication maps from $\mathcal{H}(A) \otimes \mathcal{H}(A^t)$ and $\mathcal{H}(A^t) \otimes \mathcal{H}(A)$ to $\mathcal{H}^2_{\mathrm{DR}}(A \times A^t)$ differ by a sign. That is we have shown:

9.1. Proposition. *Let $A \to S$ be an abelian scheme, $A^t \to S$ be the dual abelian scheme of A, $\rho : \Omega_{A^t} \otimes \Omega_A \to \Omega_S^1$ (resp. $\rho^t : \Omega_A \otimes \Omega_{A^t} \to \Omega_S^1$) be the Kodaira-Spencer mapping for $A \to S$ (resp. A^t). Then ρ^t is equal to ρ precomposed with the "flipping isomorphism" $\Omega_A \otimes \Omega_{A^t} \xrightarrow{\sim} \Omega_{A^t} \otimes \Omega_A$.*

We derive the following statement about the Kodaira-Spencer mapping of a polarized abelian variety:

9.2. Proposition. *Let $A \to S$ be an abelian scheme, ρ the Kodaira-Spencer mapping, $\mathcal{L}$ a rigidified invertible $\mathcal{O}_A$-module. Let $\lambda = \lambda(\mathcal{L}) : A \to A^t$ denote the S-homomorphism associated to $\mathcal{L}$ which sends a to $T_a^*\mathcal{L} \otimes \mathcal{L}^{\otimes -1}$ for any functorial point $a \in A$. Then the composition*

$$\rho(A, \mathcal{L}) = \rho \circ (\mathrm{id} \otimes \lambda^*) : \Omega_{A^t} \otimes \Omega_{A^t} \to \Omega_{A^t} \otimes \Omega_A \to \Omega_S^1$$

is a symmetric bilinear pairing, where $\lambda^* : \Omega_{A^t} \to \Omega_A$ is the map induced by λ on invariant relative 1-forms.

This follows as λ is symmetric.

We need some preliminary discussions about universal vector extensions. Let S be a base scheme. By a vector-group over S we mean a group scheme $W \to S$ which is locally isomorphic to $(G_a)^r$ for some $r \geq 0$. In other words, the local sections of W form a locally free coherent sheaf over S. Suppose G/S is a group scheme. A vector extension of G is an extension of group schemes $0 \to W \to E \to G \to 0$. It is known that sometimes there exists a universal vector extension $0 \to L_G \to E_G \to G \to 0$, so that any vector extension $0 \to W \to E \to G \to 0$ is the push-out of the universal one by a unique homomorphism $L_G \to W$. An important example is the case $G \to S$ is an abelian scheme, in which case L_G corresponds to $\Omega_{G^t} = \mathbf{Lie}(G^t/S)^* =$ sheaf of invariant relative 1-forms on the dual of G/S. We refer the reader to [Me], [Ma–Me] for general discussion of universal extensions.

When $G \to S$ is an abelian scheme, the universal extension E_G of G is related to the Poincaré sheaf $\mathcal{P}_G$ over $G \times_S G^t$ ($G^t =$ dual of G), as follows: Restricting

$\mathcal{P}_G$ to the first-order infinitesimal neighborhood of $G \times \{0\} \subset G \times_S G^t$, one gets an invertible sheaf over $G \times_S \operatorname{Spec}(\mathcal{O}_S \oplus \varepsilon\Omega_{G^t})$, which is a first order deformation of the trivial invertible sheaf on G. This first order deformation corresponds to a class in $H^1(G, \Omega_{G^t}) = \operatorname{Ext}(G, L_G)$, which is (the class of) the universal vector extension. Here L_G denote the vector group corresponding to Ω_{G^t}. The equality $H^1(G, \Omega_{G^t}) = \operatorname{Ext}(G, L_G)$ above uses the important fact that any rigidified G_a-torsor over an abelian scheme has a canonical group structure.

Suppose now that $\tilde{G}$ is an extension of an abelian scheme $A \to S$ by a torus $T \to S$. (For our purpose, we only need the case $T \to S$ split.) We have an exact sequence $0 \to T \to \tilde{G} \to A \to 0$. Then the universal vector extension $E_{\tilde{G}}$ of $\tilde{G}$ exists and is (canonically isomorphic to) the pull back of the universal vector extension E_A of A via $\tilde{G} \to A$. This follows (long exact sequence) from

(a) $\mathbf{Hom}(T, G_a) = 0$

(b) $\mathbf{Ext}(T, G_a) = 0$.

For our purpose, we need to consider a variant of the above situation. Let Y be a free abelian group of finite rank, $\tilde{G}$ an extension of an abelian scheme A by a split torus T over S as above, and $\imath : Y \to \tilde{G}(S)$ a homomorphism. We would like to consider vector extensions $0 \to W \to E \to \tilde{G} \to 0$ plus liftings $\imath^{\#} : Y \to E(S)$ of $\imath$. (This is motivated by considering $\imath : Y_{/S} \to \tilde{G}$ as a complex of fppf sheaves in degrees -1 and 0.) For convenience, we call $(E, \imath^{\#})$ a vector extension of $(\tilde{G}, \imath)$. If $\tilde{G}$ has a universal vector extension $E_{\tilde{G}}$ then there also exists a universal extension $(E_{(\tilde{G},\imath)}, \imath^{\#})$ of $(\tilde{G}, \imath)$. $E_{\tilde{G}}$ injects into $E_{(\tilde{G},\imath)}$, and the quotient is isomorphic to the vector group given by $Y \otimes \mathcal{O}_S$, via the map induced by $\imath^{\#}$.

Let us give another description the universal vector extension $(E_{(\tilde{G},\imath)}, \imath^{\#})$ of $(\tilde{G}, \imath)$. We need some notation, following that used in chapter III. Let X be the character group of T, $\mathcal{P}_A$ the Poincaré sheaf over $A \times_S A^t$, $c^t : Y \to A(S)$ the composition of $\imath : Y \to \tilde{G}(S)$ with the projection $\pi(S) : \tilde{G}(S) \to A(S)$, $c : X \to A^t(S)$ be the homomorphism attached to the extension $0 \to T \to \tilde{G} \to A \to 0$, $0 \to T^t \to \tilde{G}^t \to A^t \to 0$ be the extension attached to c^t, $\tau : 1_{Y \times X} \xrightarrow{\sim} (c^t \times c)^* \mathcal{P}_A^{-1}$ the trivialization corresponding to $\imath$, and $\tilde{\mathcal{P}}$ be the pull-back of $\mathcal{P}_A$ to $\tilde{G} \times_S \tilde{G}^t$. Let $L_{\tilde{G}} = L_A$ be the vector group over S associated to Ω_{A^t}, $\Omega_{\tilde{G}^t} = \mathbf{Lie}(\tilde{G}^t/S)^* =$ the sheaf of invariant relative 1-forms on $\tilde{G}^t$, $L_{(\tilde{G},\imath)} =$ the vector group attached to $\Omega_{\tilde{G}^t}$.

Now $E_{(\tilde{G},\imath)}$ is easy to describe: It will be the push-forward of $0 \to L_{\tilde{G}} \to E_{\tilde{G}} \to \tilde{G} \to 0$ via the canonical monomorphism $L_{\tilde{G}} \to L_{(\tilde{G},\imath)}$. To define the lifting $\imath^{\#}$, it is most convenient to use the Poincaré sheaf. If we restrict $\tilde{\mathcal{P}}$ to the first order infinitesimal neighborhood of $\tilde{G} \times \{0\} \subset \tilde{G} \times_S \tilde{G}^t$, we get an invertible sheaf over $\tilde{G} \times_S \operatorname{Spec}(\mathcal{O}_S \oplus \varepsilon\Omega_{\tilde{G}^t})$, which is a first order deformation of the trivial invertible sheaf on $\tilde{G}$. This first order deformation corresponds to a class in $= \operatorname{Ext}(\tilde{G}, L_{(\tilde{G},\imath)}) \subset H^1(G, \Omega_{\tilde{G}^t})$, which gives rise to an extension of $\tilde{G}$ by $L_{(\tilde{G},\imath)}$, which can be easily seen to be the push-forward $0 \to L_{(\tilde{G},\imath)} \to E_{(\tilde{G},\imath)} \to \tilde{G} \to 0$ of $0 \to L_{\tilde{G}} \to E_{\tilde{G}} \to \tilde{G} \to 0$ considered before. Now $Y \times X$ acts on $\tilde{\mathcal{P}}$ via the trivialization τ. Restricting to $Y \times \{0\} \subset Y \times X$, we obtain an action of Y on $\tilde{P}$, which induces an action of Y on the restriction of $\tilde{P}$ to the first order infinitesimal neighborhood of $\tilde{G} \times \{0\} \subset \tilde{G} \times_S \tilde{G}^t$, hence an action of Y on $E_{(\tilde{G},\imath)}$. Clearly such an action is just the translation action via a homomorphism $\imath^\# : Y \to E_{(\tilde{G},\imath)}(S)$, because this action commutes with translations. It is easy to see that $\imath^\#$ is a lifting of $\imath$, and that the map of Y into the quotient $E_{(\tilde{G},\imath)}/E_{\tilde{G}} = L_{(\tilde{G},\imath)}/L_{\tilde{G}} = \Omega_{T^t} = Y \otimes \mathcal{O}_S$ is the canonical one (this follows from the commutation rule for the actions of $Y \times 0$ and $0 \times T^t$ on $\tilde{\mathcal{P}}$). It follows that $(E_{(\tilde{G},\imath)}, \imath^\#)$ is indeed the universal vector extension of $(\tilde{G}, \imath)$, and in addition that $L_{(\tilde{G},\imath)} = \Omega_{\tilde{G}^t}$.

The Kodaira-Spencer mapping for abelian schemes can be defined in terms of universal vector extensions, as follows. Assume for simplicity that $S = \mathrm{Spec}(R)$ is affine. Let $f : S \to \Sigma$ be smooth, $A \to S$ be an abelian scheme, $0 \to L_A \to E_A \to A \to 0$ the universal vector extension of A. We have an exact sequence $0 \to f^*\Omega^1_{S/\Sigma} \to \Omega^1_{A/\Sigma} \to \Omega^1_{A/S} = f^*\Omega_{A/S} \to 0$. For any invariant 1-form $\alpha \in \Gamma(A, \Omega_{A/S})$, we can pull back the above extension by the homomorphism $\mathcal{O}_A \to \Omega^1_{A/S}$ corresponding to α to obtain an extension $0 \to f^*\Omega^1_{S/\Sigma} \to \alpha^*(\Omega^1_{A/\Sigma}) \to \mathcal{O}_A \to 0$. The extension $\alpha^*(\Omega^1_{A/\Sigma})$ corresponds to a class in $H^1(A, f^*\Omega^1_{S/\Sigma}) = \mathrm{Ext}(A, V(\Omega^1_{S/\Sigma}))$, where $V(\Omega^1_{S/\Sigma})$ is the vector group over S attached to $\Omega^1_{S/\Sigma}$. Denote it by $0 \to V(\Omega^1_{S/\Sigma}) \to E_\alpha \to A \to 0$. (Equivalently: Pull back the exact sequence $0 \to V(f^*\Omega^1_{S/\Sigma}) \to V(\Omega^1_{A/\Sigma}) \to V(\Omega^1_{A/S}) \to 0$ of vector bundles on A by the section α to get a rigidified $V(f^*\Omega^1_{S/\Sigma})$-torsor E_α, which has a canonical group structure.) E_α is the push-forward of E_A via a (unique) homomorphism $L_A \to V(\Omega^1_{S/\Sigma})$, by the universality of E_A. The above considerations define a homomorphism $\Gamma(A, \Omega_A) \to \mathrm{Hom}_{S\text{-}gp}(L_A \to V(\Omega^1_{S/\Sigma}) = \mathrm{Hom}(\Omega_{A^t}, \Omega^1_{S/\Sigma})$, where A^t is the dual of A. Transposing and sheafifying the above homomorphism, we obtain a homomorphism $\Omega_{A^t} \otimes \Omega_A \to \Omega^1_{S/\Sigma}$, which can be nothing else but the Kodaira-Spencer mapping of $A \to S \to \Sigma$. The proof is a tautology, which we leave to the reader.

The above construction can be given a somewhat more geometric interpretation, which will be convenient for our purpose. Consider the vector extension $0 \to V(\Omega^1_{S/\Sigma}) \to E_\alpha \to A \to 0$ of A. For any homomorphism $h : H \to A$ of smooth S-group schemes, we have an invariant 1-form $h^*(\alpha) \in \Gamma(H, \Omega^1_{H/S})$. One can ask whether $h^*(\alpha)$ can be lifted to a section $\beta \in \Gamma(H, \Omega^1_{H/\Sigma})$ which is invariant in the sense that $m^*(\beta) = \mathrm{pr}_1^*(\beta) + \mathrm{pr}_2^*(\beta)$, where m, pr_1, pr_2 : $H \times_S H \to H$ are the group law and the two projections respectively. The answer is easy: to give an invariant lifting $\beta \in \Gamma(H, \Omega^1_{H/\Sigma})$ is the same as giving an S-homomorphism $H \to E_\alpha$ lifting h. By the universal property of E_A, if we take h to be the canonical homomorphism $E_A \to A$, it can be uniquely lifted to a homomorphism $E_A \to E_\alpha$, which defines the Kodaira-Spencer map as explained before. In particular, any invariant 1-form $\alpha \in \Gamma(A, \Omega^1_{A/S}) = \Gamma(S, \Omega_{A/S})$ can be uniquely lifted to an invariant section $\beta \in \Gamma(E_A, \Omega^1_{E_A/\Sigma})$. The Kodaira-

Spencer map is determined by β in the following explicit way: The restriction of β to L_A is an invariant 1-form on L_A/Σ, whose image in $\Gamma(L_A, \Omega^1_{L_A/S})$ is 0. But the space of all invariant 1-forms on L_A/Σ with image 0 in $\Gamma(L_A, \Omega^1_{L_A/S})$ is naturally isomorphic to $L_A^* \otimes_R \Omega^1_{R/\Sigma}$, and thus the restriction of β to L_A defines the homomorphism $L_A \to V(\Omega^1_{S/\Sigma})$ in question.

The same construction works for semi-abelian extensions $0 \to T \to \tilde{G} \to A \to 0$ and defines a Kodaira-Spencer map $\rho_{\tilde{G}} : \Omega_{\tilde{G}} \otimes_R \Omega_{A^t} \to \Omega_{S/\Sigma}$. Its restriction to $\Omega_A \otimes_R \Omega_{A^t}$ is ρ_A. More interesting is the construction for the pair $(\tilde{G}, \imath)$: If $\alpha \in \Omega_{\tilde{G}}$ is an invariant differential form on $\tilde{G}$, consider the vector extension $0 \to V(\Omega^1_{S/\Sigma}) \to E_\alpha \to \tilde{G} \to 0$ as before, and the lift $\tilde{\alpha}$ to an invariant section of $\Omega^1_{E_\alpha/\Sigma}$. There exists a unique lift of $\imath$ to $\imath' : Y \to E_\alpha$ such that $\imath'^*(\tilde{\alpha})$ is trivial: Lift somehow and then correct by a suitable $Y \to \Omega^1_{S/\Sigma}$. Hence we get a map from $E_{(\tilde{G},\imath)}$ to E_α, and a Kodaira-Spencer homomorphism $\Omega_{\tilde{G}} \otimes \Omega_{\tilde{G}^t} \to \Omega_{S/\Sigma}$.

As an example consider the case where $\tilde{G} = T$ is a split torus with character group X, so $\imath$ is given by a bilinear form $b(y, \mu)$ on $Y \times X$, with values in R^*. Also $\Omega_T = X \otimes R$ and $\Omega_{T^t} = Y \otimes R$. In this case all vector extensions split canonically. For example $E_{(T,\imath)} = (Y \otimes R) \times T$, and the map from Y is $(\mathrm{id}, \imath)$. Any $\mu \in X$ defines an invariant form $d\log(\mu)$ on T, and one checks easily that the map $\imath' : Y \to V(\Omega^1_{S/\Sigma}) \times T$ sends $y \in Y$ to $-d\log(b(y,\mu)) \times \imath(y)$. It follows that $\rho_{(T,\imath)}$ is given by the bilinear form $-d\log(b(y,\mu))$ on $Y \times X$.

Later we shall need the following variant: Suppose that $\imath = \imath_1 + \imath_2$, where $\imath_1$ maps Y into $\tilde{G}(R)$ and $\imath_2$ into $T(R)$. Consider $(\imath_1, \imath_2) : Y \to \tilde{G} \times T$, which maps (via projections) to $(\tilde{G}, \imath_1)$ and $(T, \imath_2)$, and (via addition) to $(\tilde{G}, \imath)$. Via the projections $L_{(\tilde{G}\times T,(\imath_1,\imath_2))}$ becomes a subspace of $L_{(\tilde{G},\imath_1)} \oplus L_{(T,\imath_2)} = \Omega_{\tilde{G}^t} \oplus \Omega_{T^t}$, and it must consist of all pairs $(\omega^t, \omega^t|T^t)$, for invariant forms $\omega^t \in \Omega_{\tilde{G}^t}$. Furthermore the map into $L_{(\tilde{G},\imath)}$ induced by addition is the projection onto the first factor. By functoriality we derive:

9.3. Lemma. *In this situation, for any pair*

$$(\alpha, \alpha^t) \in \Omega_{\tilde{G}} \times \Omega_{\tilde{G}^t}, \quad \rho_{(\tilde{G},\imath)}(\alpha \otimes \alpha^t) = \rho_{(\tilde{G},\imath_1)}(\alpha \otimes \alpha^t) + \rho_{(T,\imath_2)}(\alpha|T \otimes \alpha^t|T^t).$$

We shall use this in the situation encountered in chapter II, where after étale localization $\imath_1$ and $\tilde{G}$ are defined over a domain R, but $\imath_2$ is only given over the quotient-field K of R. So a priori $\rho_{(\tilde{G},\imath)}$ takes values in Ω^1_K, but by our calculations these actually lie in the R-submodule spanned by Ω^1_R and the $d\log(b(y,\mu))$. That is, the values are differentials with logarithmic poles.

Again it is not necessary to assume that $S \to \Sigma$ is smooth or even of finite type. The reader might worry that for general Σ we cannot use vector extensions by $V(\Omega^1_{S/\Sigma})$, but this can be circumvented by formulating everything in terms of extensions of quasicoherent sheaves (Y-equivariant if we use $i : Y \to \tilde{G}$) $0 \to f^*\Omega^1_{S/\Sigma} \to ? \to \mathcal{O}_{\tilde{G}} \to 0$: We did not choose this approach right away because the geometric way seems to be more transparent. So we shall concentrate on the most informative case, where $\Sigma = \mathrm{Spec}(\mathbb{Z})$. Another variant is to choose once and for all a map $\sigma : \Omega_R \to M$, with M a finitely generated free R-module, and to consider Kodaira-Spencer maps with values in M.

Now we can formulate the main result of this section, namely the computation of the Kodaira-Spencer map for a semi-abelian degeneration $G \to S$ which is given by degeneration data $(\tilde{G}, \imath : \underline{Y} \to \tilde{G}_\eta)$. We assume for simplicity that $\underline{Y}$ is constant, with value Y, and that $S = \mathrm{Spec}(R)$ is affine, with R normal and I-adically complete as always. The Kodaira-Spencer map ρ_{G_η} for the generic fibre G_η is a homomorphism from $\Omega_G \otimes \Omega_{G^t} = \Omega_{\tilde{G}} \otimes \Omega_{\tilde{G}^t}$ to $\Omega_K = \Omega_R \otimes K$. To compute it we note that $\rho(\tilde{G}_\eta, \imath)$ is another such map, and so it takes very little imagination to conjecture that these two maps are related. We can show:

9.4. Theorem. *For any homomorphism of σ of Ω^1_R into a finitely generated R-module M the two compositions of ρ_{G_η} and $\rho_{(\tilde{G}_\eta,\imath)}$ with $\Omega^1_R \otimes K \to M \otimes K$ coincide.*

Proof. Choose a relatively complete model $\tilde{P}$ for $(G, \imath)$, with projection $\pi : \tilde{P} \to A$. We may assume that M is free, by dividing out torsion and injecting in such a module. It follows that the kernel of the map $\Omega_R \otimes_R \mathcal{O}_P \to \Omega_P$ (which is torsion) is contained in the kernel of $\Omega_R \otimes_R \mathcal{O}_P \to M \otimes_R \mathcal{O}_P$, so we obtain by push-out an extension $0 \to M \otimes_R \mathcal{O}_P \to ? \to \Omega_{P/R} \to 0$ on P. So any global section α of Ω_P defines an extension $0 \to M \otimes_R \mathcal{O}_P \to \mathcal{E}_\alpha \to \mathcal{O}_P \to 0$. By the same methods any Y-invariant section of $\Omega_{\tilde{P}}$ defines a Y-equivariant extension $0 \to M \otimes_R \mathcal{O}_{\tilde{P}} \to \tilde{\mathcal{E}}_\alpha \to \mathcal{O}_{\tilde{P}} \to 0$.

The vector extension $E_{(A,c)}$ (following the previous notation) defines over A a Y-equivariant extension of $\mathcal{O}_A$ by $\Omega_{\tilde{G}^t} \otimes \mathcal{O}_A$, which via π pulls back to a Y-equivariant extension (over $\tilde{P}$) $0 \to \Omega_{\tilde{G}^t} \otimes \mathcal{O}_{\tilde{P}} \to \tilde{\mathcal{E}} \to \mathcal{O}_{\tilde{P}} \to 0$. As usual by passing to P_{for} and forming the quotient we obtain an extension $\mathcal{E}$ over P whose restriction to G is independent of all choices.

We claim that over the generic fibre G_η this extension is equal to the one defined by E_{G_η}: To see this note that E_{G_η} has been constructed from the Poincaré sheaf $\mathcal{P}_G$ over $G \times_S G^t$, via the infinitesimal variation of rigidified line bundles on G obtained by restriction to the first infinitesimal neighborhood $\tilde{G}^{t(1)}$ of the identity in $\tilde{G}^t$. Now $\mathcal{P}_G$ is given by the quotient construction applied to the pullback of $\tilde{\mathcal{P}}$ of $\mathcal{P}_A$ via $\tilde{G} \times_S \tilde{G}^t \to A \times_S A^t$, with its $Y \times X$-action. We can choose a relatively complete model $\tilde{P}_1$ for $\tilde{G} \times_S \tilde{G}^t$ such that $\tilde{\mathcal{P}}$ extends to a $Y \times X$-equivariant line bundle $\tilde{\mathcal{L}}$ on $\tilde{P}_1$, whose quotient on P_1 defines $\mathcal{P}_G$. The Zariski closure of $\tilde{G}$ in $\tilde{P}_1$ defines a relatively complete model $\tilde{P} \subset \tilde{P}_1$ for $\tilde{G}$, and the action of $\tilde{G}^t$ gives a Y-equivariant map $\tilde{P} \times_S \tilde{G}^t \to \tilde{P}_1$. Now form the Y-quotient of the pullback $\tilde{\mathcal{L}}^{(1)}$ of $\tilde{\mathcal{L}}$ under the induced map $\tilde{P} \times_S \tilde{G}^{t(1)} \to \tilde{P}_1$, which gives a rigidified line bundle on $P \times_S \tilde{G}^{t(1)} = P \times_S G^{t(1)}$ whose restriction to $G \times G^{t(1)}$ is equal to that of $\mathcal{P}_G$ by construction.

On the other hand, over $\tilde{G}^* =$ union of all Y-translates of G, $\tilde{\mathcal{L}}^{(1)}$ defines the same Y-equivariant infinitesimal deformation of rigidified line bundles as the π-pull-back of $E_{(A,c)}$. This isomorphism and its inverse have poles of bounded order along the rest of $\tilde{P}$, as $\tilde{P}$ is of finite type modulo the Y-action. If we pass to formal schemes and divide by Y the assertion follows.

Finally choose an invariant form $\alpha \in \Omega_G = \Omega_{\tilde{G}}$. We try to show that $\sigma \cdot \rho_{G_\eta}(\alpha \otimes \alpha^t) = \sigma \cdot \rho_{(\tilde{G}_\eta, \imath)}(\alpha \otimes \alpha^t)$ for any $\alpha^t \in \Omega_{G^t}$, which follows if the extension $\mathcal{E}_{\alpha,\eta}$ of $\mathcal{O}_{G_\eta}$ by $M \otimes_R \mathcal{O}_{G_\eta}$ defined by α is isomorphic to the push-out of $\mathcal{E}_\eta$ via $\sigma \cdot \rho_{(\tilde{G}_\eta, \imath)}(\alpha^*(.))$. Multiplying α by a non-zero element of R we may assume that it extends to a global section of $\Omega^1_{P/R}$ over P (or of $\Omega^1_{\tilde{P}/R}$ over $\tilde{P}$, if we consider α as an invariant differential form on $\tilde{G}$), and that $\sigma \cdot \rho_{(\tilde{G}_\eta, \imath)}(\alpha^*(.))$ maps Ω_{G^t} into M. So on $\tilde{P}$ α defines a Y-equivariant extension $\tilde{\mathcal{E}}_\alpha$ of $\mathcal{O}_{\tilde{G}}$ by $M \otimes \mathcal{O}_{\tilde{P}}$ whose restriction to $\tilde{P}_\eta$ is obtained by push-out from $\tilde{\mathcal{E}}_\eta$ via the map $\Omega_{\tilde{G}^t} \to M$ defined by the Kodaira-Spencer map $\sigma \cdot \rho_{(\tilde{G}_\eta, \imath)}(\alpha \otimes (.))$. Multiplying again by a non-trivial element of R we may assume that these two Y-equivariant extensions are isomorphic over $\tilde{P}$, as $\tilde{P}$ is of finite type modulo the Y-action. Passing to formal schemes and dividing by Y we obtain the assertion, as $\tilde{\mathcal{E}}_\alpha$ descends to $\mathcal{E}_\alpha$ and $\tilde{\mathcal{E}}$ to $\mathcal{E}$. □

The following assertions follow for Kodaira-Spencer maps after composing with the map M:

9.5. Corollary. *The restriction of $\sigma \circ \rho_G$ to $\Omega_{\tilde{G}} \otimes_R \Omega_{A^t}$ is the Kodaira-Spencer map of $\tilde{G}$.*

9.6. Corollary. *Assume that as before $\imath = \imath_1 \cdot \imath_2$, where $\imath_1$ takes values in $\tilde{G}(R)$, and $\imath_2$ takes values in $T(K)$ and is associated to $b : Y \times X \to K^*$. Then ρ_G is the sum of $\sigma \circ \rho_{(\tilde{G}, \imath_1)}$ (which takes its values in Ω_R) and the composition of the restriction map $\Omega_{\tilde{G}} \otimes_R \Omega_{\tilde{G}^t} \to \Omega_T \otimes_R \Omega_{T^t}$ with the bilinear form given by $\sigma(-d\log(b(y, \mu)))$.*

9.7. Corollary. *The logarithmic derivative $d\log : K^*/R^* \to \Omega^1_K/\Omega^1_R$ induces (by gluing in the Zariski-topology) a map $d\log : \mathrm{Div}(S) \to \Omega^1_K/\Omega^1_R$. Composing with the bilinear form $B : Y \times X \to \mathrm{Div}(S)$ associated to G we obtain maps $\Omega_{G^t} \otimes_R \Omega_G \to \Omega_{T^t} \otimes_R \Omega_T \to \Omega^1_K/\Omega^1_R$. Their composition coincides with $-\rho_G$. In particular, the values of ρ_G are differentials with logarithmic poles.*

9.8. Corollary. *Let S be a smooth scheme over an excellent normal scheme Σ, $D \subset S$ a relative divisor with normal crossings, $G \to S$ a semi-abelian scheme over $U = S - D$. Then the Kodaira-Spencer map $\rho_U : \Omega_{G_U} \otimes \Omega_{G^t_U} \to \Omega^1_{U/\Sigma}$ of G_U extends to a morphism $\rho : \Omega_{G/S} \otimes \Omega_{G^t/S} \to \Omega^1_{S/\Sigma}[d\log D]$.*

10. Globalization and 2-Step Constructions

Suppose $G \to S$ is a semi-abelian scheme, with the base S normal and irreducible. We assume that the generic fibre G_η is abelian. It follows that there exists a semi-abelian $G^t \to S$ extending the dual G^t_η of G_η. Associated to G and G^t are constructible étale sheaves $\underline{X}(G)$ and $\underline{X}(G^t)$ over S (chapter I, section 2). We intend to define a bilinear map of étale sheaves $\underline{B} : \underline{X}(G) \times \underline{X}(G^t) \to \mathrm{Div}(S)$ (= Cartier-divisors) which generalizes the B defined before, for degenerations over I-adically complete rings. Also $\underline{B}$ should commute with base change by dominant morphisms $S' \to S$. It then follows by an inductive limit argument that we may assume that S is of finite type over $\mathrm{Spec}(\mathbb{Z})$, so that S is noetherian and excellent.

Assume first that $S = \mathrm{Spec}(R)$ with R a complete local ring, such that the toral part of the fibre of G over the closed point is split. Our constructions associate to G free abelian groups X and Y, an injection $\phi : X \to Y$, extensions $0 \to T \to \tilde{G} \to A \to 0$ and $0 \to T^t \to G^t \to A^t \to 0$, and a map $\imath : Y \to \tilde{G}(K)$, which up to units is given by a bilinear form $b : Y \times X \to K^*$ (K = quotient-field of R). Now $\underline{X}(G)$ is a quotient of X_S and $\underline{X}(G^t)$ a quotient of Y_S, so for any $s \in S$ the groups $X(s)$ and $Y(s)$ are quotients of X and Y respectively. To determine them we denote by $Y_s \subset Y$ the sublattice of elements y such that $b(y, \phi(y))$ is invertible in s, and by X_s the sublattice of X generated by $\phi(Y_s)$. It follows that $X(s) = X/X(s)$ and $Y(s) = Y/Y(s)$: It suffices to treat $X(s)$, and there the assertion follows from the the determination of torsion points in G_s and G^t_s, see corollary 7.5. Furthermore if R_1 denotes the completion $\mathcal{O}_{S,s}$ the induced degeneration over R_1 determines a form $b_1 : Y(s) \times X(s) \to K_1^*$, well-determined up to units. On the other hand $b(Y \times X_s)$ and $b(Y_s \times X)$ are both contained in R_1^* (this follows from definiteness), and so b also induces a bilinear form $Y(s) \times X(s) \to K_1^*/R_1^*$. These two forms coincide: Choose a number n which is invertible in R. Over K, $G_\eta[n]$ is an extension of $\tilde{G}_\eta[n]$ by Y/nY. This extension is the sum of the (push-out of the) extension of Y/nY by $T_\eta[n]$ which is determined by b, and an extension which extends to a finite étale group scheme over R. This fact determines the divisor of b modulo n. After base extension to R_1 it follows that the two forms in question coincide modulo n, for each n invertible in R, hence that they are equal. All this discussion implies that in our special situation the form $b : Y \times X \to K^*$ induces a bilinear form $B : \underline{X}(G^t) \times \underline{X}(G) \to \mathbf{Div}(S)$. By descent this also works if the torus T is not split.

In the general case we have to define for each étale $S' \to S$ a bilinear form $\underline{B}$ from $\underline{X}(G^t)(S') \times \underline{X}(G)(S')$ into $\mathrm{Div}(S')$. We first give a map $\underline{\tilde{B}}$ into Weil-divisors, that is into linear combinations of cycles of codimension one. To do this we have to exhibit for a pair of global sections $x \in \underline{X}(G)(S')$, $y \in \underline{Y}(G^t)(S')$, and for each $s' \in S'$ of height one, the coefficient of s' in $\underline{\tilde{B}}(y, x)$. This is determined by completing S' in s' and using the previous construction (for a discrete valuation ring). Note that there are only finitely many s' for which $X(s')$ is not trivial.

Now if we denote by R the completion of $\mathcal{O}_{S',s'}$, for an arbitrary $s' \in S'$ (not necessarily of dimension one), the pullback of the Weil-divisor $\underline{\tilde{B}}(y,x)$ to $\mathrm{Spec}(R)$ is the Cartier-divisor $\underline{B}(y,x)$ defined before. By descent it follows that $\underline{\tilde{B}}(y,x)$ is a Cartier-divisor, and that $\underline{\tilde{B}}$ has all the required properties. Hence:

10.1. Theorem. *Suppose S is normal and irreducible, $G \to S$ a semi-abelian scheme. There exists a bilinear map of étale sheaves $\underline{B} = \underline{B}(G) : \underline{X}(G^t) \times \underline{X}(G) \to \mathbf{Div}(S)$ such that*

i) Formation of $\underline{B}$ commutes with dominant base change $S' \to S$

ii) If G is given by the Mumford construction over an I-adically complete ring R, $\underline{B}$ is induced by the bilinear form b.

iii) If $\phi : \underline{X}(G^t) \to \underline{X}(G)$ is induced by a $G \to G^t$ which is a polarization on the generic fibre G_η, $\underline{B}(\cdot,\phi(\cdot))$ is symmetric and definite; that is, for local sections y, y_1 and y_2 of $\underline{X}(G^t)$ we have $\underline{B}(y_1,\phi(y_2)) = \underline{B}(y_2,\phi(y_1))$, and $\underline{B}(y,\phi(y))$ is effective and has the same support as y.

iv) If G is an abelian variety in codimension one then $\underline{B} \equiv 0$, and G is an abelian variety.

This result is essentially all we need in what follows. However, for the sake of completeness, we include some more precise considerations, trying to explain how degenerations can be carried out in two steps. Let us consider the following situation: Suppose R is a complete normal local (hence excellent) domain, $G \to S = \mathrm{Spec}\, R$ a (polarized) semi-abelian scheme, whose closed fibre is an extension of a semi-abelian variety A_s by a split torus T_s over the closed point $s \in S$. Let η be the generic point of S, t be a point in S (so that s lies in the closure S_1 of $\{t\}$, and we assume, for simplicity, that S_1 is normal. Then G_t is an extension of an abelian variety A_t by a torus T_t over t. T_t is a split torus over t because $\mathrm{cl}(T_t)$ = the (scheme-theoretic) closure of T_t in $G \times_S S_1$ is a torus and T_s is split over s. We denote the completion of the localization of S at t by Σ.

By the theory of degeneration, G arises from an extension $\tilde{G}$ over S as the quotient of $\tilde{G}$ by a period subgroup Y of $\tilde{G}(\eta)$. On the other hand, there are two (polarized) abelian schemes, $G \times_S \Sigma$ and $G \times_S S_1$ over Σ and S_1 respectively, which can also be analyzed by the theory of degeneration. The obvious problem then arises:

What is the relation between the period of G and the periods of $G \times_S \Sigma$ and $G \times_S S_1/\mathrm{cl}(T_t)$? In other words, how can the degeneration data of $G \times_S \Sigma$ and $G \times_S S_1$ be described in terms of the degeneration data of G? Essentially theorem 10.1. already provides the answer "up to units."

The answer to the question above can be summarized roughly in the following picture: $G \times_S S_1$ is an algebraic form of "$\tilde{G} \times_S S_1/Y_1$", where Y_1 is a subgroup of Y with Y/Y_1 free. The semi-abelian scheme $G \times_S S_1/\mathrm{cl}(T_t)$ is the Mumford quotient of $\tilde{G} \times_S S_1/\mathrm{cl}(T_t)$ by the image of Y_1 in $\tilde{G} \times_S S_1/\mathrm{cl}(T_t)(\eta)$. $G \times_S \Sigma$ is the quotient of an algebraic form of "$\tilde{G}/Y_1$" over Σ by a period subgroup canonically isomorphic to Y/Y_1.

To accurately formulate this picture and give necessary proofs requires some attention to details which makes this section look technical. If we work in the complex-analytic category (i.e. assume that S is a $\mathbb{C}$-scheme and consider the associated analytic spaces), things are much easier than in the formal case: We have a degenerating family of abelian varieties G over a base $S = \Delta^n$, which is given as a quotient of an extension $\tilde{G}$ over S by a (degenerating) family of period subgroups Y. Here Δ is the unit disk, G is assumed to be abelian over the product of punctured disks $(\Delta^*)^n$ and the periods Y degenerate (intuitively, disappear) over $s =$ the origin of S. Let t be a point of S, and let Y_1 be the subgroup of periods in Y which do not degenerate over t. Then we can also view it as the (Y/Y_1)-quotient of $\tilde{G}/Y_1$ with Y_1 a direct summand of Y, so that the family of periods (Y/Y_1) in $\tilde{G}/Y_1$ degenerates over t.

The above pictures (both formal and complex analytic) explain the title of this section. We would like to remark at this point that in terms of period matrices, the theory of degeneration and Mumford's construction gives modular meaning to "writing a period matrix in block-form in 4 blocks." This section explains relations between different type of blocks and gives modular meaning to "writing a period matrix in 9 blocks." From the point of view of classical Fourier-Jacobi expansion of modular form this section is about the relation between Fourier-Jacobi expansions of a modular form along two different cusps, one in the closure of the other.

Notation and Assumptions. (i) $R =$ a complete normal (hence excellent) local domain, $\mathbf{m} =$ the maximal ideal of R, $\kappa(s) =$ the residue field of R, $K =$ the fraction field of R, $S = \operatorname{Spec} R$, $\eta =$ the generic point of S, $s =$ the closed point of S.

(ii) $t =$ a chosen (fixed) point of S such that $S_1 = \operatorname{Spec} R_1 = \operatorname{Spec} R/\mathbf{p} = \mathrm{cl}\{t\}_{\mathrm{red}} =$ the closure of t with reduced structure is *normal*. $R' =$ the completion of $\mathcal{O}_{S,t}$ at t, $\Sigma = \operatorname{Spec} R'$, $K' =$ fraction field of R', $K_1 = \kappa(t) =$ fraction field of $R_1 =$ residue field of t. Notice that R is naturally contained in R' and K in K'.

(iii) $G \to S$ is a semi-abelian scheme (which is polarizable by a result of Raynaud). $G_s = G \times_S s$, $G_t = G \times_S t$, $G_\Sigma = G \times_S \Sigma$. G_s and G_t define extensions $0 \to T_s \to G_s \to A_s \to 0$ and $0 \to T_t \to G_t \to A_t \to 0$ of abelian varieties by tori.

(iv) $\overline{T} =$ the closure of T_t in G, which is a torus over S_1 because S_1 is normal by assumption. *We assume that T_s is a split torus over s*, which implies that $\overline{T}$ and hence T_t are split over S_1 and t respectively by density of torsion points.

(v) $G^{/s} =$ formal completion of G along G_s, $G^{/t}_\Sigma =$ formal completion of G_Σ along G_t. $G^{/s}$ fits into an exact sequence $0 \to T^{/s} \to \tilde{G}^{/s} \to A^{/s} \to 0$ with $T^{/s}$ a formal split torus and $A^{/s}$ a formal abelian scheme over $\operatorname{Spf} R$. This exact sequence determines another exact sequence $0 \to T \to \tilde{G} \to A \to 0$ with T a torus and A an abelian scheme over S as in chap. II. Similarly we have exact sequences $0 \to T^{/\Sigma} \to \tilde{G}^{/\Sigma} \to A^{/\Sigma} \to 0$ and $0 \to T_{/\Sigma} \to \tilde{G}_{/\Sigma} \to A_{/\Sigma} \to$

0 derived from G_t, where $T^{/\Sigma}$, $\tilde{G}^{/\Sigma}$, $A^{/\Sigma}$ are formal (commutative) group schemes over $\operatorname{Spf} R'$ and $T_{/\Sigma}$, $\tilde{G}_{/\Sigma}$, $A_{/\Sigma}$ are commutative group schemes over Σ.

(vi) Let $G^t \to S$ and $G^t_\Sigma \to \Sigma$ be the semi-abelian schemes over S and Σ respectively whose generic fibres are the duals G^t_K and $G^t_{K'}$ respectively. We have exact sequences $0 \to T^{t/s} \to \tilde{G}^{t/s} \to A^{t/s} \to 0$, $0 \to T^t \to \tilde{G}^t \to A^t \to 0$, $0 \to T^{t/\Sigma} \to \tilde{G}^{t/\Sigma} \to A^{t/\Sigma} \to 0$, $0 \to T^t_{/\Sigma} \to \tilde{G}^t_{/\Sigma} \to A^t_{/\Sigma} \to 0$ as before. Let X be the character group of T, Y be the character group of T^t, $\overline{X}$ be the character group of $T_{/\Sigma}$, $\overline{Y}$ be the character group of $T^t_{/\Sigma}$. $\overline{X}$ is canonically a quotient of X, because it is the character group of the split torus $\overline{T} = \operatorname{cl}(T_t)$ over S_1. The kernel of the natural surjection $X \to \overline{X}$ will be denoted by X_1. Similarly $\overline{Y}$ is canonically a quotient of Y, and the kernel of the natural surjection $Y \to \overline{Y}$ will be denoted by Y_1. Let $\tilde{G}_2$ be the quotient of $\tilde{G}$ by the subtorus $\overline{T}$ of T with character group $\overline{X}$. We have an extension $0 \to T_2 \to \tilde{G}_2 \to A \to 0$, where T_2 is a torus with character group X_1.

(vii) Let τ (resp. τ_Σ) be the semi-abelian degeneration data for $G \to S$ (resp. $G_\Sigma \to \Sigma$). It is a trivialization of the biextension $(c^t \times c)^*\mathcal{P}$ (resp. $(c^t_\Sigma \times c_\Sigma)^*\mathcal{P}_\Sigma$) of $Y \times X$ (resp. $\overline{Y} \times \overline{X}$) over K (resp. K') which defines periods subgroups inside $\tilde{G}(K)$ (resp. $\tilde{G}_{/\Sigma}(K')$). Here we use $\mathcal{P}$, $\mathcal{P}_\Sigma$ as abbreviations for $\mathcal{P}_A$ and $\mathcal{P}_{A_\Sigma}$. Let $b : Y \times X \to K^*/R^*$ (resp. $b_\Sigma : \overline{Y} \times \overline{X} \to K'^*/R'^*$) be the bimultiplicative map attached to τ (resp. τ_Σ) obtained from τ (resp. τ_Σ) by first trivializing $(c^t \times c)^*\mathcal{P}$ (resp. $(c^t_\Sigma \times c_\Sigma)^*\mathcal{P}_\Sigma$) over S (resp. Σ), then representing τ (resp. τ_Σ) by a bimultiplicative map from $Y \times X$ (resp. $\overline{Y} \times \overline{X}$) to K^* (resp. K'^*). This map depends on the choice of trivialization over S but is well-defined modulo units. This way we recover our old bilinear form B with values in $\operatorname{Div}(S)$, so we name these forms B respectively B_Σ. Note that the definitions of c^t, c, c^t_Σ, c_Σ are incorporated in the definitions of τ and τ_Σ.

(viii) Consider the semi-abelian scheme $G_1 = G \times_S S_1/\operatorname{cl}(T_t)$, whose generic fibre is abelian. Then $G_1^{/s}$ (the formal completion of G_1 along its special fibre) sits in an exact sequence $0 \to T_1^{/s} \to G_1^{/s} \to A_1^{/s} \to 0$. As before, let G_1^t be the semi-abelian scheme over S_1 whose generic fibre is the dual abelian variety of the generic fibre of G_1. $G_1^{t/s}$ (the formal completion of G_1^t along its special fibre) sits in an exact sequence $0 \to T_1^{t/s} \to G_1^{t/s} \to A_1^{t/s} \to 0$. It is easy to see that the character group of the formal torus $T_1^{/s}$ (resp. $T_1^{t/s}$) is X_1 (resp. Y_1), where X_1, Y_1 are define in (vi). Let $0 \to T_{1/S_1} \to \tilde{G}_{1/S_1} \to A_{1/S_1} \to 0$ (resp. $0 \to T^t_{1/S_1} \to \tilde{G}^t_{1/S_1} \to A^t_{1/S_1} \to 0$) be the algebraic extension corresponding to $0 \to T_1^{/s} \to G_1^{/s} \to A_1^{/s} \to 0$ (resp. $0 \to T_1^{t/s} \to G_1^{t/s} \to A_1^{t/s} \to 0$). The degeneration data τ_1 of $G_1 = G \times_S S_1/\operatorname{cl}(T_t)$ is a trivialization of the biextension $(c_1^t \times c_1)^*\mathcal{P}_1$ of $Y_1 \times X_1$ over K_1, where $\mathcal{P}_1 = \mathcal{P}_{A_1}$. As in (vii), we get a cruder form $B_1 : Y_1 \times X_1 \to K_1^*/R_1^*$ "by forgetting twists".

(ix) Let $G^1 = G^{/S_1}$ be the formal completion of G along the $G \times_S S_1$. By the rigidity of tori, T_t extends to a formal torus T^1 in G^1. Let G^2 be the quotient of G^1 by T^1, so we have an exact sequence $0 \to T^1 \to G^1 \to G^2 \to 0$ of formal groups smooth over S. G^1 is the bridge between $G^{/s}$ and $G^{/t}$ in the obvious way. Although G^2 is not proper over S and therefore we cannot directly apply Grothendieck's algebraization theorem, we shall see later that G^2 is algebraic in the sense that it is the formal completion of a unique semi-abelian scheme G^2_{alg} over S whose generic fibre is abelian. Moreover, $G^2_{\text{alg}} \times_S \Sigma = A_{/\Sigma}$, $G^2_{\text{alg}} \times_S S_1 = G_1$.

We choose and fix a rigidified ample invertible sheaf $\mathcal{L}$ on G, which defines a polarization $\lambda : G \to G^t$. Then we choose and fix a cubical trivialization of $\mathcal{L}$ along the formal torus T, which determines a cubical trivialization of $\mathcal{L}$ along the formal torus $T^{/\Sigma}$. (As in chapter 2, we make these choices in order to explore the Fourier-Jacobi expansion of theta functions, but our final result is certainly independent of the choice.) Let $\mathcal{L}^{/s}$ (resp. $\mathcal{L}^{/\Sigma}$) be the pull-back of $\mathcal{L}$ to $G^{/s}$ (resp. $G^{/\Sigma}$), which defines a cubical ample invertible sheaf $\mathcal{M}$ (resp. $\mathcal{M}_{/\Sigma}$) on A (resp. $A_{/\Sigma}$) via descent and algebraization. If we complete $A^{/\Sigma}$, $\mathcal{M}^{/\Sigma}$ along the closed fibre, we see that $\mathcal{M}^{/\Sigma}$ can be descended to $\mathcal{M}$.

For any section $s \in \Gamma(G, \mathcal{L})$, its image in $\Gamma(G^{/S_1}, \mathcal{L}^{/S_1})$ can be written as a sum of T^1-eigensections: $s = \sum_{\mu \in \overline{X}} \sigma'_\mu(s)$ in the notation of chapter 2, where $\sigma'_\mu(s) \in \Gamma(G^2, \mathcal{M}^2_\mu)$ $\forall \mu \in \overline{X}$. (Here it is understood that the formal cubical sheaf $\mathcal{L}^{/S_1}$ descends to $\mathcal{M}^2$ on the formal semi-abelian scheme G^2.) Similarly, the image of s in $\Gamma(G^{/s}, \mathcal{L}^{/s})$ can be written as a sum of $T^{/s}$-eigensections: $s = \sum_{\xi \in X} \sigma_\xi(s)$, where $\sigma_\xi(s) \in \Gamma(A, \mathcal{M}_\xi)$ $\forall \xi \in X$. Clearly we have

$$(\diamondsuit) \qquad \sigma'_\mu(s) = \sum_{\xi \in \mu} \sigma_\xi(s) \qquad \forall s \in \Gamma(G, \mathcal{L}),\ \forall \mu \in \overline{X}$$

in the obvious sense, because the completion of $A^{/\Sigma}$ along the closed fibre is the quotient of $G^{/s}$ by the completion of $T^{/\Sigma}$. This relation, while simple to state and easy to prove, is the key to the following calculations.

It remains only to keep track of various canonical isomorphisms and suitably interpret $(\diamondsuit)$.

10.2. Theorem. *(i) The restriction to $Y_1 \times X_1$, Y_1 of the degeneration data (τ, ψ) of $(G, \mathcal{L})$ defines degeneration data for the extension $0 \to T_2 \to \tilde{G}_2 \to A \to 0$. The corresponding Y_1-quotient of $\tilde{G}_2$, denoted by G^2_{alg}, is the (unique semi-abelian) algebraization of G^2.*

(ii) $G^2_{\text{alg}} \times_S \Sigma$ is canonically isomorphic to $A_{/\Sigma}$, $G^2_{\text{alg}} \times_S S_1$ is canonically isomorphic to G_1.

(iii) The extension $0 \to T^1 \to G^1 \to G^2 \to 0$ of formal group schemes defines an extension $0 \to T^1_{\text{alg}} \to G^1_{\text{alg}} \to G^2_{\text{alg}} \to 0$ of semi-abelian schemes.

(iv) τ defines an object τ^1 of $\mathrm{DD}(G^1_{\text{alg}} \times_S \Sigma)$. ψ, together with τ, defines an object ψ^1 of $\mathrm{DD}_{\text{ample}}(G^1_{\text{alg}} \times_S \Sigma)$.

(v) $\tau^1 = \tau_\Sigma$, $\psi^1 = \psi_\Sigma$. In other words, G_Σ is the $\overline{Y}$-quotient of $G^1_{\text{alg}} \times_S \Sigma$ as defined by the degeneration data in (iv).

Remarks. This theorem gives precise meaning of the second half of the title of this section. Namely G is realized in some sense as the Y/Y_1-quotient of $\tilde{G}_2$, while $\tilde{G}_2$ can be regarded as the Y_1-quotient of $\tilde{G}$. In fact one can generalize Mumford's construction and the theory of degeneration without difficulty to the case where G is an extension of a semi-abelian scheme with abelian generic fibre by a torus. Once that is carried out, we can just say that G is the Y/Y_1-quotient of $\tilde{G}_2$. Lack of incentive prevents us from pursuing this generalization.

Proof of 10.2. The first part of (i) is obvious. The second part of (i) follows from ($\diamondsuit$) by taking $\mu = 0$ in $\overline{X}$. (ii) and (iii) are immediate.

(iv) is essentially trivial if one keeps track of the relevant isomorphisms. For example, the equation $\psi(y+y_1) = \psi(y_1) \cdot \tau(y_1, \phi(y)) \cdot \psi(y)$ for $y \in Y,\ y_1 \in Y_1$ says that $\psi(y+y1)$ and $\psi(y)$ define the same element of the stalk of $\mathcal{M}_2^{-1}$ over $c_2^t(y) \in G_2(K)$, bearing in mind that $\psi(y_1) \cdot \tau(y_1, \phi(y))$ is the correct cocycle in question.

(v) follows from ($\diamondsuit$) by comparing $\sigma'_\mu(s)$ and $\sigma'_\nu(s)$ for $\mu, \nu \in \overline{X},\ \mu \equiv \nu$ (mod $\phi(\overline{Y})$), using the relation

$$\sigma_{\xi+\phi(y)}(s) = \psi(y) \cdot \tau(y, \xi) \cdot T^*_{c^t(y)} \sigma_\xi(s):$$

$$\sigma'_\mu(s) = \sum_{\xi \in \mu} \sigma_\xi(s)$$

$$\begin{aligned}\sigma'_\nu(s) &= \sum_{\zeta \in \nu} \sigma_\zeta(s)\\ &= \sum_{\xi \in \mu} \psi(\zeta_0 - \xi_0) \cdot \tau(\phi^{-1}(\zeta_0 - \xi_0), \xi) \cdot T^*_{c^t(\phi^{-1}(\zeta_0 - \xi_0))} \sigma_\xi(s)\end{aligned}$$

By (iv), $\tau(\phi^{-1}(\zeta_0 - \xi_0), \xi)$ for $\xi \in \mu$ determines a well-defined element $\tau^1(\phi^{-1}(\zeta_0 - \xi_0), \xi)$ of the stalk at the $c^{t1}(\phi^{-1}(\zeta_0 - \xi_0)) \times c^1(\xi)$ of $\mathcal{P}^{-1}_{G_2 \times G_2^t}$.

(v) now follows from the theory of degeneration.

Final Remarks. Everything in the last two chapters carries over to the case when the base is a (not necessarily affine) formal scheme satisfying the running condition of this chapter. One only has to use the language of relative schemes over formal schemes, documented in [Hak]. (In practice, a relative scheme Z over a formal scheme $\mathcal{S}$ associates a scheme $Z_{\mathcal{U}}$ over Spec(R) to every affine open $\mathcal{U} = \mathrm{Spf}(R)$ in $\mathcal{S}$, and these schemes satisfy the obvious cocycle condition. Similarly for semi-abelian schemes over $\mathcal{S}$, etc.) We leave this extra generality to the reader. This will be useful in chapter IV to get a global picture of (the formal completion of) the boundary of the arithmetic toroidal compactification.

Chapter IV. Toroidal Compactification of A_g

1. Strategy of the Construction of $\overline{\mathsf{A}}_g$

We first outline in general terms the various steps of our construction of the toroidal compactification of A_g, denoted by $\overline{\mathsf{A}}_g$. Precise definitions will be given in due course. These steps are:

(a) Let $X = \mathbb{Z}^g$. Choose a $\mathrm{GL}(X)$-invariant rational polyhedral cone decomposition $\{\sigma_\alpha\}$ of the cone $C(X)$ of positive semi-definite quadratic forms on $X \otimes \mathbb{R}$ with rational radicals such that there are only finitely many cones modulo the $\mathrm{GL}(X)$-action. (Such cone decompositions are called "admissible".)

(b) Use the machinery of Mumford's construction to produce a collection of semi-abelian degenerate families of principally polarized abelian varieties over complete rings, which is adapted to the combinatorial data chosen and fixed in (a).

(c) By M. Artin's method, approximate (instead of directly algebraizing) the complete base rings by algebraic ones to produce semi-abelian schemes over algebraic base rings. Then the theory of degeneration of polarized abelian varieties tells us that the resulting approximations can be regarded as algebraizations of those constructed in (b). The new semi-abelian schemes are still adapted to the combinatorial data chosen in (a); we will refer to them as "local models".

(d) The theory of degeneration allows us to glue together the local models constructed in (c) in the étale topology, because they are adapted to data (a). The result is an algebraic stack $\overline{\mathsf{A}}_g$ and a semi-abelian scheme $G \to \overline{\mathsf{A}}_g$ which extends the universal abelian scheme over A_g.

Several remarks are in order:

1.1. Remarks. i) Implicit in the above outline is the fact that the arithmetic toroidal compactification of A_g depends on the choice of the combinatorial data (a). This is certainly not a surprise, for the construction of toroidal compactifications of arithmetic quotients of general bounded symmetric domains also depends on the choice of an admissible rational polyhedral cone decomposition. Although different choices of such combinatorial data change $\overline{\mathsf{A}}_g$ (so we really should have written $\mathsf{A}_g(\{\sigma_\alpha\})$, such differences are clearly of an auxiliary nature. In all known application any $\overline{\mathsf{A}}_g$ "works" – it remains to be seen whether there exist situations in which one needs to use special properties of a particular cone decomposition.

ii) The existence of admissible polyhedral cone decompositions in step (a) is essentially the classical Minkowski reduction theory for positive definite quadratic forms. We will discuss admissible cone decompositions in more detail in the next section.

iii) Because of the non-canonical nature of such auxiliary choices, no one has been able to define a functor (or moduli stack) in terms of simple geometry, which is represented by $\overline{\mathsf{A}}_g$. This is quite different from the case of curves, and is responsible for the long delay in the construction of $\overline{\mathsf{A}}_g$. What is canonical is the (projective) *tower* of all such $\overline{\mathsf{A}}_g$'s.

We mention that $\overline{\mathsf{A}}_g$ also arises when one considers degeneration of polarized pure Hodge structures of weight 1 to mixed Hodge structures. The cone decompositions show up in connection with the weight filtration in degenerations. The readers interested in this line of research are referred to [C-C-K], where $\overline{\mathsf{A}}_g$ is constructed as a topological compactification of A_g (although one clearly should be able to put holomorphic structures on it).

As is obvious from the outline, the two pillars of the construction of $\overline{\mathsf{A}}_g$ are Mumford's construction and the theory of degeneration. The former is used to construction local charts at the boundary of $\overline{\mathsf{A}}_g$ and the latter is the key to gluing the local charts.

To show that the local models constructed in step (c) are really adapted to gluing when we have two local models corresponding to two cones σ, σ', one contained in the closure of another, we need to use the results of chap. III, §§ 9, 10 (to show openness of versality).

This outline is very sketchy indeed. We would like to flesh it out here.

Associated with the chosen cone decomposition in (a) above there is a torus embedding $\overline{E}$ of the torus E with character group $S^2(X)$. $\overline{E}$ is smooth over $\operatorname{Spec}\mathbb{Z}$, and the complement is a divisor with normal crossings (provided the cone decomposition is nice, which we assume). Locally in the étale topology the compactification $\overline{\mathsf{A}}_g$ of A_g will be isomorphic to $E \subset \overline{E}$. More precisely we have étale maps $U_i \to \overline{E}$ and semi-abelian schemes G_i over U_i, such that the bilinear form $\underline{B}$ given by the theory of degeneration (taking formal completion along the boundary of A_g) is given *up to units* by (the pull-back of) the universal (tautological) bilinear form (on suitable quotients of X) over E. This way $\overline{\mathsf{A}}_g$ has a natural stratification, whose dual complex is the $\mathrm{GL}(X)$-quotient of our chosen cone decomposition.

To construct the semi-abelian schemes $G_i \to U_i$ above, we first construct formal models (step (b) above) and then make them algebraic (step (c).) These two steps will occupy sections §3 and §4. The way we "make formal models algebraic" is complicated: We carefully approximate the formal models algebraically in such a way if we recomplete the approximation, the result can be obtained by base change from an automorphism of the base of the original formal model.

We define U as the disjoint union of enough U_i's (so that the boundary is covered). U has a natural stratification, with strata (not necessarily connected) parametrized by the classes of the cones σ modulo $\mathrm{GL}(X)$-action. For instance, the cone $\{0\}$ corresponds to the open stratum U° and the top-dimensional cones correspond to 0-dimensional strata. Over U we have a semi-abelian scheme G which is abelian (and principally polarized) exactly over U°.

How do we define an étale groupoid R to get the desired compactification $\overline{\mathsf{A}}_g$? Well, over U° the groupoid must be $R^\circ = \mathbf{Isom}_{U^\circ \times U^\circ}(\mathrm{pr}_1^* G_{U^\circ}, \mathrm{pr}_2^* G_{U^\circ})$, because U° is an étale covering of A_g. So we define R to be the normalization of R° over $U \times U$. We must show that R is étale over U via either projections. This is done in §5, which uses the theory of degeneration and the fact that the cones are disjoint (and properties of the algebraic models). The rest is rather easy: R is a groupoid because R° is dense in R. The properness of $\overline{\mathsf{A}}_g = U/R$ follows from the valuative criterion and the theory of degenerations: For (principally) polarized semi-abelian degeneration data over a complete discrete valuation ring, the value of the bilinear symmetric form defines a positive definite quadratic form on a free abelian group of rank $\leq g$, and therefore corresponds to a well-defined point in $C(X)/\mathrm{GL}(X)$, hence lies in some cone. Here the fact that the cones $\{\sigma_\alpha\}$ cover $C(X)$ is used.

In the subsequent sections we will implement the plan outlined above. Inexperienced readers are advised to first read the definition of admissible polyhedral cone decomposition and the beginnings of §§ 3, 4 and 5 before going on to more technical details.

We summarize some properties of $\overline{\mathsf{A}}_g$ here. Later on we shall have more information than is listed below.

(i) $\overline{\mathsf{A}}_g$ is an algebraic stack smooth and proper over $\mathbb{Z}$ which contains A_g as a dense open algebraic substack.

(ii) $\overline{\mathsf{A}}_g$ has a stratification with strata parametrized by $\{\sigma_\alpha\}/\mathrm{GL}(X)$ as dual complex. A_g is the open stratum.

(iii) Locally in the étale topology, $(\overline{\mathsf{A}}_g, \mathsf{A}_g)$ is isomorphic to $(\overline{E}, E)$, where $E \subset \overline{E}$ is the torus embedding attached to the cone decomposition $\{\sigma_\alpha\}$.

(iv) There is a semi-abelian scheme $G \to \overline{\mathsf{A}}_g$ extending the universal abelian scheme over A_g.

As a corollary we deduce the following statement: Let S be a regular 1-dimensional scheme and $G \to S$ a semi-abelian scheme whose generic fibre is a principally polarized abelian variety. Then there exists a proper S-scheme $\overline{G} \to S$ containing G is an open dense subscheme, such that the group law of G extends to an action of G on $\overline{G}$ over S. This gives a partial answer to the question of the existence of "good completion of the Néron model" formulated in [L2], Chapter 12, §4, p. 318.

The method used in this chapter works for moduli spaces of abelian varieties with assigned polarization, endomorphisms, and level structures ("Shimura varieties of PEL type.") One obtains smooth arithmetic toroidal compactifications outside the discriminant loci (for polarization, endomorphism ring, and level.) The case of unitary groups has been worked out in [Lar].

2. Admissible Polyhedral Cone Decompositions

The goal of this section is to recall the notion of admissible polyhedral cone decomposition, defined in the case of the cone of positive definite quadratic forms, in a form suited for our purpose.

2.1. Notation. We fix a free abelian group X of rank g, $g \geq 1$. $X_{\mathbb{R}} = X \otimes_{\mathbb{Z}} \mathbb{R}$ is a g-dimensional vector space with integral structure given by X.

Denote by $C = C(X)$ the convex cone of all positive semi-definite symmetric bilinear forms on $X_{\mathbb{R}}$ whose radicals (i.e. the kernel of $X_{\mathbb{R}} \to X^*_{\mathbb{R}} = \mathrm{Hom}_{\mathbb{R}}(X_{\mathbb{R}} \to \mathbb{R})$ defined by this symmetric bilinear form) are defined over $\mathbb{Q}$. C lies inside $B(X_{\mathbb{R}})$ = dual of $S^2(X_{\mathbb{R}})$ = space of symmetric bilinear forms $X_{\mathbb{R}} \times X_{\mathbb{R}} \to \mathbb{R}$. Denote by $C^\circ = C^\circ(X)$ the interior of C, i.e. the convex open cone of all positive definite symmetric bilinear forms on $X_{\mathbb{R}}$.

We give $B(X_{\mathbb{R}})$ the integral structure determined by the lattice $B(X)$ of integral valued symmetric bilinear forms on $X \times X$. ($B(X)$ is canonically isomorphic to $\Gamma^2(X^*)$, the second divided power of X^*.) C is rational w.r.t. this integral structure in the sense that $\overline{C}$ is the closure in $B(X_{\mathbb{R}})$ of the convex cone generated by $C \cap B(X)$, while C is the smallest convex subset of $B(X_{\mathbb{R}})$ generated by $C \cap B(X)$.

The natural action of $\mathrm{GL}(X)$ on X induces actions of $\mathrm{GL}(X)$ on $B(X)$ and $B(X_{\mathbb{R}})$, which stabilize C, C° and preserve the integral structure.

2.2. Definition. A $\mathrm{GL}(X)$-*admissible polyhedral cone decomposition* of C is a collection $\{\sigma_\alpha\}_{\alpha \in J}$ with an indexing set J such that

(i) Each σ_α is a non-degenerate rational polyhedral cone which is open in the smallest $\mathbb{R}$-subspace of $X \otimes_{\mathbb{Z}} \mathbb{R}$ containing itself. In other words there exists $v_1, \ldots, v_k \in B(X) \otimes_{\mathbb{Z}} \mathbb{Q}$ such that $\sigma_\alpha = \mathbb{R}_{>0} \cdot v_1 + \cdots + \mathbb{R}_{>0} \cdot v_k$, and σ_α does not contain any line. ($\sigma_\alpha = \{0\}$ is allowed.)

(ii) C is the disjoint union of the σ_α's. The closure of each σ_α is (disjoint) union of finitely many σ_β's. In other words $C = \coprod_{\alpha \in J} \sigma_\alpha$ is a stratification of C.

(iii) $\{\sigma_\alpha\}$ is $\mathrm{GL}(X)$-invariant under the natural action of $\mathrm{GL}(X)$ on $B(X_{\mathbb{R}})$, and $\{\sigma_\alpha\}/\mathrm{GL}(X)$ is finite (i.e. there are only finitely many cones modulo the action of $\mathrm{GL}(X)$).

2.3. Definition. A $\mathrm{GL}(X)$-admissible polyhedral cone decomposition $\{\sigma_\alpha\}_{\alpha \in J}$ of $C(X)$ as above is called *smooth* if each σ_α is generated by part of a $\mathbb{Z}$-basis of $B(X)$: $\sigma_\alpha = \mathbb{R}_{>0} \cdot v_1 + \cdots + \mathbb{R}_{>0} \cdot v_k$ where $v_1, \ldots, v_k \in B(X)$ and can be extended to a basis of $B(X)$.

Clearly this definition makes sense for any given integral structure on $B(X)$. When one talks about smoothness it must refer to a specific integral structure.

2.4. Definition. A *polarization function* of a $\mathrm{GL}(X)$-admissible polyhedral cone decomposition $\{\sigma_\alpha\}_{\alpha \in J}$ of $C(X)$ is a real-valued continuous function $\phi : C(X) \to \mathbb{R}_{\geq 0}$ such that

(a) $\phi(t \cdot x) = t\phi(x)\ \forall t \in \mathbb{R}_{\geq 0},\ \forall x \in C(X)$.

(b) $\phi(x) > 0$ if $x > 0$.

(c) $\phi(x) \in \mathbb{Q}$ if x is rational, i.e. if $x \in B(X) \otimes \mathbb{Q}$.

(d) ϕ is convex in the sense that $\phi(tx + (1-t)y) \geq t\phi(x) + (1-t)\phi(y)$ $\forall 0 \leq t \leq 1,\ \forall x, y \in C(X)$.

(e) ϕ is linear on each cone σ_α, i.e. $\phi|\sigma_\alpha$ is the restriction of a linear function for each cone σ_α.

(f) the closures of top-dimensional cones are exactly the maximal closed polyhedral cones in $C(X)$ on which ϕ is linear.

We remark that, unfortunately, the notion of convexity used in the literature of toroidal embeddings is exactly the notion of concavity used in elementary calculus. Thus for ϕ above, the subset $\{x \in C(X) | \phi(x) \geq t_0\}$ is convex $\forall t_0 \in \mathbb{R}_{\geq 0}$.

A polarization function ϕ as above is *integral* if $\phi(x) \in \mathbb{Z}$ for every integral element x in $C(X)$. Clearly for any $\mathrm{GL}(X)$-invariant polarization function ϕ on $C(X)$, some non-zero integral multiple of it is integral.

The existence of a $\mathrm{GL}(X)$-admissible polyhedral cone decomposition of $C(X)$ is essentially the content of classical (Minkowski) reduction theory of positive definite quadratic forms. Historically, when the reduction theory was generalized to arbitrary reductive groups, the definition of fundamental domain was relaxed (to require only the Siegel property). The issue of exact polyhedral fundamental domains was picked up in chapter 2 of [AMRT] (by A. Ash), where the notions of cores and co-cores were introduced. It is a modern treatment of reduction theory and serves as our basic reference for all assertions made in this section (although they are proved for general non-degenerate self-adjoint cones with integral structures).

We caution the reader that there is a minor error in chapter 2 of [AMRT] in that (in our special case) C was incorrectly defined to be the cone of all positive semi-definite quadratic forms on X and was asserted to be the closure in $B(X)_{\mathbb{R}}$ of the convex cone generated by $C \cap B(X)$. This error was pointed out and corrected by E. Looijenga, and the main results in [AMRT] are in no way affected.

We collect some relevant existence statements about $\mathrm{GL}(X)$-admissible polyhedral cone decompositions. Proofs can be found in [AMRT] and [KKMS].

i) $\mathrm{GL}(X)$-admissible polyhedral cone decompositions of $C(X)$ exist. Any two $\mathrm{GL}(X)$-admissible polyhedral cone decompositions of $C(X)$ have a common refinement by a $\mathrm{GL}(X)$-admissible polyhedral cone decomposition of $C(X)$.

ii) Any $\mathrm{GL}(X)$-admissible polyhedral cone decomposition $\{\sigma_\alpha\}_{\alpha \in J}$ can be subdivided (in a $\mathrm{GL}(X)$-equivariant way) into another $\mathrm{GL}(X)$-admissible polyhedral cone decomposition $\{\sigma_\kappa\}_{\kappa\in K}$ such that $\overline{\sigma}_\kappa \cap \gamma\cdot\overline{\sigma}_\kappa \neq \{0\}$ for some γ in $\mathrm{GL}(X)$ implies that γ acts as the identity on the smallest (rational) boundary component of $C(X)$ containing σ_κ.

iii) Any $\mathrm{GL}(X)$-admissible polyhedral cone decomposition $\{\sigma_\alpha\}_{\alpha\in J}$ can be subdivided (in an $\mathrm{GL}(X)$-equivariant way) into another $\mathrm{GL}(X)$-admissible polyhedral cone decomposition $\{\sigma_\kappa\}_{\kappa\in K}$ which is smooth.

iv) There exist $\mathrm{GL}(X)$-admissible polyhedral cone decompositions of $C(X)$ which admit $\mathrm{GL}(X)$-invariant polarization functions.

v) Let $\{\sigma_\alpha\}_{\alpha\in J}$ be a $\mathrm{GL}(X)$-admissible polyhedral cone decomposition of $C(X)$ admitting a polarization function. Then any $\mathrm{GL}(X)$-equivariant subdivision of $\{\sigma_\alpha\}_{\alpha\in J}$ also admits a polarization function.

The above assertion does not appear explicitly in the literature, but the proof is elementary. One just has to change a given polarization function of $\{\sigma_\alpha\}_{\alpha\in J}$ very slightly in a way suited to the given subdivision to obtain a polarization function for the subdivision. The reader is invited to supply the details of the proof of statement v).

Putting together the above statements, we know that there exists a smooth $\mathrm{GL}(X)$-admissible polyhedral cone decomposition $\{\sigma_\alpha\}_{\alpha\in J}$ of $C(X)$ with the property that $\overline{\sigma}_\alpha \cap \gamma\cdot\overline{\sigma}_\alpha \neq \{0\}$ for some γ in $\mathrm{GL}(X)$ implies that γ acts as the identity on the smallest (rational) boundary component of $C(X)$ containing σ_α. Moreover we may assume that $\{\sigma_\alpha\}_{\alpha\in J}$ admits a polarization function if we like.

From now on, *we assume that we have chosen and fixed a smooth* $\mathrm{GL}(X)$-*admissible polyhedral cone decomposition* $\{\sigma_\alpha\}_{\alpha\in J}$ *of* $C(X)$. For psychological comfort we assume that $\overline{\sigma}_\alpha \cap \gamma\cdot\overline{\sigma}_\alpha \neq \{0\}$ for some γ in $\mathrm{GL}(X)$ implies that γ acts as the identity on the smallest (rational) boundary component of $C(X)$ containing σ_α. Strictly speaking the last assumption is not logically necessary for the purpose of compactifying A_g. In fact even the smoothness assumption can perhaps be dispensed with, but we do want to assume it to avoid unnecessary difficulties.

Anticipating the next section, let us try to explain why we want to think of C as lying inside $B(X_{\mathbb{R}})$ (rather than as a quotient of $X^*_{\mathbb{R}} \otimes X^*_{\mathbb{R}}$), and why we are interested in $C(X)$ at all. The reason comes from the theory of degeneration and Mumford's construction. If we think of parametrizing quotients of a split torus $\tilde{G}$ with character group X we need to parametrize all period subgroups. Period subgroups are determined by bimultiplicative functions $b: Y\times X \to K^*$, using our usual notation. Thus each $b(y,\chi)$ is a function of the parameter space, and these functions should determine the parameter space. Since we are interested in the case of principal polarization, Y is identified with X via ϕ, and the symmetry condition says that we have functions $b(\phi^{-1}(\mu),\nu)$'s with the property that $b(\phi^{-1}(\mu),\nu) = b(\phi^{-1}(\nu),\mu)\ \forall \mu,\nu\in X$. Therefore if B is a base ring, ignoring the positivity condition, the periods are parametrized by the $\operatorname{Spec} B[X\otimes X/(\mu\otimes\nu-\nu\otimes\mu)]_{\mu,\nu\in X}$ = spectrum of the group ring of the symmetric quotient of $S^2(X)$ of $X\otimes X$. Since $B(X)$ is the dual of $S^2(X)$ the theory of torus embedding tells us that we have to use cones in $B(X)\otimes\mathbb{R}$ to produce affine torus embeddings of Spec $B[X \otimes X/(\mu \otimes \nu - \nu \otimes \mu)]_{\mu,\nu \in X}$. The cone C shows up because of the positivity condition, of course. This line of reasoning also tells us how to handle the case of non-principally polarized abelian varieties, that is when ϕ is no longer an isomorphism. For instance, $B(X)$ should be replaced by $\{B \in \mathrm{Hom}(Y \otimes X, \mathbb{Z}) | B(y, \phi(z)) = B(z, \phi(y)) \ \forall y, z \in Y\}$ = the inverse image of $B(Y)$ in $Y^* \otimes X^*$. Similarly, the symmetric quotient of $X \otimes X$ should be replaced by $Y \otimes X/(y \otimes \phi(z) - z \otimes \phi(y))_{y,z \in Y}$.

Digression: Review of Torus Embeddings

We now recall the notion of torus embedding. The standard reference is chapter one of [KKMS].

Notation. Let S denote a fixed base scheme. We will deviate from our established habit of using $X(T)$ to denote the character group of a split torus T over S. Instead, we use $M(T)$ to conform with the notations in [KKMS] and avoid possible confusions with the situations arising in the compactification theory (in which the character group of our torus embedding will be $S^2(X)$ instead of X). The cocharacter group (= group of 1-parameter subgroups) of a split torus T over S will be denoted by $N(T)$.

We will use rational polyhedral cones in $N(T)_{\mathbb{R}} = N(T) \otimes_{\mathbb{Z}} \mathbb{R}$. As a rule, a cone will always be open in its closure in the ambient real vector space, so a *rational polyhedral cone* in $N(T) \otimes_{\mathbb{Z}} \mathbb{R}$ has the form

$$\sigma = \mathbb{R}_{>0} v_1 + \cdots + \mathbb{R}_{>0} v_n \text{ with } v_1, \ldots, v_n \in N(T) \text{ (or, } N(T) \otimes \mathbb{Q}).$$

$\overline{\sigma}$ will denote the closure of σ in the underlying real vector space, so $\overline{\sigma} = \mathbb{R}_{\geq 0} v_1 + \cdots + \mathbb{R}_{\geq 0} v_n$.

The intersection of a supporting hyperplane H of a rational polyhedral cone $\sigma (\sigma \cap H = \emptyset)$ with $\overline{\sigma}$ is itself a rational polyhedral cone of the form $\overline{\tau}$ for some face τ of σ. This defines faces of σ.

A rational polyhedral cone σ in $N(T) \otimes_{\mathbb{Z}} \mathbb{R}$ is *smooth* w.r.t. the integral structure given by $N(T)$ if σ is generated by part of a $\mathbb{Z}$-basis of $N(T)$.

If σ is a rational polyhedral cone σ in $N(T) \otimes_{\mathbb{Z}} \mathbb{R}$, then its dual closed cone $\sigma^\vee \subset M(T) \otimes_{\mathbb{Z}} \mathbb{R}$ is defined by

$$\sigma^\vee = \{x \in M(T) \otimes_{\mathbb{Z}} \mathbb{R} \,|\, \langle x, y \rangle \geq 0 \ \forall y \in \sigma\}$$

where $\langle \ , \ \rangle$ is the natural pairing of $M(T) \otimes_{\mathbb{Z}} \mathbb{R}$ with $N(T) \otimes_{\mathbb{Z}} \mathbb{R}$.

A *rational partial polyhedral (abbreviated r.p.p.) cone decomposition* in $N(T)_{\mathbb{R}}$ is a mutually disjoint collection C of rational polyhedral cones $\{\sigma_\alpha\}$ in $N(T)_{\mathbb{R}}$ such that any face σ_β of a cone σ_α in this collection is itself in this collection.

Thus $\{\sigma_\alpha\}$ is a stratification of $|C| = \cup \sigma_\alpha$ by rational polyhedral cones.

The notion of rational polyhedral cones, smoothness, dual closed cones and r.p.p. decompositions clearly makes sense w.r.t. any finite dimensional real vector space with integral structure. r.p.p. decompositions in finite dimensional real vector spaces with integral structures form a category R.P.P. in the obvious way, with morphisms given by subdivisions.

A *torus embedding* over S consists of a T-equivariant open immersion $T \hookrightarrow Z$ of a split torus T over S into a separated S-scheme Z with a T-action extending the translation action of T, which is fibrewise an open dense immersion. Torus embeddings over S form a category T.E.

We can formulate the main result of torus embeddings as follows:

2.5. Theorem. *(i) There is a functor Φ : R.P.P. $\rightsquigarrow$ T.E. from the category of r.p.p. decompositions in finite dimensional real vector spaces with integral structures to the category of torus embeddings over S.*

(ii) If an r.p.p decomposition C in a finite dimensional real vector space $N \otimes_{\mathbb{Z}} \mathbb{R}$ with integral structure is given by a cone σ and all its faces, then Φ sends C to $T \hookrightarrow Z = Z(\sigma)$ with

$$\begin{aligned} T &= N \otimes_{\mathbb{Z}} G_m = \mathbf{Spec}_S \mathcal{O}_S[M], \quad M = \mathrm{Hom}_{\mathbb{Z}}(N, \mathbb{Z}), \\ Z &= \mathbf{Spec}_S \mathcal{O}_S[M \cap \sigma^\vee] \quad \textit{with natural T-action.} \end{aligned}$$

We call $T \hookrightarrow Z$ the affine torus embedding associated to σ.

(iii) If τ is a face of σ (notations as in (ii)), $T \hookrightarrow Z(\tau)$, $T \hookrightarrow Z(\sigma)$ the affine torus embedding associated to τ, σ respectively, then Φ transforms the obvious morphism between the two r.p.p. decompositions to the natural morphism from $T \hookrightarrow Z_\tau$ to $T \hookrightarrow Z_\sigma$, which identifies Z_τ as an S-affine open subscheme of Z_σ.

(iv) Notations as in (ii). There is a natural stratification $Z = \coprod Z_\tau$ of Z by locally closed subschemes, where τ runs through the faces of σ. Each stratum Z_τ is a T-orbit in Z, which can be described as follows:

Let $\tau_\circ^\vee = \{x \in M(T) \otimes_{\mathbb{Z}} \mathbb{R} \mid \langle x, y \rangle > 0\ \forall y \in \tau\}$ be the interior of $\tau^\vee$, $\tau^\perp = \{x \in M(T) \otimes_{\mathbb{Z}} \mathbb{R} \mid \langle x, y \rangle = 0\ \forall y \in \tau\}$ be the annihilator of τ. Let $\langle M \cap \sigma^\vee / M \cap \sigma^\vee \cap \tau_\circ^\vee \rangle$ be the group generated by the semigroup $M \cap \sigma^\vee / M \cap \sigma^\vee \cap \tau_\circ^\vee$. There is a natural isomorphism $M \cap \tau^\perp \xrightarrow{\sim} \langle M \cap \sigma^\vee / M \cap \sigma^\vee \cap \tau_\circ^\vee \rangle$. Then $\mathcal{O}_S[M \cap \sigma^\vee \cap \tau_\circ^\vee]$ is a sheaf of ideals of $\mathcal{O}_S[M \cap \sigma^\vee]$, the subscheme $Z_{\sigma,\tau}$ defined by which is canonically isomorphic to $\mathbf{Spec}_S \mathcal{O}_S[M \cap \sigma^\vee / M \cap \sigma^\vee \cap \tau_\circ^\vee]$. The stratum Z_τ is canonically identified with the subscheme

$$\mathbf{Spec}_S \mathcal{O}_S[\langle M \cap \sigma^\vee / M \cap \sigma^\vee \cap \tau_\circ^\vee \rangle] (\xleftarrow{\sim} \mathbf{Spec}_S \mathcal{O}_S[\tau^\perp])$$

of $Z_{\sigma,\tau}$.

Intuitively, Z_τ consists of limits (as $s \to 0$ in G_m) of orbits $y(s) \cdot t_0$, for 1-parameter subgroups $y : G_m \to T$ with $y \in \tau \cap N$.

We remark that the scheme $Z_{\sigma,\tau}$ above is itself an affine torus embedding, with $\mathbf{Spec}_S \mathcal{O}_S[\langle M \cap \sigma^\vee / M \cap \sigma^\vee \cap \tau_\circ^\vee \rangle] \xleftarrow{\sim} \mathbf{Spec}_S \mathcal{O}_S[\tau^\perp]$ as the torus, and this quotient torus of T acts on $Z_{\sigma,\tau}$ in the natural way since $Z_{\sigma,\tau}$ is a T-orbit in $Z(\sigma)$. It corresponds to the r.p.p. decomposition of the cone τ and its faces inside $\langle \tau \rangle \otimes \mathbb{R}$ = the vector subspace of $N \otimes_{\mathbb{Z}} \mathbb{R}$ spanned by τ, with the integral structure given by the lattice $N \cap \langle \tau \rangle \otimes \mathbb{R}$.

(v) If $Z(\tau) \subset Z(\sigma)$ is the immersion corresponding to the inclusion of a face τ of σ in $\overline{\sigma}$ as in (ii), this inclusion preserves the stratifications of $Z(\tau)$ and $Z(\sigma)$ and induces isomorphisms on strata of $Z(\tau)$.

(vi) Φ sends a general r.p.p. decomposition $C = \{\sigma_\alpha\}$ in a finite dimensional real vector space $N \otimes_{\mathbb{Z}} \mathbb{R}$ with integral structure given by a lattice N to the torus embedding $T \hookrightarrow Z(C)$, where $Z(C)$ is the union of S-affine open $Z(\sigma)$'s with σ running over the cones in C, and the gluing is given by (iii). $Z(\tau) = Z(\sigma_\alpha) \cap Z(\sigma_\beta)$ if $\overline{\tau} = \overline{\sigma}_\alpha \cap \overline{\sigma}_\beta$ (i.e. τ is the largest common face of σ_α and σ_β).

(vii) Notations as in (vi). $Z(C)$ has a natural stratification $Z(C) = \amalg_{\sigma \in C} Z_\sigma$ given by T-orbits compatible with the stratifications of its affine open subschemes $Z(\sigma)$ in (iv). $Z_\sigma \subset \overline{Z}_\tau$ iff τ is a face of σ.

(viii) Notations as in (vii). $Z(C)$ is smooth over S iff the r.p.p. decomposition C is smooth. $Z(C)$ is proper over S iff $|C| = N \otimes_{\mathbb{Z}} \mathbb{R}$.

The relative version of the above statement for morphisms of torus embeddings coming from morphisms of r.p.p. decompositions (with possibly different tori) is left to the reader as an easy exercise.

(ix) When S is the spectrum of an algebraically closed field k, Φ is an equivalence of categories.

2.6. Remark. In the documented literature one works over an algebraically closed field, in which case one can classify torus embeddings in term of rational partial polyhedral cone decompositions. (statement (ix) above). When one works over an arbitrary base the classification part is not available but the construction part is intact, as expressed by the functor Φ above. The proof of statements (i) to (viii) above are elementary, and the proof in [KKMS] carries over without modification.

One can also construct T-equivariant invertible sheaves on torus embeddings in terms of combinatorial data. We refer the reader to [KKMS] for that.

As an illustration of the theory of torus embeddings, let $\{\sigma_\alpha\}_{\alpha \in J}$ be the smooth $\mathrm{GL}(X)$-admissible polyhedral cone decomposition of $C(X)$ chosen above. Let $E \hookrightarrow \overline{E}$ be the corresponding torus embedding over $\operatorname{Spec} \mathbb{Z}$. We list some of its properties and establishes notations to be used in the next section:

(i) E is the split torus over $\operatorname{Spec}(\mathbb{Z})$ with character group $Q = S^2(X) =$ the symmetric quotient of $X \otimes X$. Elements of Q will be identified with linear forms on $B(X)$.

(ii) $\overline{E}$ is a scheme separated over $\operatorname{Spec}(\mathbb{Z})$ which contains E as a dense open subscheme. The translation action of E on itself extends to $\overline{E}$.

(iii) For each σ_α there is an affine E-invariant dense open subscheme $E(\sigma_\alpha)$ of $\overline{E}$ containing E. $\overline{E}$ is the union of various $E(\sigma_\alpha)$'s.

(iv) For each cone $\sigma =$ some σ_α, $E(\sigma) = \operatorname{Spec}(B_\alpha)$ with $B_\alpha = \mathbb{Z}[q]_{q \in Q, q \geq 0 \text{ on } \sigma}$. Notice that $\{q \in Q,\ q \geq 0 \text{ on } \sigma\}$ is a subsemigroup of Q containing zero, which generates Q. The coordinate ring B_α of $E(\sigma_\alpha)$ is just the group ring of this semigroup over $\mathbb{Z}$. When $\sigma = \{0\}$, $E(\sigma) = E$.

(v) If σ_α is a face of σ_β then $E(\sigma_\alpha) \subset E(\sigma_\beta)$.

(vi) If $\overline{\sigma}_\alpha = \overline{\sigma}_\beta \cap \overline{\sigma}_\gamma$ (so that σ_α is the largest common face of σ_α and σ_β) then $E(\sigma_\alpha) = E(\sigma_\beta) \cap E(\sigma_\gamma)$.

(vii) $\overline{E}$ has a natural stratification by locally closed subschemes $Z(\sigma_\alpha)$, $\alpha \in J$. Each stratum $Z(\sigma_\alpha)$ is an orbit of E on $\overline{E}$ which is canonically isomorphic to a quotient torus of E. $\sigma_\alpha \subset \overline{\sigma}_\beta$ iff $\overline{Z}(\sigma_\alpha) \supset \overline{Z}(\sigma_\beta)$.

(viii) $\forall \alpha \in J,\ Z(\sigma_\alpha) = \mathrm{Spec}(\mathbb{Z}[q]_{q\in Q,\ q=0 \text{ on } \sigma})$. $Z(\sigma_\alpha)$ is a closed subscheme of $E(\sigma_\alpha)$ via $\mathbb{Z}[q]_{q\in Q,\ q=0 \text{ on } \sigma} \overset{\sim}{\rightarrow} B_\alpha/I_\alpha$, where $I_\alpha =$ the ideal of B_α generated by $\{q | q \in Q,\ q > 0 \text{ on } \sigma\}$.

(ix) $\overline{E}$ is smooth over $\mathrm{Spec}(\mathbb{Z})$ by the assumption that each σ_α is generated by part of a $\mathbb{Z}$-basis.

(x) $\mathrm{GL}(X)$ acts on $\overline{E}$ by functoriality of the construction of torus embeddings. This action preserves the stratification: $\gamma \cdot Z(\sigma) = Z(\gamma \cdot \sigma)$, $\gamma \cdot E(\sigma) = E(\gamma \cdot \sigma)\ \forall \gamma \in \mathrm{GL}(X)$. In particular we may assume that if σ has maximal dimension and $\gamma \cdot \overline{Z}(\sigma) \cap \overline{Z}(\sigma) \neq \emptyset$, then $\gamma = \pm \mathrm{id}_X$.

(xi) There exists a constructible $\mathrm{GL}(X)$-equivariant étale sheaf $\underline{X}$ on $\overline{E}$ which is a quotient of the constant sheaf defined by X. Over $Z(\sigma_\alpha)$ $\underline{X}$ is constant and equal to the smallest quotient X_α of X such that σ_α is contained in $B(X_\alpha \otimes \mathbb{R})$.

(xii) There exists a bilinear symmetric form $\underline{B} : \underline{X} \times \underline{X} \to \mathrm{Div}(\overline{E})$ induced by the tautological form on $X \times X$ which sends (μ, ν) to the divisor of $[\mu \otimes \nu] \in S^2(X)$. $\underline{B}$ is $\mathrm{GL}(X)$-equivariant and invariant under E. For any local section χ of $\underline{X}$ the support of $\underline{B}(\chi, \chi)$ is effective, with the same support as χ.

(xiii) $\overline{E}$ is a "moduli space" for certain types of bilinear forms: More precisely there exists a symmetric bilinear form $b : X \times X \to \Gamma(E, \mathcal{O}_E^*)$ which sends $\mu, \nu \in X$ to $b(\mu, \nu) = [\mu \otimes \nu] =$ class of $\mu \otimes \nu$ in $S^2(X)$, the group of characters of E. Note that $b(\mu, \mu)$ extends to a regular function on $\overline{E}$, as C consists of semi-definite forms. If R is a normal local integral domain with field of quotients K, and $b' : X \times X \to K^*$ a symmetric bilinear form such that $b'(\mu, \mu) \in R$ for all μ, then b' defines a K-rational point of E, which extends to an R-rational point of $\overline{E}$ if and only if the following condition holds: For each height one valuation $v : K^* \to \mathbb{Z}$, consider the positive semi-definite form $v \cdot b$ on $X \times X$ (which vanishes for almost all v). Then an extension exists if and only if there exists a σ_α whose closure contains all $v \cdot b$. The extension maps the closed point of $\mathrm{Spec}(R)$ into the stratum $Z(\sigma_\alpha)$ parametrized by the smallest such σ_α. Equivalently a K-valued point $t : \mathrm{Spec}(K) \to E$ extends to an R-valued point $\mathrm{Spec}(R) \to E(\sigma)$ of the affine torus embedding defined by σ and its faces if and only if for every discrete valuation $v : K^* \to \mathbb{Z}$ defined by a prime ideal of height one, the element $v \circ t^* \in S^2(X)$ lies in the closure of σ. This statement is obvious: whether an element of K is in R can be checked at height-one primes, because R is normal.

3. Construction of Local Charts: Formal Theory

Introduction. The goal of this section is to carry out step (b) of the outline in section 1. Let us first sketch how to construct local charts formally "near the 0-dimensional cusp", i.e. construct totally degenerate semi-abelian schemes.

At the end of the last section we constructed a torus embedding $E \subset \overline{E}$ corresponding to the $\mathrm{GL}(X)$-admissible polyhedral cone decomposition chosen and fixed in 2. As explained there E can be thought of as a parameter space of periods in the split torus $\tilde{G}$ with character group X when we disregard the positivity condition. If we complete $\overline{E}$ along a stratum Z_σ with $\sigma \subset C(X)^\circ$, we get a complete regular ring R_σ formally smooth over $\mathbb{Z}$, and the pull-back of the period subgroup to R_σ satisfies the positivity condition. Then we can apply Mumford's construction to get a semi-abelian scheme ${}^\heartsuit G_\sigma \to \operatorname{Spec} R_\sigma$. It will turn out that $\operatorname{Spf} R_\sigma$ is the formal completion of the toroidal compactification of A_g along a stratum of the boundary over the 0-dimensional cusp.

We have seen that using cones in $C(X)^\circ$ one can construct semi-abelian schemes which degenerate to tori. What about the general case? Clearly the cones at the boundary of $C(X)$ correspond to degenerations with non-trivial abelian part. Since the abelian part has moduli, torus embeddings alone cannot produce good base schemes. Instead one should expect to use fibre bundles with torus embeddings as fibres and moduli of abelian varieties as bases.

More precisely, what we should do is dictated by the theory of degeneration and Mumford's construction: The "universal" $\tilde{G}$ lives over some moduli space B related to A_{g-r} with $r > 0$, and the "periods without positivity condition" are parametrized by a suitable torus bundle over B. The combinatorial data chosen in section 2 tell us how to complete in order to get the right base over which the period subgroup satisfies the positivity condition. (This is where one uses the combinatorial data!) Then we proceed as before: apply Mumford's construction to obtain semi-abelian varieties over complete rings, the base of which will turn out to be local charts at the boundary of A_g. (This time over higher dimensional cusps.)

For later use we will prove various properties of semi-abelian schemes ${}^\heartsuit G \to \operatorname{Spec} R$ obtained above. We will put them together and formulate a notion "good formal model" over a strict complete local ring R', for semi-abelian schemes over $\operatorname{Spec} R'$ sharing the nice properties of the local charts ${}^\heartsuit G \to \operatorname{Spec} R$. These properties are formulated in such a way that they are preserved under the approximation process. In the next section we will formulate the notion "good algebraic model" by direct analogue, so that the good algebraic models serve as local charts of $\overline{\mathsf{A}}_g$.

Local Charts Over the 0-Dimensional Cusp

For motivation, we first treat the totally degenerate case. Recall once more that throughout this chapter a $\mathrm{GL}(X)$-admissible polyhedral cone decomposition $\mathfrak{C}$ satisfying the requirement of section 2 has been picked. We will also use the notations of section 2.

For each cone $\mathfrak{C} \ni \sigma_\alpha \subset C(X)^\circ$, let $R_\alpha = I_\alpha$-adic completion of B_α, $S_\alpha = \operatorname{Spec}(B_\alpha)$, $S_{\alpha,0} = \operatorname{Spec}(B_\alpha/I_\alpha \cdot B_\alpha) \cong Z_\alpha$, $K_\alpha =$ fraction field of R_α. We would like to construct for each cone $\sigma_\alpha \subset C(X)^\circ$ a semi-abelian scheme ${}^\heartsuit G_\alpha \to S_\alpha$ whose generic fibre is principally polarized. In view of Mumford's construction we only have to define a period subgroup $\imath : Y = X \to \tilde{G}(K_\sigma)$ with $\phi =$

id : $Y = X \to X$ as the polarization and $\tilde{G} =$ the split torus over S_α, which satisfies the symmetry and positivity conditions. But giving such an $\imath$ satisfying the symmetry condition is the same as giving a symmetric bimultiplicative map $b : X \times X \to K_\alpha^*$, and we simply use the rational function $\mu \otimes \nu \in$ the symmetric quotient of $X \otimes X =$ character group of E as $b(\mu, \nu)$. In brief, $b(\mu, \nu) = \mu \otimes \nu \ \forall \mu, \nu \in X$. The positivity condition requires that $b(\mu, \mu) \in I_\alpha \ \forall 0 \neq \mu \in X$. This condition is equivalent to : $\mu \otimes \mu > 0$ on $\sigma_\alpha \ \forall 0 \neq \mu \in X$ $\Leftrightarrow$ elements of σ_α are positive definite quadratic forms $\Leftrightarrow \sigma_\alpha \subset C(X)^\circ$.
The positivity condition is indeed satisfied, so we have the desired semi-abelian scheme ${}^\heartsuit G_\alpha \to S_\alpha$.

It is useful to reformulate this so as to have a global picture of this construction. The global statement is that the construction described above defines a "semi-abelian scheme" over the formal completion $\overline{E}^\wedge$ of $\overline{E}$ along the locally closed subscheme whose support is $\cup Z_\alpha$, where α runs through those with $\sigma_\alpha \subset C^\circ$. By this we mean not just a formal group scheme but the following: To give a semi-abelian scheme ${}^\heartsuit G \to \overline{E}^\wedge$ just means that for each affine open $U \subset \overline{E}^\wedge$, we have a semi-abelian scheme ${}^\heartsuit G_{\mathcal{U}} \to \operatorname{Spec} \Gamma(\mathcal{U}, \overline{E}^\wedge)$, and that this assignment is compatible with all pull-backs coming from inclusion $\mathcal{V} \hookrightarrow \mathcal{U}$ of affine opens.

Now for each cone $\sigma_\alpha \subset C^\circ$ we have a semi-abelian scheme ${}^\heartsuit G_\alpha \to S_\alpha$ over the affine open S_α of $\overline{E}^\wedge$, and by functoriality of Mumford's construction this assignment is compatible with all pull-backs $S_\beta \to S_\alpha$ induced by $E(\sigma_\beta) \subset E(\sigma_\alpha)$, $C^\circ \supset \sigma_\beta \subset \overline{\sigma}_\alpha$. These data clearly define a semi-abelian scheme ${}^\heartsuit G \to \overline{E}^\wedge$ over $\overline{E}^\wedge$. Notice that there is a natural $\mathrm{GL}(X)$-action on ${}^\heartsuit G \to \overline{E}^\wedge$, and the quotient is a semi-abelian scheme ${}^\heartsuit G \to \overline{E}^\wedge/\mathrm{GL}(X)$.

Later we will see that $\overline{E}^\wedge/\mathrm{GL}(X)$ is isomorphic to the formal completion of the boundary of $\overline{\mathsf{A}}_g$ along the union of those strata lying over the 0-dimensional cusp.

Local Charts: General Case

We have dealt with those cones $\mathfrak{C} \ni \sigma_\alpha \subset C^\circ$. What should we do for those σ_α lying in the boundary $C - C^\circ$ of C? Clearly we want to use Mumford's construction in the general (mixed) case. To do this we have to construct a suitable base scheme, over which there lies an extension $\tilde{G}_\sigma$ and a set of periods Y satisfying the symmetry and polarization conditions. If $\tilde{G}_\sigma$ is an extension of an abelian scheme A_σ by a split torus T_σ, then A_σ and Y should both be principally polarized.

Fix a general cone σ in the chosen decomposition $\mathfrak{C}$, and let $X \to X_\sigma$ be the smallest quotient of X such that all elements of σ (thought of as quadratic forms on $X \otimes \mathbb{R}$) factor through $X \otimes \mathbb{R} \to X_\sigma \otimes \mathbb{R}$. $B(X_\sigma \otimes \mathbb{R}) \subset B(X \otimes \mathbb{R})$ contains the rational subspace spanned by σ. Let $Q_\sigma = S^2(X_\sigma)$ be the symmetric quotient of X_σ, $C(X_\sigma) = C \cap B(X_\sigma \otimes \mathbb{R})$ the cone of all positive semi-definite symmetric bilinear forms on X_σ whose radicals are rational, $C(X_\sigma)^\circ$ the interior of $C(X_\sigma)$ consisting of all positive definite symmetric bilinear forms on X_σ. By reduction theory ([AMRT] chap. 2), $\mathfrak{C}_\sigma = \{\sigma_\beta | \sigma_\beta \subset B(X_\sigma \otimes \mathbb{R})\}$ is a $\mathrm{GL}(X_\sigma)$-admissible polyhedral decomposition of $C(X_\sigma)$. Clearly $C(X_\sigma)$, $C(X_\sigma)^\circ$ and $\mathfrak{C}_\sigma$ depend only on the quotient X_σ of S, not on the (top dimensional) cone σ in $C(X_\sigma)^\circ$. In the terminology of [AMRT], $C(X_\sigma)$ is a rational boundary component of $C(X)$. We will use the notation $\mathfrak{C}_\xi$ for a general rational boundary component of $C(X)$, $\mathfrak{C}_\xi^\circ$ for its interior. Each rational boundary component $\mathfrak{C}_\xi$ is of the form $C(X_\sigma)$ for some (any) top dimensional cone σ in $\mathfrak{C}_\xi^\circ$. X_σ as an X-quotient depends only on $\xi = \xi(\sigma)$ so we will denote it by X_ξ. $\mathrm{rk}(X_\xi)$ is called the *rank* of $\mathfrak{C}_\xi$ in [AMRT]. (Here we do not have to specify whether we mean $\mathbb{Q}$-rank or $\mathbb{R}$-rank, as they are equal.) Q_ξ will denote the symmetric quotient of $X_\xi \otimes X_\xi$ as before.

Let $\mathfrak{C}_\xi$ be a rational boundary component of $\mathfrak{C}_\xi$, $E_\xi \hookrightarrow \overline{E}_\xi$ the torus embedding associated to $\mathfrak{C}_\xi$. We have $E_\xi = B(X_\xi) \otimes G_m$, $\overline{E}_\xi$ = union of affine E_ξ-invariant open subschemes $E_\xi(\sigma_\beta)$'s for $\sigma_\beta \in \mathfrak{C}_\xi$; $\overline{E}_\xi$ is stratified by E_ξ-orbits $Z_\xi(\sigma_\beta)$ for $\sigma_\beta \subset \mathfrak{C}_\xi$.

This time the (formal completion of the) torus embedding $\overline{E}_\xi$ itself will not provide us base schemes for Mumford's construction for several reasons. For one thing the dimension is wrong: it should be $1/2 \cdot g \cdot (g+1)$, but the relative dimension of $\overline{E}_\xi$ over $\mathbb{Z}$ is $1/2 \cdot r \cdot (r+1)$ if $r = \mathrm{rk}(X_\xi)$. Second, we have to have an extension of an abelian scheme by a split torus of dimension r over the base. The universal family of such extensions has dimension $(1/2) \cdot (g-r) \cdot (g-r+1) + r \cdot (g-r)$, where $(1/2) \cdot (g-r) \cdot (g-r+1)$ is the dimension of the moduli of abelian schemes over $\mathbb{Z}$ and $r \cdot (g-r)$ is the relative dimension of moduli of extensions. In fact, if we let $A \to \mathsf{A}_{g-r}$ be the universal abelian scheme with principal polarization $\lambda : A \xrightarrow{\simeq} A^t$ over A_{g-r}, then over $\mathbf{Hom}_{\mathbb{Z}}(X_\xi, A)$ (which is isomorphic to the fibred product of r copies of A) lives a tautological (or universal) extension $0 \to T_\xi \to \tilde{G}_\xi \to A \to 0$, where A here is the pull-back of $A \to \mathsf{A}_{g-r}$ to $\mathbf{Hom}_{\mathbb{Z}}(X_\xi, A)$ and T_ξ is the split torus with character group X_ξ. This tautological extension is defined by requiring that the push-forward via $X_\xi \ni -\chi : T_\xi \to G_m$ be classified by the tautological morphism $\mathbf{Hom}_{\mathbb{Z}}(X_\xi, A) \to A$ given by χ.

We will construct an E_ξ-torsor Ξ_ξ over $\mathbf{Hom}_{\mathbb{Z}}(X_\xi, A)$. As we are dealing with principally polarized abelian schemes here we may identify A^t with A (over A_{g-r}) via the principal polarization, and can therefore can regard the Poincaré G_m-torsor $\mathcal{P}_A \to A \times_{\mathsf{A}_{g-r}} A^t$ as a G_m-torsor $\mathcal{P}_A \to A \times_{\mathsf{A}_{g-r}} A$. Ξ_ξ is defined by requiring its push-forward by $[\mu \otimes \nu] : E_\xi \to G_m$ to be the rigidified (along the zero-section) G_m-torsor $(c(\mu) \times c(\nu))^* \mathcal{P}_A^{\otimes -1}$ on $\mathbf{Hom}_{\mathbb{Z}}(X_\xi, A)$, for every $\mu, \nu \in X_\xi$. Here $[\mu \otimes \nu]$ denote the image of $\mu \otimes \nu$ in Q_ξ and $c : X_\xi \to A$ is the tautological map over $\mathbf{Hom}_{\mathbb{Z}}(X_\xi, A)$. This is symmetric in μ and ν, by the symmetry of $\mathcal{P}_A$. Also over Ξ_ξ the pullback of each $(c(\mu) \times c(\nu))^* \mathcal{P}_A^{\otimes -1}$ is canonically trivial. Define an $\overline{E}_\xi$-bundle $\overline{\Xi}_\xi$ over $\mathbf{Hom}_{\mathbb{Z}}(X_\xi, A)$ by $\overline{\Xi}_\xi = \Xi_\xi \times^{E_\xi} \overline{E}_\xi$ (contraction product). $\overline{E}_\xi$ has a locally closed subscheme W_ξ which is the union of all E_ξ-orbits $Z_\xi(\sigma_\beta)$ with $\sigma_\beta \subset C(X_\xi)^\circ$. Similarly, define $\Xi(\sigma_\beta) = \Xi_\xi \times^{E_\xi} E_\xi(\sigma_\beta)$ for $\sigma_\beta \in \mathfrak{C}_\xi$. There also exists an étale constructible sheaf $\underline{X}_\xi$ on $\overline{\Xi}_\xi$, together with a symmetric bilinear form $\underline{B}_\xi : \underline{X}_\xi \times \underline{X}_\xi \to \mathbf{Div}(\overline{\Xi}_\xi)$. Over $\Xi(\sigma_\beta)$ $\underline{X}_\xi$ is equal to the quotient X_{σ_β} of X_ξ.

We define $\mathcal{S}_\xi$ to be the formal completion of $\overline{\Xi}_\xi$ along the locally closed subscheme $\Xi_\xi \times_{E_\xi} W_\xi$. This is the base we wanted. Note that $\overline{\Xi}_\xi$ has a natural stratification by $\Xi_\xi \times^{E_\xi} Z_\xi(\sigma_\beta)$'s for $\sigma_\beta \subset C(X_\xi)$, and $\Xi_\xi \times^{E_\xi} W_\xi$ is just the union of $\Xi_\xi \times^{E_\xi} Z_\xi(\sigma_\beta)$'s with $\sigma_\beta \subset C(X_\xi)^\circ$. Notice that $\mathrm{GL}(X_\xi)$ acts on $\mathcal{S}_\xi$ naturally.

Similarly, for each cone $\sigma \subset C(X_\xi)^\circ$, define $\mathcal{S}_\sigma$ to be the formal completion of $\overline{\Xi}_\xi$ along the locally closed subscheme $\Xi_\xi \times^{E_\xi} Z_\xi(\sigma)$ (contraction product again). It is obtained from $\mathcal{S}_\xi$ by localization and further completion.

Let us examine the data at hand. First we have an extension $0 \to T_\xi \to \tilde{G}_\xi \to A \to 0$ over $\mathbf{Hom}_{\mathbb{Z}}(X_\xi, A)$, which we can pull back to $\overline{\Xi}_\xi$. Since we are dealing with principal polarizations we will identify Y_ξ with X_ξ and $\tilde{G}_\xi$ with $\tilde{G}^t_\xi$ if we want to stick to the notations of chapters 2, 3. (Strictly speaking, here we were simply defining Y_ξ, $\tilde{G}^t_\xi$ as X_ξ, $\tilde{G}_\xi$ respectively.) We have the tautological homomorphism $c : X_\xi \to A$ over $\mathbf{Hom}_{\mathbb{Z}}(X_\xi, A)$ and also $c^t = c : Y_\xi = X_\xi \to A = A^t$ over $\mathbf{Hom}_{\mathbb{Z}}(X_\xi, A^t) = \mathbf{Hom}_{\mathbb{Z}}(Y_\xi, A)$. Moreover we have a tautological trivialization τ of $(c^t \times c)^* \mathcal{P}_A^{\otimes -1}$ over Ξ_ξ, which is in fact a trivialization of biextensions as is easily checked. When we pull everything back to $\overline{\Xi}_\xi$, we see that $\tau(\chi, \chi)$ extends to a section of $(c^t(\chi) \times c(\chi))^* \mathcal{P}_A^{\otimes -1}$ over $\overline{\Xi}_\xi$ $\forall \chi \in X_\chi$, again denoted by $\tau(\chi, \chi)$. Moreover the collection of sections $\{\tau(\chi, \chi) | \chi \neq 0\}$ vanishes simultaneously exactly on $\Xi_\xi \times^{E_\xi} W_\xi$. This assertion is a direct generalization of the corresponding statement in in the totally degenerate case, and is left as an exercise. Therefore when we pull back everything to the formal completion $\mathcal{S}_\xi$ of $\overline{\Xi}_\xi$ we get principally polarized semi-abelian degeneration data. In fact $\mathcal{S}_\xi$ can serve as a moduli space for certain types of degeneration-data: Homomorphisms into it correspond to data where the toral part of $\tilde{G}$ is the torus T_ξ with character group X_ξ, and where the bilinear form b on $X_\xi \times X_\xi$ (well determined up to units) can be obtained by pull-back from the universal form on $\overline{E}_\xi$. This last condition is equivalent to the fact that locally in the base there exists some cone σ_α of the decomposition such that the form "$(\mu, \nu) \to$ Weil-divisor of $b(\mu, \nu)$" is a linear combination of elements in the closure of σ_α. If the minimal such σ_α is always equal to σ we obtain maps into $\mathcal{S}_\sigma$.

Applying Mumford's construction we obtain a principally polarized semi-abelian scheme ${}^\heartsuit G \to \mathcal{S}_\xi$ over the formal scheme $\mathcal{S}_\xi$, and $\mathrm{GL}(X_\xi)$ acts on it by functoriality of Mumford's construction. The quotient is a semi-abelian scheme ${}^\heartsuit G \to \mathcal{S}_\xi/\mathrm{GL}(X_\xi)$ over the formal scheme $\mathcal{S}_\xi/\mathrm{GL}(X_\xi)$.

More precisely the semi-abelian scheme ${}^\heartsuit G \to \mathcal{S}_\xi$ is a functorial assignment which to every affine open formal subscheme $\mathcal{U}$ of $\mathcal{S}_\xi$ assigns a semi-abelian scheme ${}^\heartsuit G \to S_{\mathcal{U}} = \mathrm{Spec}(\Gamma(\mathcal{U}, \mathcal{O}_{\mathcal{U}}))$ which is abelian and principally polarized over the generic points. Note that $R_{\mathcal{U}} = \Gamma(\mathcal{U}, \mathcal{O}_{\mathcal{U}})$ is complete w.r.t. its natural adic topology; its (largest) ideal of definition will be denoted by $I_{\mathcal{U}}$.

If we take a completed local ring of $\mathcal{S}_\xi$ we get a semi-abelian scheme over (the spectrum of) a complete local ring. Clearly to each such complete local ring there is a cone $\sigma \in C(X_\xi)^\circ/\mathrm{GL}(X_\xi) \subset C(X)/\mathrm{GL}(X)$ naturally attached. (Its closed point is located at the stratum corresponding to σ mod $\mathrm{GL}(X)$.)

We will refer to the semi-abelian schemes (denoted by ${}^{\heartsuit}G$, over various bases) constructed above as *Mumford families.*

Properties of Mumford Families

We collect the important properties of ${}^{\heartsuit}G$ in the following proposition. For convenience we shall state it over an affine base.

3.1. Proposition. *Let* $S = \mathrm{Spf}(R, I)$ *be an irreducible affine open formal subscheme of* $\mathcal{S}_\xi$ *and* ${}^{\heartsuit}G \to \mathrm{Spec}(R)$ *the pull-back of* ${}^{\heartsuit}G$. *R is an I-complete excellent ring, formally smooth over* $\mathbf{Hom}_{\mathbb{Z}}(X_\xi, A)$, *hence also over* $\mathbb{Z}$. *Let K be the quotient field of R,* $S_0 = \mathrm{Spec}(R/I)$.

(i) The stratification of $\overline{\Xi}_\xi$ *determines a stratification of S parametrized by (i.e. with dual graph) a subcomplex of* $C(X_\xi)^\circ/\mathrm{GL}(X_\xi) \subset C(X)/\mathrm{GL}(X)$ *depending on R. Thus every point of S uniquely determines a cone in* $C(X_\xi)/\mathrm{GL}(X_\xi) \subset C(X)/\mathrm{GL}(X)$.

(ii) The completion of any strict localization of S in a point of S_0 *is isomorphic to the completion of a strict local ring of the torus embedding* $\overline{E}$ *attached to* $\{\sigma_\alpha\}$, *via a strata-preserving isomorphism.*

(iii) Under the above stratification, the cone attached to a point of S_0 *lies in* $C(X_\xi)^\circ/\mathrm{GL}(X_\xi)$. *The cone attached to a point* $s \in S$ *is* $\{0\}$ *iff* ${}^{\heartsuit}G_s$ *is abelian. More generally, if the dimension of the torus part of* ${}^{\heartsuit}G_s$ *is* r_1, *then the cone attached to s lie in a boundary component* $C(X_\zeta)^\circ\ \mathrm{mod}\mathrm{GL}(X)$ *with dimension* $X_\zeta = r_1$.

(iv) The formal completion of ${}^{\heartsuit}G$ *along the inverse image of* S_0 *is canonically isomorphic to the pull-back of* $\tilde{G}$ *to S.*

(v) The étale sheaf $\underline{X}({}^{\heartsuit}G)$ *is the quotient of the constant sheaf* X_ξ *such that on the* σ*-stratum* $\underline{X}({}^{\heartsuit}G)$ *is equal to* X_ξ/X_σ. *The tautological degeneration data with which we perform the Mumford construction defines a bilinear symmetric pairing* $B : X_\xi \times X_\xi \to \mathrm{Div}(S)$ *which induces the form* $\underline{B}({}^{\heartsuit}G)$ *associated to* ${}^{\heartsuit}G$. *In other words* $X({}^{\heartsuit}G)$ *and* $\underline{B}({}^{\heartsuit}G)$ *are induced by the* $\underline{X}_\xi$ *respectively* $\underline{B}_\xi$ *on* $\overline{\Xi}_\xi$ *defined before.*

(vi) Let $\Omega = \underline{\Omega}({}^{\heartsuit}G_{/S})$ *be the dual of the relative Lie algebra of* ${}^{\heartsuit}G$, $\Omega^1_S[d\log\infty] = \Omega^1_{S/\mathbb{Z}}[d\log\infty]$ *the universal finite module of differentials on S w.r.t the divisor* S_0 *with normal crossings. Then the Kodaira-Spencer mapping defines an isomorphism* $\rho : S^2(\Omega) \xrightarrow{\sim} \Omega^1_S[d\log\infty]$.

The proof of prop. 3.1 is easy at this stage. (i)–(iv) are immediate from the definition of ${}^{\heartsuit}G$ (and the theory of Mumford's construction). (v) follows from chapter III, §10. (vi) follows from chapter III, §9: 9.4. shows that ρ is equal to $\rho_{(\tilde{G},\imath)}$ and maps $S^2(\Omega)$ into $\Omega^1_S[d\log\infty]$. The restriction of ρ to the image of $\Omega_{\tilde{G}/S} \otimes \Omega_{A/S}$ in $S^2(\Omega)$ defines the Kodaira-Spencer map $\rho_{\tilde{G}}$, and maps this image isomorphically onto $\Omega^1_{\mathrm{Hom}(X_\xi,A)/S}$, because of the construction of $\tilde{G}$. III, 9.4 shows that ρ induces an isomorphism modulo the two isomorphic subbundles on both sides as above.

3.2. Definition. Let $\sigma \in C(X_\xi)^\circ$ be a cone in $\{\sigma_\alpha\}$. A *good formal σ-model* consists of

(a) a strict local ring R, complete w.r.t. an ideal $I = \mathrm{rad}(I)$,
(b) a stratification of $\mathrm{Spec}(R)$ with strata parametrized by faces of σ,
(c) a semi-abelian scheme G over $\mathrm{Spec}(R)$ which is abelian over the generic points,
(d) a principal polarization on the abelian part of G,
(e) a strata-preserving isomorphism between R and the completion of a strict local ring of $\overline{E}$ w.r.t. the σ-stratum,
(f) an isomorphism between $\underline{X}(G)$ and the pull-back of $\underline{X}_\xi$ under the isomorphism in (e)

which satisfy the following properties:

(i) The form $\underline{B}(G)$ is the pull-back of $\underline{B}_\xi$. In particular, G is abelian precisely over the open stratum.

(ii) The Kodaira-Spencer mapping defines an isomorphism $\rho : S^2(\Omega_{G/R}) \xrightarrow{\sim} \Omega^1_{R/\mathbb{Z}}[d\log\infty]$, where $\Omega_{G/R}$ is the dual of the relative Lie algebra of $G_{/R}$ and $\Omega^1_{R/\mathbb{Z}}[d\log\infty]$ is the module of logarithmic 1-differentials on R w.r.t. the stratification on $\mathrm{Spec}(R)$.

The result of this section can be summarized as follows:

3.3. Proposition. *(i) If s is a point of $\mathcal{S}_\xi$ which lies in a stratum corresponding to σ, then the pull-back of ${}^\heartsuit G$ to the completion of the strict local ring of s (along the σ-stratum) is a good formal σ-model.*

(ii) Let $G_{/R}$ be a good formal σ-model as in 3.2. Then it is induced from ${}^\heartsuit G \to \mathcal{S}_\sigma$ via a canonically defined morphism $\mathrm{Spf}(R) \to \mathcal{S}_\sigma$. *This morphism* $\mathrm{Spf}(R) \to \mathcal{S}_\sigma$ *is formally étale.*

Proof. (i) is immediate from prop. 3.1 (and the construction of ${}^\heartsuit G$). The first part of (ii) follows easily from the theory of degeneration, using the universal property of $\mathcal{S}_\sigma$. The morphism $\mathrm{Spf}(R) \to \mathcal{S}_\sigma$ respects the ideals which define the closures of the strata, since these ideals are determined by the bilinear forms. Therefore it is formally étale because it induces an isomorphism between the Ω^1's (consider determinants).

3.4. Remarks. (i) We have shown that the maps $\mathrm{Spec}(R/I)$ to the support of $\mathcal{S}_\sigma$ deduced from $\mathrm{Spf}(R) \to \mathcal{S}_\sigma$, for the various good formal σ-models, cover the whole σ-stratum.

(ii) The group $\mathrm{GL}(X)$ operates naturally on the set of good formal local models, compatibly with the natural $\mathrm{GL}(X)$-action on the set of ${}^\heartsuit G \to \mathcal{S}_\sigma$'s.

4. Construction of Local Charts: Algebraization

Introduction. In this section, we shall carry out step (c) of the outline in part 1. In the previous section we have constructed semi-abelian degenerations of principally polarized abelian varieties over *complete* base rings. In order to use them as pieces of the boundary of the moduli stack A_g, one has to *algebraize* them. Unfortunately, we cannot directly quote M. Artin. The reason is that we do not have an easily definable functor which gives rise to the Mumford families. (In fact, the non-existence of a "compactifying functor" is exactly what makes the compactification problem for A_g more difficult than for M_g.)

Instead of trying to directly algebraizing these Mumford families, we shall proceed by first approximating them. Great care has to be taken to control the approximation. Then in some sense the approximation gives an algebraization. More precisely, let ${}^{\heartsuit}G \to \mathrm{Spec}(R^{\wedge})$ be a Mumford family, where $R^{\wedge}$ is the formal completion of a strict henselian local ring R along the closed stratum. It is not easy (or even possible) to descend G to R w.r.t. the inclusion $R \subset R^{\wedge}$. However, it is easy to approximate ${}^{\heartsuit}G_{/R^{\wedge}}$ by another semi-abelian scheme $G_{/R}$ "as close as we want". Although $G \times_R R^{\wedge}$ is not isomorphic to ${}^{\heartsuit}G$ over $\mathrm{Spec}(R^{\wedge})$ in general, it turns out that it is isomorphic to the pull-back of $G \times_R R^{\wedge}$ via an automorphism $R^{\wedge} \to R^{\wedge}$ close to the identity. Therefore we can algebraize ${}^{\heartsuit}G \to \mathrm{Spec}(R^{\wedge})$ after twisting the base by an automorphism close to the identity. This serves our purpose, *because this way the relevant cone does not change.*

We shall preserve most of the notation employed in the previous sections, *except that R will denote the strict henselization of $\overline{E}$ or $\overline{\Xi}_{\xi}$ at a geometric point.* If the point lies in the σ-orbit of $\overline{E}$, and if I defines the σ-stratum in $\mathrm{Spec}(R)$, $R^{\wedge}$ denotes the I-adic completion of R. Usually we prefer to consider R not as a strict local ring of $\overline{E}$, but of $\overline{\Xi}_{\xi}$, where $\sigma \subset C(X_{\xi})^{\circ}$.

$\mathrm{Spec}(R^{\wedge})$ is the base of a good formal σ-model, as defined in the last section. Over $\mathrm{Spec}(R^{\wedge})$ we have a semi-abelian variety ${}^{\heartsuit}G$ such that the generic fibre of ${}^{\heartsuit}G$ is a principally polarized abelian variety, and ${}^{\heartsuit}G \times_{\mathrm{Spec}(R^{\wedge})} \mathrm{Spec}(R/I)$ is isomorphic to the universal $\tilde{G}$. Furthermore $\underline{X}({}^{\heartsuit}G)$ and $\underline{B}({}^{\heartsuit}G)$ are given by the universal $\underline{X}_{\xi}$ and $\underline{B}_{\xi}$.

Now $\hat{R}$ is the filtering direct union of its normal subalgebras R_{γ} which are of finite type over R. Furthermore by M. Artin's fundamental result [Ar 3], for any such R_{γ} the inclusion $R \subset R_{\gamma}$ has sections $R_{\gamma} \to R$, such that the composition $R_{\gamma} \to R \subset \hat{R}$ is arbitrarily close to $R_{\gamma} \subset \hat{R}$ in the I-adic topology. Since ${}^{\heartsuit}G \to \mathrm{Spec}(R^{\wedge})$ is of finite presentation, by [EGA IV] we can find an R_{δ} such that the ${}^{\heartsuit}G$, and the isomorphism ${}^{\heartsuit}\lambda : {}^{\heartsuit}G \xrightarrow{\sim} {}^{\heartsuit}G^{t}$ induced by the principal polarization are already defined over R_{δ}. Moreover we may assume that the isomorphism between the pairs $(\underline{X}({}^{\heartsuit}G), \underline{B}({}^{\heartsuit}G))$ and $(\underline{X}_{\xi}, \underline{B}_{\xi})$ is already defined over R_{δ}. By pull-back with a section $R_{\delta} \to R$ of $R \subset R_{\delta}$ with $R_{\delta} \to R \hookrightarrow R^{\wedge}$ close to $R_{\delta} \hookrightarrow R^{\wedge}$, we obtain a pair (G, λ) consisting of a semi-abelian scheme G over R with abelian generic fibre and a principal polarization λ of the generic fibre, such that $\underline{X}(G) = \underline{X}_{\xi}$, and $\underline{B}(G) = \underline{B}_{\xi}$.

Of course the extended pair $(G, \lambda) \otimes_R R^\wedge$ over $\hat{R}$ is not identical to the original $({}^\heartsuit G, {}^\heartsuit \lambda)$, but they are close in the I-adic topology. More precisely we claim we can achieve the following:

4.1. Claim. (i) over $\mathrm{Spec}(R/I)$, both G's are isomorphic: ${}^\heartsuit G \otimes_{R^\wedge} R/I \cong G \otimes_R R/I$.

(ii) the degeneration data for $(G, \lambda) \otimes_R R^\wedge$ and $({}^\heartsuit G, {}^\heartsuit \lambda)$ are close $IR^\wedge$-adically.

Explicitly, (ii) means: $\exists n > 0$, which can be made arbitrarily large by choosing R_δ sufficiently large and $R_\delta \to R \hookrightarrow R^\wedge$ close enough to $R_\delta \hookrightarrow R^\wedge$, that the abelian parts ${}^\heartsuit A$, A of ${}^\heartsuit G$, $G \otimes_R R^\wedge$ respectively are isomorphic modulo $I^n R^\wedge$ (as principally polarized abelian schemes), and the maps $c : X_\sigma \to {}^\heartsuit A(R^\wedge)$, $c : X_\sigma \to A(R^\wedge)$ are congruent modulo $I^n R^\wedge$. Moreover, one can choose symmetric trivializations of the biextensions of the two Poincaré sheaves over $X \times X$ which are equal modulo I^n, such that the two resulting bimultiplicative symmetric forms (which coincide up to units) ${}^\heartsuit b$, b on X_σ are multiplicatively congruent modulo $1 + I^n R^\wedge$. (Notice that we have implicitly identified A with its dual using the principal polarization.)

Proof of the Claim 4.1. We shall need a lemma:

4.2. Lemma. *Suppose R is a noetherian ring, $I \subset R$ an ideal such that $\mathrm{gr}_I(R)$ is an integral domain and such that the I-adic topology on R is separated. Suppose furthermore that $f, g \neq 0$ are elements of R such that f/g lies in R. If $f_n, g_n \neq 0$ are sequences of elements of R which converge I-adically to f and g respectively, and if all the quotients f_n/g_n lie in R, then the quotients f_n/g_n converge I-adically to f/g.*

Proof. $g \cdot g_n \cdot (f/g - f_n/g_n) = f \cdot g_n - f_n \cdot g$ converges to 0. Clearly for big n the I-adic order of $g \cdot g_n$ is constant. Since $\mathrm{gr}_I(A)$ is an integral domain, the I-adic order of a product is the sum of the orders of its terms. In particular, $\mathrm{ord}_I(f/g - f_n/g_n) = \mathrm{ord}_I(f \cdot g_n - f_n \cdot g) - \mathrm{ord}_I(g \cdot g_n) \to \infty$ as $n \to \infty$. The assertion follows.

Let us first prove the following statement:
(*) If R_δ is sufficiently large, by choosing the section $R_\delta \to R$ with $R_\delta \to R \hookrightarrow R^\wedge$ sufficiently close to $R_\delta \hookrightarrow R^\wedge$, we can make b and ${}^\heartsuit b$ close additively, in the following sense: $\exists R^\wedge$-valued quadratic functions a, ${}^\heartsuit a$ on X_σ with b, ${}^\heartsuit b$ as their associated bimultiplicative functions, such that $a(\mu) \equiv {}^\heartsuit a(\mu)$ modulo some big power of I, $\forall \mu \in X_\sigma$.

Note that by lemma 4.2 this assertion implies that for any given linear combination $\sum \mu_i \otimes \nu_i \in S^2(X_\sigma)$ (= the symmetric quotient of $X_\sigma \times X_\sigma$) which lies in the dual of the cone of positive semi-definite forms, the corresponding products $\prod b(\mu_i, \nu_i)$ (which are integral by the assumption) are I-adically close in $\hat{R}$. As they coincide up to units we are done.

To prove (*) it suffices to choose an ample invertible sheaf ${}^\heartsuit \mathcal{L}$ on ${}^\heartsuit G$ giving rise to the principal polarization (which always exists), and a non-zero section

${}^{\heartsuit}s$ of $\mathcal{L}$ over $R^\wedge$. After choosing a cubical trivialization of the restriction of ${}^{\heartsuit}\tilde{\mathcal{L}}$ to a lifting of c to ${}^{\heartsuit}\tilde{G}(R^\wedge)$, we obtain an $R^\wedge$-valued quadratic functions ${}^{\heartsuit}a$ on X_σ with ${}^{\heartsuit}b$ as its associated bimultiplicative function by expanding ${}^{\heartsuit}s$ as sum of T_σ-eigenfunctions. We get $a : X_\sigma \to R^\wedge$ in a similar way, using a cubical trivialization of the restriction of $\tilde{\mathcal{L}}$ to a lifting of c to $\tilde{G}(R^\wedge)$ congruent to the previous one modulo a high power of $IR^\wedge$. Of course we use $\mathcal{L} = (R_\delta \to R)^*({}^{\heartsuit}\mathcal{L})$, $s = (R_\delta \to R)^*({}^{\heartsuit}s)$, provided ${}^{\heartsuit}\mathcal{L}$ and ${}^{\heartsuit}s$ are already defined over R_δ. Since s and ${}^{\heartsuit}s$ are $IR^\wedge$-adically close, so are a and ${}^{\heartsuit}a$. This proves the assertion.

Combining results of chap. III, §§ 9, 10 and (part of) what has been proved above, we obtain the following

4.3. Proposition. *Suppose $\hat{R}$ is the I-adic completion of the strict local ring R of $\overline{\Xi}_\xi$ at a geometric point of the σ-stratum, $I \subset R$ is the ideal defining the σ-stratum, and ${}^{\heartsuit}G$, ${}^{\heartsuit}\lambda : {}^{\heartsuit}G \xrightarrow{\sim} {}^{\heartsuit}G^t$ together with the stratification of* Spec$(R^\wedge)$ *induced by that of $\overline{E}$ define a good formal model over $\hat{R}$. Then we may find a pair (G, λ) consisting of a semi-abelian scheme G over R with abelian generic fibre and a principal polarization $\lambda : G \xrightarrow{\sim} G^t$, which approximates the original pair $({}^{\heartsuit}G, {}^{\heartsuit}\lambda)$ in the following sense:*

(a) Over Spec(R/I) *both pairs are isomorphic.*

(b) $\underline{X}(G)$ and $\underline{B}(G)$ induce $\underline{X}({}^{\heartsuit}G)$ respectively $\underline{B}({}^{\heartsuit}G)$.

(c) The maps from Spec(R/I) *into the σ-stratum of S_σ (defined because of (b)) coincide.*

Moreover the following holds:

(d) The Kodaira-Spencer class of (G, λ) induces an isomorphism $S^2(\Omega_{G/R}) \cong \Omega^1_{R/\mathbb{Z}}[d\log\infty]$.

(d) follows from III, §9, after passing to a larger R_δ if necessary.

Consider a geometric point $\overline{x}$ of the σ-stratum of $\overline{\Xi}_\xi$. Its strict local ring R is the inductive limit of the coordinate ring R_α of all affine étale neighborhoods of $\overline{x}$. Now we can play the same trick again: A pair (G, λ) over R as in 4.3 is already defined over some R_α etc.. After passing to larger R_α (i.e. smaller étale neighborhood) if necessary, we obtain:

4.4. Proposition. *Let R be the strict local ring of a geometric point $\overline{x}$ of the σ-stratum of $\overline{\Xi}_\xi$, and $R^\wedge$ the completion of R along the σ-stratum. There exist an étale neighborhood* Spec(R_α) *of $\overline{x}$, a pair (G, λ) over* Spec(R) *consisting of a semi-abelian scheme with abelian generic fibre and a principal polarization, and an embedding $R \hookrightarrow R^\wedge$ close to the canonical inclusion, such that:*

(o) The stratification on Spec(R) *is parametrized by faces τ of σ.*

(i) $\underline{X}(G)$ is equal to the pull-back of $\underline{X}_\xi$, and $\underline{B}(G)$ to the pullback of $\underline{B}_\xi$.

(ii) The Kodaira-Spencer map induces an isomorphism $S^2(\Omega_G) \cong \Omega^1_R[d\log\infty]$.

(iii) The universal map Spf$(R^\wedge) \to S_\sigma$ *coincides on the σ-stratum with the map induced by the inclusion $R \subset \hat{R}$.*

4.5. Definition. We call such an étale neighborhood, together with the stratification and (G, λ) as above, a *good algebraic σ-model.*

We have shown that the good algebraic σ-models cover the σ-stratum in $\mathcal{S}_\sigma$. For any geometric point $\overline{x}$ in the τ-stratum of a good algebraic σ-model, we obtain by pull-back a good formal τ-model over the completion along the τ-stratum of the strict local ring at $\overline{x}$ by III §§ 9, 10. More generally, for any point y in the τ-stratum, the semi-abelian variety induced by G over the completion of the strict henselization $R^{\wedge}_{\overline{y}}$ at a geometric point $\overline{y}$ over y gives degeneration data, and as we already know the bilinear form $\underline{B}$ we see that we obtain a morphism from $\mathrm{Spf}(R^{\wedge}_{\overline{y}})$ to $\mathcal{S}_\tau$. By III §§ 9, 10, this morphism is formally étale, and $(G, \lambda)_{/\mathrm{Spec}(R^{\wedge}_{\overline{y}})}$ is the pull-back of the Mumford family $({}^{\heartsuit}G, {}^{\heartsuit}\lambda)$ over $\mathcal{S}_\tau$.

Remark. Again everything behaves well under the natural action of the group $\mathrm{GL}(X)$.

5. Construction of $\overline{\mathsf{A}}_g$: Gluing Good Algebraic Models

The compactification $\overline{\mathsf{A}}_g$ will be constructed as the quotient of an étale covering U of $\overline{\mathsf{A}}_g$ by an étale equivalence relation R. For U we take the disjoint union of finitely many good algebraic σ-models $\mathrm{Spec}(R_i)$. More precisely, we choose finitely many σ's such that any simplex in the chosen rational polyhedral cone decomposition is conjugate to one of them under $\mathrm{GL}(X)$, and for each of them we choose finitely many good algebraic σ-models $\mathrm{Spec}(R_i)$ such that the corresponding tale maps from $\mathrm{Spec}(R_i/I)$ (notation as in §4) to the σ-stratum in $\mathcal{S}_\sigma$ cover this stratum. U is their disjoint union. Our convention is that the stratum $\{0\}$ corresponds to A_g, hence U contains an étale covering of A_g by finitely many affines. We remark that the compactification we construct will not depend on the choice of U (although it does depend on the chosen rational polyhedral cone decomposition), as the reader can easily check as we proceed.

By construction U is stratified with strata parametrized by classes σ modulo $\mathrm{GL}(X)$. Over U we have a semi-abelian variety G which is abelian over the open dense stratum U_0 corresponding to $\{0\}$. The abelian part has a principal polarization $\lambda : G_{/U_0} \xrightarrow{\sim} G^t_{/U_0}$, which extends to a homomorphism $G \xrightarrow{\sim} G^t$ by I.2.7. Furthermore if we form the completion (w.r.t. the maximal ideal) of a strict local ring of U in a point of the σ-stratum, this is isomorphic to the completion of a strict local ring of $\tilde{\mathcal{S}}_\sigma$ in a geometric point of its σ-stratum, the isomorphism being such that G with its principal polarization is given by the Mumford construction applied to the universal semi-abelian degeneration data with principal polarization on $\mathcal{S}_\sigma$.

The construction of U is motivated by the following proposition 5.1. Strictly speaking, it is not logically necessary for the construction of the compactification $\overline{\mathsf{A}}_g$. Given the compactification, it can be viewed as a property of (the complete strict local rings of) the boundary points of $\overline{\mathsf{A}}_g$, which characterizes the (chosen) rational polyhedral cone decomposition and hence the compactification $\overline{\mathsf{A}}_g$.

5.1. Proposition. *Suppose R is a normal complete local ring which is strictly henselian, K its fraction field and k its residue field. Assume furthermore that $G^\dagger$ is a semi-abelian variety over R whose generic fibre $G^\dagger_K$ is an abelian variety with a principal polarization λ_K. Associated to this we have a symmetric bimultiplicative form b on the character group of the torus part of the special fibre of G, defined up to units in R, with values in K^*. The following statements are equivalent:*

i) There exists a morphism $\mathrm{Spec}(R) \to U$*, sending the generic point into the 0-stratum, such that $G^\dagger$ is isomorphic to the pull-back of the G defined over U, with the isomorphism respecting the polarization on the generic fibre.*

ii) Write the character group of the torus part of the special fibre in some way as a quotient lattice of $X = \mathbb{Z}^g$. If $v : K^* \to \mathbb{Z}$ is any discrete valuation defined by a prime of R of height one, $v \circ b$ defines a positive semi-definite symmetric bilinear form on X. There exists a (closed) simplex in the decomposition which contains all $v \circ b$ obtained this way.

If the above two conditions hold, the closed point of $\mathrm{Spec}(R)$ *maps to a point of the σ-stratum in U iff σ is minimal.*

5.2. Lemma. *Let $G^\dagger_{/R}$ be as in 5.1. Then statement (ii) in 5.1 is equivalent to:*

(iii) $(G^\dagger, \lambda)$ is the pull-back of a Mumford family $({}^\heartsuit G, {}^\heartsuit\lambda) \to \mathcal{S}_\sigma$ via a map $\mathrm{Spf}(R) \to \mathcal{S}_\sigma$ *(equivalently, a map* $\mathrm{Spec}(R) \to \Xi(\sigma)$*) for some cone σ in the chosen rational polyhedral cone decomposition. In fact, the two σ's in (ii) and (iii) are in the same* $\mathrm{GL}(X)$*-orbit, hence can be chosen to be the same.*

The proof of 5.2 is standard.

Proof of Proposition 5.1. It is easy to see that i) implies ii): The pull-back of $G \to U$ to the completion of a strict local ring at a point of the σ-spectrum is isomorphic to a Mumford family centered at a point over $\mathcal{S}_\sigma$. Hence (i) implies (iii) $\Leftrightarrow$ (ii).

Now for the converse (ii) $\Rightarrow$ (i): Choose σ minimal such that all $v \circ b$ lie inside its closure. Then some positive linear combination of the $v \circ b$'s must lie in σ (= interior of $\overline{\sigma}$). As such a linear combination is positive definite on the character group of the torus part of the special fibre of $G^\dagger$, this group must be equal to X_σ. Changing the way we write the character group of the torus part as a quotient of X, we may (and do) assume that σ occurs in the finite selection used in the construction of U.

The semi-abelian degeneration data attached to $(G^\dagger, \lambda)$ defines a map $\mathrm{Spec}(K) \to \Xi(\sigma)$. The argument of lemma 5.2 shows that it extends to a morphism $\mathrm{Spec}(R) \to \Xi(\sigma)$. In fact one derives that the closed point of $\mathrm{Spec}(R)$ is mapped to a point z of the σ-stratum, since otherwise σ would not be minimal. By construction there exists a good algebraic σ-model over some $\mathrm{Spec}(R_\alpha)$

such that the image under the étale map of $\mathrm{Spec}(R_\alpha/I_{\alpha,\sigma})$ into the σ-stratum of $\Xi(\sigma)$ contains the point z, where $I_{\alpha,\sigma}$ is the ideal defining the σ-stratum of $\mathrm{Spec}(R_\alpha)$. The étale map naturally extends to a morphism from the formal spectrum of the completion $R_\alpha^\wedge$ of $\mathrm{Spec}(R_\alpha)$ along $I_{\alpha,\sigma}$ to $\mathcal{S}_\sigma$, as the data defining it do, and the result is formally étale. It follows that the morphism $\mathrm{Spf}(R) \to \mathcal{S}_\sigma$ lifts (uniquely) to $\mathrm{Spf}(R) \to \mathrm{Spf}(R_\alpha^\wedge)$, and the corresponding morphism $\mathrm{Spec}(R) \to \mathrm{Spec}(R_\alpha^\wedge)$ identifies the degeneration data of $(G^\dagger, \lambda)_{/R}$ with the pull-back of the degeneration of $(G, \lambda)_{/R_\alpha^\wedge}$. It follows that $(G^\dagger, \lambda)_{/R}$ is isomorphic to the pullback of $(G, \lambda)_{/R_\alpha^\wedge}$, via $\mathrm{Spec}(R) \to \mathrm{Spec}(R_\alpha^\wedge) \to \mathrm{Spec}(R_\alpha)$. This finishes the proof of prop. 5.1.

5.3. Remark. The assumption (ii) of prop. 5.1 is always fulfilled if R is a (complete) discrete valuation-ring (because there is only one choice of v, and the union of the cones cover $C(X)$).

The next step is constructing the étale equivalence relation R on U. There is a natural candidate: Over the dense open stratum U_0, $G \to U_0$ becomes a principally polarized abelian scheme, and there is an étale groupoid $R_0 = \mathrm{Isom}_{U_0 \times_S U_0, \mathrm{pol}}(\mathrm{pr}_1^* G, \mathrm{pr}_2^* G)$ over $U_0 \times_S U_0$ ($S = \mathrm{Spec}(\mathbb{Z})$). Since U/R should extend $\mathsf{A}_g = U_0/R_0$, R should extend R_0. Moreover, since R is expected to be an étale equivalence on U, it should be the normalization of $R_0 \to U \times_S U$. (i.e. normalization of the closure Z of the image of R_0 in $U \times_S U$ w.r.t. the finite extensions of the function fields at the maximal points of Z.) *So we define R as the normalization of $R_0 \to U \times_S U$.* By definition R is normal, and so the universal isomorphism on R_0 extends to a (polarization-preserving) isomorphism $h : (\mathrm{pr}_1^* G)_{/R} \xrightarrow{\sim} (\mathrm{pr}_2^* G)_{/R}$ of the two pull-backs of G on R by I.2.7.

The next result contains almost everything we need for compactification. The key point, rather trivial as it seems, is that *the cones do not overlap.* (Otherwise we would not be able to glue.)

5.4. Proposition. *The two projections from R onto U are étale.*

Proof. We shall show that for any geometric point z of R, the two projections pr_i ($i = 1, 2$) induce an isomorphism between the completion of the strict henselization of R at z and the completion of the strict henselization of U at $\mathrm{pr}_i(z)$. More precisely, let $x = \mathrm{pr}_1(z)$, $y = \mathrm{pr}_2(z)$, and R_{12} (resp. R_1, R_2) be the completions of the strict henselizations of R (resp. U) at z (resp. x, y). As any irreducible component of R dominates a component of U via either projection, R_1 and R_2 are naturally subrings of R_{12}. Trivially, the maximal ideal of R_{12} intersects both R_1 and R_2 in their maximal ideals (i.e. $\mathrm{pr}_i^* : R_i \to R_{12}$ is a local homomorphism $\forall i = 1, 2$.) We shall show that $R_{12} = R_1 = R_2$ via pr_1^* and pr_2^* respectively. This clearly suffices.

There are simplices σ, τ in the chosen rational polyhedral cone decomposition such that R_1 and R_2 are isomorphic to completions of strict local rings of $\mathcal{S}_\sigma$ (resp. $\mathcal{S}_\tau$) at points of the σ-stratum (resp. τ-stratum). The semi-abelian varieties G_1 over R_1 and G_2 over R_2 are given by the Mumford construction, applied to the semi-abelian degeneration data with principal polarization $(\tilde{G}_1, X_\sigma, c_1, \lambda_1, \imath_1, \phi_1, \tau_1)$ respectively $(\tilde{G}_2, X_\tau, c_2, \lambda_2, \imath_2, \phi_2, \tau_2)$. Here we use notation of chap. II. As G_1 and G_2 become isomorphic over R_{12} via the isomorphism h as in 5.3.1, f induces an isomorphism between $(\tilde{G}_1, X_\sigma, c_1, \lambda_1, \imath_1, \phi_1, \tau_1) \otimes_{R_1} R_{12}$ and $(\tilde{G}_2, X_\tau, c_2, \lambda_2, \imath_2, \phi_2, \tau_2) \otimes_{R_2} R_{12}$. In particular, there exists an isomorphism $\alpha : \tilde{G}_{1/R} \to \tilde{G}_{2/R}$ over R_{12}, which must induce an isomorphism between the torus parts, hence an isomorphism $\alpha^* : X_\tau \to X_\sigma$.

We claim that α^* identifies the simplices σ and τ in $S^2(X_\sigma)_\mathbb{R}$ respectively $S^2(X_\tau)_\mathbb{R}$. This follows easily from the fact that the two bilinear forms B_1 and B_2 (with values in $\mathrm{Div}(R_1)$, $\mathrm{Div}(R_2)$ respectively) determine the cones σ and τ, and B_1, B_2 become equal when extended to $\mathrm{Div}(R)$. Explicitly: after identifying X_σ and X_τ, if $\sigma \neq \tau$, the intersection $\sigma \cap \tau$ would be a proper face in one of them, say in σ. It follows that there exists a linear combination $\sum \mu_i \otimes \nu_i$ in $S^2(X_\sigma)$ such that $f = \prod b_1(\mu_i, \nu_i)$ is integral and even non-invertible in R_1, while f^{-1} is an element of R_2. Thus f lies in the maximal ideal of R_{12}, and is invertible in R_{12} at the same time. Contradiction!

The above claim shows that α^* induces an isomorphism between $\mathcal{S}_\sigma$ and $\mathcal{S}_\tau$, and up to this isomorphism the two mappings $\mathrm{Spec}(R_{12}) \to \mathrm{Spec}(R_1) \to \Xi(\sigma)$, $\mathrm{Spec}(R_{12}) \to \mathrm{Spec}(R_2) \to \Xi(\tau)$ defined by the two sets of degeneration data coincide. As R_1 and R_2 are completions of the strict henselizations of $\Xi(\sigma)$ and $\Xi(\tau)$ at the image of the closed point of $\mathrm{Spec}(R_{12})$, they coincide as subrings of R_{12}. So we obtain an isomorphism between R_1 and R_2 which transforms one set of semi-abelian data into the other. It follows from the theory of degeneration that the isomorphism between $G_{1/\mathrm{Spec}(R_{12})}$ and $G_{2/\mathrm{Spec}(R_{12})}$ is already defined over $R_1 = R_2$, considered as a subring of R_{12}. We derive a map from the generic point of $\mathrm{Spec}(R_1)$ into $R_0 = \mathrm{Isom}_{\mathrm{pol}}(\mathrm{pr}_1^* G, \mathrm{pr}_2^* G)$ (over $U_0 \times_S U_0$), which covers the map into $U_0 \times_S U_0$ with components given by the identity of R_1, respectively the isomorphism of R_1 with R_2. By definition of R this lifts to a regular map from $\mathrm{Spec}(R_1)$ into R, which sends the closed point of $\mathrm{Spec}(R_1)$ to the geometric point z of R. This gives a homomorphism $R_{12} \to R_1$, which is a left inverse to $\mathrm{pr}_1^* : R_1 \hookrightarrow R_{12}$, hence $R_{12} = R_1$ via pr_1^* and consequently $R_{12} = R_2$ via pr_2^* and the proposition is proved. Another way to conclude the proof is to use the fact that formation of the Isom scheme commutes with arbitrary change of base scheme, and therefore R is equal to a connected component of the normalization of $\mathrm{Isom}_{\mathrm{pol}}(\mathrm{pr}_1^* G_{/(U_0 \times_S U_0) \otimes R_1 \otimes R_2}, \mathrm{pr}_2^* G_{/(U_0 \times_S U_0) \otimes R_1 \otimes R_2}) \to \mathrm{Spec}(R_1 \otimes R_2)$ by the analytic normality of excellent rings.

In fact the above argument shows much more, but let us try first to make good use of the proposition. An immediate consequence is

5.5. Corollary. *R is an étale groupoid, hence U/R defines a separated algebraic stack.*

Proof. Any fibred product over U of copies of R is étale over U, hence normal and equal to the normalization of R_0. The necessary maps between those fibered products are obtained by extension to the normalization, and all the axioms hold by continuity. This proves that R a groupoid. Since R is finite over $U \times_S U$ by definition, U/R is a separated algebraic stack. It only depends on the cone-decomposition, as follows easily from statement (5) in theorem 5.7 below, which itself is a variant of proposition 5.1.

5.6. Definition. The separated algebraic stack U/R constructed above will be denoted $\overline{\mathsf{A}}_g$, by abuse of notation. As remarked before, it depends on the choice of the rational polyhedral cone decomposition. Therefore we shall write $\overline{\mathsf{A}}_g\ (\{\sigma_\alpha\})$ when we want to indicate the dependence on the cone decomposition. Note that $\overline{\mathsf{A}}_g$ does not depend on the choice of étale covering U.

By construction, $\overline{\mathsf{A}}_g$ contains A_g, which is open dense in $\overline{\mathsf{A}}_g$. Furthermore the semi-abelian variety G over U has descent data over R by I.2.7, and thus defines a locally quasiprojective semi-abelian scheme G over $\overline{\mathsf{A}}_g$ which extends the universal abelian scheme over A_g. Finally, $\overline{\mathsf{A}}_g$ is proper over $\mathrm{Spec}(\mathbb{Z})$, by the valuative criterion: If V is a complete discrete valuation-ring with algebraically closed residue-field, K its field of fractions, and $\mathrm{Spec}(K) \to \mathsf{A}_g$ a morphism. We would like to show that it extends to a morphism $\mathrm{Spec}(V) \to \overline{\mathsf{A}}_g$. The morphism $\mathrm{Spec}(K) \to \mathsf{A}_g$ gives a separable overfield L of K and a principally polarized abelian variety G_L over L together with descent data of G_L w.r.t. $\mathrm{Spec}(L) \to \mathrm{Spec}(K)$. By the theorem of semi-stable reduction for abelian varieties, we may assume that G has semi-stable reduction over $\mathrm{Spec}(V)$, i.e. that there is a semi-abelian scheme $G_V \to \mathrm{Spec}(V)$ extending G_K. The theory of degeneration tells us that G_V corresponds to a set of degeneration data over the complete discrete valuation ring V. The base being 1-dimensional, the cone attached to the degeneration data (as in 5.2) is a half-line. Thus proposition 5.2 tells us that G_V is a pull-back of the universal G over U. (Actually, the special case of 5.2 used here admits a shorter proof.) Thus the morphism $\mathrm{Spec}(K) \to \mathsf{A}_g$ lifts to U and extends to $\mathrm{Spec}(V)$, and the valuative criterion for properness is satisfied.

Let us formulate our main result:

5.7. Theorem. *Let X be a free abelian group of rank g. To a smooth* $\mathrm{GL}(X)$-*admissible polyhedral cone decomposition $\{\sigma_\alpha\}$ of $C(X)$, we can attach an algebraic stack $\overline{\mathsf{A}}_g\ (\{\sigma_\alpha\})$ ($= U/R$ as constructed above), which has the following properties:*

(1) $\overline{\mathsf{A}}_g = U/R$ is a proper algebraic stack over $\mathrm{Spec}(\mathbb{Z})$ *containing A_g as an open dense algebraic substack.*

(2) The complement of A_g in $\overline{\mathsf{A}}_g$ is a relative Cartier divisor D_∞ with normal crossings. $\overline{\mathsf{A}}_g$ has a stratification parametrized by $\mathrm{GL}(X)$-*orbits of simplices σ in the cone-decomposition of $C(X)$: $\overline{\mathsf{A}}_g = \amalg_\sigma Z(\sigma)$, with σ running over $\{\sigma_\alpha\}/\mathrm{GL}(X)$, and $C(X)/\mathrm{GL}(X)$ is the dual complex of this stratification. Each stratum is open dense in an intersection of components of D_∞ (including possible self-intersection). Locally in the étale topology, $\overline{\mathsf{A}}_g$ is isomorphic to $\overline{E}$, the*

isomorphism respecting stratifications and the parametrization by $\{\sigma_\alpha\}/\mathrm{GL}(X)$*. In particular the closure of each stratum is smooth over* $\mathrm{Spec}(\mathbb{Z})$*.*

(3) The universal abelian scheme over A_g *extends to a semi-abelian scheme* $G \to \overline{\mathsf{A}}_g$*. The Kodaira-Spencer mapping for* $G \to \overline{\mathsf{A}}_g$ *induces an isomorphism* $S^2(\mathbf{Lie}(G)^*) \cong \Omega_{\overline{\mathsf{A}}_g/\mathbb{Z}}[d\log D_\infty]$ *of locally free sheaves of rank* $(1/2)\cdot g\cdot(g+1)$ *on* $\overline{\mathsf{A}}_g$*.*

(4) The formal completion $\overline{\mathsf{A}}_g^{/\sigma}$ *of* $\overline{\mathsf{A}}_g$ *along the* σ*-stratum is isomorphic to the formal algebraic stack* $\mathcal{S}_\sigma/\Gamma_\sigma$ *(i.e. quotient of* $\mathcal{S}_\sigma$ *as stack under the étale equivalence relation determined by* Γ_σ*), where* $\mathcal{S}_\sigma$ *is the formal completion of the* $\overline{E}_\xi$*-bundle* $\overline{\Xi}_\xi$ *along its* σ*-stratum as in 3.2, and* Γ_σ *is the subgroup of elements of* $\mathrm{GL}(X_\sigma)$ *which stabilizes* $\sigma \subset S^2(X_\sigma \otimes \mathbb{R})$*. (Note that* Γ_σ *is finite and acts naturally on* $\mathcal{S}_\sigma$*, and the quotient exists as an algebraic stack.) The isomorphism respects stratifications. The degenerating semi-abelian scheme* $G \to \overline{\mathsf{A}}_g^{/\sigma}$ *is isomorphic to the Mumford family* $({}^{\heartsuit}G \to {}^{\heartsuit}\mathcal{S}_\sigma)/\Gamma_\sigma$ *after identifying the base stacks.*

(5) Let S *be an irreducible normal scheme,* $G^\dagger \to S$ *a semi-abelian scheme with abelian generic fibre and a principal polarization on the generic fibre. Then* $G^\dagger \to S$ *with its principal polarization is the pull-back of the universal* G *over* $\overline{\mathsf{A}}_g$ *via a (necessarily unique) morphism* $S \to \overline{\mathsf{A}}_g$ *iff for every point* s *of* S*, there is a cone* σ *in* $\{\sigma_\alpha\}$ *and a surjection* $X \to X_s$ *such that for any complete discrete valuation ring with algebraically closed residue field and any morphism* $\mathrm{Spec}(V) \to S$ *centered at* s*, the element* $\mathrm{ord}\circ b \in B(X_s) \subseteq B(X)$ *is contained in* σ*. (Here* b *is the symmetric bimultiplicative form on* X_s *defined up to unit in* V^* *by the theory of degeneration.)*

Proof. (1) has already been proved. (3) follows immediately from the construction. For (2) we only need to remark that R respects stratifications, which follows from the proof of proposition 5.3.2 although not stated explicitly. (5) is essentially an easy corollary of proposition 5.1: The abelian part of $G^\dagger$ defines a morphism from a dense open subscheme U of S to $\mathsf{A}_g \subset \overline{\mathsf{A}}_g$. Since extendibility is a local condition, prop. 5.1 shows that this morphism extends to $S \to \overline{\mathsf{A}}_g$. This means that there is an étale covering S' of S and a semi-abelian scheme $G \to S'$, together with descent data w.r.t. $f : S' \to S$, such that $G|_{f^{-1}(U)}$ descends to $G^\dagger|_U$. Then $G^\dagger \times_S S'$ is isomorphic to G by I.2.7, i.e. G descends to $G^\dagger$. (This is the what we mean by saying that $G^\dagger$ is the pull-back of the universal G.)

Let us prove (4). Essentially everything follows from the considerations in the proof of prop. 5.4. By (5) (which was just proved) we get a morphism from $\mathcal{S}_\sigma$ into $\overline{\mathsf{A}}_g$. This defines a surjective formally étale morphism from $\mathcal{S}_\sigma$ into the formal completion of $\overline{\mathsf{A}}_g$, along the σ-stratum, which is easily seen to factor through the algebraic stack quotient under the stabilizer Γ_σ of σ in $\mathrm{GL}(X_\sigma)$. For an inverse morphism we first use the fact that in the construction of U we naturally obtain a morphism from its formal completion along the σ-stratum to $\mathcal{S}_\sigma$. In the proof of prop. 5.4. we saw that any geometric point in the σ-stratum of R is related to elements in $\mathrm{GL}(X_\sigma)$ which stabilizes σ. (We mean the isomorphism $\alpha^* : X_\tau \xrightarrow{\sim} X_\sigma$ there which is unique up to Γ_σ.) One easily derives a map from the formal completion of $\overline{\mathsf{A}}_g$ along the σ-stratum into the quotient stack of $\mathcal{S}_\sigma$ under the finite group Γ_σ. The two morphisms are clearly inverse to each other, and assertion (4) follows.

5.8. Remarks. (a) When we do not assume that $\gamma \cdot \overline{\sigma} \cap \overline{\sigma} \neq \{0\}$ for $\gamma \in \mathrm{GL}(X)$ implies that γ acts as identity on the smallest (rational) boundary component of $C(X)$ containing σ, the strata of $\overline{\mathsf{A}}_g$ may have self-intersections. One important feature of the toroidal compactification is that these self-intersections are *without monodromy* in the sense of [AMRT], p.16. This partially explains why these self-intersections never give rise to serious trouble in applications.

(b) In statement (5) of theorem 5.7, it (obviously) suffices to use completion of localization at height-one prime ideals of $\mathcal{O}_s$.

Statement (4) in theorem 5.7 has a more global analogue/generalization: Define (intersections of) relative Cartier divisors with normal crossings D_r to be the unions (with reduced structure) of strata parametrized by cones σ whose elements when considered as belonging to $S^2(X_{\mathbb{R}})$ have radical of dimension r. (In other words, cones which lie in some $C(X_\xi)^\circ$ with $\dim(X_\xi) = g - r$.)

5.9. Proposition. *The formal completion of $\overline{\mathsf{A}}_g$ along D_r is isomorphic to the stack quotient of $\mathcal{S}_\xi$ by $\mathrm{GL}(X_\xi)$, where X_ξ is any quotient of X of rank $g - r$ and $\mathcal{S}_\xi$ is the formal completion of the $\overline{E}_\xi$-bundle $\overline{\Xi}_\xi$ along $\Xi_\xi \times_{E_\xi} W_\xi$ as in 3.2.4.*

The proof is entirely similar to that of thm. 5.7 (4), hence omitted.

5.10. Corollary 1 (to Thm 5.7). *The geometric fibres of A_g over* $\mathrm{Spec}(\mathbb{Z})$ *are all irreducible.*

Proof. As in Deligne-Mumford [De-Mum], §5, $\overline{\mathsf{A}}_g$ is geometrically connected in characteristic 0, because by complex uniformization $\mathsf{A}_g(\mathbb{C}) \cong \mathrm{Sp}(2g, \mathbb{Z}) \backslash \mathbb{H}_g$ is connected. Applying Zariski's connectedness theorem, we conclude that the geometric fibres of $\overline{\mathsf{A}}_g$ over $\mathrm{Spec}(\mathbb{Z})$ are connected. Since the fibres of $\overline{\mathsf{A}}_g$ are smooth, hence geometrically unibranch, we conclude that the geometric fibres of A_g over $\mathrm{Spec}(\mathbb{Z})$ are irreducible.

5.11. Corollary 2. *(a) If k is an algebraically closed field, the k-rational points of $\overline{\mathsf{A}}_g$ are naturally given as follows: They constitute the disjoint union of the sets of k-rational points which lie in σ-strata $Z(\sigma)$, over the* $\mathrm{GL}(X)$*-orbits of simplices σ in the decomposition of $C(X)$. In other words, they form the disjoint union of the k-rational points of $\Xi(\sigma)/\mathrm{Gl}(X_\sigma)$, with σ running through* $\mathrm{GL}(X)$*-orbits of $C(X)$ and $\Xi(\sigma)$ as in §3. Note that for $\sigma = \{0\}$, $\Xi(\sigma) = \mathsf{A}_g$ by our convention.*

(b) For any such σ, $Z(\sigma)(k) \cong (\Xi(\sigma)/\mathrm{GL}(X_\sigma))(k) = \Xi(\sigma)(k)$ can be explicitly described as follows:

(i) They are in 1–1 correspondence with isomorphism classes $(\tilde{G}, \lambda, \imath)$ *where* $\tilde{G}$ *is a g-dimensional semi-abelian variety over k which is an extension* $0 \to T_\sigma \to \tilde{G} \to A \to 0$, $T_\sigma =$ *torus with character group* X_σ, (A, λ) *a principally polarized abelian variety, and* $\imath$ *a k-rational point of the σ-stratum of a certain "*E_σ*-equivariant* $E(\sigma)$*-torsor"* $\Xi(\sigma)$. *Here* E_σ *denotes the torus with character group* $S^2(X_\sigma)$, $E(\sigma)$ *the torus embedding of* E_σ *determined by σ and its faces. Notice that* $\Xi(\sigma)$ *depends on the* $(\tilde{G}, \lambda)$, *and the subgroup* Γ_σ *(= stabilizer of σ in* $\mathrm{GL}(X_\sigma)$*) operates on such triples.*

(ii) In calling $\Xi(\sigma)$ *a "*E_σ*-equivariant* $E(\sigma)$*-torsor" we mean that* E_σ *acts on it and it is non-canonically isomorphic to* $E(\sigma)$ *via an* E_σ*-equivariant map. Thus any two choices of such isomorphisms differ by translation via an element of* $E_\sigma(k)$. $\Xi(\sigma)$ *has a natural stratification parametrized by σ and its faces, and any isomorphism between* $\Xi(\sigma)$ *and* $E(\sigma)$ *as above preserves stratification. The open stratum* Ξ *is an* E_σ*-torsor, and* $\Xi(\sigma)$ *is determined by* Ξ.

(iii) An isomorphism between $\Xi(\sigma)$ *and* $E(\sigma)$ *corresponds to a trivialization of the* $E(\sigma)$*-torsor* $\Xi(\sigma)$. *To get a trivialization, it suffices to give a lifting* $\tilde{c} : X_\sigma \to \tilde{G}(k)$ *of the map* $c : X_\sigma \to A(k)$ *classifying the extension* $0 \to T_\sigma \to \tilde{G} \to A \to 0$. *After choosing such a lifting, triples* $(\tilde{G}, \lambda, \imath)$ *correspond to k-rational points b of the σ-stratum of* $E(\sigma)$. *Given* $\tilde{G}$, *the pair* $(\tilde{c}, b)$ *is only determined up to action of* $E_\sigma(k)$, *and isomorphisms of such triples* $(\tilde{G}, \tilde{c}, b)$ *should act on* T_σ *via elements* α^* *of* $\mathrm{GL}(X_\sigma)$ *which stabilize σ (Note that* α^* *also acts on the b's).*

(iv) The E_σ*-torsor* Ξ *can be canonically described as in §3: for any homomorphism* $[\mu \otimes \nu] : E_\sigma \to G_m$, *where* $[\mu \otimes \nu] \in S^2(X_\sigma)$ *with* $\mu, \nu \in (X_\sigma)$, *the push-forward of* Ξ *by* $[\mu \otimes \nu]$ *is the* G_m*-torsor* $(c(\mu) \times c(\nu))^* \mathcal{P}_A^{\otimes -1}$, *where* $\mathcal{P}_A$ *is the Poincaré sheaf over* $A \times A$, *with A identified with its dual via the principal polarization λ.*

Proof. Immediate from part (4) of theorem 5.8.

5.12. Corollary 3. *The geometric fibres over* $\mathrm{Spec}(\mathbb{Z})$ *of the σ-stratum are irreducible, for each cone σ in* $C(X)$.

This follows immediately from thm. 5.7 (4) and cor. 1.

5.13. Remark. In [F3] the theorem of degeneration for principally polarized abelian varieties was proved under the extra assumption that the fraction field K of the complete local ring R has characteristic $\neq 2$. Then G. Faltings remarked that one can derive the assertions about degenerations of principally polarized abelian varieties even in characteristic 2: The only point which might be in doubt is the non-vanishing and the multiplicatively quadratic character of the $a(\,)$-function. For this one reduces to the case of a discrete valuation ring. As $\overline{\mathsf{A}}_g$ is proper over $\mathbb{Z}$ such a degeneration can be obtained from it by pullback, and the required property holds for $\overline{\mathsf{A}}_g$ because its degenerations already come from the Mumford construction. Notice that for the construction of $\overline{\mathsf{A}}_g$ we only used degenerations where the generic point has characteristic 0.

Hence the theory of degeneration for generic characteristic 0 suffices for the construction of $\overline{\mathsf{A}}_g$. So that there is no circle in this argument. This provides an alternative (and historically earlier) proof of the theory of degeneration for principally polarized abelian varieties over any complete local ring.

5.14. Remark. A word on how the compactification $\overline{\mathsf{A}}_g$ depends on the smooth rational polyhedral cone-decomposition of $C(X)$. If we have a second such decomposition which refines the first, we obtain a map from the second compactification onto the first. It sends the σ-stratum, for σ a simplex in the second decomposition, into the τ-stratum of the first decomposition, τ being the minimal simplex in the first decomposition which contains σ. It follows that $X_\sigma = X_\tau$, and that we obtain a canonical map from $\mathcal{S}_\sigma$ to $\mathcal{S}_\tau$ which gives a local description of the refinement-map, by part (4) of theorem 5.7. All these assertions follow easily from part (5) of theorem 5.7 (and proposition 5.1).

The fact that $\overline{\mathsf{A}}_g$ is non-canonical causes no problem in practice, and usually any choice will do. To get a canonical object, notice that we have the filtering projective system of the smooth rational polyhedral cone decomposition of $C(X)$, and the corresponding filtering projective system of smooth compactifications is clearly canonical.

5.15. Proposition. *$\overline{\mathsf{A}}_g(\mathbb{C})$ is canonically isomorphic to the toroidal compactification of $\mathsf{A}_g(\mathbb{C})$ (which is a Moisheson complex analytic space) associated in [AMRT] to the cone-decomposition of $C(X)$ which we used to construct $\overline{\mathsf{A}}_g$. Even better, the construction of of the toroidal compactification constructed in [AMRT] gives a proper smooth algebraic stack via GAGA, which is canonically isomorphic to $\overline{\mathsf{A}}_g \otimes_{\mathbb{Z}} \mathbb{C}$.*

This follows easily from the description of $\overline{\mathsf{A}}_g$ above, by comparing to [AMRT].

Thus the compactification $\overline{\mathsf{A}}_g$ is an arithmetic (over $\mathbb{Z}$) version of Mumford's toroidal compactification, which we may refer to as the *arithmetic toroidal compactification* of A_g.

6. Principal Level-Structures

The goal of this section is to construct arithmetic toroidal compactifications of the moduli space of principally polarized abelian varieties with level-n structures, for every positive integer $n \geq 1$. Of course the method is exactly the same as before, so we shall only indicated changes posed by the presence of level-structures. We begin by reviewing the notion of level structures. For simplicity, we shall concentrate on principal level-n structures to illustrate the method. Other level-structures can be treated in analogous fashion, but we omit details.

Fix an integer $n \geq 1$. For principal level-n structures, the natural base to work over is not the integers $\mathbb{Z}$, but the ring $\mathbb{Z}[e^{2\pi i/n}, 1/n]$. Writing $\zeta_n =$

$e^{2\pi i/n}$, we have a chosen primitive root of unity in this base ring. Sometimes we prefer to think of $\mathbb{Z}[e^{2\pi i/n}, 1/n]$ as an abstract ring, and then we write it $\mathbb{Z}[\zeta_n, 1/n]$, defined to be $\mathbb{Z}[1/n][T]/\Phi_n(T)$, where $\Phi_n(T)$ is the n-th cyclotomic polynomial. We also let $\mathbb{Z}[\zeta_n] = \mathbb{Z}[T]/\Phi_n(T)$, so $\mathbb{Z}[\zeta_n, 1/n] = \mathbb{Z}[\zeta_n] \otimes \mathbb{Z}[1/n]$. By our choice of the primitive root of unity ζ_n, the group scheme $\mu_{n/\mathbb{Z}[\zeta_n,1/n]}$ of n-th roots of unity is canonically identified with the constant group $\mathbb{Z}/n\mathbb{Z}$ over $\mathrm{Spec}(\mathbb{Z}[\zeta_n, 1/n])$ (1 corresponds to ζ_n).

We recall the following:

6.1. Definition. (i) A *symplectic principal level-n structure* on a principally polarized abelian scheme $(A \to S, \lambda)$ of relative dimension g over a $\mathbb{Z}[\zeta_n, 1/n]$-scheme S is a symplectic isomorphism $\alpha : A[n] \cong (\mathbb{Z}/n\mathbb{Z})^{2g}$. Here the Weil pairing is thought of as a (non-degenerate) skew-symmetric pairing $A[n] \times A[n] \to \mathbb{Z}/n\mathbb{Z}$ via the principal polarization λ and the identification of μ_n with $\mathbb{Z}/n\mathbb{Z}$. $(\mathbb{Z}/n\mathbb{Z})^{2g}$ is given the *standard* non-degenerate alternating form $(\mathbb{Z}/n\mathbb{Z})^{2g} \times (\mathbb{Z}/n\mathbb{Z})^{2g} \to \mathbb{Z}/n\mathbb{Z}$ defined by $(u, v) \times (z, w) \mapsto u \cdot {}^t w - v \cdot {}^t z$ for $u, v, z, w \in (\mathbb{Z}/n\mathbb{Z})^g$. To say α is symplectic means that α preserves the non-degenerate alternating forms on both sides.

(ii) We denote by $\mathsf{A}_{g,n\,/\mathrm{Spec}(\mathbb{Z}[\zeta_n,1/n])}$ the moduli stack over $\mathrm{Spec}(\mathbb{Z}[\zeta_n, 1/n])$ which classifies principally polarized abelian schemes over $\mathrm{Spec}(\mathbb{Z}[\zeta_n, 1/n])$ with symplectic principal level-n structure.

6.2. Remarks. (a) For an abelian scheme $A \to S$ of relative dimension g, we have defined a level-n structure on $A \to S$ as an isomorphism between $A[n]$ and $(\mathbb{Z}/n\mathbb{Z})^{2g}$. Unless n is invertible in $\mathcal{O}_S$, such an isomorphism cannot possibly exist. Assuming $A \to S$ is principally polarized, the Weil pairing $A[n] \times A[n] \to \mu_n$ induces a non-degenerate symmetric pairing $(\mathbb{Z}/n\mathbb{Z})^{2g} \times (\mathbb{Z}/n\mathbb{Z})^{2g} \to \mu_n$. In particular, n is invertible in $\mathcal{O}_S$ and $\mu_{n/S}$ is constant.

(b) If S is given a structure of scheme over $\mathrm{Spec}(\mathbb{Z}[\zeta_n, 1/n])$, then a level-$n$ structure on a principally polarized abelian scheme $A \to S$ gives rise to a non-degenerate alternating pairing $(\mathbb{Z}/n\mathbb{Z})^{2g} \times (\mathbb{Z}/n\mathbb{Z})^{2g} \to \mathbb{Z}/n\mathbb{Z}$, which can be regarded as a kind of numerical invariant on principally polarized abelian schemes over S with level-n structures. This implies that the moduli space of principally polarized abelian schemes with level-n structure over $\mathbb{Z}[\zeta_n, 1/n]$ is not connected. Elementary linear algebra tells us that any non-degenerate alternating pairing $(\mathbb{Z}/n\mathbb{Z})^{2g} \times (\mathbb{Z}/n\mathbb{Z})^{2g} \to \mathbb{Z}/n\mathbb{Z}$ can be put into the standard form by an automorphism of $(\mathbb{Z}/n\mathbb{Z})^{2g}$, therefore one may restrict to the standard non-degenerate alternating forms when studying level-n structures (for principally polarized abelian schemes) over $\mathbb{Z}[\zeta_n, 1/n]$.

(c) $\mathsf{A}_{g,n}$ is an algebraic stack smooth (but not proper) over $\mathrm{Spec}(\mathbb{Z}[\zeta_n, 1/n])$ as before. When $n = 1$, it reduces to A_g. For positive integers m, n with $m|n$, there is a natural morphism $\mathsf{A}_{g,n} \to \mathsf{A}_{g,m}$, which is finite étale over $\mathrm{Spec}(\mathbb{Z}[\zeta_n, 1/n])$. It turns out that $\mathsf{A}_{g,n}$ is connected, which is a corollary of the existence of an arithmetic toroidal compactification of $\mathsf{A}_{g,n}$ over $\mathrm{Spec}(\mathbb{Z}[\zeta_n, 1/n])$. When $n \geq 3$, a lemma of Serre says that any algebraic integer which is a unit and $\equiv 1 \pmod 3$ must be equal to 1. This implies that for any $n \geq 3$ not divisible by the characteristic of the base field, the set of automorphisms of finite order of any abelian variety, fixing all n-torsion points, is trivial. Consequently $\mathsf{A}_{g,n}$ is actually an algebraic space over $\mathrm{Spec}(\mathbb{Z}[\zeta_n, 1/n])$ if $n \geq 3$. In fact, $\mathsf{A}_{g,n}$ is quasi-projective over $\mathrm{Spec}(\mathbb{Z}[\zeta_n, 1/n])$ if $n \geq 3$ (cf. Mumford), in particular it is a scheme.

(d) There is a natural action of $\mathrm{Sp}(2g, \mathbb{Z}/n\mathbb{Z})$ on $\mathsf{A}_{g,n}$, which is defined by composing a symplectic principal level-n structures with elements of $\mathrm{Sp}(2g, \mathbb{Z}/n\mathbb{Z})$. Clearly $\mathsf{A}_{g,n}/\mathrm{Sp}(2g, \mathbb{Z}/n\mathbb{Z}) = \mathsf{A}_{g\,/\mathrm{Spec}(\mathbb{Z}[\zeta_n,1/n])}$, which is direct translation of the fact that the action of $\mathrm{Sp}(2g, \mathbb{Z}/n\mathbb{Z})$ on the set of all symplectic level-n structures for a principally polarized abelian variety (over a field containing $\mathbb{Z}[\zeta_n, 1/n]$) is free and transitive.

In this section, $(\mathbb{Z}/n\mathbb{Z})^{2g}$ will be given the standard non-degenerate alternating pairing with value in $\mathbb{Z}/n\mathbb{Z}$, such that the first g basis-vectors and the last g basis-vectors each generate a maximal isotropic subspace. When $\mathbb{Z}^{2g}$ is given the standard $\mathbb{Z}$-valued non-degenerate alternating form, the reduction map $\mathrm{Sp}(2g, \mathbb{Z}) \to \mathrm{Sp}(2g, \mathbb{Z}/n\mathbb{Z})$ is surjective. This is a consequence of the strong approximation theorem (because $\mathrm{Sp}(2g)$ is simply connected), and can also be proved by elementary methods.

We find it convenient to introduce some subgroups of $\Lambda = \Lambda(\mathbb{Z}/n\mathbb{Z}) = (\mathbb{Z}/n\mathbb{Z})^{2g}$ and related subgroups of $\mathrm{Sp}(2g, \mathbb{Z}/n\mathbb{Z})$.

6.3. Definition. (i) For $0 \leq r \leq g$, let $\Psi_{g,r}(\mathbb{Z}/n\mathbb{Z})$ (or simply $\Psi_{g,r}$) be the sublattice of $(\mathbb{Z}/n\mathbb{Z})^{2g}$ generated by the first r basis-vectors $e_1, \dots, e_r$ and $\Sigma_{g,r}(\mathbb{Z}/n\mathbb{Z})$ (or simply $\Sigma_{g,r}$) be the sublattice of $(\mathbb{Z}/n\mathbb{Z})^{2g}$ generated by the $2g - r$ basis-vectors $e_1, \dots, e_g, e_{g+r+1}, \dots, e_{2g}$. Clearly $\Psi_{g,r}$ is the radical of the restriction of the standard symplectic form to $\Sigma_{g,r}$, and $\Sigma_{g,r}/\Psi_{g,r}$ with the induced non-degenerate symplectic form is isomorphic to the $(\mathbb{Z}/n\mathbb{Z})^{2g-2r}$ with the standard form via the obvious isomorphism between $\Sigma_{g,r}/\Psi_{g,r}$ and $(\mathbb{Z}/n\mathbb{Z})^{2g-2r}$. Moreover $\Psi_{g,r}$ and $\Sigma_{g,r}$ are mutual annihilators w.r.t. the standard symplectic pairing.

(ii) Let $P_{g,r}(\mathbb{Z}/n\mathbb{Z})$ be the subgroup of $\mathrm{Sp}(2g, \mathbb{Z}/n\mathbb{Z})$ consisting of element stabilizing $\Sigma_{g,r}$ (or equivalently $\Psi_{g,r}$ by duality). It is the group of $(\mathbb{Z}/n\mathbb{Z})$-points of a parabolic subgroup $P_{g,r}$ of $\mathrm{Sp}(2g)$ over $\mathbb{Z}$.

(iii) Denote by $N_{g,r}$ the unipotent radical of $P_{g,r}$. It is easy to see that $N_{g,r}(\mathbb{Z}/n\mathbb{Z})$ consists of all elements in $P_{g,r}(\mathbb{Z}/n\mathbb{Z})$ which act trivially on $\Psi_{g,r}$ and $\Sigma_{g,r}/\Psi_{g,r}$. Again we often abbreviate $N_{g,r}(\mathbb{Z}/n\mathbb{Z})$ to $N_{g,r}$.

(iv) Let $U_{g,r}$ be the center of $N_{g,r}$. It is easy to see that $U_{g,r}(\mathbb{Z}/n\mathbb{Z})$ consists of all elements of $P_{g,r}$ acting trivially on $\Sigma_{g,r}$. One can check that it is canonically isomorphic to $\Gamma_2(\Psi_{g,r}) = B(\Psi^*_{g,r}) \subset \Psi_{g,r} \otimes_{\mathbb{Z}} \Psi_{g,r} = \Psi_{g,r} \otimes_{\mathbb{Z}} (\Lambda/\Sigma_{g,r})^*$ after identifying $\Psi_{g,r}$ with $(\Lambda/\Sigma_{g,r})^*$ via the symplectic pairing. It can also be seen from the matricial description of the relevant groups (see below.) It is also easily checked that $P_{g,r}/U_{g,r}$ is the group of all automorphisms of $\Sigma_{g,r}$ which preserve the standard (degenerate) alternating form.

(v) Let $L_{g,r}(\mathbb{Z}/n\mathbb{Z})$ be (the $(\mathbb{Z}/n\mathbb{Z})$-points of) the standard Levi-component of $P_{g,r}$. It consists of element of $P_{g,r}$ fixing the basis-vectors $e_{g+1}, \dots, e_{g+r}$ of

$(\mathbb{Z}/n\mathbb{Z})^{2g}$. Observe that elements of $L_{g,r}(\mathbb{Z}/n\mathbb{Z})$ stabilize the sublattice generated by $e_{r+1}, \ldots, e_g, e_{g+r+1}, \ldots, e_{2g}$ because they stabilize its annihilator. $L_{g,r}$ is canonically isomorphic to $\mathrm{GL}(\Psi_{g,r}) \times \mathrm{Sp}(\Sigma_{g,r}/\Psi_{g,r})$.

(vi) We give the matricial description of the above subgroups. Typical elements of the subgroups are tabulated below, written in $(r, g-r, r, g-r)$ block form. For instance, the entry in the upper-right corner is an $r \times (g-r)$-matrix.

$$P_{g,r}: \quad \begin{pmatrix} A_1 & A_{12} & B_{11} & B_{12} \\ 0 & A' & B_{21} & B' \\ 0 & 0 & D_1 & 0 \\ 0 & C' & D_{21} & D' \end{pmatrix} \qquad D_1 = {}^tA_1^{-1} + \text{ other symplectic identities}$$

$$N_{g,r}: \quad \begin{pmatrix} I_r & A_{12} & B_{11} & B_{12} \\ 0 & I_{g-r} & B_{21} & 0 \\ 0 & 0 & I_r & 0 \\ 0 & 0 & D_{21} & I_{g-r} \end{pmatrix} \qquad B_{21} = {}^tB_{12}\,,\ D_{21} = {}^tA_{12}, \quad B_{11} + B_{12} \cdot {}^tA_{12} = {}^tB_{11} + A_{12} \cdot {}^tB_{12}$$

$$U_{g,r}: \quad \begin{pmatrix} I_r & 0 & B_{11} & 0 \\ 0 & I_{g-r} & 0 & 0 \\ 0 & 0 & I_r & 0 \\ 0 & 0 & 0 & I_{g-r} \end{pmatrix} \qquad B_{11} = {}^tB_{11}$$

$$L_{g,r}: \quad \begin{pmatrix} A_1 & 0 & 0 & 0 \\ 0 & A' & 0 & B' \\ 0 & 0 & D_1 & 0 \\ 0 & C' & 0 & D' \end{pmatrix} \qquad D_1 = {}^tA_1^{-1} + \text{ sympletic identities}$$

In the above table, symplectic identities for A', B', C', D' in a typical element of $L_{g,r}$ just guarantee that they define an element of $\mathrm{Sp}(2g-2r, \mathbb{Z}/n\mathbb{Z})$. For a typical element of $P_{g,r}$, the symplectic identities imply that A', B', C', D' define an element of $\mathrm{Sp}(2g-2r, \mathbb{Z}/n\mathbb{Z})$.

6.4. Definition. Let $\tilde{G}$ be a semi-abelian scheme of relative dimension g over a $\mathbb{Z}[\zeta_n, 1/n]$-scheme S which is an extension $0 \to T \to \tilde{G} \to A \to 0$, with T a torus of relative dimension r over S and A a principally polarized abelian scheme of relative dimension $g-r$. A *symplectic principal level-n structure* on this extension $\tilde{G}$ is an isomorphism $\alpha : \tilde{G}[n] \cong \Sigma_{g,r}$, which respects the (degenerate) alternating forms, where $\tilde{G}[n]$ is given the pull-back of the Weil pairing of $A[n]$ and $\Sigma_{g,r}$ given the (restriction of the) standard alternating pairing.

To give a (symplectic principal) level-n structure on $\tilde{G}$ as above is the same as to give isomorphisms $T[n] \cong \Psi_{g,r}$, $A[n] \cong \Sigma_{g,r}/\Psi_{g,r}$ (preserving symplectic forms), and a splitting of the exact sequence $0 \to T[n] \to \tilde{G}[n] \to A[n] \to 0$. Clearly the quotient $P_{g,r}/U_{g,r}$ operates freely and transitively on the set of all symplectic principal level-n structures for $\tilde{G}$.

The key point of this section is a close analysis of symplectic principal level-n structures on the generic fibre of a semi-abelian variety G, obtained by applying the Mumford construction to principally polarized semi-abelian degeneration data over some complete base R. After this, it will become clear that the method of the previous section produces arithmetic toroidal compactifications of the algebraic stack $\mathsf{A}_{g,n}$ over $\mathbb{Z}[\zeta_n, 1/n]$.

Let R be a noetherian normal domain complete w.r.t. an ideal I with $\mathrm{rad}(I) = I$ satisfying the conditions in ch. II, section 3, K = fraction field of R, $S = \mathrm{Spec}(R)$. Moreover we assume that it is a $\mathbb{Z}[\zeta_n, 1/n]$-algebra; i.e. n is invertible in R and we have chosen a primitive n-th root of unity in R. Assume that we have semi-abelian degeneration data with principal polarization $(0 \to T \to \tilde{G} \to A \to 0,\ \imath, c, \lambda, \tau)$ over R. Here we have followed the notation scheme of chapters II, III, so λ is a principal polarization of the abelian scheme A, X = character group of T (assumed to be split), $c : X \to A^t(S) = A(S)$ defines the extension, and τ is a symmetric trivialization of the pull-back of the inverse of the Poincaré sheaf of A over $c \times c$ which defines the period subgroup $\imath : X = Y \to \tilde{G}(K)$. We would like to express symplectic principal level-n structures on G_K in terms of the degeneration data.

Assume first that we are given a symplectic principal level-n structure α on G_K. Let us analyze it. The generic fibre $G[n]_K$ of $G[n]$ has natural structure as an extension $0 \to \tilde{G}[n]_K \to G[n]_K \to (1/n) \cdot Y/Y \to 0$. Since $\tilde{G}[n]$ is finite étale over the normal base R, $\tilde{G}[n]$ is constant because $G[n]$ is. We know that the restriction of the Weil pairing on $G[n]_K$ to $\tilde{G}[n]$ has $T[n]$ as radical, and is derived from the Weil pairing on $A[n]$. Moreover we know that $T[n]$ and $\tilde{G}[n]$ are mutual annihilators with respect to the Weil pairing on $G[n]_K$, which induces the Cartier-duality pairing between $T[n]$ and $(1/n) \cdot Y/Y$.

From the above properties of $\tilde{G}[n]$ and the alternating form on $\tilde{G}[n]$, we see that after transforming the level-n structure on G_K by a suitable element of $\mathrm{Sp}(2g, \mathbb{Z}/n\mathbb{Z})$ we can assure that the symplectic isomorphism $\alpha : G[n]_K \xrightarrow{\sim} (\mathbb{Z}/n\mathbb{Z})^{2g}$ induces a symplectic isomorphism $\tilde{G}[n] \xrightarrow{\sim} \Sigma_{g,r}$ over S. We shall say such symplectic principal level-n structures on G_K are *compatible with the degeneration.* Clearly the obvious action of $P_{g,r}(\mathbb{Z}/n\mathbb{Z})$ on the set of symplectic level-n structures on G_K which are compatible with the degeneration is free and transitive. Moreover we can recover all symplectic principal level-n structures on G_K from the compatible ones by $\mathrm{Sp}(2g, \mathbb{Z}/n\mathbb{Z})$-action.

From a given a symplectic principal level-n structure α on G_K which is compatible with the degeneration we can derive a splitting of the exact sequence $0 \to \tilde{G}[n]_K \to G[n]_K \to (1/n) \cdot Y/Y \to 0$, from the canonical splitting of $0 \to \Psi_{g,r} \to (\mathbb{Z}/n\mathbb{Z})^{2g} \to (\mathbb{Z}/n\mathbb{Z})^{2g}/\Psi_{r,g} \to 0$. Moreover the image of the section $(1/n) \cdot Y/Y \to G[n]_K$ must be isotropic and perpendicular to $A[n]$. One checks that (over some extension of K) such sections always exist, and that they form a torsor under symmetric homomorphisms of $(1/n) \cdot Y/Y$ into $T[n]$, that is under $E[n]$, E the torus with character group $S^2(X)$. As one can easily see this splitting is the crucial data given by α. However it is not quite straightforward to analyze the isotropy-condition. Because of that we use an indirect approach.

A splitting of $0 \to \tilde{G}[n]_K \to G[n]_K \to (1/n) \cdot Y/Y \to 0$ obviously corresponds to a lifting of $(1/n) \cdot Y/Y$ in $G[n]_K$, which corresponds in turn to an isogeny $G_{/K}$ to an abelian variety with kernel the lifting of $(1/n) \cdot Y/Y$. Such isogenies are characterized by the fact that the intersection of their kernels with $\tilde{G}[n] \subset G[n]$ is trivial. Any such isogeny extends to a homomorphism between semi-abelian schemes over R, which by the theory of degenerations corresponds to a morphism between semi-abelian degeneration data. It follows that we can extend $\imath : Y \to \tilde{G}(K)$ to a $\imath_n : Y_n = (1/n) \cdot Y \to \tilde{G}(K)$. Conversely we remark that if $\imath_n : Y_n = (1/n) \cdot Y \to \tilde{G}(K)$ is a homomorphism such that $n \cdot \imath_n(y) = \imath(ny)$ $\forall y \in Y_n = (1/n) \cdot Y_1$, clearly $\imath_n$ defines an homomorphism from Y_n/Y_1 to $G[n]_K$ which splits the exact sequence $0 \to \tilde{G}[n]_K \to G[n]_K \to (1/n) \cdot Y/Y \to 0$. As an aside we note that the existence of a level-n structure on G_K implies that the bilinear from $\underline{B}(G) : \underline{X}(G^t) \times \underline{X}(G) \to \mathbf{Div}(S)$ is (uniquely) divisible by n. In fact by descent that holds for degenerations over any normal base S, not necessarily the spectrum of an I-adically complete ring.

6.5. Construction. A word on notation: $Y_n = X_n = (1/n) \cdot X$, since we are dealing with principally polarized abelian varieties. From the map $\imath_n$ we get $c_n : X_n = (1/n) \cdot X \to A^t(S)$ which extends the map c classifying $\tilde{G}$, that is we obtain an extension $0 \to T_n \to \tilde{G}_n \to A \to 0$, T_n the torus with character group $X_n = (1/n) \cdot X$, which induces $\tilde{G}$ under push-out by $T_n \to T$. Assume first that Y_n lifts to a period map $Y_n \to \tilde{G}_n(K)$, such that the quotient acquires a principal polarization (i.e. some pull-back of $\mathcal{P}_A$ is trivialized over K, etc.). If we divide $\tilde{G}_n$ instead by $Y = n \cdot Y_n$ we obtain a covering G_n "$= \tilde{G}_n/Y$" (with covering group Y_n/Y) which (by pull-back) has a polarization of degree n^{2r}. This polarization defines (over K) a map into the dual of G_n, whose kernel must be the subgroup $T_n[n] \oplus Y_n/Y$, with its canonical symplectic form. Thus dividing by the maximal isotropic subgroup $T_n[n]$ we obtain a principal polarization on the quotient G. Furthermore Y_n/Y injects into $G_K[n]$ and is isotropic (check by pullback to G_n). Thus we have found a way to construct isotropic sections. Note furthermore that $A[n] = \tilde{G}_n[n]/T_n[n]$ injects into $G[n]$. So if we choose an isomorphism $Y/nY \approx (\mathbb{Z}/n\mathbb{Z})^r$ and a symplectic level-n structure $A[n] \approx (\mathbb{Z}/n\mathbb{Z})^{2(g-r)}$ we obtain a level-n structure on G_K which is compatible with the degeneration. One also checks that this level-n structure only depends on the map of Y_n into $\tilde{G}_K = \tilde{G}_{n,K}/T_n[n]_K$.

Now we may ask whether we have found all possible level-n structures. Such a structure defines an isomorphism between Y/nY and $(\mathbb{Z}/n\mathbb{Z})^r$ and a symplectic level-n structure on A_K. Also it gives a covering $G_{n,K} \to G_K$ with group $T_n[n]$ (Choose $G_{n,K}$ = quotient of G_K by $A[n] \oplus (Y_n/Y)$). If we lift Y_n/Y to an isotropic subgroup of $G_{n,K}[n]$ perpendicular to $A[n]$ (these liftings form a principal homogeneous space under $E[n]$, and a lifting is possible after extending K) the quotient by Y_n/Y has a principal polarization, and we obtain a map $Y_n \to \tilde{G}_{n,K}$ as before. In short we might say that our construction gives an $E[n]$-covering of the set of pairs (principal polarization on G_K, symplectic level-n structure on G_K compatible with the degeneration).

We can also analyze the action of $P_{g,r}$: First we can change the isomorphism $Y/nY \approx (\mathbb{Z}/n\mathbb{Z})^r$ and the symplectic level-n structure on $A[n]$. These choices correspond to the action of $P_{g,r}/N_{g,r}(\mathbb{Z}/n\mathbb{Z}) = \mathrm{GL}(r, \mathbb{Z}/n\mathbb{Z}) \times \mathrm{Sp}(2g-2r, \mathbb{Z}/n\mathbb{Z})$. Then there is the ambiguity of dividing $\imath$ by n, and lifting to $\tilde{G}_n$. Ways of dividing $c : Y \to A^t(S) = A(S)$ by n form a principal homogeneous space under $\mathrm{Hom}(Y, A[n]) \approx N_{g,r}/U_{g,r}(\mathbb{Z}/n\mathbb{Z})$. If we fix such a division (which determines

$\tilde{G}_n$) what remains is a homogeneous space under symmetric homomorphisms of Y into $T_n[n^2]$. However symmetric homomorphisms into $T_n[n]$ operate trivially, and we get an operation of $E[n] \approx U_{g,r}(\mathbb{Z}/n\mathbb{Z})$. As all level-$n$ structures compatible with the degeneration form a principal homogeneous space under $P_{g,r}(\mathbb{Z}/n\mathbb{Z})$ we see again that we have obtained all of them.

Finally, if we pick a set of representatives $\{\gamma_i\}$ of $\mathrm{Sp}(2g, \mathbb{Z}/n\mathbb{Z})/P_{g,r}(\mathbb{Z}/n\mathbb{Z})$, every symplectic principal level-n structure on G_K is obtained from one which respects the degeneration by transformation with a unique γ_i.

Let us translate these discussions into the global context by constructing Mumford families with symplectic principal level-n structures. They will serve as good formal models for the arithmetic toroidal compactification $\overline{\mathsf{A}}_{g,n}$.

Let us first make some remarks about the rational polyhedral cone decomposition. Fortunately we can use the same decomposition as before, because we are dealing with symplectic principal level-n structures. Thus we just have to choose a smooth $\mathrm{GL}(X)$-invariant admissible rational polyhedral cone decomposition $\{\sigma_\alpha\}$ of $C(X)$, with $X \cong \mathbb{Z}^g$. Actually, this is more than is necessary: all one needs is an admissible rational polyhedral cone decomposition invariant under the kernel of $\mathrm{GL}(X) \to \mathrm{GL}(X/nX)$, for each maximal isotropic sublattice X in ($\mathbb{Z}^{2g}$, standard symplectic pairing). (This is what Γ-admissible rational polyhedral cone decomposition means in this case, where Γ is the principal congruence subgroup of level n.) The advantage of using a $\mathrm{GL}(X)$-invariant rational polyhedral cone decomposition is not only its simplicity (only *one* cone decomposition is needed), but also that the natural action of $\mathrm{Sp}(2g, \mathbb{Z}/n\mathbb{Z})$ on $\mathsf{A}_{g,n}$ extends to $\overline{\mathsf{A}}_{g,n}$. There is one thing to keep in mind: It is natural to change the integral structure on $C(X)$. Namely the lattice $n \cdot B(X)$ defines the new integral structure. The effect is that we have more rational functions on the corresponding torus embedding, which is what we need for dividing the period subgroup by n.

Again the rational boundary components of $C(X)$ are parametrized by quotients $\xi = (X \to X_\xi)$, where X_ξ is a torsion-free quotient X, say of rank r. First there exists a moduli stack for semi-abelian extensions $0 \to T_{\xi,n} \to \tilde{G}_n \to A \to 0$ with symplectic principal level-n structure on A, where the split torus $T_{\xi,n}$, has $X_{\xi,n} = (1/n) \cdot X_\xi$ as character group and $\tilde{G}_n$ has relative dimension g: We start with the moduli stack $\mathsf{A}_{g-r,n}$ for principally polarized abelian schemes of relative dimension $g - r$, with a (symplectic principal) level-n structure. Identify the universal abelian scheme A over $\mathsf{A}_{g-r,n}$ with its dual via the universal principal polarization. Form the r-fold product of the universal abelian variety A over $\mathsf{A}_{g,n}$, which should be thought of as $\mathbf{Hom}_{\mathbb{Z}}(X_{\xi,n}, A^t) = \mathbf{Hom}_{\mathbb{Z}}(X_{\xi,n}, A)$. This is clearly the moduli stack referred to above. As before this is the base of a semi-abelian extension $0 \to T_{\xi,n} \to \tilde{G}_n \to A \to 0$ over $\mathbf{Hom}_{\mathbb{Z}}(X_{\xi,n}, A)$, where $T_{\xi,n}$ is the split torus with character group $X_{\xi,n}$ and A is the universal principally polarized abelian scheme over $\mathsf{A}_{g,n}$. The extension is associated to the universal map $c_{\xi,n} : X_{\xi,n} \to A$. Also on A we have the universal symplectic level-n structure.

The rational polyhedral cone decomposition $\{\sigma_\alpha\}$ of $C(X)$ induces a rational polyhedral cone decomposition of $C(X_\xi)$, which is smooth and $\mathrm{GL}(X_\xi)$-invariant. Attached to this rational polyhedral cone decomposition of $C(X_\xi)$ is a torus embedding $E_\xi \subset \overline{E}_\xi$. If we denote by $E_{\xi,n}$ the torus with character-group $(1/n) \cdot S^2(X_\xi)$ we have a natural map $E_{\xi,n} \to E_\xi$ which can be thought of as multiplication by n on E_ξ. We also obtain an embedding $E_{\xi,n} \subset \overline{E}_{\xi,n}$. Following previous notation the affine torus embedding parametrized by a cone σ in $C(X_\xi)$ is denoted $E_{\xi,n}(\sigma)$, and the $E_{\xi,n}$-orbit parametrized by cone σ in $C(X_\xi)$ is denoted $Z_\xi(\sigma)$. $Z_{\xi,n}(\sigma)$ is the smallest (closed) orbit in $E_{\xi,n}(\sigma)$. Let us remark that $\mathrm{GL}(X_\xi)$ operates naturally on $E_{\xi,n}$.

The base for Mumford's construction will be the formal completion of an $\overline{E}_{\xi,n}$-bundle over $\mathbf{Hom}_\mathbb{Z}(X_{\xi,n}, A)$. Let us first describe it. As before this bundle $\overline{\Xi}_{\xi,n}$ will be defined as $\Xi_{\xi,n} \times^{E_{\xi,n}} \overline{E}_{\xi,n}$ (contraction product) for some $E_{\xi,n}$-bundle $\Xi_{\xi,n}$, so it suffices to define $\Xi_{\xi,n}$.

The character group of $E_{\xi,n}$ is $(1/n) \cdot S^2(X_\xi) = n \cdot S^2(X_{\xi,n})$. To define $\Xi_{\xi,n}$ amounts to mapping this into the group of rigidified line-bundles on $\mathbf{Hom}_\mathbb{Z}(X_{\xi,n}, A)$. We choose the map which sends the class of $n \cdot (\mu \otimes \nu)$ to $(c_{\xi,n}(\mu) \times c_{\xi,n}(\nu))^* \mathcal{P}_A^{\otimes -n}$. Instead of that we can also map to $(c_{\xi,n}(\mu) \times c_{\xi,n}(\nu))^* \mathcal{P}_A^{\otimes -1}$ and obtain another $E_{\xi,n}$-torsor $\Xi_{\xi,n,d}$. The n-th power map on $E_{\xi,n}$ induces $\Xi_{\xi,n,d} \to \Xi_{\xi,n}$. For the associated bundles we obtain $\overline{\Xi}_{\xi,n,d} \to \overline{\Xi}_{\xi,n}$ which locally looks like the map $\overline{E}_{\xi,n} \to \overline{E}_{\xi,n}$ induced by the n-th power.

Furthermore ξ defines a rational boundary component in the old sense (forgetting about level-n structures). We obtain a map from $\mathbf{Hom}_\mathbb{Z}(X_{\xi,n}, A)$ into the old $\mathbf{Hom}_\mathbb{Z}(X_\xi, A)$, by forgetting about the level-n structure of A and restricting to $X_\xi \subset X_{\xi,n}$. It follows from the definitions that this extends to $\Xi_{\xi,n} \to \Xi_\xi$ and $\overline{\Xi}_{\xi,n} \to \overline{\Xi}_\xi$, equivariant for the map $E_{\xi,n} \to E_\xi$. Formally completing along non-degenerate strata defines formal stacks $\mathcal{S}_{\xi,n,d} \to \mathcal{S}_{\xi,n} \to \mathcal{S}_\xi$. The maps between them are finite, and étale on the "open stratum" (that is over $\mathrm{Spec}(R)$, for an affine piece $\mathrm{Spf}(R)$). $\mathrm{GL}(X_\xi)$ still operates naturally on everything. Over the formal stack $\mathcal{S}_{\xi,n,d}$ we have

(i) a semi-abelian extension $0 \to T_{\xi,n} \to \tilde{G}_n \to A \to 0$ by pull-back.

(ii) a tautological homomorphism $c_{\xi,n} : Y_{\xi,n} = (1/n) \cdot X_\xi \to A$ which classifies the extension.

(iii) a tautological symmetric trivialization $\tau_{\xi,n}$ of $(c_{\xi,n} \times c_{\xi,n})^* \mathcal{P}_A^{\otimes -1}$.

By construction the above define semi-abelian degeneration data with periods subgroup $X_{\xi,n}$ (i.e. the positivity condition is satisfied). Thus Mumford's construction gives us a semi-abelian scheme over $\mathcal{S}_{\xi,n,d}$ whose generic fibre is abelian and principally polarized. If instead we divide by $Y_\xi = n \cdot Y_{\xi,n}$ we obtain a covering ${}^\heartsuit G_n$ of it. In turn dividing ${}^\heartsuit G_n$ by $T_{\xi,n}[n]$ produces the pull-back of the old ${}^\heartsuit G$ defined over $\mathcal{S}_\xi$. It follows (by construction 6.5) that, after choice of an isomorphism $Y_{\xi,n}/Y_\xi \approx (\mathbb{Z}/n\mathbb{Z})^r$, over the open stratum this pull-back has a natural level-n structure compatible with the degeneration, induced by the inclusions of $T_\xi[n]$, $A[n]$ and $Y_{\xi,n}/Y_\xi$ into it. Again this means that these properties hold over $\mathrm{Spec}(R)$ for any affine piece $\mathrm{Spf}(R)$ of the formal stack.

It is constant on the fibres of $\mathcal{S}_{\xi,n,d} \to \mathcal{S}_{\xi,n}$, so induces a level-$n$ structure on the open stratum of $\mathcal{S}_{\xi,n}$. In fact it follows from construction 6.5 that on $\mathcal{S}_{\xi,n} \times \mathrm{Isom}(Y_{\xi,n}/Y_\xi, (\mathbb{Z}/n\mathbb{Z})^r)$ wc obtain the universal level-n structure compatible with the degeneration. Note that again everything is equivariant under $\mathrm{GL}(X_\xi)$.

Construction of $\overline{\mathsf{A}}_{g,n}$

At this point we could just use the arguments of sections 3, 4, 5 to first approximate and eventually algebraize the good formal models, and obtain enough formal algebraic models. However it is easier to define the stack $\overline{\mathsf{A}}_{g,n}$ as the normalization of $\overline{\mathsf{A}}_g$ in $\mathsf{A}_{g,n}$. $\overline{\mathsf{A}}_{g,n}$ has the form U_n/R_n, where R_n is a groupoid defining an étale equivalence relation on U_n. We can choose for U_n the normalization of U_1 in the finite étale cover of its open stratum which classifies level-n structures. R_n is defined to be the normalization of $\mathrm{Isom}_{U_{n,0}\times U_{n,0}}(\mathrm{pr}_1^*(G,\lambda,\alpha),\ \mathrm{pr}_2^*(G,\lambda,\alpha)) \to U_n \times_{\mathrm{Spec}(\mathbb{Z}[\zeta_n,1/n])} U_n$. One checks that both projections of R_n to U_n are indeed étale.

The previous discussions give information about $\overline{\mathsf{A}}_{g,n}$: The formal completion of $\overline{\mathsf{A}}_g$ along the stratum where the toral part of G has dimension r is equal to the quotient $\mathcal{S}_\xi/\mathrm{GL}(X_\xi)$. The corresponding formal completion of $\overline{\mathsf{A}}_{g,n}$ is obtained by normalizing this in the covering of the open stratum defined by level-n structures. If we only consider level-n structures compatible with the degeneration, it follows from previous discussions that we obtain the quotient $\{\mathcal{S}_{\xi,n} \times \mathrm{Isom}(X_{\xi,n}/X_\xi, (\mathbb{Z}/n\mathbb{Z})^r)\}/\mathrm{GL}(X_\xi)$. The group $P_{g,r}(\mathbb{Z}/n\mathbb{Z})$ operates on this, and to obtain all level-n structures we form the induced $\mathrm{Sp}(2g,\mathbb{Z}/n\mathbb{Z})$-set. In particular, it follows that $\overline{\mathsf{A}}_{g,n}$ is smooth over $\mathrm{Spec}(\mathbb{Z}[e^{2\pi i/n}, 1/n])$, and that locally in the étale topology the projection $\overline{\mathsf{A}}_{g,n} \to \overline{\mathsf{A}}_g$ is isomorphic (preserving strata) to $\overline{E}_n \to \overline{E}$.

Similarly we can compute the formal completion along σ-strata, for any cone σ in the decomposition.

We now can state the main theorem about the arithmetic toroidal compactification of $\mathsf{A}_{g,n}$. But first we need some bookkeeping in order to describe the boundary structure of $\overline{\mathsf{A}}_{g,n}$.

Let us first discuss cusps. In our concrete situation, the rational boundary components of $\mathbb{H}_g$ with respect to $\mathrm{Sp}(2g)_{/\mathbb{Q}}$ can be parametrized explicitly as follows: They corresponds to $\mathbb{Q}$-subspaces $\Sigma_\mathbb{Q}$ of $\Lambda_\mathbb{Q}$ with $\Sigma_\mathbb{Q}^\perp \subset \Sigma_\mathbb{Q}$, where $\Lambda_\mathbb{Q} = \mathbb{Q}^{2g}$ with the standard symplectic pairing. Equivalently they are parametrized by exact sequences $0 \to \Sigma_\mathbb{Z} \to \Lambda_\mathbb{Z} \to \Lambda_\mathbb{Z}/\Sigma_\mathbb{Z} \to 0$, where $\Lambda_\mathbb{Z} = \mathbb{Z}^{2g}$ with the standard symplectic pairing, $\Sigma_\mathbb{Z}$ is a co-torsion free subgroup of $\Lambda_\mathbb{Z}$ with $\Sigma_\mathbb{Z}^\perp \subset \Sigma_\mathbb{Z}$. The group $\mathrm{Sp}(2g,\mathbb{Z})$ acts naturally on them. This motivates the following definition: (where $\mathrm{Sp}(2g,\mathbb{Z})(n) = \mathrm{Ker}(\mathrm{Sp}(2g,\mathbb{Z}) \to \mathrm{Sp}(2g,\mathbb{Z}/n\mathbb{Z}))$)

6.6. Definition. A *cusp at symplectic level-n* is a $\mathrm{Sp}(2g,\mathbb{Z})(n)$-orbit in the set of exact sequences $0 \to \Sigma_\mathbb{Z} \to \Lambda_\mathbb{Z} \to \Lambda_\mathbb{Z}/\Sigma_\mathbb{Z} \to 0$ as above.

Equivalently, a cusp at symplectic level-n is (the isomorphism class of) an exact sequence $0 \to \Sigma_{\mathbb{Z}/n\mathbb{Z}} \to \Lambda_{\mathbb{Z}/n\mathbb{Z}} \to \Lambda_{\mathbb{Z}/n\mathbb{Z}}/\Sigma_{\mathbb{Z}/n\mathbb{Z}} \to 0$ together with an isomorphism $\Lambda_{\mathbb{Z}/n\mathbb{Z}}/\Sigma_{\mathbb{Z}/n\mathbb{Z}} \cong (1/n)\cdot X_\zeta/X_\zeta$, where $\Lambda_{\mathbb{Z}/n\mathbb{Z}} = (\mathbb{Z}/n\mathbb{Z})^{2g}$ with the standard symplectic pairing, $\Sigma_{\mathbb{Z}/n\mathbb{Z}}$ is a subgroup of $\Lambda_{\mathbb{Z}/n\mathbb{Z}}$ with $\Sigma^{\perp}_{\mathbb{Z}/n\mathbb{Z}} \subset \Sigma_{\mathbb{Z}/n\mathbb{Z}}$ such that both $\Sigma_{\mathbb{Z}/n\mathbb{Z}}$ and $\Lambda_{\mathbb{Z}/n\mathbb{Z}}/\Sigma_{\mathbb{Z}/n\mathbb{Z}}$ are free over $\mathbb{Z}/n\mathbb{Z}$, and X_ζ is a free abelian group of rank $\leq g$.

Equivalently a cusp is a $\mathrm{Sp}(2g,\mathbb{Z}/n\mathbb{Z})$-transform of some $\Sigma_{g,r}$, $0 \leq r \leq g$. A cusp as above is *proper* if $\Lambda_{\mathbb{Z}}/\Sigma_{\mathbb{Z}}$ (or equivalently X_ζ) is non-trivial (that is, $r > 0$.)

We define a conical complex $\sum_n$ as follows:

(a) First consider the disjoint union of $C(X_\zeta)$'s, indexed by exact sequences $0 \to \Sigma_{\mathbb{Z}} \to \Lambda_{\mathbb{Z}} \to \Lambda_{\mathbb{Z}}/\Sigma_{\mathbb{Z}} \to 0$ plus isomorphisms $\Lambda_{\mathbb{Z}}/\Sigma_{\mathbb{Z}} \xrightarrow{\sim} X_\zeta$. (Recall that $C(X_\zeta)$ is the cone of positive semi-definite quadratic forms on X_ζ with rational radical.) In other words, the parameter set is the set of all surjections $\Lambda_{\mathbb{Z}} \to X_\zeta$ whose kernels contain their own annihilators. Denote this disjoint union by $\sum^*$. It has a natural structure as a simplicial cone induced by the cone decomposition $\{\sigma_\alpha\}$. The group $\mathrm{Sp}(2g,\mathbb{Z})$ operates naturally on $\sum^*$.

(b) There is an equivalence relation on $\sum^*$: for an exact sequence $0 \to \Sigma_{\mathbb{Z}} \to \Lambda_{\mathbb{Z}} \to X_\zeta \to 0$ and a surjection $X_\zeta \to X_{\zeta'}$, identify $C(X_{\zeta'})$ indexed by the composition $\Lambda_{\mathbb{Z}} \to X_\zeta \to X_{\zeta'}$, with a subspace of $C(X_\zeta)$ (indexed by $0 \to \Sigma_{\mathbb{Z}} \to \Lambda_{\mathbb{Z}} \to X_\zeta \to 0$) via the canonical injection $C(X_{\zeta'}) \hookrightarrow C(X_\zeta)$. This equivalence relation clearly preserves the structure of simplicial cone on $\sum^*$, and is stable under the action of $\mathrm{Sp}(2g,\mathbb{Z})$.

(c) $\sum_n$ is defined to be the quotient of $\sum^*$ under the above equivalence relation and the action of $\mathrm{Sp}(2g,\mathbb{Z})(n)$. It has the natural structure of conical complex (because there might be identification among faces of a cone). When $n \geq 3$, $\sum_n$ is a simplicial complex by Serre's lemma. Note that $\mathrm{Sp}(2g,\mathbb{Z}/n\mathbb{Z})$ operates naturally on $\sum_n$ preserving the conical structure. Note also that for $n|m$ there is a natural map $\sum_m \to \sum_n$, which preserves the conical structures and is equivariant w.r.t. $\mathrm{Sp}(2g,\mathbb{Z}/m\mathbb{Z}) \to \mathrm{Sp}(2g,\mathbb{Z}/n\mathbb{Z})$. In particular, we have a natural map $\sum_n \to \sum_1 = \sum$.

(d) Alternatively $\sum_n$ can be described as follows: First take the disjoint union of cones $C(X_\zeta)$'s, indexed by exact sequences $0 \to \Sigma_{\mathbb{Z}/n\mathbb{Z}} \to \Lambda_{\mathbb{Z}/n\mathbb{Z}} \to \Lambda_{\mathbb{Z}/n\mathbb{Z}}/\Sigma_{\mathbb{Z}/n\mathbb{Z}} \to 0$ as in 6.6 (i.e. cusps at level-n) together with isomorphism $\Lambda_{\mathbb{Z}/n\mathbb{Z}}/\Sigma_{\mathbb{Z}/n\mathbb{Z}} \xrightarrow{\sim} (1/n)\cdot X_\zeta/X_\zeta$. Then pass to the quotient by the following equivalence relation: for an exact sequence $0 \to \Sigma_{\mathbb{Z}/n\mathbb{Z}} \to \Lambda_{\mathbb{Z}/n\mathbb{Z}} \to (1/n)\cdot X_\zeta/X_\zeta \to 0$ and a surjection $X_\zeta \to X_{\zeta'}$, identify $C(X_{\zeta'})$ indexed by the composition $\Lambda_{\mathbb{Z}/n\mathbb{Z}} \to (1/n)\cdot X_\zeta/X_\zeta \to (1/n)\cdot X_{\zeta'}/X_{\zeta'}$, with a subspace of $C(X_\zeta)$ (indexed by $0 \to \Sigma_{\mathbb{Z}/n\mathbb{Z}} \to \Lambda_{\mathbb{Z}/n\mathbb{Z}} \to (1/n)\cdot X_\zeta/X_\zeta \to 0$) via the canonical injection $C(X_{\zeta'}) \hookrightarrow C(X_\zeta)$. Note that we can think of X_ζ's as quotients of a fixed X, with X the quotient $\Lambda_{\mathbb{Z}}$ by a maximal isotropic sublattice $\Sigma_{\mathbb{Z}}$ in $\Lambda_{\mathbb{Z}}$.

(e) There is a natural map from $\sum^*$ modulo the equivalence relation in (b) to the set of exact sequences $0 \to \Sigma_{\mathbb{Z}} \to \Lambda_{\mathbb{Z}} \to \Lambda_{\mathbb{Z}}/\Sigma_{\mathbb{Z}} \to 0$ as in (a). This map preserves the natural incidence relations of its source and target. Passing to $\mathrm{Sp}(2g,\mathbb{Z})(n)$-quotients (or directly from the description in (d)), we get a natural map from $\sum_n$ to the set of cusps at level-n, which preserves the natural incidence relation of both the source and the target and is equivariant w.r.t. $\mathrm{Sp}(2g,\mathbb{Z}) \to \mathrm{Sp}(2g,\mathbb{Z}/n\mathbb{Z})$. (Of course, the incidence relation on the set of exact sequences $0 \to \Sigma_{\mathbb{Z}} \to \Lambda_{\mathbb{Z}} \to \Lambda_{\mathbb{Z}}/\Sigma_{\mathbb{Z}} \to 0$ as in (a) is induced by surjections among the $\Lambda_{\mathbb{Z}}/\Sigma_{\mathbb{Z}}$'s as in (b). Similarly for the cusps at level-n.) This map from $\sum_n$ to the cusps at level-n tells us how $\sum_n$ is built up from orbit sets $C(X_\zeta)^\circ/\mathrm{Ker}(\mathrm{GL}(X_\zeta) \to \mathrm{GL}(X_\zeta/nX_\zeta))$, which are fibres of this map. For instance, the set of connected components of $\sum_n -(0)$ with natural $\mathrm{Sp}(2g,\mathbb{Z}/n\mathbb{Z})$-action is canonically isomorphic to the connected components of the set of proper cusps at level-n under the natural incidence relation.

6.7. Theorem. *Let X be a free abelian group of rank g. To a $\mathrm{GL}(X)$-admissible polyhedral cone decomposition $\{\sigma_\alpha\}$ of $C(X)$ smooth w.r.t. the lattice $B(X)$ and an integer $n \geq 1$, we can attach an algebraic stack $\overline{\mathsf{A}}_{g,n} = \mathsf{A}_{g,n}(\{\sigma_\alpha\})$ over* $\mathrm{Spec}(\mathbb{Z}[\zeta_n, 1/n])$*, which has the following properties:*

(1) $\overline{\mathsf{A}}_{g,n}$ is a proper algebraic stack over $\mathrm{Spec}(\mathbb{Z}[\zeta_n, 1/n])$ *which contains $\mathsf{A}_{g,n}$ as a dense open algebraic subscheme. In fact it is the normalization of $\mathsf{A}_{g,n} \to \overline{\mathsf{A}}_g$. The natural action of the group* $\mathrm{Sp}(2g,\mathbb{Z}/n\mathbb{Z})$ *extends to $\overline{\mathsf{A}}_{g,n}$. The complement $D_{\infty,n} = \overline{\mathsf{A}}_{g,n} - \mathsf{A}_{g,n}$ is a relative Cartier divisor with normal crossings.*

(2) $\overline{\mathsf{A}}_{g,1} = \overline{\mathsf{A}}_g$. If $n|m$, the natural morphism $\mathsf{A}_{g,m} \to \mathsf{A}_{g,n}$ over $\mathrm{Spec}(\mathbb{Z}[\zeta_m, 1/m])$ *extends to a morphism $\overline{\mathsf{A}}_{g,m} \to \overline{\mathsf{A}}_{g,n}$ over* $\mathrm{Spec}(\mathbb{Z}[\zeta_m, 1/m])$*, which is equivariant with respect to the surjection* $\mathrm{Sp}(2g,\mathbb{Z}/m\mathbb{Z}) \to \mathrm{Sp}(2g,\mathbb{Z}/n\mathbb{Z})$.

(3) The universal principally polarized abelian scheme over $\mathsf{A}_{g,n}$ extends to a semi-abelian scheme G over $\overline{\mathsf{A}}_{g,n}$, which is canonically isomorphic to the pull-back of the semi-abelian scheme over $\overline{\mathsf{A}}_g$. The Kodaira-Spencer mapping for $G \to \overline{\mathsf{A}}_{g,n}$ induces an isomorphism $S^2(\Omega_G) \cong \Omega^1_{\overline{\mathsf{A}}_{g,n}}[d\log D_\infty]$ between locally free sheaves on $\overline{\mathsf{A}}_{g,n}$.

(4) $\overline{\mathsf{A}}_{g,n}$ has a stratification with strata parametrized by the strata of the conical complex $\sum_n$, so that the dual conical complex of this stratification is $\sum_n$: $\overline{\mathsf{A}}_{g,n} = \amalg_\sigma Z_n(\sigma)$ with σ running through the cones in $\sum_n$. For instance irreducible components of D_∞ are parametrized by 1-dimensional strata (half-lines) of $\sum_n$. Each stratum of $\overline{\mathsf{A}}_{g,n}$ is irreducible and smooth over $\mathrm{Spec}(\mathbb{Z}[\zeta_n, 1/n])$*, and is open dense in an intersection of irreducible components of D_∞ (including possible self-intersections). For notation, we recall that a cone in $\sum_n$ corresponds to an exact sequence $0 \to \Sigma_{\mathbb{Z}/n\mathbb{Z}} \to \Lambda_{\mathbb{Z}/n\mathbb{Z}} \to (1/n)\cdot X_\zeta/X_\zeta \to 0$ with $\Sigma^\perp_{\mathbb{Z}/n\mathbb{Z}} \subset \Sigma_{\mathbb{Z}/n\mathbb{Z}}$ plus a* $\mathrm{Ker}(\mathrm{GL}(X_\zeta) \to \mathrm{GL}(X_\zeta/nX_\zeta))$*-orbit of cones σ in $C(X_\zeta)^\circ$. We shall sometimes write X_σ for X_ζ.*

(5) If $n|m$, the natural morphism $\overline{\mathsf{A}}_{g,m} \to \overline{\mathsf{A}}_{g,n}$ over $\mathrm{Spec}(\mathbb{Z}[\zeta_m, 1/m])$ *is compatible with the natural map $\sum_m \to \sum_n$.*

(6) Locally in the étale topology $\overline{\mathsf{A}}_{g,n}$ with its stratification parametrized by the cones in $\{\sigma_\alpha\}$ is isomorphic to the torus embedding $\overline{E} = \overline{E}(\{\sigma_\alpha\})$ attached to the cone decomposition $\{\sigma_\alpha\}$. Moreover, the $\mathrm{Sp}(2g,\mathbb{Z}/m\mathbb{Z}) \to \mathrm{Sp}(2g,\mathbb{Z}/n\mathbb{Z})$ *equivariant morphisms $\overline{\mathsf{A}}_{g,m} \to \overline{\mathsf{A}}_{g,n}$ which are compatible with the parametrization of stratifications are locally isomorphic to the map $[m/n] : \overline{E} \to \overline{E}$ induced*

by multiplication with m/n *on* E. *Especially the closures of all strata are smooth over* $\operatorname{Spec}(\mathbb{Z}[\zeta_n, 1/n])$. *The 0-stratum is isomorphic to* $\mathsf{A}_{g,n}$.

(7) For any cone σ *in* $\sum_n$ *the formal completion* $\overline{\mathsf{A}}_{g,n}^{/\sigma}$ *of* $\overline{\mathsf{A}}_{g,n}$ *along the* σ*-stratum is isomorphic to the formal algebraic stack* $\mathcal{S}_{\sigma,n}/\Gamma_{\sigma,n}$, *where* $\Gamma_{\sigma,n}$ *is the stabilizer of* σ *in* $\operatorname{Ker}(\operatorname{GL}(X_\sigma) \to \operatorname{GL}(X_\sigma/nX_\sigma))$. *This isomorphism respects stratifications and the* $\operatorname{Sp}(2g, \mathbb{Z}/n\mathbb{Z})$*-action.*

Proof. Essentially similar to that of theorem 5.7, hence omitted. The assertions about stratification are implicit in our previous descriptions.

6.8. Corollary 1. *The geometric fibres of* $\overline{\mathsf{A}}_{g,n}$ *over* $\operatorname{Spec}(\mathbb{Z}[\zeta_n, 1/n])$ *are all irreducible.*

6.9. Corollary 2. *If* $n \geq 3$ $\overline{\mathsf{A}}_{g,n}$ *is an algebraic space.*

Proof. In terms of the étale groupoid R, this just means that $R \to U \times U$ is a closed embedding. This can be checked after passing to formal completions. By statement (6) we only have to show that $\mathcal{S}_{\sigma,n}/\Gamma_{\sigma,n}$ is a formal algebraic space for each σ. This is obvious, for $\mathcal{S}_{\sigma,n}$ is a formal algebraic space because $\mathsf{A}_{g,n}$ is an algebraic space, and $\Gamma_{\sigma,n}$ is trivial if $n \geq 3$ by Serre's lemma.

6.10. Corollary 3. *(a) For an algebraically closed field* k *over* $\mathbb{Z}[\zeta_n, 1/n]$, *the* k*-rational points of* $\overline{\mathsf{A}}_{g,n}$ *are the disjoint union of the* k*-rational points which lie in the different* σ*-strata* $Z(\sigma)$, *where* σ *runs through cones in* $\sum_n$.

(b) Choose and fix a set of representatives of $\operatorname{Sp}(2g, \mathbb{Z}/n\mathbb{Z})/P_{g,r}(\mathbb{Z}/n\mathbb{Z})$, *for each* $0 \leq r \leq g$. *For a cone* σ *in* $\sum_n$, $Z(\sigma)(k) = \Xi(\sigma)/\Gamma_{\sigma,n}(k)$ *can be describes as follows: They are in 1–1 correspondence with isomorphism classes of quadruples* $(\tilde{G}, A, c_n, \imath_n)$, *where*

— $\tilde{G}$ *is a semi-abelian variety over* k *of dimension* g, *which is an extension* $0 \to T_\sigma \to \tilde{G} \to A \to 0$, *where* T_σ *is a torus of dimension* r *with character group* X_σ, A *a principally polarized abelian variety of dimension* $r = g - r$ *with a symplectic principal level-n structure.*

— $c_n : (1/n)\cdot X_\sigma \to A(k)$ *is a homomorphism whose restriction to* X_σ *classifies the extension* $0 \to T_\sigma \to \tilde{G} \to A \to 0$.

— $\imath_n$ *is a* k*-rational point of the* σ*-stratum of an "*E_σ*-equivariant* $E(\sigma)$*-torsor". The corresponding* E_σ*-torsor can be canonically described as follows: for any* $y \in (1/n) \cdot X_\sigma$ *and any* $\mu \in X_\sigma$ *its push-forward by* $y \otimes \mu \in (1/n) \cdot S^2(X_\sigma)$ *is the* G_m*-torsor* $(c_n(y) \times c_n(\mu))^* \mathcal{P}_A^{\otimes -1}$, *where* $\mathcal{P}_A$ *is the Poincaré torsor of* A, *thought of as a* G_m*-torsor over* $A \times A$ *via the principal polarization.*

The proof is immediate. The chosen sets of representatives of $\operatorname{Sp}(2g, \mathbb{Z}/n\mathbb{Z})/P_{g,r}(\mathbb{Z}/n\mathbb{Z})$ are used to trivialize relevant homogeneous spaces under $\operatorname{Sp}(2g, \mathbb{Z}/n\mathbb{Z})$ non-canonically isomorphic to $\operatorname{Sp}(2g, \mathbb{Z}/n\mathbb{Z})/P_{g,r}(\mathbb{Z}/n\mathbb{Z})$.

6.11. Corollary 4. *For each cusp* $\mathfrak{C} = (0 \to \Sigma_{\mathbb{Z}/n\mathbb{Z}} \to \Lambda_{\mathbb{Z}/n\mathbb{Z}} \to \Lambda_{\mathbb{Z}}/\Sigma_{\mathbb{Z}/n\mathbb{Z}} \to 0$, $\Lambda_{\mathbb{Z}}/\Sigma_{\mathbb{Z}/n\mathbb{Z}} \cong (1/n)\cdot X_\zeta/X_\zeta)$ *at level-n, the union of strata of* $\overline{\mathsf{A}}_{g,n}$ *lying over*

$\mathfrak{C}$ is an intersection of divisors, which we denote by $D_{\mathfrak{C}}$. The formal completion of $\overline{\mathsf{A}}_{g,n}$ along $D_{\mathfrak{C}}$ is (non-canonically) isomorphic to a disjoint union of copies of $\mathcal{S}_{\zeta,n}/\mathrm{Ker}(\mathrm{GL}(X_\zeta) \to \mathrm{GL}(X_\zeta/nX_\zeta))$, via the classifying morphism of G over the formal completion.

6.12. Remark. We can consider $\overline{\mathsf{A}}_{g,n}$ as an algebraic stack over $\mathrm{Spec}(\mathbb{Z}[1/n])$ via $\mathrm{Spec}(\mathbb{Z}[\zeta_n, 1/n]) \to \mathrm{Spec}(\mathbb{Z}[1/n])$, i.e. Grothendieck's restriction of scalars. If we do so $\overline{\mathsf{A}}_{g,n} \to \mathrm{Spec}(\mathbb{Z}[1/n])$ is not geometrically connected any more, but has $\phi(n)$-components. The advantage is that we may extend the action of the group $\mathrm{Sp}(2g, \mathbb{Z}/n\mathbb{Z})$ to $\mathrm{GSp}(2g, \mathbb{Z}/n\mathbb{Z})$ (= the group of symplectic similitudes, which acts on $\mathrm{Spec}(\mathbb{Z}[\zeta_n, 1/n])$ via the standard character $\mathrm{GSp}(2g, \mathbb{Z}/n\mathbb{Z}) \to (\mathbb{Z}/n\mathbb{Z})^*$). This corresponds to considering a modified notion of principal symplectic level-n structures: For a principally polarized abelian scheme $A \to S$, consider isomorphisms $A[n] \cong (\mathbb{Z}/n\mathbb{Z})^{2g}$ which respect symplectic forms up to a constant factor.

7. Theta Level Structures

Another sort of level structure is obtained by prescribing a symmetric Θ-divisor on a principally polarized abelian variety A, or equivalently an ample symmetric line-bundle $\mathcal{L}$ defining the polarization. It should not be confused with the notion of a level-2-structure. Over A_g it is represented by a stack N_g which is a principally homogeneous space over A_g, principally homogeneous under the group $A[2]$ of the 2-torsion-points in the universal abelian variety A. It is relatively Gorenstein, hence Cohen-Macaulay, over $\mathbb{Z}$, and even smooth away from characteristic 2. However even for the case of elliptic curves it is not normal in all points of characteristic two. There exists a map sign: $N_g \to \mu_2$ given by the parity of the unique global section of $\mathcal{O}(\Theta)$.

We can compactify N_g by normalizing $\overline{\mathsf{A}}_g$ but that would lead to singularities. In characteristics different from 2 a better way is to redo our theory, using the description of degenerations of abelian varieties with symmetric line bundles. We sketch the main ideas:

We want to construct symmetric ample line bundles $\mathcal{L}$ on degenerating abelian varieties. We normalize so all symmetries are the identity at the origin. If such a variety is the quotient of an extension $\tilde{G}$ of A by T, the restriction of the symmetry to $T[2]$ is given by a character, that is, a class $\overline{\rho}$ in $X/2 \cdot X$, and we can descend $\mathcal{L}$ to an $\mathcal{M}$ on A with $[-1]^*(\mathcal{M}) \cong \mathcal{M}_\rho$. We then expand a global section of $\mathcal{L}$ into its Fourier-series, and this leads us to the following construction:

Let $X = \mathbb{Z}^g$, and choose an element $\rho \in X$. Instead of $B(X)$ we consider for each such ρ the lattice $Q_\rho(X)$ of quadratic functions $a : X \to \mathbb{Z}$ satisfying $a(\rho - \mu) = a(\mu)$, modulo the constant functions. Up to isomorphism $Q_\rho(X)$ depends only on the class of ρ modulo $2 \cdot X$, and over the rationals all $Q_\rho(X)$

become isomorphic to $B(X)$, via the associated bilinear form. Over the integers the image of $Q_\rho(X)$ in $B(X)$ has index 2^g if $\rho \in 2\cdot X$, and index 2^{g-1} otherwise. We denote by E_ρ the toral group with character-group the dual $Q^*_\rho(X)$ (if $\rho \in 2 \cdot X$) or (non-canonically) $Q^*_\rho(X) \oplus (\mathbb{Z}/2\mathbb{Z})$ (if ρ is not in $2 \cdot X$), such that elements of $E_\rho(R)$ are functions $a(\,) : X \to R^*$ satisfying $a(\rho - \mu) = s \cdot a(\mu)$ (For some fixed choice of $s \in \mu_2$, which obviously must be $+1$ if $\rho \in 2\cdot X$). Note that $T[2]$ injects into E_ρ, sending a local section t to the family $a(\mu) = \mu(t)$ with sign $s = \rho(t)$ (this corresponds to the translation-action on symmetric Θ-divisors.) The quotient $E_\rho/T[2]$ is our old torus E with character-group $S^2(X)$, under the homomorphism $E_\rho \to E$ which sends $a(\,)$ to its associated bilinear form. In $Q_\rho(X)_\mathbb{R}$ we have the cone C_ρ of functions which are bounded below, corresponding to the cone $C \subset B(X)_\mathbb{R}$ under the identifications above. The cone decomposition of C is replaced by smooth cone decompositions of each C_ρ, well behaved under the group $\mathrm{GL}(X)$ and under translations (Transferring the cone decomposition of C by the identifications above would correspond to normalizing $\overline{\mathsf{A}}_g$ in N_g, but the resulting cone decompositions are not smooth in general). This defines embeddings $E_\rho \subset \overline{E}_\rho$. For each simplex $\sigma \subset C_\rho$ of the decomposition we obtain a quotient $X \to X_\sigma$, an embedding $E_{\rho,\sigma} \subset \overline{E}_{\rho,\sigma}$ and a stack T_ρ, as follows: Start with the stack which classifies extensions $0 \to T_\sigma \to \tilde{G} \to A \to 0$ (given by $c : X_\sigma \to A^t = A$), an ample line-bundle $\mathcal{M}$ on A defining a principal polarization such that $[-1]^*(\mathcal{M}) \cong \mathcal{M}_\rho$, and a trivialization of $c^*(\mathcal{M})$ compatible (up to sign) with its symmetry under $\mu \to \rho - \mu$. This is a principal homogeneous space under the torus $E_{\rho,\sigma}$, which acts on the trivializations above, and T_σ is the associated bundle with fibre $\overline{E}_{\rho,\sigma}$. Over the formal completion $\mathcal{T}_\sigma$ we construct a degenerating abelian variety with a symmetric ample line bundle: The a's define the Fourier-coefficients of the theta-function, and the sign is its parity at the origin.

Let us first remove characteristic two. Then after algebraization we obtain a stack $\overline{N}_g$ over $\mathrm{Spec}(\mathbb{Z}[1/2])$ satisfying:

7.1. Theorem. *a)* $\overline{N}_g$ *is a proper over* $\mathrm{Spec}(\mathbb{Z}[1/2])$, *with a stratification parametrized by simplices* σ *modulo the group* $\mathrm{GL}(X)$. *Locally in the étale topology* $\overline{N}_g$ *is isomorphic to* $\overline{F}_\rho$, *as a stratified space. The 0-stratum is isomorphic to* N_g.

b) The formal completion of $\overline{N}_g$ *along the* σ*-stratum is isomorphic to* $\mathcal{T}_\sigma$ *divided by the subgroup of* $\mathrm{GL}(X_\sigma)$ *which fixes* ρ *modulo* $2 \cdot X$ *and* σ.

c) $\overline{N}_g$ *is the disjoint union* $\overline{N}_g = \overline{N}_g^+ \amalg \overline{N}_g^-$, *according to the subdivision into even and odd theta-characteristics. The fibres of* $\overline{N}_g^+$ *and* $\overline{N}_g^-$ *over* $\mathrm{Spec}(\mathbb{Z}[1/2])$ *are geometrically irreducible.*

We leave the proof as an exercise to the reader. Similarly for geometric points etc. One may also construct stacks $\overline{N}_{g,n}$ compactifying the stack $N_{g,n}$ which classifies principally polarized abelian varieties with level-n structure and a symmetric theta-divisor, for n odd. (For even n there is some relation between the two structures which makes things more involved.)

However our construction also gives information about the (much more interesting) behavior of N_g in characteristic two.

7.2. Theorem. *a) $N_g \otimes \mathbb{F}_2$ is has two irreducible components which are both geometrically irreducible.*

b) Let $\mathcal{N}_g^{\mathrm{ord}}$ denote the formal stack over $\mathbb{Z}_2$ obtained by completing the ordinary locus of N_g. Then $\mathcal{N}_g^{\mathrm{ord}}$ is the disjoint union of two stacks $\mathcal{N}_g^{\mathrm{ord},1}$ and $\mathcal{N}_g^{\mathrm{ord},2}$.

i) $\mathcal{N}_g^{\mathrm{ord},1}$ is smooth over $\mathbb{Z}_2$, with geometrically irreducible fibres. The sign map is identically one on it. It has degree 2^g over the formal completion of A_g.

ii) $\mathcal{N}_g^{\mathrm{ord},2}$ is smooth over μ_2, again with geometrically irreducible fibres. Its degree over the formal completion of A_g is $2^g \cdot (2^g - 1)$.

Proof. It is clear that b) implies a), as N_g is finite flat over A_g and as the ordinary locus is dense in $\mathsf{A}_g \otimes \mathbb{F}_2$. To see b) let us study symmetric line bundles on ordinary abelian varieties. Suppose A is such an abelian variety over a complete local ring R, with algebraically closed residue-field k of characteristic two. The 2-divisible group $A[2^\infty]$ is an extension $0 \to \mathrm{Hom}(X, \mu_{2^\infty}) \to A[2^\infty] \to Y \otimes (\mathbb{Q}_2/\mathbb{Z}_2) \to 0$, for free $\mathbb{Z}_2$-modules $X, Y \approx \mathbb{Z}_2^g$. This extension is classified by a bilinear form $b : Y \times X \to \mathrm{lim.proj.}\ R^*/(R^*)^{2^n}$. It is characterized by the fact that for $\mu \in Y$, $\nu \in X$ the fibre over $\mu/2^n$ of $A[2^n] \to 2^{-n} \cdot X/X$ is the $\mathrm{Hom}(X, \mu_{2^n})$-torsor whose push-out under ν is the μ_{2^n}-torsor of n-th roots of $b(\mu, \nu)$. If A has a principal polarization then $X = Y$ (by the Weil-pairing) and b is symmetric, and deformations of A as polarized abelian variety correspond to deformations of this form (use the Serre-Tate theorem). It follows that the space of first order deformations of $A \otimes k$ is $\mathrm{Hom}(S^2(X), k)$, which also can be obtained using the Kodaira-Spencer map.

Suppose A admits a symmetric line-bundle $\mathcal{L}$ which induces the principal polarization. Normalize the isomorphism $[-1]^*(\mathcal{L}) \cong \mathcal{L}$ so that it induces the identity at the origin. Then its restriction to the 2-division points $A[2]$ defines a regular map $e : A[2] \to \mu_2$, and it is known that $e(x+y)/(e(x) \cdot e(y))$ defines the Weil-pairing on $A[2]$. In particular, the restriction of e to $\mathrm{Hom}(X, \mu_2)$ is a character, given by some $\rho \in X$ well determined up to $2 \cdot X$. Translation by a section t of $\mathrm{Hom}(X, \mu_2)$ changes the parity of the unique global section of $\mathcal{L}$ by $\rho(t)$. Furthermore, e restricted to the pre-image of $\mu/2$ under $A[2] \to 1/2 \cdot Y/Y$ is a map of torsors equivariant under $(\rho - \mu) : \mathrm{Hom}(X, \mu_2) \to \mu_2$. It follows that $b(\mu, \rho - \mu)$ is a square. We derive that not all first order deformations of $A \otimes k$ can have such an $\mathcal{L}$: It is necessary that the corresponding bilinear form $X \times X \to k$ vanishes on elements of the form $\mu \otimes (\rho - \mu)$. If $\rho \in 2 \cdot X$ this determines a subspace of codimension g, otherwise of codimension $g - 1$. It follows $N_g^{\mathrm{ord}} \otimes \mathbb{F}_2$ is the disjoint union of $N_g^{\mathrm{ord},1} \otimes \mathbb{F}_2$ and $N_g^{\mathrm{ord},2} \otimes \mathbb{F}_2$, where $N_g^{\mathrm{ord},1} \otimes \mathbb{F}_2$ classifies symmetric $\mathcal{L}$ with $e \equiv 1$ on the toral part (or $\rho \in 2 \cdot X$), and $N_g^{\mathrm{ord},2} \otimes \mathbb{F}_2$ the rest. This induces the decomposition of the formal scheme $\mathcal{N}^g$. Now consider the projection map $N_g^{\mathrm{ord}} \otimes \mathbb{F}_2 \to \mathsf{A}_g^{\mathrm{ord}} \otimes \mathbb{F}_2$: At each point the kernel of the associated tangent map has dimension g, while the cokernel has dimension $\geq g$ (over $N_g^{\mathrm{ord},1} \otimes \mathbb{F}_2$) respectively $\geq g-1$ (over $N_g^{\mathrm{ord},2} \otimes \mathbb{F}_2$). If $d = g\cdot(g+1)/2$ is the dimension of all these schemes it follows that $N_g^{\mathrm{ord},1} \otimes \mathbb{F}_2$ is smooth, and that at each point of $N_g^{\mathrm{ord},2} \otimes \mathbb{F}_2$ the tangent-space can be generated by $\leq d+1$ elements. However the map sign: $N_g^{\mathrm{ord},2} \otimes \mathbb{F}_2 \to \mu_2$ is flat and surjective on tangent-spaces: Using previous notation, locally the toral part μ_2^g of $A[2]$ operates freely on $N_g^{\mathrm{ord},2}$, and sign is equivariant for the surjection $\mu_2^g \to \mu_2$ induced by ρ (which now is odd!). As above it follows that sign: $N_g^{\mathrm{ord},2} \otimes \mathbb{F}_2 \to \mu_2$ is smooth. Also the assertion about degrees works out easily.

Finally we are at a point where we can use torus embeddings: Choose as usual an admissible cone-decomposition on symmetric bilinear forms on $\mathbb{Z}^g$. We derive a formal scheme $\mathcal{S}$ whose fibres over Spec($\mathbb{Z}$) are smooth and geometrically connected, and a fibrewise dominant map $\mathcal{S}/\mathrm{GL}(g,\mathbb{Z}) \to \overline{\mathsf{A}}_g$. More precisely for each affine piece Spf(R) of $\mathcal{S}$ we have a semi-abelian $G \to \mathrm{Spec}(R)$, etc. Using the same cone-decomposition for each Q_ρ we get formal schemes $\mathcal{T}_\rho \to \mathcal{S}$, for $\rho \in (\mathbb{Z}/2\mathbb{Z})^g$, such that the $\mathcal{T}_\rho$ are geometrically normal and connected over μ_2 (if $\rho \neq 0$) respectively Spec($\mathbb{Z}$) (if $\rho = 0$). Let $\mathcal{T}$ denote the disjoint union of all $\mathcal{T}_\rho$. $\mathrm{GL}(g,\mathbb{Z})$ operates on $\mathcal{T}$ (permuting the ρ's). Furthermore over the open stratum of $\mathcal{T}$ the pullback of G admits a symmetric theta-divisor, and this is the universal one (One checks that the open stratum is a $G[2]$-torsor, etc.). Finally everything is equivariant under $\mathrm{GL}(g,\mathbb{Z})$.

For any affine piece Spf(R) of $\mathcal{T}_\rho$ we thus obtain a map from the open stratum of Spec(R) into N_g, and the collection of all these maps is dominant fibre by fibre. Furthermore the maps $\mathcal{T}_\rho \to \mu_2$ correspond to the sign map. In characteristic two the open strata of the affine pieces of $\mathcal{T}_0 \otimes \mathbb{F}_2$ dominate $N_g^{\mathrm{ord},1} \otimes \mathbb{F}_2$ and those of the other $\mathcal{T}_\rho \otimes \mathbb{F}_2$ dominate $N_g^{\mathrm{ord},1} \otimes \mathbb{F}_2$. It follows that on $\mathcal{N}_g^{\mathrm{ord},1}$ the sign-map is identically one.

It remains to show the assertions about irreducibility. For this we provisionally denote by $\overline{N}_g$ the normalization of $\mathsf{A}_g \otimes \mathbb{F}_2$ in $(N_g \otimes \mathbb{F}_2)_{\mathrm{red}}$ or in $(N_g^{\mathrm{ord}} \otimes \mathbb{F}_2)_{\mathrm{red}}$. According to the decomposition of $N_g^{\mathrm{ord}} \otimes \mathbb{F}_2$ this is the disjoint union of stacks $\overline{N}_g^1 \supset N_g^{\mathrm{ord},1} \otimes \mathbb{F}_2$ and $\overline{N}_g^2 \supset (N_g^{\mathrm{ord},2} \otimes \mathbb{F}_2)_{\mathrm{red}}$. Also the reduced fibre over $\mathbb{F}_2$ of the formal stack $\mathcal{T}/\mathrm{GL}(g,\mathbb{Z})$ dominates $\overline{N}_g$. This fibre has two connected components, one given by $\mathcal{T}_0/\mathrm{GL}(g,\mathbb{Z})$ and the second by the rest. It follows that $\overline{N}_g$ also has only two connected components, namely $\overline{N}_g^1$ and $\overline{N}_g^2$. As $\overline{N}_g$ is normal they must be irreducible. The same reasoning over finite extensions of $\mathbb{F}_2$ shows that these components are geometrically irreducible, and we are done.

7.3. Corollary. *Let N_g^{norm} denote the normalization of N_g (over $\mathbb{Z}$). Then N_g^{norm} is the disjoint union $N_g^{\mathrm{norm},+}$ and $N_g^{\mathrm{norm},-}$, distinguished by the value of sign. The fibres of $N_g^{\mathrm{norm},+}$ and $N_g^{\mathrm{norm},-}$ over* Spec($\mathbb{Z}$) *are geometrically reduced, and geometrically irreducible except in characteristic two. Furthermore $N_g^{\mathrm{norm},-} \otimes \mathbb{F}_2$ is geometrically irreducible, while $N_g^{\mathrm{norm},+} \otimes \mathbb{F}_2$ has two (geometrically) irreducible components extending $\mathcal{N}_g^{\mathrm{ord},1} \otimes \mathbb{F}_2$ and $(\mathcal{N}_g^{\mathrm{ord},2} \otimes \mathbb{F}_2)_{\mathrm{red}}$.*

This follows as all irreducible components show up over the ordinary locus.

Chapter V. Modular Forms and the Minimal Compactification

0. Introduction

The major goal of this section is to lay the foundation for an arithmetic theory of Siegel modular forms and to construct an arithmetic minimal compactification of (the coarse moduli space of) A_g, which mimics the Satake-Baily-Borel compactification of arithmetic quotients of bounded symmetric domains. These two are closely intertwined, as they must be.

We shall define modular forms as global sections of powers of ω over (any) $\overline{\mathsf{A}}_g$. This definition is independent of the choice of the cone decomposition. Then we establish the q-expansion principle and the Koecher principle, to the effect that modular forms are determined by their q-expansions and that any meromorphic modular form regular on A_g is automatically holomorphic at the boundary, if the genus g is > 1. The minimal compactification A_g^* is defined as the projective spectrum of the graded ring of modular forms. By its construction, A_g^* is a blowing-down of $\overline{\mathsf{A}}_g$. Roughly speaking, in terms of the semi-abelian scheme G over A_g, the natural map $\overline{\mathsf{A}}_g \to \mathsf{A}_g^*$ associates to a semi-abelian variety its abelian part. Some power of ω descends to an ample invertible sheaf on the minimal compactification. The necessary power comes from the non-trivial automorphisms of principally polarized abelian varieties of dimension $\leq g$. There are obvious analogues if we throw in principal level-n structure. If we do so for $n \geq 3$ then ω itself descends to the minimal compactification. As a byproduct, it follows that the ring of integral Siegel modular forms is finitely generated over $\mathbb{Z}$.

As an application we prove a theorem that the p-adic monodromy of the "physical p-torsion points" of generic principally polarized abelian varieties in characteristic $p > 0$ is "as large as possible", namely $\mathrm{GL}_g(\mathbb{Z}_p)$. Of course, this same method can be applied to show that the p-adic monodromy is big for modular varieties coming from abelian varieties, polarizations and endomorphism structures.

In chapter IV it was shown that for $n \geq 3$, $\overline{\mathsf{A}}_{g,n}$ is in fact an algebraic space. The reader may ask whether it is indeed a scheme. Although we do not give a complete answer, we do supply a sufficient condition under which $\overline{\mathsf{A}}_{g,n}$ is a projective scheme over $\mathbb{Z}[\zeta_n, 1/n]$. We need is a polarization function (ch. IV, def. 2.4) for the cone decomposition $\{\sigma_\alpha\}$. If it exists then $\overline{\mathsf{A}}_{g,n}$ is the blow-up of the minimal compactification $\mathsf{A}^*_{g,n}$ along an explicitly defined coherent sheaf of ideal of $\mathsf{A}^*_{g,n}$ with the boundary as support (i.e. the ideal of cusp forms with specified order of vanishing along the boundary). This is a generalization of a theorem of Y.-S. Tai in the last chapter of [AMRT].

As in the previous chapter, the method in this chapter can be used to handle Shimura varieties of PEL type. For instance, the resulting arithmetic minimal compactification for the Hilbert-Blumenthal moduli spaces can be used to improve the results of Brylinski-Labesse [Br-La] on the Ramanujan-Petersson conjecture from "almost all p" to "all p outside the discriminant".

1. Modular Forms and Their q-Expansions

Let us fix a $\mathrm{GL}(X)$-admissible polyhedral cone decomposition $\{\sigma_\alpha\}$ of $C(X)$ as in chap. IV §2, and let $\overline{\mathsf{A}}_g$ be the associated arithmetic toroidal compactification. Over $\overline{\mathsf{A}}_g$ we have a canonical invertible sheaf $\omega = \det(\mathcal{T}_G^*) = \det(\Omega_{G/\overline{\mathsf{A}}_g})$, corresponding to the highest order translation-invariant relative differentials of the semi-abelian variety G over $\overline{\mathsf{A}}_g$. Modular forms of weight k are just global sections of ω^k:

1.1. Definition. Suppose M is an abelian group. A *(Siegel) modular form of genus g and weight k with coefficients in M regular at infinity* is an element of $\Gamma(\overline{\mathsf{A}}_g, \omega^k \otimes_{\mathbb{Z}} M)$.

1.2. Remarks. (a) In most applications M will be a commutative ring, but for the proof of the q-expansion principle we need the full generality of the definition 1.1. As a functor in M, modular forms are left exact and commute with filtering inductive limits.

(b) The readers surely have noticed that we have made a choice of a $\mathrm{GL}(X)$-admissible polyhedral cone decomposition $\{\sigma_\alpha\}$ of $C(X)$, so it seems that definition 1.1. depends on that choice, which is undesirable. *In fact this definition is really independent of the choice of the cone-decomposition leading to* $\overline{\mathsf{A}}_g$, here is one way to see it: By taking refinements, we only have to show that for two $\mathrm{GL}(X)$-admissible polyhedral cone decompositions $\{\sigma_\alpha\}, \{\sigma'_\zeta\}$, one being a refinement of the other, we have $\Gamma(\overline{\mathsf{A}}_g\{\sigma_\alpha\}, \omega^{\otimes k} \otimes M) = \Gamma(\overline{\mathsf{A}}_g\{\sigma'_\zeta\}, \omega^{\otimes k} \otimes M)$. The point is that the natural morphism map π from $\overline{\mathsf{A}}_g\{\sigma_\alpha\}$ to $\overline{\mathsf{A}}_g\{\sigma'_\zeta\}$ induces a natural identification between the corresponding sets of global sections of $\omega^{\otimes k}$. This is a consequence of the facts that $\pi_*\mathcal{O} = \mathcal{O}$ and $R^i\pi_*\mathcal{O} = 0\ \forall i > 0$ universally. The first is immediate, and the second is because the locally in the étale topology π can be explicitly described. In fact locally in the étale topology π is isomorphic to the map between the torus embeddings attached to $\{\sigma_\alpha\}$ and $\{\sigma'_\zeta\}$: say $\{\sigma_\alpha\}$ is a refinement of $\{\sigma'_\zeta\}$, and $\pi_T : \mathrm{Temb}\{\sigma_\alpha\} \to \mathrm{Temb}\{\sigma'_\zeta\}$

is the natural morphism between the corresponding torus embeddings (c. [TE chap. 1]). We know that $\overline{\mathsf{A}}_g\{\sigma'_\zeta\}$ and $\mathrm{Temb}\{\sigma'_\zeta\}$ are isomorphic locally in the étale topology. Moreover locally in the étale topology of the target, π_T and π are isomorphic. In particular the irreducible components of the fibres of π are complete rational varieties, and the cones tell us how the components intersect.

(c) Later on in this section we shall see that for $g \geq 2$, the natural injective map $\Gamma(\overline{\mathsf{A}}_g, \omega^{\otimes k} \otimes_{\mathbb{Z}} M) \to \Gamma(\mathsf{A}_g, \omega^{\otimes k} \otimes_{\mathbb{Z}} M)$ is an isomorphism (Koecher's principle). Since for $g = 1$ there is only one cone decomposition, namely $\{0\} \cup \mathbb{R}_{>0}$, this gives another proof of independence of choice of $\{\sigma_\alpha\}$ in the definition of Siegel modular forms.

q-Expansion of Elements of $\Gamma(\mathsf{A}_g, \omega^{\otimes k} \otimes_{\mathbb{Z}} M)$

Let us first study $\Gamma(\mathsf{A}_g, \omega^{\otimes k} \otimes_{\mathbb{Z}} M)$. Any element f in this space has a *q-expansion*, defined as follows:

(i) Giving f amounts to a functorial assignment, which associates to any principally polarized abelian scheme $A \to \mathrm{Spec}(R)$ of relative dimension g over a base ring R, an element $f(A) \in \Gamma(A, \omega_{A/R}^{\otimes k} \otimes_{\mathbb{Z}} M)$, and this association is compatible with arbitrary pull-backs and isomorphisms.

(ii) We may construct special A's via the Mumford construction, and evaluate f there. In particular, we can construct (principally polarized) abelian varieties as quotients of split tori by multiplicative periods. The invertible sheaves ω for A and the split torus are canonically isomorphic to $\Lambda^g(X)$, hence trivial. Evaluating f via this canonical trivialization gives the q-expansion.

Let us carry out the above idea more carefully. So let $X = \mathbb{Z}^g$, and let $s_1, \dots, s_r$ $(r = g(g+1)/2)$ denote a $\mathbb{Z}$-basis of $B(X)$ = the space of all $\mathbb{Z}$-valued symmetric bilinear forms on X, with each s_i positive semi-definite. Let $t_1, \dots, t_r$ be the dual basis of $s_1, \dots, s_r$ in $S^2(X)$ = the symmetric quotient of $X \otimes_{\mathbb{Z}} X$. Over the base ring $R = \mathbb{Z}[\![t_1, \dots, t_r]\!]$ we have the split torus $\tilde{G}$ with character group X. Let K = fraction field of R, and $b : X \times X \to K^*$ the positive definite bilinear form b on X given by $b(\mu, \nu) = [\mu \otimes \nu]$, or explicitly $b(\mu \otimes \nu) = \prod_i t_i^{s_i(\mu,\nu)}$. Then b defines a period subgroup $Y = X \subset \tilde{G}(K)$, polarized by $Y = X$, so Mumford's construction can be applied to give a principally polarized abelian variety $G_{/K}$ which extends to a semi-abelian scheme G over R. By construction, $\omega_{G/R} \cong \omega_{\tilde{G}/\tilde{R}} \cong R \otimes_{\mathbb{Z}} \Lambda^g(X)$ canonically. Over $\mathrm{Spec}\, R[1/(t_1 \dots t_r)]$, G is a principally polarized abelian scheme, so we obtain $f(G) \in R[1/(t_1 \dots t_r)] \otimes \Lambda^g(X)^{\otimes k} \otimes_{\mathbb{Z}} M$.

To describe the dependence on the basis $s_1, \dots, s_r$ of $B(X)$ we identify R with the ring of power series $\mathbb{Z}[\![S^2(X) \cap \sigma^\vee]\!]$ associated to the monoid $S^2(X) \cap \sigma^\vee$, where $S^2(X)$ is the symmetric quotient of $X \otimes_{\mathbb{Z}} X$ dual to $B(X)$ = symmetric bilinear forms on X, $\sigma \subset B(X) \otimes_{\mathbb{Z}} \mathbb{R}$ (= the space of all $\mathbb{R}$-valued symmetric bilinear forms on X) is the cone generated by $s_1, \dots, s_r$, and $\sigma^\vee \subset S^2(X) \otimes_{\mathbb{Z}} \mathbb{R}$ is the dual cone of σ and is generated by $t_1, \dots, t_r$. Now it is evident that b is given by $b(\mu, \nu)$ = the image of $\mu \otimes \nu$ in $S^2(X)$. In this notation $f(G)$ becomes a power-series $f(G) = \sum_{\lambda \in S^2(X)} f(\lambda) \cdot \lambda$, with coefficients $f(\lambda) \in M \otimes_{\mathbb{Z}} \Lambda^g(X)^{\otimes k}$ such that the support of $f(\cdot)$ is contained in a translate of the cone $\sigma^\vee$. In fact it is not necessary to assume that σ is generated by a basis of $B(X)$. The construction works for any rational polyhedral cone in $C(X)^\circ$.

Now it is easy to show that the q-expansion is independent of the choice of the cone σ. Indeed if we have two cones σ_1, σ_2 with $\sigma_1 \subset \sigma_2$ the independence is clear, because the pull-back of $G_1 \to \operatorname{Spec} \mathbb{Z}[\![S^2(X) \cap \sigma_2^\vee]\!]$ by $\operatorname{Spec} \mathbb{Z}[\![S^2(X) \cap \sigma_1^\vee]\!] \to \operatorname{Spec} \mathbb{Z}[\![S^2(X) \cap \sigma_2^\vee]\!]$ is the semi-abelian scheme G_1 over $\operatorname{Spec} \mathbb{Z}[\![S^2(X) \cap \sigma_1^\vee]\!]$, say by functoriality of Mumford's construction. The general case follows, because any two cones are contained in a third one. At this point it should be clear that we made the assumption that σ is a top-dimensional cone generated by a $\mathbb{Z}$-basis only for simplicity. In fact any rational cone σ with $\sigma^\circ \subset C(X)^\circ$ can be used.

Moreover, the coefficients $f(\lambda)$ are invariant under $\mathrm{GL}(X)$: an element γ in $\mathrm{GL}(X)$ induces a natural isomorphism $\gamma^* : \mathbb{Z}[\![S^2(X) \cap \gamma\sigma^\vee]\!] \xrightarrow{\sim} \mathbb{Z}[\![S^2(X) \cap \sigma^\vee]\!]$. By the functoriality of Mumford's construction and q-expansion and the independence of cone which we just proved, we see that $\gamma^* f(\gamma^*(\lambda)) = f(\lambda)$.

We have seen that for any finitely generated rational subcone σ of $C(X)^\circ$ the support of $f(\cdot)$ is contained in a translate of $\sigma^\vee$. We may ask whether it is actually contained in all such $\sigma^\vee$, that is in their intersection $C(X)^\vee$. This holds for $g \geq 2$:

1.3. Lemma. *Suppose $L \subset S^2(X) \otimes \mathbb{Q}$ is a subset invariant under* $\mathrm{GL}(X)$, *such that for any finitely generated rational subcone $\sigma \subset C(X)^\circ$ L is contained in a translate of $\sigma^\vee$. If $g = \mathrm{rk}(X) \geq 2$, then $L \subset C(X)^\vee$.*

Proof. Let $\lambda \in L$ denote an element. The assumptions imply that for any $b \in C(X)$ the set $\{\langle b, \gamma(\lambda)\rangle | \gamma \in \mathrm{GL}(X)\}$ has a finite infimum. Let us translate into matricial notation: λ is represented by a symmetric $g \times g$ matrix M with rational entries, such that for any positive semi-definite real symmetric matrix N $\operatorname{Inf}\{\operatorname{Trace}(\mathrm{MTNT}^t) | T \in \mathrm{GL}(g, \mathbb{Z})\}$ is finite. We must show that M is positive semi-definite.

We may change the basis of $\mathbb{Q}^g$, so we choose a basis $e_1, \ldots, e_g$ of $\mathbb{Q}^g$ in which the quadratic form given by M becomes diagonal, with diagonal elements $m_1, \ldots, m_g \in \mathbb{Q}$. The change of basis also acts on the N's, but with the inverse matrix, because of duality. We have to show that none of the m_i is negative. If for example $m_1 < 0$, we choose for N the diagonal-matrix with entries $(0, 1, 0, \ldots, 0)$ (using that $g \geq 2$), and for T unipotent transformations of the form $T(e_1) = e_1$, $T(e_2) = n \cdot e_1 + e_2$, $T(e_i) = e_i$ for $i \geq 3$. For n sufficiently divisible T is integral not only in the basis $e_1, \ldots, e_g$ but also in the standard basis, and $\operatorname{Trace}(\mathrm{MTNT}^t) = n^2 \cdot m_1 + m_2$. As this is bounded below, $m_1 \geq 0$. Thus all $m_i \geq 0$ and the lemma is proved.

This result is of interest because of the following:

1.4. Lemma. *(a) The q-expansion map $\Gamma(\mathsf{A}_g, \omega^{\otimes k} \otimes_{\mathbb{Z}} M) \to \mathbb{Z}[\![S^2(X)]\!] \otimes_{\mathbb{Z}} M \otimes_{\mathbb{Z}} \Lambda^g(X)^{\otimes k}$ is injective.*

(b) An element of $\Gamma(\mathsf{A}_g, \omega^{\otimes k} \otimes_{\mathbb{Z}} M)$ lies in $\Gamma(\overline{\mathsf{A}}_g, \omega^{\otimes k} \otimes_{\mathbb{Z}} M)$ if and only if its q-expansion-coefficients are supported in $C(X)^\vee$.

Proof. (a) *Injectivity.* M is the direct limit of finitely generated $\mathbb{Z}$-modules. Since cohomology and tensor product both commute with direct limits, we are reduced to the case when M is a finitely generated $\mathbb{Z}$-module, then to $M = \mathbb{Z}$ or $= \mathbb{Z}/n\mathbb{Z}$ by additivity in M. In both cases M is actually a ring.

Now assume M is a (commutative) ring. Choose $\sigma \subset C(X)^\circ$ to be a top-dimensional cone in the $\mathrm{GL}(X)$-admissible polyhedral cone decomposition $\{\sigma_\alpha\}$ of $C(X)$ which we used to construct $\overline{\mathsf{A}}_g$. We have seen that the q-expansion of an element f of $\Gamma(\mathsf{A}_g \otimes M, \omega^{\otimes k} \otimes_{\mathbb{Z}} M)$ can be computed using σ. σ corresponds to the formal completion $\mathcal{S}_\sigma$ of $\overline{\mathsf{A}}_g$ along a closed stratum, and the injectivity follows from the fact that $\overline{\mathsf{A}}_g \otimes M$ is smooth over the ring M with geometrically irreducible fibres.

(b) $f \in \Gamma(\mathsf{A}_g, \omega^{\otimes k} \otimes_{\mathbb{Z}} M)$ lies in $\Gamma(\overline{\mathsf{A}}_g, \omega^{\otimes k} \otimes_{\mathbb{Z}} M)$ if and only if its q-expansion-coefficients are supported in $C(X)^\vee$.

One implication is clear: evaluation at Mumford test families shows the "only if" part. For the other one, arguing as before, we may and do assume that M is either $\mathbb{Z}$ or $\mathbb{Z}/n\mathbb{Z}$, hence a ring. Suppose that the coefficients of the q-expansion of f are supported in $C(X)^\vee$. f is a rational section of ω^k, regular on A_g, so its poles must be supported at the boundary $\overline{\mathsf{A}}_g - \mathsf{A}_g$. By fpqc descent, the assumption means that the polar locus $Z(f)$ of f is disjoint from strata corresponding to cones in $C(X)^\circ/\mathrm{GL}(X)$. In particular, $Z(f)$ is disjoint from strata of $\overline{\mathsf{A}}_g - \mathsf{A}_g$ corresponding to the top-dimensional cones in $C(X)$ (i.e. relative dimension is zero). Since every cone in $C(X)$ is the face of a top-dimensional cone, the closure of every stratum in $\overline{\mathsf{A}}_g - \mathsf{A}_g$ contains a stratum corresponding to a top-dimensional cone, therefore $Z(f)$ must be empty. In other words f is regular on $\overline{\mathsf{A}}_g$.

Let us summarize our discussion in the following

1.5. Proposition. *(i) The q-expansion of a modular form f is $\mathrm{GL}(X)$-invariant, and can be computed using any cone in $C(X)^\circ$, in particular any top-dimensional cone in $C(X)$.*

(ii) (Koecher principle). If $g \geq 2$, $\Gamma(\mathsf{A}_g, \omega^{\otimes k} \otimes_{\mathbb{Z}} M) = \Gamma(\overline{\mathsf{A}}_g, \omega^{\otimes k} \otimes_{\mathbb{Z}} M)$ for any $\mathbb{Z}$-module M, and the q-expansion-coefficients are supported in $C(X)^\vee$.

(iii) (q-expansion principle). If $M_1 \subset M_2$ is a subgroup of M_2, a modular form with coefficients in M_2 and q-expansion in $M_1[\![S^2(X)]\!] \otimes \Lambda^g(X)^{\otimes k}$ actually is a modular form with coefficients in M_1 (with the same weight).

Proof. (i) has been shown, (ii) is the combination of 1.3 and 1.4(b), (iii) follows from 1.4 (a) and left-exactness in M.

We remark that one usually applies the q-expansion principle for the case M_1, M_2 are both rings and $M_1 \subset M_2$ is an inclusion of rings, in which case the q-expansion principle reads: if f is a modular form defined over a ring M_2 whose q-expansion has coefficients in a subring M_1 of M_2, then f is actually defined over M_1.

Siegel Modular Forms and Their q-Expansions over $\mathbb{C}$

Over the field of complex numbers $\mathbb{C}$, we can consider Siegel modular forms with coefficients in $\mathbb{C}$, and we recover the usual notion of Siegel modular forms, as for example defined in [Fr]: As a complex analytic stack $\mathsf{A}_{g/\mathbb{C}}$ is the quotient of the Siegel upper half-plane $\mathbb{H}_g$ by the group $\mathrm{Sp}(2g,\mathbb{Z})$ (with the usual action), via the family of principally polarized abelian varieties over $\mathbb{H}_g$ given by $A(\Omega) = \mathbb{C}^g/(\mathbb{Z}^g \oplus \Omega \cdot \mathbb{Z}^g)$ for $\Omega \in \mathbb{H}_g$. Note that the tangent space to $A(\Omega)$ is canonically isomorphic to $\mathbb{C}^g$, so the ω is canonically isomorphic to $\mathcal{O}_{\mathbb{H}_g} \otimes_{\mathbb{C}} \Lambda^g(\mathbb{C}^g)$, and a complex modular form s of weight k becomes an expression of the form $s = f(\Omega)\cdot(dz_1 \wedge \cdots \wedge dz_g)^{\otimes k}$ where $f(\Omega)$ is a complex holomorphic function on $\mathbb{H}_g$ holomorphic at ∞, dz_i's are (fibre) coordinates of $\mathbb{C}^g$ and $f(\Omega)\cdot(dz_1 \wedge \cdots \wedge dz_g)^{\otimes k}$ is invariant under the natural $\mathrm{Sp}(2g,\mathbb{Z})$-action. In other words after trivializing ω on $\mathbb{H}_g$, complex modular forms of weight k are identified with complex holomorphic functions $f(\Omega)$ on $\mathbb{H}_g$ holomorphic at ∞, satisfying the transformation rule $f(\gamma\cdot\Omega) = \det(C\Omega+D)^k \cdot f(\Omega)$ for any element $\gamma \in \mathrm{Sp}(2g,\mathbb{Z})$ with entries A, B, C, D, where $\gamma \cdot \Omega = (A\Omega + B)\cdot(C\Omega + D)^{-1}$. Notice that the condition "holomorphic at ∞" is automatically satisfied if $g \geq 2$, by the classical Koecher principle.

We have shown that a modular form of weight k can be thought of (after trivializing ω on $\mathbb{H}_g$) as a holomorphic function $f(\Omega)$ on $\mathbb{H}_g$ holomorphic at ∞ satisfying the correct transformation rule. In fact the converse is also true: any holomorphic function $f(\Omega)$ on $\mathbb{H}_g$ holomorphic at ∞ satisfying the correct transformation rule (i.e. a holomorphic section of $\omega^{\otimes k}$ on $\mathsf{A}_g(\mathbb{C}) = \mathrm{Sp}(2g,\mathbb{Z})\backslash\mathbb{H}_g$) extends to a section of $\omega^{\otimes k}$ over $\overline{\mathsf{A}}_g(\mathbb{C})$, hence is algebraic by GAGA.

The classical Fourier-expansion of $f(\Omega)$ (using invariance of f under $\Omega \to \Omega + S$, $\forall S \in B(X)$, identified with symmetric $g \times g$ matrices with integral coefficients) coincides with the q-expansion defined in 1.2.3 above, maybe up to factor $(2\pi i)^{gk}$ where k is the weight of f, depending on the normalizations one chooses to make. Indeed if we use $(dz_1 \wedge \cdots \wedge dz_g)$ to trivialize ω, then we have the factor $(2\pi i)^{gk}$. If we use $(2\pi i\, dz_1 \wedge \cdots \wedge 2\pi i\, dz_g)$ instead, then we can eliminate this factor.

To see this assertion, take a top-dimensional cone σ in the $\mathrm{GL}(X)$-admissible polyhedral cone decomposition $\{\sigma_\alpha\}$ of $C(X)$. On the one hand σ corresponds to a point s_σ of the boundary of $\overline{\mathsf{A}}_g(\mathbb{C})$ such that the holomorphic local coordinate is described as follows: we have $\mathbb{H}_g = B(X)\otimes_{\mathbb{Z}}\mathbb{R} + i\cdot C(X)^\circ$, and the action of the unipotent subgroup $B(X) \subset \mathrm{Sp}(2g,\mathbb{Z})$ is just the translation. $B(X)\backslash\mathbb{H}_g$ is canonically embedded in $B(X)\backslash B(X)\otimes_{\mathbb{Z}}\mathbb{C} = B(X)\otimes_{\mathbb{Z}}\mathbb{C}^*$ via exponentiation ($\exp(2\pi i\cdot)$), and σ defines a torus embedding $B(X)\otimes_{\mathbb{Z}}\mathbb{C}^* \subset \mathrm{Temb}(\sigma)$. Let S_σ be the interior of the closure of $B(X)\backslash\mathbb{H}_g$ in $\mathrm{Temb}(\sigma)$, which contains the closed $B(X)\otimes_{\mathbb{Z}}\mathbb{C}^*$-orbit $\tilde{s}_\sigma$ (in fact a point). There is an étale map (in the complex analytic category) from S_σ to $\overline{\mathsf{A}}_g(\mathbb{C})$ sending $\tilde{s}_\sigma$ to s_σ, so that S_σ gives local holomorphic coordinates near s_σ. The holomorphic function $f(\Omega)$ defines a holomorphic function on $B(X)\backslash\mathbb{H}_g$ because of $B(X)$-invariance, in fact defines a holomorphic function on S_σ (either by 1.5 or the classical Koecher principle), and can therefore be expanded as a convergent power series $\sum_{\lambda\in S^2(X)} a(\lambda)\cdot\exp(2\pi i\cdot\langle\Omega,\lambda\rangle)$. Here $S^2(X)$ = the symmetric quotient of $X\otimes_{\mathbb{Z}} X = \mathbb{Z}$-dual of $B(X)$.

Over S_σ we can perform a complex analytic quotient construction analogous to Mumford's construction: dividing the constant torus $B(X)\otimes_{\mathbb{Z}}\mathbb{C}^*$ over S_σ by the degenerating set of periods $\exp(2\pi i\,\mathbb{Z}^g\cdot\Omega)$ to get a semi-abelian family, which is nothing but $G(\mathbb{C})$ restricted to S_σ. Thus canonically a modular form of weight k has a convergent Fourier expansion of the form $\sum_{\lambda\in S^2(X)} a(\lambda)\cdot\exp(2\pi i\cdot\langle\Omega,\lambda\rangle)\cdot(dz_1\wedge\cdots\wedge dz_g)^{\otimes k}$.

By now it is clear that what we did in the algebraic context is nothing but passing to the formal completion of S_σ at $\tilde{s}_\sigma$, so we certainly get the same series. The only point to note is that algebraically we used $d\chi_1/\chi_1\wedge\cdots\wedge d\chi_g/\chi_g$ to trivialize ω over (the formal completion of) S_σ, where $\chi_1,\cdots\chi_g$ are the standard basis of $X=\mathbb{Z}^g$ = character group of the torus $(\mathbb{C}^*)^g$, and $(\mathbb{C}^*)^g$ is identified with $\mathbb{C}^g/\mathbb{Z}^g$ via $\exp(2\pi i\cdot)$. Therefore $d\chi_1/\chi_1\wedge\cdots\wedge d\chi_g/\chi_g = 2\pi i\,dz_1\wedge\cdots\wedge 2\pi i\,dz_g$, where the z_i's are the standard coordinates of $\mathbb{C}^g$ ($\chi_j=\exp(2\pi i\cdot z_j)$). This explains the factor $(2\pi i)^{kg}$.

Fourier-Jacobi Expansion

When discussing the q-expansion principle we have used only the strata of $\overline{\mathsf{A}}_g$ corresponding to cones in $C(X)^\circ$, i.e. where the abelian part of G is trivial. There is no reason to restrict ourselves to these strata and not to look at the general ones. In fact we can apply the Mumford construction not just to tori but to general semi-abelian extensions, thus obtaining generalized q-expansions, which in the theory of holomorphic automorphic forms are called Fourier-Jacobi expansions.

Let us fix a quotient-lattice $X\to X_\xi$ of rank r, and a smooth positive definite simplex $\sigma\subset C(X_\xi)^\circ_{\mathbb{R}}$, not necessarily of maximal dimension (but $X_\sigma = X_\xi$). (Later on we shall apply our discussion to a cone σ in the $\mathrm{GL}(X)$-admissible polyhedral cone decomposition used to construct $\overline{\mathsf{A}}_g$.) We can perform the Mumford construction on the formal completion $\mathcal{S}_\sigma$ of a stack $\overline{\Xi}_\sigma$ along its closed stratum as in chapter IV,3, and we shall follow the notation there pretty closely. $\overline{\Xi}_\sigma$ is a bundle over the algebraic stack $\mathbf{Hom}_{\mathbb{Z}}(X_\xi, A_{g-r})$ which classifies semi-abelian extensions $0\to T_\xi\to\tilde{G}\to A\to 0$, with T_ξ the torus with character group X_ξ, and A a principally polarized abelian scheme of relative dimension $g-r$. (Of course A_{g-r} is the universal abelian scheme over A_{g-r}.) The fibre of the bundle $\overline{\Xi}_\sigma\to\mathbf{Hom}_{\mathbb{Z}}(X_\xi, A_{g-r})$ is the torus embedding $\overline{E}_\sigma$ defined by the smooth cone $\sigma\subset C(X_\xi)^\circ_{\mathbb{R}}$. Mumford's construction gives a semi-abelian scheme ${}^\heartsuit G_\sigma\to\mathcal{S}_\sigma$, so we can evaluate a modular form on ${}^\heartsuit G_\sigma\to\mathcal{S}_\sigma$. To do this properly we have to analyze the structure of $\overline{\Xi}_\sigma$, $\mathcal{S}_\sigma$, and $\omega({}^\heartsuit G_\sigma/\mathcal{S}_\sigma)$ more carefully.

It is clear that $\omega({}^\heartsuit G_\sigma/\mathcal{S}_\sigma)$ is canonically isomorphic to $\omega_A\otimes_{\mathbb{Z}}\Lambda^r(X_\xi)$, where ω_A is shorthand for the pull-back of $\omega(A_{g-r}/\mathsf{A}_{g-r})$ to $\mathcal{S}_\sigma$ via $\mathcal{S}_\sigma\to\overline{\Xi}_\sigma\to\mathbf{Hom}_{\mathbb{Z}}(X_\xi, A_{g-r})\to\mathsf{A}_{g-r}$. Let us turn to $\overline{\Xi}_\sigma$. Recall that $\overline{\Xi}_\sigma = \Xi_\sigma\times^{E_\sigma}\overline{E}_\sigma$, and Ξ_σ is a E_σ ($=\mathbf{Hom}_{\mathbb{Z}}(S^2(X_\sigma), G_m)$)-bundle defined by $\lambda_*\Xi_\sigma = \mathbf{ev}(\lambda)^{\otimes -1}$, where $\mathbf{ev}$ is the homomorphism from $S^2(X_\sigma)$ = (the character group of E_σ) to the group (or rather Picard category) of G_m-torsors over $\mathbf{Hom}_{\mathbb{Z}}(X_\xi, A_{g-r})$ such that $\mathbf{ev}$ sends the class $[\mu \otimes \nu]$ of $\mu \otimes \nu$ to $(c(\mu), c(\nu))^* \mathcal{P}_A$. Here $(c(\mu), c(\nu))$: $\mathbf{Hom}_{\mathbb{Z}}(X_\xi, A_{g-r}) \to A_{g-r} \times_{\mathsf{A}_{g-r}} A_{g-r}$ is the obvious (tautological) morphism, $\mathcal{P}_A$ is the Poincaré G_m-torsor over $A_{g-r} \times_{\mathsf{A}_{g-r}} A_{g-r}$ with A_{g-r} identified with its dual via the principal polarization. From the canonical symmetric biextension structure of the Poincaré torsor it follows that $\mathbf{ev}$ is well-defined. We denote by $\mathcal{L}(\lambda)$ the line bundle associated to $\mathbf{ev}(\lambda)$. If $\lambda = [\mu \otimes \mu]$, for some $\mu \in X_\xi$, then $\mathcal{L}(\lambda)$ is pull–back of an ample line bundle on the quotient $\mathbf{Hom}_{\mathbb{Z}}(\mathbb{Z}\cdot\mu, A_{g-r})$ of $\mathbf{Hom}_{\mathbb{Z}}(X_\xi, A_{g-r})$. If λ is positive semi-definite some multiple of it is a positive linear combination of such elements $[\mu\otimes\mu]$, so it follows that $\mathcal{L}(\lambda)$ is pullback of an ample line bundle on the quotient $\mathbf{Hom}_{\mathbb{Z}}(X(\lambda), A_{g-r})$ of $\mathbf{Hom}_{\mathbb{Z}}(X_\xi, A_{g-r})$, where $X(\lambda)$ is the smallest sublattice of X_ξ such that λ is contained in $S^2(X(\lambda))$. In particular, we deduce that formation of direct images (under the projection to A_{g-r}) of $\mathcal{L}(\lambda)$ commutes with base change.

The split torus E_σ acts naturally on $(\overline{\Xi}_\sigma, \omega({}^\heartsuit G_\sigma/\mathcal{S}_\sigma))$, so any section of $\omega({}^\heartsuit G_\sigma/\mathcal{S}_\sigma)$ over $\overline{\Xi}_\sigma$ can be expanded as a (formally convergent) infinite series according to the characters of E_σ. Especially we can expand (the pull-back of) any modular form $f \in \Gamma(\overline{\mathsf{A}}_g, \omega^k \otimes_{\mathbb{Z}} M)$ according to the characters of E_σ. The result is that every modular form $f \in \Gamma(\overline{\mathsf{A}}_g, \omega^{\otimes k} \otimes_{\mathbb{Z}} M)$ has a q-expansion

$$f = \sum_{\lambda \in S^2(X_\xi)} f(\lambda)$$

with the $f(\lambda)$ global algebraic sections of $(\omega_A \otimes \wedge^r(X_\xi))^{\otimes k} \otimes_{\mathbb{Z}} M$ over $\mathcal{S}_\sigma$ which transform according to λ under E_σ. It follows that the $f(\lambda)$ are already algebraic sections defined over $\overline{\Xi}_\sigma$ (before completion). To be more precise, $f(\lambda)$ is canonically identified as a global section of the coherent sheaf $\mathcal{L}^k(\lambda; M) = \mathcal{L}^k_{g,r}(\lambda; M) = \mathcal{L}(\lambda) \otimes (\omega_A \otimes \wedge^r(X_\xi))^{\otimes k} \otimes_{\mathbb{Z}} M$ on $\mathbf{Hom}_{\mathbb{Z}}(X_\xi, A_{g-r})$.

By argument similar to the totally degenerate case as before, one checks that they are independent of σ and supported in the dual of the cone of positive semi-definite forms if $r > 1$. Moreover, we claim that the same is true for all r. In fact, we may assume that $r = 1$ and σ is a cone in the $\mathrm{GL}(X)$-admissible cone decomposition $\{\sigma_\alpha\}$ used to construct $\overline{\mathsf{A}}_g$. If the Fourier-Jacobi expansion $f = \sum_{\lambda \in \mathbb{Z}} f(\lambda)$ is not supported in $\{\lambda \geq 0\}$, it means that the global section f of $\omega^{\otimes k}$ has a pole along the σ-stratum, contradicting Koecher's principle. Since f defines a $\mathrm{GL}(X_\xi)$-invariant global section of $(\omega_A \otimes \Lambda^r(X_\xi))^{\otimes k}$ over $\mathcal{S}_\xi$, it follows that $\gamma \cdot f(\lambda) = f(\gamma \cdot \lambda)$ under the natural $\mathrm{GL}(X_\xi)$-actions.

As σ is positive definite it follows that all $f(\lambda)$ vanish on the σ-stratum, except for $f(0)$ which is constant along the fibres of $\overline{\Xi}_\sigma \to \mathbf{Hom}_{\mathbb{Z}}(X_\xi, A_{g-r})$, which reflects the fact that $\mathcal{L}^k(0)$ is the pull-back of the coherent sheaf $(\omega_A \otimes \wedge^r(X_\xi))^{\otimes k} \otimes_{\mathbb{Z}} M$ on $\mathbf{Hom}_{\mathbb{Z}}(X_\xi, A_{g-r})$.

As $\mathbf{Hom}_{\mathbb{Z}}(X_\xi, A_{g-r})$ is fibred over A_{g-r}, with the fibres complete, geometrically irreducible varieties (namely r-fold fibre products of the $(g-r)$-dimensional universal abelian varieties) and the invertible sheaf in question is trivial along the fibres, it follows that $f(0)$ defines a global section in $\Gamma(\mathsf{A}_{g-r}, (\omega \otimes \wedge^r(X_\xi))^{\otimes k} \otimes_{\mathbb{Z}} M)$. Notice that $\Gamma(\mathsf{A}_{g-r}, (\omega \otimes \wedge^r(X_\xi))^{\otimes k} \otimes_{\mathbb{Z}} M) = 0$ for k

odd except when we are in characteristic 2 (i.e. 2 is a zero divisor in M), by invariance of the q-expansion under $\mathrm{GL}(X_\xi)$. Thus barring the characteristic 2 case, k must be even for $f(0)$ not to vanish, in which case $\wedge^r(X_\xi)^{\otimes k}$ is canonically trivial and $(\omega \otimes \wedge^r(X_\xi))^{\otimes k} = \omega^{\otimes k}$. Note that this is a special feature of A_g, and is no longer valid when level structures are present.

Assume M *is a ring in which 2 is not a zero divisor* (for simplicity, otherwise we have to carry the term $\Lambda^g(X)^{\otimes k}$ around), which we denote by R. In other words we extend base to $\mathrm{Spec}(R)$ and forget about the coefficient-modules. Under the above assumptions there are no non-trivial modular forms of odd weights. The assignment $f \to f(0)$ is called the *Φ-operator*, mapping modular forms of genus g into modular forms of genus $g - r$. When $R = \mathbb{C}$, we recover the classical Siegel Φ-operator via complex uniformization. Elements in the kernels of all Φ-operators are called *cusp-forms*. They are characterized as global sections of $\omega^{\otimes k}$ which vanish on the boundary of $\overline{\mathsf{A}}_g$. (This follows immediately from Fourier-Jacobi expansion.) Note that since the Fourier-Jacobi expansion is independent of the choice of cones, so are the Φ-operators, and the definition of cusp form is independent of the choice of $\overline{\mathsf{A}}_g$.

By letting the abelian part A of $\tilde{G}$ degenerate we can compute the q-expansion of $f(0)$: Its coefficients are indexed by $S^2(\mathrm{Ker}(X \to X_\xi)) \subset S^2(X)$, and coincide with those of f on this subgroup. In particular, it is holomorphic at infinity; that is, $f(0)$ is an element of $\Gamma(\overline{\mathsf{A}}_{g-r}, \omega^{\otimes k})$. We leave it to the reader to make everything precise using the 2-step quotients discussed in chapter III, §10.

The expansion $f = \sum_{\lambda \in S^2(X_\xi)} f(\lambda)$ as explained above will be called the Fourier-Jacobi expansion of f. The homomorphism

$$\begin{aligned} \mathrm{FJ}_\xi : \Gamma(\overline{\mathsf{A}}_g, \omega^{\otimes k} \otimes_{\mathbb{Z}} M) \to& \\ &\{\mathrm{GL}(X_\xi)\text{-invariant families } f(\lambda) \in \Gamma(\mathbf{Hom}_{\mathbb{Z}}(X_\xi, A_{g-r}), \mathcal{L}^k(\lambda; M)), \\ &\text{for } \lambda \in C(X_\xi)^\vee \cap S^2(X_\xi)\}\ , \end{aligned}$$

called the Fourier-Jacobi homomorphism, is injective. Combined with the left-exactness in M in the formation of $\Gamma(\overline{\mathsf{A}}_g, \omega^{\otimes k} \otimes_{\mathbb{Z}} M)$, we obtain the "Fourier-Jacobi expansion principle": if $M_1 \subset M_2$, $f \in \Gamma(\overline{\mathsf{A}}_g, \omega^{\otimes k} \otimes_{\mathbb{Z}} M_2)$ and $\mathrm{FJ}_\xi(f)$ "takes values in M_1", then in fact f is in $\Gamma(\overline{\mathsf{A}}_g, \omega^{\otimes k} \otimes_{\mathbb{Z}} M_1) \subset \Gamma(\overline{\mathsf{A}}_g, \omega^{\otimes k} \otimes_{\mathbb{Z}} M_2)$.

If we work in the category of complex analytic spaces and take $M = \mathbb{C}$, since each term $f(\lambda)$ comes from a section of an invertible sheaf $\mathcal{L}^k(\lambda)$ on an abelian scheme $\mathbf{Hom}_{\mathbb{Z}}(X_\xi, A)$, it can be represented as a θ-function. Thus a modular form f is represented as an infinite sum of θ-functions, which suggested the name Fourier-Jacobi expansion. Of course it is just the classical Fourier-Jacobi expansion used, for example, by Baily, Igusa, Piatetskii-Shapiro, Shimura, etc. when considering holomorphic automorphic forms on bounded symmetric domains.

We have seen that the constant term $f(0)$ of a Siegel modular form f determines the restriction of f to any σ-stratum with $\sigma \subset C(X_\xi)^\circ$ (notation as above), because all other terms $f(\lambda)$ vanish on any such stratum. In fact, we have proved a stronger statement:

1.6. Proposition. *On the boundary of $\overline{\mathsf{A}}_g$, the value of a Siegel modular form $f \in \Gamma(\overline{\mathsf{A}}_g, \omega^{\otimes k})$ depends only on the abelian part of the semi-abelian scheme G (and not on the choice of cone decomposition or the cone σ), and it is given $\Phi(f)$ (= the constant term $f(0)$ in the Fourier-Jacobi expansion of f).*

We recall that the part of the boundary of $\overline{\mathsf{A}}_g$ lying over the component corresponding to $X \to X_\xi$ (i.e. the union of all strata corresponding to cones in $C(X_\xi)^\circ$) can be described as follows: It is the $\mathrm{GL}(X_\xi)$-quotient of the union of non-degenerate strata in $\overline{\Xi}_\xi - \Xi_\xi = \Xi_\xi \times^{E_\xi} (\overline{E}_\xi - E_\xi) \to \mathrm{Hom}_{\mathbb{Z}}(X_\xi, A_{g-r})$, which sits over A_{g-r}. The proposition follows from the fact that $\Phi(f)$ is the section of the pull-back of the invertible sheaf $\omega^{\otimes k}$ on A_{g-r} and is constant on the fibres of the projection to $\mathrm{Hom}_{\mathbb{Z}}(X_\xi, A_{g-r})$, while in turn the geometric fibres of $\mathrm{Hom}_{\mathbb{Z}}(X_\xi, A_{g-r})$ over A_{g-r} are all connected.

q-Expansion and the Fourier-Jacobi Expansion for $\mathsf{A}_{g,n}$

Our discussion about the q-expansion and Fourier expansion naturally extends to the stacks $\mathsf{A}_{g,n}$.

1.7. Definition. Let M be a module over $\mathbb{Z}[\zeta_n, 1/n]$. A *Siegel modular form of genus g, weight k and (principal symplectic) level n* is an element of $\Gamma(\overline{\mathsf{A}}_{g,n}, \omega^{\otimes k} \otimes_{\mathbb{Z}[\zeta_n,1/n]} M)$. The q-expansion is the homomorphism

$$\Gamma(\overline{\mathsf{A}}_{g,n}, \omega^{\otimes k} \otimes_{\mathbb{Z}[\zeta_n,1/n]} M) \to \mathbb{Z}[\zeta_n, 1/n][\![(1/n) \cdot S^2(X)]\!] \otimes (\Lambda^g X)^{\otimes k} \otimes M$$

obtained by the methods above, where the two tensor products (excluding the one before k) are over $\mathbb{Z}$ and $\mathbb{Z}[\zeta_n, 1/n]$ respectively, and the image lies in the subspace of $\mathrm{GL}(X)(n)$-invariants (again $\mathrm{GL}(X)(n)$ denotes the congruence subgroup).

We should note that in this case there is in fact more than one q-expansion, corresponding to the different 0-dimensional cusps. (Among these cusps there is one favored by many people, usually referred to as $\sqrt{-1}\,\infty$, which in our previous notation corresponds to the coset $P_{g,g}(\mathbb{Z}/n\mathbb{Z})'$ in $\mathrm{Sp}(2g, \mathbb{Z}/n\mathbb{Z})/P_{g,g}(\mathbb{Z}/n\mathbb{Z})'$. Here $G(\mathbb{Z}/n\mathbb{Z})'$ denotes the image of $G(\mathbb{Z})$ in $G(\mathbb{Z}/n\mathbb{Z})$.) The same argument as before (with obvious modifications) shows that for any 0-dimensional cusp, if $g > 1$ the elements in the image of the q-expansion homomorphism are all supported in the dual cone $C(X)^\vee$ of $C(X)$. This means that the natural map $\Gamma(\overline{\mathsf{A}}_{g,n}, \omega^{\otimes k} \otimes_{\mathbb{Z}[\zeta_n,1/n]} M) \to \Gamma(\mathsf{A}_{g,n}, \omega^{\otimes k} \otimes_{\mathbb{Z}[\zeta_n,1/n]} M)$ is an isomorphism (Koecher's principle). The q-expansion homomorphisms

$$\Gamma(\overline{\mathsf{A}}_{g,n}, \omega^{\otimes k} \otimes_{\mathbb{Z}[\zeta_n,1/n]} M) \to \mathbb{Z}[\zeta_n, 1/n][\![(1/n) \cdot S^2(X)]\!] \otimes (\Lambda^g X)^{\otimes k} \otimes M$$

are all injective because $\overline{\mathsf{A}}_{g,n}$ is smooth over $\mathbb{Z}[\zeta_n, 1/n]$ with geometrically irreducible fibres. From this and the left exactness in M in the formation of $\Gamma(\overline{\mathsf{A}}_{g,n}, \omega^{\otimes k} \otimes_{\mathbb{Z}[\zeta_n,1/n]} M)$ we obtain the q-expansion principle: If $M_1 \subset M_2$ are two $\mathbb{Z}[\zeta_n, 1/n]$-modules and the q-expansion of an element of $\Gamma(\overline{\mathsf{A}}_{g,n}, \omega^{\otimes k} \otimes_{\mathbb{Z}[\zeta_n,1/n]} M_2)$ lies in $\mathbb{Z}[\zeta_n, 1/n][\![(1/n) \cdot S^2(X)]\!] \otimes (\Lambda^g X)^{\otimes k} \otimes M_1$, then this element belongs to $\Gamma(\overline{\mathsf{A}}_{g,n}, \omega^{\otimes k} \otimes_{\mathbb{Z}[\zeta_n,1/n]} M_1) \subset \Gamma(\overline{\mathsf{A}}_{g,n}, \omega^{\otimes k} \otimes_{\mathbb{Z}[\zeta_n,1/n]} M_2)$. We summarize the above discussion in

1.8. Proposition. *(i) For any $\mathbb{Z}[\zeta_n, 1/n]$-module M and any 0-dimensional cusp c of $\overline{\mathsf{A}}_{g,n}$ there is a q-expansion homomorphism*

$$q\text{-}exp(c; M) : \Gamma(\overline{\mathsf{A}}_{g,n}, \omega^{\otimes k} \otimes_{\mathbb{Z}[\zeta_n,1/n]} M) \to$$
$$\to \mathbb{Z}[\zeta_n, 1/n][\![(1/n) \cdot S^2(X)]\!] \otimes_{\mathbb{Z}} (\Lambda^g X)^{\otimes k} \otimes_{\mathbb{Z}[\zeta_n,1/n]} M$$

obtained from totally degenerate Mumford families with level-n structures over the cusp c.

(ii) Each q-$exp(c; M)$ is injective and is independent of the cone(s) used for Mumford's construction, and its image lies in the subspace of $\mathrm{Ker}(\mathrm{GL}(X) \to \mathrm{GL}(X/nX))$*-invariants of $\mathbb{Z}[\zeta_n, 1/n][\![(1/n) \cdot S^2(X)]\!] \otimes (\Lambda^g X)^{\otimes k} \otimes M$.*

(iii) Koecher's principle is true for any 0-dimensional cusp c.

(iv) If $M_1 \subset M_2$ are two $\mathbb{Z}[\zeta_n, 1/n]$-modules, then a modular form $f \in \Gamma(\overline{\mathsf{A}}_{g,n}, \omega^{\otimes k} \otimes_{\mathbb{Z}[\zeta_n,1/n]} M_2)$ lies in $\Gamma(\overline{\mathsf{A}}_{g,n}, \omega^{\otimes k} \otimes_{\mathbb{Z}[\zeta_n,1/n]} M_1)$ if its q-expansion (at one cusp) lies in $\mathbb{Z}[\zeta_n, 1/n][\![(1/n) \cdot S^2(X)]\!] \otimes (\Lambda^g X)^{\otimes k} \otimes M_1$.

More generally, we consider the Fourier-Jacobi expansion over a cusp of $\mathsf{A}_{g,n}$ corresponding to a surjection $X \twoheadrightarrow X_\xi$ and a representative of $\mathrm{Sp}(2g, \mathbb{Z}/n\mathbb{Z})'/P_{g,r}(\mathbb{Z}/n\mathbb{Z})$, with coefficient in a $\mathbb{Z}[\zeta_n, 1/n]$-module M. It is a homomorphism $\mathrm{FJ}(c; M)$ which associates to an $f \in \Gamma(\overline{\mathsf{A}}_{g,n}, \omega^{\otimes k} \otimes_{\mathbb{Z}[\zeta_n,1/n]} M)$ a family of $f(\lambda) \in \Gamma(\mathbf{Hom}_{\mathbb{Z}}(X_{\xi,n}, A_{g-r,n}), \mathcal{L}^k(\lambda; M))$, for $\lambda \in C(X_\xi)^\vee \cap (1/n) \cdot S^2(X_\xi)$, invariant under $\mathrm{Ker}(\mathrm{GL}(X_\xi) \to \mathrm{GL}(X_\xi/nX_\xi))$.

Here $\mathcal{L}^k(\lambda; M) = \mathcal{L}^k_{g,r,n}(\xi, \lambda; M) = \mathbf{ev}(\lambda)^*\mathcal{P}(A_{g-r,n})^{\otimes n} \otimes (\underline{\omega} \otimes \Lambda^r(X_\xi))^{\otimes k} \otimes M, \mathcal{P}(A_{g-r,n})$ stands for the Poincaré sheaf on the abelian scheme $A_{g-r,n}$, $X_{\xi,n} = (1/n) \cdot X_\xi$, $\mathcal{L}(n \cdot [\mu \otimes \nu]) = \mathbf{ev}(n \cdot [\mu \otimes \nu])^* \mathcal{P}(A_{g-r,n})^{\otimes n} (\mu, \nu \in X_{\xi,n})$ is the pull-back of $\mathcal{P}(A_{g-r,n})^{\otimes n}$ via $(c_{\xi,n}(\mu), c_{\xi,n}(\nu)) : \mathbf{Hom}_{\mathbb{Z}}(X_\xi, A_{g-r}) \to A_{g-r,n}$, and $\mathrm{Ker}(\mathrm{GL}(X_\xi) \to \mathrm{GL}(X_\xi/nX_\xi))$ operates naturally on λ and $\mathbf{Hom}_{\mathbb{Z}}(X_\xi, A_{g-r})$.

The Fourier-Jacobi expansion over a cusp c of an element $f \in \Gamma(\overline{\mathsf{A}}_{g,n}, \omega^{\otimes k} \otimes_{\mathbb{Z}[\zeta_n,1/n]} M)$ shall be abbreviated as $\mathrm{FJ}(c; M)(f) = \sum f(\lambda)$ (the sum is over $C(X_\xi)^\vee \cap (1/n) \cdot S^2(X_\xi)$). We define a version of the Φ-operator

$$\Phi_c : \Gamma(\overline{\mathsf{A}}_{g,n}, \omega^{\otimes k} \otimes_{\mathbb{Z}[\zeta_n,1/n]} M) \to \Gamma(\mathsf{A}_{g-r,n}, \omega^{\otimes k} \otimes_{\mathbb{Z}} \Lambda^r(X_\xi)^{\otimes k} \otimes_{\mathbb{Z}[\zeta_n,1/n]} M)$$

by $\Phi_c(f) = f(0) \in \Gamma(\mathbf{Hom}_{\mathbb{Z}}(X_{\xi,n}, A_{g-r}), \mathcal{L}^k(0; M)) =$

$\Gamma(\mathsf{A}_{g-r,n}, \omega^{\otimes k} \otimes_{\mathbb{Z}} \Lambda^r(X_\xi)^{\otimes k} \otimes_{\mathbb{Z}[\zeta_n,1/n]} M)$. Again the equality holds because $\mathbf{Hom}_{\mathbb{Z}}(X_{\xi,n}, A_{g-r}) \to \mathsf{A}_{g-r,n}$ has geometrically connected fibres. We also remark that the Fourier-Jacobi expansion homomorphism $\mathrm{FJ}(c, M)$ is injective for every $\mathbb{Z}[\zeta_n, 1/n]$-module M and is independent of the cone-decomposition. From the injectivity and the left-exactness in M in the formation of $\Gamma(\overline{\mathsf{A}}_{g,n}, \omega^{\otimes k} \otimes_{\mathbb{Z}[\zeta_n,1/n]} M)$ we derive a Fourier-Jacobi expansion principle. We summarize our discussion in

1.9. Proposition. *(i) For any cusp c of $\mathsf{A}_{g,n}$ indexed by a quotient X_ξ of X and any $\mathbb{Z}[\zeta_n, 1/n]$-module M, we have a well-defined Fourier-Jacobi expansion, invariant under* $\mathrm{Ker}(\mathrm{GL}(X_\xi) \to \mathrm{GL}(X_\xi/nX_\xi))$*:* $\mathrm{FJ}(c; M)(f) = \sum_\lambda f(\lambda)$, *with $f(\lambda) \in \Gamma(\mathbf{Hom}_{\mathbb{Z}}(X_{\xi,n}, A_{g-r}), \mathcal{L}^k(\lambda; M))$ for $\lambda \in C(X_\xi)^\vee \cap (1/n) \cdot S^2(X_\xi)$. Here $X_{\xi,n} = (1/n) \cdot X_\xi$ and $\mathcal{L}^k(\lambda; M) = \mathcal{L}^k_{g,r,n}(\xi, \lambda; M) = \mathbf{ev}(\lambda)^* \mathcal{P}^{\otimes n}(A_{g-r,n}) \otimes (\omega \otimes_{\mathbb{Z}} \Lambda^r(X_\xi))^{\otimes k} \otimes_{\mathbb{Z}[\zeta_n,1/n]} M$. $\mathrm{FJ}(c : M)$ is always injective.*

(ii) (Fourier-Jacobi expansion principle). If $M_1 \subset M_2$ are two $\mathbb{Z}[\zeta_n, 1/n]$-modules, then a section of $\Gamma(\overline{\mathsf{A}}_{g,n}, \omega^{\otimes k} \otimes_{\mathbb{Z}[\zeta_n,1/n]} M_2)$ lies in $\Gamma(\overline{\mathsf{A}}_{g,n}, \omega^{\otimes k} \otimes_{\mathbb{Z}[\zeta_n,1/n]} M_1)$ if for its Fourier-Jacobi expansion at one cusp all its coefficients "lie in M_1".

(iii) The restriction of $f \in \Gamma(\overline{\mathsf{A}}_{g,n}, \omega^{\otimes k} \otimes_{\mathbb{Z}[\zeta_n,1/n]} M)$ to the part of the boundary of $\overline{\mathsf{A}}_{g,n}$ lying over the cusp c (with reduced structure) is determined by the constant term $f(0)$ of $\mathrm{FJ}(c : M)(f)$. The assignment $f \mapsto f(0)$ defines a map, called the Φ-operator at c

$$\Phi_c : \Gamma(\overline{\mathsf{A}}_{g,n}, \omega^{\otimes k} \otimes_{\mathbb{Z}[\zeta_n,1/n]} M) \to \Gamma(\overline{\mathsf{A}}_{g-r,n}, \omega^{\otimes k} \otimes_{\mathbb{Z}} \Lambda^r(X_\xi)^{\otimes k} \otimes_{\mathbb{Z}[\zeta_n,1/n]} M) .$$

1.10. Remark. The construction in chapter IV, proof of theorem 7.2 allows us to define q-expansions for modular forms on N_g and its normalization. However to establish the q-expansion principle we have to consider all cusps, as the geometric fibres are no longer irreducible. For one distinguished cusp (corresponding to $\rho = 0$) the coefficients of the q-expansion are parametrized by $Q_0(X)$ = dual of the space of quadratic symmetric functions on X, modulo constants). Similarly for level-n structures (n odd).

2. Construction of the Minimal Compactification A_g^*

Introduction. In the category of complex analytic spaces, any arithmetic quotient of a bounded symmetric domain $\Gamma\backslash D$ has a minimal (Satake-Baily-Borel) compactification $\Gamma\backslash D^*$. $\Gamma\backslash D^*$ is the quotient of a Hausdorff topological space D^* containing D as a dense open subset. D^* is obtained from D by adding all (proper) rational boundary components of D. It does not have a natural complex analytic structure. However, $\Gamma\backslash D^*$ has the structure of a normal compact complex analytic variety extending the complex structure of $\Gamma\backslash D$. In fact $\Gamma\backslash D^*$ is projective, with the projective embedding given by Γ-invariant pluricanonical forms. (Thus some positive power k of the canonical bundle descends and extends to an ample line bundle on $\Gamma\backslash D^*$, where k can be any number divisible by the order of all torsion elements in Γ.) Moreover $\Gamma\backslash D^*$ has a natural stratification by (Zariski) locally closed subvarieties, each of which is again an arithmetic quotient. Apparently Satake [Sa 1] was the first person to construct $\Gamma\backslash D^*$ as a topological space. For $D = \mathbb{H}_g$ the analytic structure on $\Gamma\backslash D^*$ was furnished by Baily and Shimura independently, cf. [B], [Sh 1]. (Baily's paper is earlier but contains a number of mistakes. Shimura also treated the more delicate arithmetic structure on $\Gamma\backslash D^*$, and considered more general D's.) The general minimal compactification $\Gamma\backslash D^*$ was treated by Baily and Borel in their joint work [B-B]. A similar construction was also done independently by Pijatetski-Shapiro. We remark that the adjective "minimal" is supported by the following facts: $\Gamma\backslash D^*$ not only lies under any toroidal compactification of $\Gamma\backslash D$, any smooth (as complex manifold) compactification of $\Gamma\backslash D$, but also under any (general) Satake compactification of $\Gamma\backslash D$, in particular the Borel-Serre compactification of $\Gamma\backslash D$.

When $D = \mathbb{H}_g$ is the Siegel upper-half space and $\Gamma = \mathrm{Sp}(2g, \mathbb{Z})$ is the full modular group, the natural stratification on $\Gamma\backslash D^*$ is easy to describe:

$$\mathrm{Sp}(2g,\mathbb{Z})\backslash\mathbb{H}_g^* = \mathrm{Sp}(2g,\mathbb{Z})\backslash\mathbb{H}_g \amalg \mathrm{Sp}(2g-2,\mathbb{Z})\backslash\mathbb{H}_{g-1} \amalg \cdots \amalg \mathrm{Sp}(2,\mathbb{Z})\backslash\mathbb{H} \amalg \{\mathrm{pt.}\} .$$

The geometric meaning of this stratification is that the compactification $\mathrm{Sp}(2g,\mathbb{Z})\backslash\mathbb{H}_g^*$ is obtained from $\mathrm{Sp}(2g,\mathbb{Z})\backslash\mathbb{H}_g$ by adding abelian parts of all degenerations. (Thus points of $\mathrm{Sp}(2g,\mathbb{Z})\backslash\mathbb{H}_g^*$ are in one-to-one correspondence with isomorphism classed of principally polarized abelian varieties of dimension $\leq g$.) More precisely, assume that $A^* \to \Delta^*$ is an analytic family of principally polarized abelian varieties over the punctured unit disk Δ^*. In view of the semi-stable reduction theorem we may assume that $A^* \to \Delta^*$ extends to a semi-abelian family $A \to \Delta$ over the unit disk, so that the special fibre A_0 is an extension of a (principally polarized) abelian variety B_0 by an algebraic torus T_0. Then the classifying map $\Delta^* \to \mathrm{Sp}(2g,\mathbb{Z})\backslash\mathbb{H}_g$ of the family $A^* \to \Delta^*$ extends to a holomorphic map $\Delta \to \mathrm{Sp}(2g,\mathbb{Z})\backslash\mathbb{H}_g^*$, which sends $0 \in \Delta$ to the modular point $[B_0] \in \mathrm{Sp}(2g,\mathbb{Z})\backslash\mathbb{H}_g^*$ of B_0.

In this section we would like to carry out a parallel arithmetic theory of the minimal compactification over $\mathbb{Z}$. In particular we shall construct a scheme A_g^* projective over $\mathbb{Z}$, together with a canonical morphism $\overline{\mathsf{A}}_g \to \mathsf{A}_g^*$ for any (smooth) toroidal compactification $\overline{\mathsf{A}}_g$ of A_g. Moreover we shall see that the (geometric) points of A_g^* over an algebraically closed field k are in natural one-to-one correspondence with the isomorphism classes of principally polarized abelian varieties over k of dimension $\leq g$. In fact A_g^* contains the coarse moduli scheme $[\mathsf{A}_i]$ of A_i for $i \leq g$, and $\mathsf{A}_g^* = \coprod_{0\leq i\leq g}[\mathsf{A}_i]$. We shall also give the level-n version of the minimal compactification. Later on in this chapter we shall see some applications of the theory of minimal compactification.

The construction of the minimal compactification of A_g is based on the following two propositions to be stated below. All known proofs so far use the idea that theta constants are modular forms of weight 1/2 (chapter I, theorem 5.1). Arithmetically this translates into the statement that ω is equal up to torsion to the square of the invertible sheaf "$\mathcal{O}(1)$" given by algebraic theta constants (à la Mumford). Thus ω is "positive" on $\overline{\mathsf{A}}_g$ in some sense, in particular ample over $\mathsf{A}_{g,n}$ for $n \geq 3$.

We shall quote existing literature for the proof of these two propositions in this section and indicate the major lines of argument. In the next section we shall prove them directly to make this book more self-contained. It can be skipped if the reader is willing to believe these two propositions.

2.1. Proposition. *Let $\overline{\mathsf{A}}_g$ be a smooth arithmetic toroidal compactification of A_g. There exists an integer $m > 0$ such that $\omega^{\otimes m}$ is generated by its global sections over $\overline{\mathsf{A}}_g$.*

Proof. A quick way is to quote [MB 1] chap. IX thm. 2.1, p.208, which says that for any semi-stable commutative group scheme $G \to S$ (i.e. the identity component of each fibre is semi-abelian) over a normal excellent noetherian scheme S, there exists a positive integer $m > 0$ such that $\omega_{G/S}^{\otimes m}$ is generated by its global sections. The proof applies to the algebraic space $\overline{\mathsf{A}}_{g,n}$ if $n \geq 3$, and thus also to the stack $\overline{\mathsf{A}}_g$. (Moret-Bailly's result is slightly stronger than, but essentially equivalent, to prop. 2.1.)

The main step has already been indicated in the preceding discussion. Since theta constants determine the moduli (basically because theta functions give projective embeddings of abelian varieties), we are essentially through.

2.2. Proposition. *Let C be a proper smooth irreducible curve over an algebraically closed field k, $f: G \to C$ a semi-abelian scheme over C. Suppose that* $\deg(\omega_{G/C}) \leq 0$, *then*

(i) $\omega_{G/C}$ is a torsion-bundle (i.e. some positive power of it is trivial), hence $\deg(\omega_{G/C}) = 0$.

(ii) G is an extension of an isotrivial abelian scheme (over C) by a torus (over C).

Here we recall that an abelian scheme $A \to C$ is isotrivial if it becomes constant on a finite étale covering of C. Note that if $A \times_C C'$ is constant for some proper smooth curve C' sitting over C, then A is isotrivial: For any $n \geq 3$ prime to char.(k), *$S := A[n]$ is finite étale over C and $A \times_C S \to S$ is constant because $A \times_C S \times_C C' \to S \times_C C'$ is.*

Proof. This is an immediate consequence of [MB 1] chap. IX thm. 4.5, p.235. This result probably is probably in the category of "well-known" facts. Let us indicate a proof here.

By the theory of the torus part (chap. I §2) G is an extension by a torus over C of a semi-abelian scheme over C which is generically abelian. This reduces us immediately to the case that G is generically abelian. Replacing G_η by $G_\eta^4 \times (G_\eta^t)^4$ if necessary, we can assume that the generic fibre G_η is principally polarized (Zarhin's trick, see ch. I, 5.5). We must show that G has good reduction everywhere and is isotrivial. Notice that we are free to replace C by finite coverings.

The key point is that $\deg(\omega^{\otimes n}) \leq 0 \; \forall n \geq 0$, hence all global sections of $\omega^{\otimes n}$ are constant in the sense that they are either zero or nowhere zero. Therefore *all theta constants of* G_η *are constants.* It follows that G_η has potential good reduction everywhere and is isotrivial, because theta constants determine the moduli.

As we have said, the above propositions are the basic ingredient in proving the main result of this chapter:

2.3. Theorem. *There exists a normal scheme* A_g^* *proper and of finite type over* $\mathrm{Spec}(\mathbb{Z})$, *a positive integer* $m > 0$, *and a projective embedding* $\mathsf{A}_g^* \hookrightarrow \mathbb{P}^N$ *over* $\mathrm{Spec}(\mathbb{Z})$ *given by a very ample invertible sheaf* $\mathcal{L}$ *on* A_g^*, *such that the following assertions hold:*

(1) For any (smooth) arithmetic toroidal compactification $\overline{\mathsf{A}}_g$ *of* A_g, *there is a canonical morphism* $\overline{\pi} : \overline{\mathsf{A}}_g \to \mathsf{A}_g^*$ *such that* $\overline{\pi}^*\mathcal{L} = \omega^{\otimes m}$ *on* $\overline{\mathsf{A}}_g$. *Moreover, the morphisms* $\overline{\pi} : \overline{\mathsf{A}}_g \to \mathsf{A}_g^*$ *(for different choices of admissible polyhedral cone decompositions) are compatible in the sense that if* $\overline{\mathsf{A}}_g'$ *is another (smooth) arithmetic toroidal compactification of* A_g *dominating* $\overline{\mathsf{A}}_g$, *then the composition* $\overline{\mathsf{A}}_g' \to \overline{\mathsf{A}}_g \to \mathsf{A}_g^*$ *is equal to* $\overline{\pi}' : \overline{\mathsf{A}}_g' \to \mathsf{A}_g^*$.

(2) The graded ring $\oplus_{k\geq 0} \Gamma(\overline{\mathsf{A}}_g, \omega^{\otimes k})$ *is a finitely generated* $\mathbb{Z}$*-algebra, which is independent of the choice of* $\overline{\mathsf{A}}_g$ *by results of* §1. A_g^* *is canonically isomorphic to* $\mathrm{Proj}(\oplus_{k\geq 0} \Gamma(\overline{\mathsf{A}}_g, \omega^{\otimes k}))$ *and the morphism* $\overline{\pi} : \overline{\mathsf{A}}_g \to \mathsf{A}_g^*$ *is the canonical one defined by* $\mathcal{L}$ *(which is a morphism by prop. 2.1).*

(3) A_g^* *has a natural stratification by locally closed normal subschemes flat over* $\mathrm{Spec}(\mathbb{Z})$, *each of which is (canonically isomorphic to) the coarse moduli scheme* $[\mathsf{A}_i]$ *of* A_i *for a unique* i, $0 \leq i \leq g$: $\mathsf{A}_g^* = \coprod_{0\leq i\leq g} [\mathsf{A}_i]$. *Consequently the geometric points of* A_g^* *are in 1–1 correspondence with those of the disjoint union of the* A_i, $0 \leq i \leq g$. $[\mathsf{A}_g]$ *is dense open in* A_g^*, *the closure of* $[\mathsf{A}_i]$ *is* $\coprod_{0\leq j\leq i} [\mathsf{A}_j]$ *and is isomorphic to* A_i^* *(and hence normal).*

(4) The map $\overline{\pi}$ *associates to each geometric point of* $\overline{\mathsf{A}}_g$ *the classifying point of the abelian part of the fibre of the universal semi-abelian variety* $G \to \overline{\mathsf{A}}_g$ *over this point.*

(5) $\overline{\pi}$ *is compatible with the stratifications of its source and target, where* $\overline{\mathsf{A}}_g$ *is given the stratification by dimension of fibres of the abelian part of the universal semi-abelian variety* $G \to \overline{\mathsf{A}}_g$ *and* A_g^* *is given its natural stratification as in (3).*

(6) The restriction $\pi : \mathsf{A}_g \to [\mathsf{A}_g]$ *of* $\overline{\pi}$ *to* $\mathsf{A}_g \subset \overline{\mathsf{A}}_g$ *identifies* $[\mathsf{A}_g]$ *as the coarse moduli scheme of* A_g. *More generally for a rank-r quotient* X_ξ *of* X, *let* π_ξ *be the restriction of* $\overline{\pi}$ *to the part of the boundary corresponding to* X_ξ *(i.e. where the abelian part of* G *has dimension* $g - r$*). This is the* $\mathrm{GL}(X_\xi)$*-quotient of a bundle over* $\mathbf{Hom}_{\mathbb{Z}}(X_\xi, A_{g-r})$, *and* π_ξ *is induced by the composition of* $\mathbf{Hom}_{\mathbb{Z}}(X_\xi, A_{g-r}) \to \mathsf{A}_{g-r}$ *and the canonical morphism* $\mathsf{A}_{g-r} \to [\mathsf{A}_{g-r}]$.

2.4. Remarks. (i) By analogy with the complex case, the $[\mathsf{A}_i]$'s will be called *cusps* of A_g^*. The codimension of the boundary $\coprod_{0\le i\le g-1}[\mathsf{A}_i]$ of A_g^* is equal to g. This already indicates that the singularity of A_g^* at the boundary is not very pleasant.

(ii) Taking $\mathbb{C}$-points, we recover the usual picture in the analytic category: $\overline{\mathsf{A}}_g(\mathbb{C})$ blows down to the Satake-Baily-Borel compactification $\mathsf{A}_g^*(\mathbb{C}) = \mathrm{Sp}(2g,\mathbb{Z})\backslash\mathbb{H}^*$, as explained in [AMRT].

Proof. We may and do assume that $g \ge 2$, because the elliptic modular case is well-understood (A_1^* is the coarse moduli space of $\overline{\mathsf{A}}_1$). By prop. 2.1 there exists a positive integer m such that $\omega^{\otimes m}$ is generated by its global sections over $\overline{\mathsf{A}}_g$, for *any* (smooth) arithmetic toroidal compactification $\overline{\mathsf{A}}_g$. The reader may be worried that proposition 2.1 only asserts the existence of m for one $\overline{\mathsf{A}}_g$ at a time, but this suffices by the usual arguments.

Let us choose and fix one $\overline{\mathsf{A}}_g$. As $\omega^{\otimes m}$ is generated by its global sections, it defines a morphism from $\overline{\mathsf{A}}_g$ into some projective space $\mathbb{P}^n$, which we denote by f. The theory of Stein factorization ([EGA III 4.3.3], with the obvious generalization to algebraic stacks) tells us that $f : \overline{\mathsf{A}}_g \to \mathbb{P}^n$ factors as $\overline{\mathsf{A}}_g \to \mathsf{A}_g^* \to \mathbb{P}^n$ such that the first factor $\overline{\pi} : \overline{\mathsf{A}}_g \to \mathsf{A}_g^*$ has connected geometric fibres and the second factor $\mathsf{A}_g^* \to \mathbb{P}^n$ is finite. (The domain A_g^* of the second factor will be shown to be what we want. In particular it is independent of the choice of $\overline{\mathsf{A}}_g$.) Also $\overline{\pi}_*(\mathcal{O}_{\overline{\mathsf{A}}_g}) = \mathcal{O}_{\mathsf{A}_g^*}$. Moreover after replacing m by a multiple there is a very ample invertible sheaf $\mathcal{L}$ on A_g^* which pulls back to $\omega^{\otimes m}$. This shows that $\oplus_{k\ge0}\Gamma(\overline{\mathsf{A}}_g,\omega^{\otimes mk})$ is a normal (by smoothness of $\overline{\mathsf{A}}_g$) integral finitely generated $\mathbb{Z}$-algebra independent of the choice of $\overline{\mathsf{A}}_g$. Also $\mathsf{A}_g^* = \mathrm{Proj}(\oplus_{k\ge0}\Gamma(\overline{\mathsf{A}}_g,\omega^{\otimes mk}))$, hence is normal, integral and independent of the choice of $\overline{\mathsf{A}}_g$. Obviously $\oplus_{k\ge0}\Gamma(\overline{\mathsf{A}}_g,\omega^{\otimes k})$ is integral over $\oplus_{k\ge0}\Gamma(\overline{\mathsf{A}}_g,\omega^{\otimes mk})$ and integrally closed in its field of fractions, which is a finite extension of that of $\oplus_{k\ge0}\Gamma(\overline{\mathsf{A}}_g,\omega^{\otimes mk})$. The ring $\oplus_{k\ge0}\Gamma(\overline{\mathsf{A}}_g,\omega^{\otimes k})$ is of finite type over $\mathbb{Z}$ since $\Gamma(\overline{\mathsf{A}}_g,\omega^{\otimes mk})$ is excellent (as is any finitely generated $\mathbb{Z}$-algebra), and for many other reasons. By now we have proved statements (1) and (2) in the theorem.

Look now at the fibres of $\overline{\pi} : \overline{\mathsf{A}}_g \to \mathsf{A}_g^*$. They are all geometrically connected. Since $\omega^{\otimes m}$ = the pullback of $\mathcal{L}$ to $\overline{\mathsf{A}}_g$, the restriction of $\omega^{\otimes m}$ to any fibre is trivial. So we may apply prop. 2.2 to morphisms of complete curves into those fibres (there always exist plenty, except if the fibre is discrete): It follows that the isomorphism class of the abelian part of G is constant on the fibres. So first if a fibre meets A_g it has only one geometric point. The coarse module scheme $[\mathsf{A}_g]$ being a categorical quotient of A_g, the morphism $\pi : \mathsf{A}_g \to \mathsf{A}_g^*$ factors through $\mathsf{A}_g \to [\mathsf{A}_g]$ and induces a morphism $[\mathsf{A}_g] \to \mathsf{A}_g^*$, each of whose geometric fibres is a single point. Hence $[\mathsf{A}_g] \to \mathsf{A}_g^*$ is an open immersion, by Zariski's main theorem.

In general, for a quotient X_ξ of X as in statement (6), consider the restriction π_ξ of $\overline{\pi}$ to the $(g-r)$-th stratum, and denote its image by Z_{g-r}. Then prop. 1.5 shows that π_ξ factors through A_{g-r}, and therefore also through $[\mathsf{A}_{g-r}] \to Z_{g-r} \subset \mathsf{A}_g^*$. Furthermore we know that the morphism $[\mathsf{A}_{g-r}] \to Z_{g-r}$ is bijective on geometric points, hence radicial. Since $[\mathsf{A}_{g-r}]$ is normal $[\mathsf{A}_{g-r}] \to Z_{g-r}$ is an isomorphism if we can show Z_{g-r} to be normal. (This is not so obvious because Stein factorization does not commute with arbitrary base change, although it commutes with flat base change.) Let us first note that statements (4) and (5) have already been proved and the proofs of (3) and (6) will be completed with the proof that the closure of Z_{g-r} is normal.

The normality of the closure of Z_{g-r} can be shown by a direct calculation of the completed strict local rings of A_g^*: by [EGA III thm. 4.1.5], the completed strict local ring at a geometric point $\overline{x}$ of A_g^* is the set of global sections of the completion of $\overline{\mathsf{A}}_g$ along the inverse image of $\overline{x}$. We can proceed as all the necessary information about this completion can be obtained from the study of $\mathcal{S}_\xi/\mathrm{GL}(X_\xi)$, which we know rather explicitly. However it takes quite a few lines to do it properly, and we defer the proof until 2.7. Once we know that this closure is normal it follows easily that it is isomorphic to A_{g-r}^* (it already has the same geometric points). Fortunately the normality of Z_{g-r} itself does not require the explicit structure of the completed local ring as in 2.7: From what we just said and the structure of the boundary of $\overline{\mathsf{A}}_g$ it is clear that the completed strict local rings of $[\mathsf{A}_{g-r}]$ are contained in the completed strict local rings of A_g^*, using pull-back via the natural map $\mathcal{S}_\xi \to \mathsf{A}_{g-r}$. The reader is encouraged to look at 2.7 for a better view of the situation. □

There is a corresponding (symplectic principal) level-n version of theorem 2.3. As usual we obtain a little bit more if we *assume* $n \geq 3$. The case $n = 2$ is left to the reader.

2.5. Theorem. *Assume $n \geq 3$. There exists a normal scheme $\mathsf{A}_{g,n}^*$ proper and of finite type over* $\mathrm{Spec}(\mathbb{Z}[\zeta_n, 1/n])$ *containing $\mathsf{A}_{g,n}$ as a dense open subscheme such that the following assertions hold:*

(1) The invertible sheaf $\omega(G \to \mathsf{A}_{g,n})$ on $\mathsf{A}_{g,n}$ extends to an invertible sheaf on $\mathsf{A}_{g,n}^*$ relatively ample over $\mathrm{Spec}(\mathbb{Z})$*, which we again denote by ω.*

(2) For any (smooth) arithmetic toroidal compactification $\overline{\mathsf{A}}_{g,n}$ of $\mathsf{A}_{g,n}$, there is a canonical morphism $\overline{\pi} : \overline{\mathsf{A}}_{g,n} \to \mathsf{A}_{g,n}^*$ with $\overline{\pi}^*\omega = \omega(G \to \overline{\mathsf{A}}_{g,n})$ on $\overline{\mathsf{A}}_{g,n}$. Moreover, the morphisms $\overline{\pi} : \overline{\mathsf{A}}_{g,n} \to \mathsf{A}_{g,n}^*$ (for different choices of admissible polyhedral cone decompositions) are compatible as usual: For any map $\overline{\mathsf{A}}'_{g,n} \to \overline{\mathsf{A}}_{g,n}$ the composition $\overline{\mathsf{A}}'_{g,n} \to \overline{\mathsf{A}}_{g,n} \to \overline{\mathsf{A}}_{g,n}^*$ is equal to $\overline{\pi}' : \overline{\mathsf{A}}'_{g,n} \to \mathsf{A}_{g,n}^*$.

(3) The graded ring $\oplus_{k\geq 0} \Gamma(\overline{\mathsf{A}}_{g,n}, \omega^{\otimes k})$ is a finitely generated $\mathbb{Z}[\zeta_n, 1/n]$-algebra, which is independent of the choice of $\overline{\mathsf{A}}_{g,n}$ by results of §1. $\mathsf{A}_{g,n}^*$ is canonically isomorphic to $\mathrm{Proj}(\oplus_{k\geq 0} \Gamma(\overline{\mathsf{A}}_{g,n}, \omega^{\otimes k}))$ *and the morphism $\overline{\pi} : \overline{\mathsf{A}}_{g,n} \to \mathsf{A}_{g,n}^*$ is the canonical one defined by ω. The natural* $\mathrm{Sp}(2g, \mathbb{Z}/n\mathbb{Z})$*-action on $\mathsf{A}_{g,n}$ extends to $\mathsf{A}_{g,n}^*$, and $\overline{\pi}$ is equivariant w.r.t. this action.*

(4) $\mathsf{A}_{g,n}^*$ has a natural stratification by locally closed subschemes geometrically normal and flat over $\mathrm{Spec}(\mathbb{Z})$*, each of which is (canonically isomorphic to) a (fine) moduli scheme $\mathsf{A}_{i,n}$ for some i between 0 and g. Furthermore their closures are isomorphic to $\mathsf{A}_{i,n}^*$. We call these subschemes* cusps. *There are* $[\mathrm{Sp}(2g, \mathbb{Z}/n\mathbb{Z}) : P_{g,r}(\mathbb{Z}/n\mathbb{Z})']$ *(disjoint) copies of $\mathsf{A}_{g-r,n}$ for each r*

between 0 and g. One way to parametrize these strata is to use direct summands $\Sigma^\perp \subset \Sigma \subset \mathbb{Z}^{2g}$ (with the standard symplectic pairing), modulo the level-n subgroup of $\mathrm{Sp}(2g,\mathbb{Z})$, as in chap. IV §6: $\mathsf{A}_{g,n}^* = \coprod Z_c$ where c runs over all such classes. Z_c is isomorphic to $\mathsf{A}_{g-r,n}$ if c is of rank r. More canonically, Z_c classifies abelian schemes A of relative dimension $g-r$ plus isomorphisms $A[n] \xrightarrow{\sim} \Sigma/\Sigma^\perp \bmod n$. The incidence relation among the cusps is determined by the incidence relation among these totally isotropic subspaces inside $\mathbb{Z}^{2g}$. Thus $\overline{\pi}$ is compatible with the stratifications of its source and target in an obvious sense.

(5) The map $\overline{\pi}$ associates to each geometric point $\overline{x}$ of $\overline{\mathsf{A}}_{g,n}$ the point of $\mathsf{A}_{g,n}^*$ determined as follows: if $G_{\overline{x}}$ is the fibre of the universal semi-abelian variety $G \to \overline{\mathsf{A}}_{g,n}$ over $\overline{x}$, and $T_{\overline{x}}$ its maximal torus, we have a canonical symplectic embedding $T[n] \subset G[n] \hookrightarrow (\mathbb{Z}/n\mathbb{Z})^{2g}$, and a generator of $\det(T[n]) = \det(X(T))^*$ well determined up to ± 1. Lift this to $\Sigma^\perp \subset \Sigma \subset \mathbb{Z}^{2g}$, and the canonical generator (up to ± 1) of $\det(\Sigma^\perp)$. This determines the cusp c and an isomorphism $A[n] \xrightarrow{\sim} \Sigma/\Sigma^\perp \bmod n$, where A is the abelian part of G. $\overline{\pi}(\overline{x})$ is the corresponding point in $\mathsf{A}_{g-r,n}$.

(6) The restriction of $\overline{\pi}$ to $\mathsf{A}_{g,n} \subset \overline{\mathsf{A}}_{g,n}$ identifies $\mathsf{A}_{g,n}$ as a dense open subscheme of $\mathsf{A}_{g,n}^*$. More generally for a rank-r quotient X_ξ of X and a coset $\overline{c}$ in $\mathrm{Sp}(2g,\mathbb{Z}/n\mathbb{Z})/P_{g,r}(\mathbb{Z}/\mathbb{Z})'$, let $\pi_{\xi,\overline{c}}$ be the restriction of $\overline{\pi}$ to the part of the boundary corresponding to X_ξ and $\overline{c}$ (according to IV §6). Then $\pi_{\xi,\overline{c}}$ factors through the projection onto $\mathsf{A}_{g-r,n}$ and induces an isomorphism $\mathsf{A}_{g-r,n} \to Z_c$. (This isomorphism depends on the choice of the isomorphism between $\Sigma/\Sigma^\perp \bmod n$ and $(\mathbb{Z}/n\mathbb{Z})^{2g-2r}$, hence strictly speaking is determined only up to $\mathrm{Sp}(2g-2r,\mathbb{Z}/n\mathbb{Z})$-action.)

Proof. The proof is essentially the same as in 2.3: Assertions (2) and (3) follow as before. The abelian part of G is constant on each fibre, and on each stratum $\overline{\pi}$ factors over the projection onto $\mathsf{A}_{g-r,n}$. As each fibre of $\overline{\pi}$ is geometrically connected it must be contained in the image of the closed stratum of $\mathcal{S}_{\xi,n}$, for some quotient X_ξ of X. Now careful bookkeeping shows assertions (4) to (6). We leave this as an exercise to the reader. There remains statement (1) that ω on $\mathsf{A}_{g,n}$ extends to $\mathsf{A}_{g,n}^*$, or equivalently that $(\mathsf{A}_{g,n} \to \mathsf{A}_{g,n}^*)_* \omega$ is invertible (since $\mathsf{A}_{g,n}^*$ is normal and $\mathrm{codim}(\mathsf{A}_{g,n}^* - \mathsf{A}_{g,n}) = g \geq 2$. By fpqc descent we can pass to the completion. Now the assertion that ω extends is easy to check by the explicit description/construction of the boundary. Over a cusp we have the following picture: Quotient of $\mathcal{S}_{\xi,n}$ under the kernel of $\mathrm{GL}(X_\xi) \to \mathrm{GL}(X_\xi/n \cdot X_\xi) =$ (formal completion of $\overline{\mathsf{A}}_{g,n}$ along the part over a cusp) $\to$ (formal completion of $\mathsf{A}_{g,n}^*$ along this cusp) $\to \mathsf{A}_{g-r,n}$. Because of the level-n structure, $\omega(G \to \mathcal{S}_{\xi,n})$ is the pull-back of the ω on $A_{g-r,n}$, therefore the pull-back of the ω on $\mathsf{A}_{g-r,n}$ to the formal completion of $\mathsf{A}_{g,n}^*$ along this cusp obviously gives the extension we are looking for. □

2.6. Remarks. (i) Over the complex numbers we obtain, by totally algebraic methods, the classical Satake compactification ([Sa1]). Our construction shows that it extends over the integers.

(ii) The cone decomposition has disappeared, as it should. More precisely, in chapter IV §6, we described the stratification of $\overline{\mathsf{A}}_{g,n}$ by certain conical complexes denoted by $\sum_n$, obtained by gluing the cones $C(\Lambda_{\mathbb{Z}}/\Sigma_{\mathbb{Z}})$'s indexed by direct summands $\Sigma_{\mathbb{Z}}$ of the standard symplectic lattice $\Lambda_{\mathbb{Z}}$ of rank $2g$ with $\Sigma_{\mathbb{Z}}^{\perp} \subset \Sigma_{\mathbb{Z}}$, then dividing out by $\mathrm{Sp}(2g, \mathbb{Z})(n)$. The cones are given by an admissible polyhedral cone decomposition, which gives $\sum_n$ the structure of a conical complex. If we forget about the cone decompositions, then we obtain a CW-complex $\sum_n^*$ with the same underlying topological space as $\sum_n$ but different combinatorial structure. The global structure of the stratification of $\mathsf{A}_{g,n}^*$ coded in $\sum_n^*$ ("via the dual complex construction"), and the map between strata of $\overline{\mathsf{A}}_{g,n}$ and $\mathsf{A}_{g,n}^*$ is best understood via the obvious map $\sum_n \to \sum_n^*$.

Local Structure of the Minimal Compactification at the Boundary

We can determine the completion of the strict local rings of $\mathsf{A}_{g,n}^*$ and A_g^*. (For $\mathsf{A}_{g,n}$, $n \geq 3$, we can even get the completed local rings, as will become clear.) The method is to use [EGA III] thm. 4.1.5 together with the description of the blowing-down map between the boundary of $\overline{\mathsf{A}}_g$ and A_g^*, as explained near the end of the proof of theorem 2.3. We shall do the $\mathsf{A}_{g,n}$ case first ($n \geq 3$), because there we do not have to worry about coarse moduli spaces. Let $\overline{x}$ be a geometric point of a cusp c of $\mathsf{A}_{g,n}^*$. Using the previous notation, we have $\overline{\Xi}_{\xi,n} \to \mathbf{Hom}_{\mathbb{Z}}(X_{\xi,n}, A_{g-r}) \to \mathsf{A}_{g-r}$ and its formal completion (along the union of non-degenerate strata in $\overline{\Xi}_{\xi,n} - \Xi_{\xi,n}$) $\mathcal{S}_{\xi,n} \to \mathbf{Hom}_{\mathbb{Z}}(X_{\xi,n}, A_{g-r}) \to \mathsf{A}_{g-r}$. Identify $\overline{x}$ with a geometric point of A_{g-r}, which we still denote by $\overline{x}$. The completed strict local ring $\mathcal{O}_{\overline{x}}^{\wedge}$ of $\mathsf{A}_{g,n}^*$ at $\overline{x}$ is the ring of global sections, invariant under $\mathrm{GL}(X_{\xi})(n)$, on the formal completion of $\mathcal{S}_{\xi,n}$ along the inverse image of $\overline{x}$.

We have exhibited the structure of the torus bundle $\overline{\Xi}_{\xi,n}$ in ch. IV. From this it is clear that $\mathcal{O}_{\overline{x}}^{\wedge}$ is equal to the invariant elements of the product (over $\lambda \in C(X_{\xi})^{\wedge} \bigcap (1/n) \cdot S^2(X_{\xi})$) of all $\Gamma(\mathbf{Hom}_{\mathbb{Z}}(X_{\xi,n}, A_{(\overline{x})}), \mathcal{L}(\lambda)_{(\overline{x})})$, where $A_{(\overline{x})}$ is the fibre over $\mathrm{Spec}(\mathcal{O}_{\overline{x}}^{\wedge})$ of the universal abelian scheme $A_{g-r,n} \to A_{g-r,n}$ and $\mathcal{L}(\lambda)_{(\overline{x})}$ the same line bundle as in §1: Any such family defines a global section, and conversely the q-expansion of a global section is such a family. Thus we have computed the ring $\mathcal{O}_{\overline{x}}^{\wedge}$.

Furthermore we promised to compute the completed strict local rings at the $\overline{x}$ (thought of as points in the closures of strata A_{g-s} ($s \leq r$)) to show that they are normal. Choose a direct summand $Y_{\xi} \subset X_{\xi}$ of rank $r-s$, and consider the invariant elements (under the congruence subgroup in $\mathrm{GL}(Y_{\xi})$, which is the image of the stabilizer of Y_{ξ} in the congruence subgroup of $\mathrm{GL}(X_{\xi})$) in the product of $\Gamma(\mathbf{Hom}_{\mathbb{Z}}(X_{\xi,n}, A_{(\overline{x})}), \mathcal{L}(\lambda)_{(\overline{x})})$ for $\lambda \in C(X_{\xi})^{\vee} \bigcap (1/n) \cdot S^2(Y_{\xi})$. For such λ $\Gamma(\mathbf{Hom}_{\mathbb{Z}}(X_{\xi,n}, A_{(\overline{x})}), \mathcal{L}(\lambda)_{(\overline{x})})$ and $\Gamma(\mathbf{Hom}_{\mathbb{Z}}(Y_{\xi,n}, A_{(\overline{x})}), \mathcal{L}(\lambda)_{(\overline{x})})$ coincide (as $X(\lambda) \subset Y_{\xi}$), and two such λ's conjugate under $\mathrm{GL}(X_{\xi})(n)$ are already conjugate under $\mathrm{GL}(Y_{\xi})(n)$. It follows that any $\mathrm{GL}(Y_{\xi})(n)$-invariant family extends to a $\mathrm{GL}(X_{\xi})(n)$-invariant one, so that we obtain the completed strict local ring in A_{g-s}^*.

Recall that the invertible sheaves $\mathcal{L}(\lambda)$'s on $\mathbf{Hom}_{\mathbb{Z}}(X_{\xi,n}, A_{(\overline{x})})$ are pull-backs of relatively ample (w.r.t. $A_{g-r,n}$) line bundles on quotients $\mathbf{Hom}_{\mathbb{Z}}(X(\lambda)_n, A_{(\overline{x})})$. So formation of their global sections commutes with arbitrary base change by the Riemann-Roch theorem for non-degenerate invertible sheaves on abelian schemes. Furthermore the stabilizer of λ in the group $\mathrm{GL}(X_\xi)(n)$ operates trivially on $X(\lambda)$, so formation of $\mathrm{GL}(X_\xi)(n)$-invariants also commutes with base change. Hence the local rings $\mathcal{O}_{\overline{x}}^{\wedge}$ are geometrically normal over $\mathbb{Z}$.

If we do not assume $n \geq 3$ (i.e. consider A_g^* or $\mathsf{A}_{g,2}^*$), then we have to take into account non-trivial automorphisms of the closed fibre A_0, and of $X(\lambda)$. More precisely, we keep the notation as above, let $\overline{x}$ be a geometric point of $\mathsf{A}_{g,n}^*$, and denote by $\mathrm{Aut}_{\overline{x}}$ the (finite) group of automorphisms of the closed fibre $A_{\overline{x}}$ of $A_{(\overline{x})}$ which preserve the principal polarization and the principal level-n structure. (Of course $\mathrm{Aut}_{\overline{x}}$ is trivial if $n \geq 3$.) $\mathrm{Aut}_{\overline{x}}$ operates on the ring product of all $\Gamma(\mathbf{Hom}_{\mathbb{Z}}(X_{\xi,n}, A_{(\overline{x})}), \mathcal{L}(\lambda)_{(\overline{x})})$ and this action commutes with that of $\mathrm{GL}(X_\xi)(n)$. Now the completed strict local ring $\mathcal{O}_{\overline{x}}^{\wedge}$ consists of the $\mathrm{Aut}_{\overline{x}} \times \mathrm{GL}(X_\xi)(n)$-invariants. (This time formation of invariants need not commute with base change.) This is just a matter of definition involving algebraic stacks, and we leave the necessary checking to the reader. For example if $n = 1$ we obtain the product, over all λ modulo $\mathrm{GL}(X_\xi)$, of the $\mathrm{Aut}_{\overline{x}} \times \mathrm{Aut}(X(\lambda), \lambda)$-invariants in $\Gamma(\mathbf{Hom}_{\mathbb{Z}}(X(\lambda), A_{(\overline{x})}), \mathcal{L}(\lambda)_{(\overline{x})})$. Again it follows that closures of strata are normal. So now we have computed the completed strict local rings of $\mathsf{A}_{g,n}$ for all $n \geq 1$.

The stabilizer in $\mathrm{Sp}(2g, \mathbb{Z}/n\mathbb{Z})$ of a cusp c as above is isomorphic to the finite group $P_{g,r}(\mathbb{Z}/n\mathbb{Z})' =$ image of $P_{g,r}(\mathbb{Z})$ in $P_{g,r}(\mathbb{Z}/n\mathbb{Z})$. It is a semi-direct product of $\mathrm{GL}(r, \mathbb{Z}/n\mathbb{Z})' \times \mathrm{Sp}(2g-2r, \mathbb{Z}/n\mathbb{Z})$ with $N_{g,r}(\mathbb{Z}/n\mathbb{Z})$. The action of $\mathrm{GL}(r, \mathbb{Z}/n\mathbb{Z})' \times \mathrm{Sp}(2g-2r, \mathbb{Z}/n\mathbb{Z})$ is easy: $\mathrm{Sp}(2g-2r, \mathbb{Z}/n\mathbb{Z})$ operates by changing level structure of the abelian part, while elements of $\mathrm{GL}(r, \mathbb{Z}/n\mathbb{Z})' \cong \mathrm{GL}(X_\xi/n \cdot X_\xi)'$ act naturally on $\mathcal{O}_{\overline{x}}^{\wedge}$ by lifting them to $\mathrm{GL}(X_\xi)$. $N_{g,r}(\mathbb{Z}/\mathbb{Z})$ is isomorphic to an extension of $\mathrm{Hom}(X_{\xi,n}, A[n])$ by $E_\xi[n]$. $E_\xi[n]$ acts on $\mathcal{O}_{\overline{x}}^{\wedge}$ via the various characters λ of E_ξ on $(\mathbf{Hom}_{\mathbb{Z}}(X_{\xi,n}, A_{(\overline{x})}), \mathcal{L}(\lambda)_{(\overline{x})})$. $\mathrm{Hom}(X_{\xi,n}, A[n])$ acts "by translation on $\mathrm{Hom}(X_{\xi,n}, A)$". The description can be made more canonical with patience, a task which we leave to the reader.

It is instructive to look at the morphisms $\pi_{m,n}^* : A_{g,n}^* \to A_{g,m}^*$ for $m|n$ at the boundary in terms of completed strict local rings. Thus let $\overline{x}$ be a geometric point of $A_{g,n}^*$, and $\overline{y} = \pi_{m,n}^*(\overline{x})$. We have

$$\mathcal{O}_{\overline{x}}^{\wedge} = \{\mathrm{Ker}(\mathrm{GL}(X_\xi) \to \mathrm{GL}(X_\xi/nX_\xi))\}\text{-invariants in} \prod_{\lambda \in C(X_\xi)^\vee \cap ((1/n)\cdot S^2(X_\xi))} \Gamma(\mathbf{Hom}_{\mathbb{Z}}(X_{\xi,n}, A_{(\overline{x})}), \mathcal{L}(\lambda)_{(\overline{x})}), \text{ while}$$

$$\mathcal{O}_{\overline{y}}^{\wedge} = \{\mathrm{Ker}(\mathrm{GL}(X_\xi) \to \mathrm{GL}(X_\xi/mX_\xi))\}\text{-invariants in} \prod_{\lambda \in C(X_\xi)^\vee \cap ((1/m)\cdot S^2(X_\xi))} \Gamma(\mathbf{Hom}_{\mathbb{Z}}(X_{\xi,m}, A_{(\overline{y})}), \mathcal{L}(\lambda)_{(\overline{y})}).$$

Recall that for $\lambda \in (1/m) \cdot S^2(X_\xi) \subset (1/n) \cdot S^2(X_\xi)$, the invertible sheaf $\mathcal{L}(\lambda)$ on $\mathbf{Hom}_{\mathbb{Z}}(X_{\xi,n}, A_{(\overline{x})})$ is canonically a pull-back of an invertible sheaf on

$\mathbf{Hom}_{\mathbb{Z}}(X_{\xi,m}, A_{(\overline{y})})$, which is also denoted $\mathcal{L}(\lambda)$. This defines a natural inclusion map on the products above, which is invariant under the natural $\mathrm{GL}(X_\xi)$-action. Taking invariants, we obtain the natural map from $\mathcal{O}^{\wedge}_{\overline{y}}$ to $\mathcal{O}^{\wedge}_{\overline{x}}$. This map is also equivariant w.r.t. $P_{g,r}(\mathbb{Z}/n\mathbb{Z})' \to P_{g,r}(\mathbb{Z}/m\mathbb{Z})'$.

To summarize:

2.7. Theorem. *Notation as above.*

(i) If $n \geq 3$, the completion of the strict localization of a geometric point $\overline{x}$ at a cusp c of $\mathsf{A}^*_{g,n}$ is (canonically isomorphic to) the ring of $\mathrm{GL}(X_\xi)(n)$-invariants in $\prod_{\lambda \in C(X_\xi)^\vee \cap ((1/n)\cdot S^2(X_\xi))} \Gamma(\mathbf{Hom}_{\mathbb{Z}}(X_{\xi,n}, A_{(\overline{x})}), \mathcal{L}(\lambda)_{(\overline{x})})$. $\mathsf{A}^*_{g,n}$ $(n \geq 3)$ is geometrically normal over $\mathrm{Spec}(\mathbb{Z}[\zeta, 1/n])$.

(ii) More generally, for any integer $n \geq 1$, the completion of the strict localization of a geometric point $\overline{x}$ at a cusp c of $\mathsf{A}^*_{g,n}$ is (canonically isomorphic to) the ring of $\mathrm{Aut}_{\overline{x}} \times \mathrm{GL}(X_\xi)(n)$-invariants in $\prod_{\lambda \in C(X_\xi)^\vee \cap ((1/n)\cdot S^2(X_\xi))} \Gamma(\mathbf{Hom}_{\mathbb{Z}}(X_{\xi,n}, A_{(\overline{x})}), \mathcal{L}(\lambda)_{(\overline{x})})$. The action of the stabilizer of the cusp c is explicitly described above (the "obvious one").

(iii) In terms of the above description of the completed local rings, the canonical morphism $\pi^*_{m,n} : \mathsf{A}^*_{g,n} \to \mathsf{A}^*_{g,m}$ for $m|n$ can be described as follows: for a geometric point $\overline{x}$ of $\mathsf{A}^*_{g,n}$ and $\overline{y} = \pi^*_{m,n}(\overline{x})$, we know that that $\forall \lambda \in (1/m)\cdot S^2(X)$, the invertible sheaf $\mathcal{L}(\lambda)$ on $\mathbf{Hom}_{\mathbb{Z}}(X_{\xi,n}, A_{(\overline{x})})$ is canonically a pull-back of an invertible sheaf (also denoted by $\mathcal{L}(\lambda)$) on $\mathbf{Hom}_{\mathbb{Z}}(X_{\xi,m}, A_{(\overline{y})})$. The homomorphism from $\mathcal{O}^{\wedge}_{\overline{y}}$ to $\mathcal{O}^{\wedge}_{\overline{x}}$ is obtained by taking invariants in the induced inclusion of $\prod_{\lambda \in C(X_\xi)^\vee \cap ((1/m)\cdot S^2(X_\xi))} \Gamma(\mathbf{Hom}_{\mathbb{Z}}(X_{\xi,m}, A_{(\overline{y})}), \mathcal{L}(\lambda)_{(\overline{y})})$ into $\prod_{\lambda \in C(X_\xi)^\vee \cap (1/n)\cdot S^2(X_\xi)} \Gamma(\mathbf{Hom}_{\mathbb{Z}}(X_{\xi,n}, A_{(\overline{x})}), \mathcal{L}(\lambda)_{(\overline{x})})$.

It is instructive to formulate a variant of theorem 2.7 over the 0-dimensional cusps. The statement is more concrete in some sense because we use only the q-expansion. The method is exactly the same as before, so we shall not repeat the argument.

2.8. Corollary (of the Proof). *For $n \geq 1$, the formal completion of $\mathsf{A}^*_{g,n}$ at a 0-dimensional cusp is isomorphic to the ring of all possible q-expansions, that is to $\mathbb{Z}[1/n, \zeta_n][\![n^{-1}S^2(X) \cap C(X)]\!]^{\mathrm{GL}(X)(n)}$ = the ring of $\mathrm{GL}(X)(n)$-invariants in the formal power series ring $\mathbb{Z}[1/n, \zeta_n][\![n^{-1}S^2(X) \cap C(X)]\!]$, where $\mathrm{GL}(X)(n) = \mathrm{Ker}\{\mathrm{GL}(X) \to \mathrm{GL}(\mathbb{Z}/n\mathbb{Z})\}$ is the principal congruence subgroup of $\mathrm{GL}(X)$ of level-n.*

3. Modular Forms and Classical Θ-Series

Two themes are interwoven in this section. The first is the relation between the Hodge line bundle ω and the line bundle corresponding to theta constants. The second is the positivity of these two line bundles and various phenomena associated with the positivity. In particular, we shall discuss ideas behind the

the quoted proof of prop. 2.1, 2.2, upon which the theory of minimal compactification was built. We also offer somewhat more direct proof of these two propositions as an alternative for the reader.

As discussed in ch. I, 5, the Hodge line bundle ω is equal up to torsion to the square of "the line bundle attached to theta constants". Strictly speaking we can define "the line bundles attached to theta constants" only when we have a theta structure. But if we are willing to ignore torsion (i.e. operate in $\mathrm{Pic} \otimes \mathbb{Q}$), then it is well-defined, and in fact equal to $1/2 \cdot \omega$. We shall bypass theta structures since introducing them would unduly lengthen our exposition. When we have a *symmetric* relatively ample invertible sheaf $\mathcal{L}$ on an abelian scheme $f : A \to S$ rigidified along the zero section which defines a principal polarization, $f_*(\mathcal{L})^{\otimes -1}$ is by definition the invertible sheaf attached to theta constants w.r.t. $\mathcal{L}$. More generally if $\mathcal{L}$ is a *symmetric* relatively ample invertible sheaf on an abelian scheme $f : A \to S$ rigidified along the zero section, the invertible sheaf attached to theta constants w.r.t. $\mathcal{L}$ is $-\mathrm{rk.}(f_*(\mathcal{L}))^{-1} \cdot \Lambda^{\mathrm{top}}(f_*(\mathcal{L}))$, thought of as an element of $\mathrm{Pic}(S) \otimes \mathbb{Q}$. (Note that $\mathrm{rk.}(f_*(\mathcal{L}))$ is locally constant on S w.r.t the Zariski topology, hence the above formula makes sense.)

In this section we first give the reader a tour of the main ideas behind the proof of the positivity of ω which is found in the literature. We later give a second proof with more details.

Recall that for any abelian scheme $A \to S$ and relatively ample invertible sheaf $\mathcal{L}$ on A, $\mathcal{L}^{\otimes n}$ is very ample if $n \geq 3$. (c. [Mum 1] §17, p.163.) Exploring the method of theta groups and the method of variable pencils, D. Mumford proved

3.1. Theorem. *If $f : A \to S$ is an abelian scheme and $\mathcal{L}$ a relatively ample invertible sheaf of degree d on A, then*

(i) $f_*(\mathcal{L}^{\otimes m}) \otimes f_*(\mathcal{L}^{\otimes m}) \to f_*(\mathcal{L}^{\otimes m+n})$ is surjective $\forall m, n \geq 4$.

(ii) Fix $m \geq 4$, let $\mathbb{P} = \mathbb{P}(f_*(\mathcal{L}^{\otimes m})) = \mathbf{Proj}_S(\mathrm{Sym}^\bullet f_*(\mathcal{L}^{\otimes m}))$ and consider the projective embedding $\phi = \phi_{\mathcal{L}^{\otimes m}} : A \hookrightarrow \mathbb{P}$. Denote by $\mathcal{J} = \mathcal{J}_A$ the sheaf of ideals in $\mathcal{O}_{\mathbb{P}}$ vanishing on A. Then the natural map $f^* f_* \mathcal{J}(2) \to \mathcal{J}(2)$ is surjective. Here as usual $\mathcal{J}(2) = \mathcal{J} \otimes \mathcal{O}_{\mathbb{P}}(2)$.

(iii) In (ii), if N is a positive integer such that $N^2 > 2m^g\, d \cdot g!$, then the natural map $f_* \mathcal{J}_A(2) \to f_* \mathcal{J}_{A(N)}(2)$ is bijective (injectivity is trivial), where $\mathcal{J}_{A(N)}$ is the sheaf of ideals of $\mathcal{O}_{\mathbb{P}}$ of functions vanishing on $A[N]$.

The statement (i) is proved in [Mum 4] §3, (ii) is proved in [Mum 4] §4, thm. 10, p.80. (iii) is a crude estimate using intersection theory. (It is clear that such an N exists, because torsion points are dense in an abelian variety.) Although Mumford stated his results over fields, the general case follows easily. See also [MB 1] chap. VII. In statement (iii), one can in fact replace $A[N]$ with $N^2 > 2m^g\, d \cdot g!$ by the much smaller torsion subgroup $K(\mathcal{L}^{\otimes m})$ if $4|m$. This result is in [Mum 2] I §4 when 2 is invertible in $\mathcal{O}_S$. It is certainly true in general, but nobody has taken the trouble to write it down. The point is that the theta constants give the structural constants for multiplication of sections of powers of $\mathcal{L}$.

The significance of the above theorem (especially part (ii) and (iii)) is that for the projective embedding given by $\mathcal{L}^{\otimes m}$, $m \geq 4$, the abelian scheme A can be explicitly recovered as the (scheme theoretic) intersection of all quadrics in $\mathbb{P}$ containing $A[N]$ (or, $K(\mathcal{L}^{\otimes m})$ if $4|m$ in view of the above). From this it follows that theta constants determine the moduli. Combined with I, th. 5.1 it follows for instance that ω is ample on $A_{g,n}$ for any $n \geq 3$. Prop. 2.2 follows from this and the theory of Néron models.

The results of Moret-Bailly (c. [MB 1]) generalize these considerations to semi-stable group schemes with abelian generic fibre. (These are essentially semi-abelian schemes, but fibres are not necessarily connected.) Besides the method of theta groups, he used extension results for cubical invertible sheaves and the groups $K(\mathcal{L})$ to deal with degeneration. To prove prop. 2.1 he used Grassmannian embeddings obtained by evaluating global sections of powers of $\mathcal{L}$ on $A[N]$.

Excluding characteristic 2, Mumford obtained very delicate results in [Mum 3]: He showed that an abelian scheme can be explicitly defined by universal quadratic equations with coefficients given by theta constants. Moreover, using theta constants as coordinates, the moduli scheme (with suitably chosen level structure) is defined by quartic equations (in the sense that one gets an open immersion of the moduli into the variety defined by these equations). These quartic equations are nothing but the classical Riemann theta relations. If we include characteristic 2 such precise results are not available, but the general qualitative statements still hold as we have already seen.

The theta constants not only give moduli of abelian varieties, but they essentially give the minimal compactification. Mumford showed this outside of characteristic 2 in [Mum 3] III §11 for the whole 2-adic tower using his theory of algebraic theta constants and theta measures. In view of §2 this is what must happen. Notice that in general at a finite level, the theta constants do *not* give an embedding of the minimal compactification: they only give a finite birational morphism, and we have to take the normalization of the closure of the moduli scheme (in the projective space) to get the minimal compactification. This discrepancy was first noticed by Mumford, then studied by Igusa over $\mathbb{C}$ (c. [I 6]). His results still hold outside of characteristic 2 ([Ch 1] Appendix III).

The rest of this section is geared toward proving prop. 2.1 and prop. 2.2. We shall use I, th. 5.1 to produce modular forms from theta constants. Classical theta series together with the q-expansion principle help us to control the modular forms so obtained. Since we need a symmetric ample invertible sheaf $\mathcal{L}$ to get started, we naturally work on the algebraic stack $N_{g,n}$ (n odd, ≥ 1) which classifies abelian varieties A of dimension g, with a level-n structure, n odd, and a symmetric relatively ample invertible sheaf $\mathcal{L}$ which defines a principal polarization. We usually normalize these $\mathcal{L}$'s so that they are trivial along the 0-section, unless stated otherwise.

We intend to apply I, th. 5.1 for the universal data over $N_{g,n}$ (n odd). It complicates matters that in general $N_{g,n}$ is just an algebraic stack. However we can use PGL(N)-torsors as in the proof of theorem 5.1 of chapter I, and everything follows. Let us also mention that for $n \geq 3$, $N_{g,n}$ is actually a scheme, which leaves only $n = 1$ as an exceptional case. The fact that $N_{g,n}$ is a scheme requires either Mumford's geometric invariant theory or his results described above.

Now the ideas in chapter I,5 give modular forms: Evaluating $f_*(\mathcal{L})$ at the n-torsion sections of the universal abelian scheme gives global sections of $\omega^{\otimes 1/2}$, up to torsion, and suitable powers of these (the exponent $8n$ suffices) are genuine modular forms on $N_{g,n}$. Finally traces and norms of these relative to A_g (resp. $\mathsf{A}_{g,n}$) are modular forms on A_g (resp. $\mathsf{A}_{g,n}$). This is easy to see because $N_{g,n}$ is finite flat over $\mathsf{A}_{g,n}$. Unfortunately we see no easy direct arguments which give the q-expansions of these modular forms, so we must take a detour about classical theta series. The references are [I 3], [I 5], [Mum 7].

Review of Classical Theta Series and Complex Uniformization of $N_{g,n}$

The classical Riemann theta-series associates to $\Omega \in \mathbb{H}_g$ and $z \in \mathbb{C}^g$ the sum

$$\theta(\Omega, z) = \sum_{n\in\mathbb{Z}^g} \exp(\pi\sqrt{-1}\,{}^t n\Omega n)\cdot\exp(2\pi\sqrt{-1}\,{}^t nz)\ .$$

(Elements of $\mathbb{Z}^g$ are thought of as column vectors.) As a function in z, it satisfies the following functional equations: it is invariant under translation by $\mathbb{Z}^g$, and

$$\theta(\Omega, z+\Omega m) = \exp(-\pi\sqrt{-1}\,{}^t m\Omega m)\cdot\exp(-2\pi\sqrt{-1}\,{}^t mz)\cdot\theta(\Omega, z)\ \forall m\in\mathbb{Z}^g\ .$$

More interestingly, it satisfies the following transformation-rules

(i) $\theta(-\Omega^{-1}, \Omega^{-1}z) = \det(-\sqrt{-1}\,\Omega)^{1/2}\cdot\exp(\pi\sqrt{-1}\,{}^t z\Omega^{-1}z)\cdot\theta(\Omega, z)$, where the branch of the square root has positive value when Ω is purely imaginary.

(ii) $\theta(\Omega+B, z) = \theta(\Omega, z+\Delta(B)/2)$, where B denotes a symmetric integral $g\times g$ matrix, and $\Delta(B)$ its diagonal-vector.

(iii) $\theta(A\Omega\,{}^t A, Az) = \theta(\Omega, z)\ \forall A\in \mathrm{GL}_g(\mathbb{Z})$.

Of the above rules, (ii) and (iii) are easy, while (i) follows from the Poisson summation formula. They can also be proved by exploiting the heat equation satisfied by theta functions. These rules describe the behavior of the theta-series under generators of the group $\mathrm{Sp}(2g,\mathbb{Z})$ acting on $\mathbb{H}_g$, and hence under the whole group. Due to the term $\Delta(B)/2$ in (iii), the transformation rule for a general element of $\mathrm{Sp}(2g,\mathbb{Z})$ is a little bit more complicated. However if we take the subgroup of $\mathrm{Sp}(2g,\mathbb{Z})$ generated by elements of type (i), (iii) and type (ii) with $\Delta(B)$ even, then it is easy to write down a uniform transformation formula. This is the subgroup which Igusa called $\Gamma(1,2)$. In fact $\Gamma(1,2)$ consists of elements which in usual block form with terms A, B, C, D such that $\Delta(A\,{}^t B)$, $\Delta(C\,{}^t D)$ are even (equivalently, $\Delta({}^t AC)$, $\Delta({}^t DB)$ are even). For such an element, the transformation formula can be written as

(iv) $\theta((A\Omega+B)\cdot(C\Omega+D)^{-1}, {}^t(C\Omega+D)^{-1}\cdot z) =$ (element in μ_8 depending only on γ) $\cdot\ \det(C\Omega+D)^{1/2}\cdot\exp(\pi\sqrt{-1}\,{}^t z(C\Omega+D)^{-1}Cz)\cdot\theta(\Omega, z)$.

The transformation rules imply that for special choices of the values of z we obtain modular forms of weight 1/2. More precisely define for $a, b \in \mathbb{Q}^g$,

$$\begin{aligned}\theta(\Omega, a, b) &= \sum_{n\in\mathbb{Z}^g} \exp(\pi\sqrt{-1}\,{}^t(n+a)\Omega(n+a) + 2\pi\sqrt{-1}\,{}^t(n+a)b) \\ &= \exp(2\pi\sqrt{-1}\,{}^tab)\cdot\exp(\pi\sqrt{-1}\,{}^ta\Omega a)\cdot\theta(\Omega, \Omega a + b)\ .\end{aligned}$$

These are the classical theta constants. Note that the factor $\exp(2\pi\sqrt{-1}\,{}^tab)$ is simply a root of unity, we include it here to conform with the usual notation. It can be omitted with no consequence for our purpose. It is clear from the definition that if a, b are changed by some elements in $\mathbb{Z}^g$, the theta constants change only by a root of unity. We can reformulate the above the transformation formula in terms of these theta constants:

(i)′ $\theta(-\Omega^{-1}, -b, a) = \det(-\sqrt{-1}\,\Omega)^{1/2}\cdot\exp(-2\pi\sqrt{-1}\,{}^tab)\cdot\theta(Z, a, b)$, where the branch of the square root has positive value when Ω is purely imaginary,

(ii)′ $\theta(\Omega + B, a, -Ba + b + \Delta(B)) = \theta(\Omega, a, b)$ where B denotes a symmetric integral $g\times g$ matrix,

(iii)′ $\theta(A\cdot\Omega\cdot{}^tA,\ {}^tA^{-1}\cdot a,\ Ab) = \theta(\Omega, a, b)\ \forall A \in \mathrm{GL}_g(\mathbb{Z})$.

The above formula in principle gives the transformation rule of $\mathrm{Sp}(2g, \mathbb{Z})$ on these theta constants. If we ignore the precise roots of unity, it can be described as follows: for a general element $\gamma \in \mathrm{Sp}(2g, \mathbb{Z})$ with entries A, B, C, D when written in block-form as usual, we have

(iv)′ $\theta(\Omega', a', b') =$ (root of unity depending on γ, a, b) $\cdot \det(C\Omega + D)^{1/2}\cdot \theta(\Omega, a, b)$, where $\Omega' = \gamma\cdot\Omega = (A\Omega + B)(C\Omega + D)^{-1}$, $a' = Da - Cb + \Delta(C\,{}^tD)/2$, $b' = -Ba + Ab + \Delta(A\,{}^tB)/2$. In matricial notation,

$$\begin{pmatrix} a' \\ b' \end{pmatrix} = \begin{pmatrix} D & -C \\ -B & A \end{pmatrix}\begin{pmatrix} a \\ b \end{pmatrix} + \frac{1}{2}\begin{pmatrix} \Delta(C\,{}^tD) \\ \Delta(A\,{}^tB) \end{pmatrix}\ ;$$

$$\text{note that}\ \begin{pmatrix} D & -C \\ -B & A \end{pmatrix} = {}^t\begin{pmatrix} A & B \\ C & D \end{pmatrix}^{-1}\ .$$

When $\gamma \in \Gamma(1,2)$, we can control the root of unity, and it is best to introduce a modified theta constant $\theta^{\alpha}(\Omega, a, b) := \exp(-\pi\sqrt{-1}\,{}^tab)\theta(\Omega, a, b)$. Then we have

(v)′ $\theta^{\alpha}(\Omega', Da - Cb, -Ba + Ab) =$ (element of μ_8 depending only on γ) $\cdot \det(C\Omega + D)^{1/2}\cdot\theta^{\alpha}(\Omega, a, b)$, where $\Omega' = \gamma\cdot\Omega = (A\Omega + B)(C\Omega + D)^{-1}$.

It is also possible to give the transformation rule for a general element of $\mathrm{Sp}(2g, \mathbb{Z})$. For this it is convenient to introduce the classical theta series with characteristics: define for $a, b \in \mathbb{Q}^g$,

$$\begin{aligned}\theta(\Omega, z, a, b) &= \sum_{n\in\mathbb{Z}^g} \exp(\pi\sqrt{-1}\,{}^t(n+a)\Omega(n+a) + 2\pi\sqrt{-1}\,{}^t(n+a)(z+b)) \\ &= \exp(2\pi\sqrt{-1}\,{}^tab)\cdot\exp(\pi\sqrt{-1}\,{}^ta\Omega a)\cdot\exp(2\pi\sqrt{-1}\,{}^taz)\cdot\theta(\Omega, z + \Omega a + b)\ ,\end{aligned}$$

which is a holomorphic function on $\mathbb{H}_g \times \mathbb{C}^g$. Note that $\theta(\Omega, 0, a, b) = \theta(\Omega, a, b)$ as defined before. For a general element element $\gamma \in \mathrm{Sp}(2g, \mathbb{Z})$ with entries A, B, C, D in block-form, we have $\theta(\Omega', {}^t(C\Omega+D)^{-1}z, a', b') = \kappa(\gamma)\cdot\zeta(\gamma, a', b')\cdot \exp(\pi\sqrt{-1}\,{}^tz(C\Omega + D)^{-1}Cz) \cdot \det(C\Omega + D)^{1/2}\theta(\Omega, z, a, b)$, where $\Omega' = \gamma \cdot \Omega = (A\Omega + B)(C\Omega + D)^{-1}$, $a' = Da - Cb + \Delta(C \cdot {}^tD)/2$, $b' = -Ba + Ab + \Delta(A \cdot {}^tB)/2$ as before, $\kappa(\gamma)$ is an element of μ_8 depending only on γ, $\zeta(\gamma, a', b')$ is a root of unity whose precise form is complicated: $\zeta(\gamma, a', b') = \exp(-\pi\sqrt{-1}\,\{{}^ta\,{}^tB\,D\,b + {}^tb\,{}^tA\,C\,b - 2\,{}^ta\,{}^tB\,C\,b - {}^t\Delta(A\cdot{}^tB)(D\,a - C\,b)\})$ (c. Igusa [I 3], II, §2).

Some remarks about formulae (iv)$'$, (v)$'$ are in order. Consider the map from $\mathrm{Sp}(2g, \mathbb{Z})$ to $(\mathbb{Q}/\mathbb{Z})^{2g}$ sending a general element γ with entries A, B, C, D in block-form to the element δ_γ of $(1/2\cdot\mathbb{Z}/\mathbb{Z})^{2g}$ whose first g entries are represented by $\Delta(C \cdot {}^tD)/2$ and second g entries by $\Delta(A \cdot {}^tB)/2$. We regard $(1/2 \cdot \mathbb{Z}/\mathbb{Z})^{2g}$ as a $\mathrm{Sp}(2g, \mathbb{Z})$-module via the composition of the inner automorphism $\gamma \mapsto {}^t\gamma^{-1}$ and the standard representation of $\mathrm{Sp}(2g, \mathbb{Z})$. (This inner automorphism is always present if one follows the classical formula.) Then $\gamma \mapsto \delta_\gamma$ is a 1-cocycle. Equivalently if we twist (i.e. compose with translation) the linear action of $\mathrm{Sp}(2g, \mathbb{Z})$ on $(1/2\cdot\mathbb{Z}/\mathbb{Z})^{2g}$ just explained by this 1-cocycle, we obtain a twisted action of $\mathrm{Sp}(2g, \mathbb{Z})$ on $(1/2 \cdot \mathbb{Z}/\mathbb{Z})^{2g}$. This twisted action preserves the $(\mathbb{Z}/2\mathbb{Z})$-valued quadratic form $q({}^t({}^ta, {}^tb)) = {}^ta\cdot b$ on $(1/2\cdot\mathbb{Z}/\mathbb{Z})^{2g}$, and the 1-cocycle δ_γ is characterized by this property. The assertion that δ_γ is a cocycle can be checked by easy computation, or as an immediate consequence of (v)$'$. The uniqueness can be easily checked. The kernel of this homomorphism from $\mathrm{Sp}(2g, \mathbb{Z})$ to the group of (affine) isometries of $(1/2 \cdot \mathbb{Z}/\mathbb{Z})^{2g}$ is just $\Gamma(2)$ = the principal congruence subgroup of level-2. A final remark: the 1-cocycle δ_γ describes how the Riemann theta divisor (i.e. the zero-locus of $\Theta(\Omega, z)$) transforms under $\mathrm{Sp}(2g, \mathbb{Z})$. This is why elements of the group $(1/2 \cdot \mathbb{Z}/\mathbb{Z})^{2g}$ are called theta characteristics.

We first use the theta-series to describe the complex analytic stack associated to $N_{g,n}$ over the complex numbers. Let the group $\mathrm{Sp}(2g, \mathbb{Z})$ operate on the product $\mathbb{H}_g \times (1/2 \cdot \mathbb{Z}/\mathbb{Z})^{2g}$ as follows: On the first factor we have the usual action and on the second, we twist the action on $(1/2\cdot\mathbb{Z}/\mathbb{Z})^{2g}$ by δ_γ, as described above.

Now we can identify the complex analytic stack associated to N_g with the quotient of $\mathbb{H}_g \times (1/2 \cdot \mathbb{Z}/\mathbb{Z})^{2g}$ under $\mathrm{Sp}(2g, \mathbb{Z})$. A typical element of $\mathbb{H}_g \times (1/2 \cdot \mathbb{Z}/\mathbb{Z})^{2g}$ will be denoted $(\Omega, (x, y))$, where an element of $(1/2 \cdot \mathbb{Z}/\mathbb{Z})^{2g}$ is represented by two column vectors consisting of the first half and the second half of its entries respectively. A pair $(\Omega, (x, y))$ corresponds to the abelian variety $A(\Omega) = \mathbb{C}^g/(\mathbb{Z}^g \oplus \Omega\cdot\mathbb{Z}^g)$ with the zero-locus of $\theta(\Omega, z+\Omega x+y)$ as its symmetric theta-divisor. It follows from the definitions and the transformation rules (i) – (iii) for $\theta(\Omega, z)$ (or from (v)$'$) that $\mathrm{Sp}(2g, \mathbb{Z})$ operates on this analytic family of principally polarized abelian varieties with symmetric theta divisor. (Recall that $\gamma \in \mathrm{Sp}(2g, \mathbb{Z})$ sends (Ω, z) to $(A\Omega+B)\cdot(C\Omega+D)^{-1}$, ${}^t(C\Omega+D)^{-1}z)$.) Thus the $\mathrm{Sp}(2g, \mathbb{Z})$-quotient of $\mathbb{H}_g \times (1/2\cdot\mathbb{Z}/\mathbb{Z})^{2g}$ is indeed N_g^{an}. Since the stabilizer at $(0, 0)$ is $\Gamma(1, 2)$, $\mathrm{Sp}(2g, \mathbb{Z})\backslash\mathbb{H}_g \times (1/2\cdot\mathbb{Z}/\mathbb{Z})^{2g}$ contains $\Gamma(1, 2)\backslash\mathbb{H}_g$ as a connected component. But $\mathrm{Sp}(2g, \mathbb{Z})$ does *not* operate transitively on $(1/2\cdot\mathbb{Z}/\mathbb{Z})^{2g}$. In fact there are two orbits, distinguished by the value of the $(\mathbb{Z}/2\mathbb{Z})$-valued quadratic form $q({}^t({}^ta,\ {}^tb)) = 4 \cdot {}^ta \cdot b \bmod 2\mathbb{Z}$. (Classically, an element of $(1/2 \cdot \mathbb{Z}/\mathbb{Z})^{2g}$ is called an even (resp. odd) theta characteristic if the quadratic form has value 0 (resp. 1) mod $2\mathbb{Z}$ on it. There are $2^{g-1} \cdot (2^g + 1)$ even theta characteristics and $2^{g-1} \cdot (2^g - 1)$ odd ones.) So $\mathrm{Sp}(2g, \mathbb{Z}) \backslash \{\mathbb{H}_g \times (1/2 \cdot \mathbb{Z}/\mathbb{Z})^{2g}\}$ has exactly two connected components.

Similarly we obtain the complex analytic version of $N_{g,n}$ (n odd as always) by dividing out the principal congruence subgroup of level-n in $\mathrm{Sp}(2g, \mathbb{Z})$. $\Gamma(n)\backslash\mathbb{H}_g \times (1/2 \cdot \mathbb{Z}/\mathbb{Z})^{2g}$ has two connected components, one of which is $\Gamma(n, 2n)\backslash\mathbb{H}_g$, where $\Gamma(n, 2n) = \Gamma(n) \cap \Gamma(1, 2)$. The transformation rules above imply that for $m = 4n^2$ and $a, b \in (n^{-1}\mathbb{Z})^g$ with $a, b \equiv 0 \pmod 2$, the functions $(Z, x, y) \to \theta(Z, a + x, b + y)^{2m}$ define sections of $\omega^{\otimes m}$ over the complex analytic stack associated to $N_{g,n}$. Notice that $(Z, x, y) \to \Theta(Z, a + x, b + y)^{2m}$ depends only on a, b modulo $\mathbb{Z}$, although $\theta(Z, a + x, b + y)$ does not share this property. (We require $a, b \equiv 0 \pmod 2$ just to separate 2 from other odd primes.) Thus we get n^{2g} modular forms on $N^{\mathrm{an}}_{g,n}$ of weight $m = 4n^2$, one for each element of $(n^{-1}\mathbb{Z}/\mathbb{Z})^{2g}$; a typical one will be denoted $g(a, b)^{2m}$. Taking symmetric functions we obtain modular forms on $\mathsf{A}_{g,n}$ or A_g. (Since $N^{\mathrm{an}}_{g,n}$ has two connected components, it suffices to use one component.) The q-expansions (or better, the generalizations to $N_{g,n}$ of the q-expansions) of these modular forms at the standard cusps can be easily computed: By the formula above, $\theta(Z, a + x, b + y) = \sum_{\mu \in X} \exp\{2\pi\sqrt{-1}\,{}^t(b + y)(\mu + a + x)\} \cdot (\mu + a + x) \otimes (\mu + a + x)/2$, where $\exp\{2\pi\sqrt{-1}\,{}^t(b + y)(\mu + a + x)\} \in \mathbb{C}$ (remember that $X = \mathbb{Z}^g$), $(\mu + a + x) \otimes (\mu + a + x)/2 \in (4n^2)^{-1} \cdot S^2(X)$ (i.e. "the λ's"). To get the expansion at the other cusps, transform with $\mathrm{Sp}(2g, \mathbb{Z}/n\mathbb{Z})$. Then we just have to take the $2m$-th power of the right side, and form symmetric functions with respect to either orbit in $(1/2 \cdot \mathbb{Z}/2\mathbb{Z})^{2g}$. Note that all coefficients of the resulting q-expansions lie in the ring of integers of some cyclotomic field, so that these modular forms are not only algebraic, but even defined over a ring of cyclotomic integers, after inverting n. Let $\overline{N}_{g,n}$ denote the normalization of $\overline{\mathsf{A}}_{g,n}$ in $N_{g,n}$. It follows that the sections of $\omega^{\otimes m}$ ($m = 4n^2$) over $\overline{N}_{g,n}$ given by $\theta(Z, a, b)^{2m}$ are already defined over the ring of integers of some (cyclotomic) number field, again after inverting n.

We intend to compare the two sets of modular forms on $N_{g,n}$ (n odd as always) constructed above. Fix a pair (a, b) of elements of $(1/n\mathbb{Z}/\mathbb{Z})^g$. From the universal level-n structure we obtain a corresponding torsion section of the universal abelian scheme over $N_{g,n}$. Evaluation of the global section of $\mathcal{L}$ (unique up to units, which we can normalize) at these torsion sections gives an algebraically defined theta constant $f(a, b)$ such that for any multiple m of $4n$ $f(a, b)^{2m}$ is a global section of $\omega^{\otimes m}$ over $N_{g,n}$. We shall use $m = 4n^2$, as before. On the other hand the classical θ-series defines an analytically defined theta constant $g(a, b)$ such that over the integers of some (cyclotomic) number field, $g(a, b)^{2m}$ defines a global section of $\omega^{\otimes m}$ over $\overline{N}_{g,n}$ via the q-expansion described above. (We hope the reader will pardon this abuse of notation: $g(a, b)$ itself is not quite well defined; only its $8n^2$-th power is.)

3.2. Lemma. *Notation as above, if $(a,b) \neq (0,0)$, the section $g(a,b)^{2m}$ of $\omega^{\otimes m}$ divides $f(a,b)^{2m}$. (This makes sense over the ring of integers of a cyclotomic field over which $g(a,b)^{2m}$ is defined, after inverting n.)*

Proof. If $a \neq 0$ the λ's in the q-expansion of $\theta(Z, a+x, b+y) = \sum_{\mu\in X} \exp\{2\pi\cdot \sqrt{-1}\,{}^t(b+y)(\mu+a+x)\} \cdot (\mu+a+x) \otimes (\mu+a+x)/2$ are all different, so all coefficients in the q-expansion are roots of unity. If $a = 0$ we must pair the terms indexed by μ and $2x-\mu$. Up to a root of unity the resulting coefficient is $1+\exp(4\pi\sqrt{-1}\,{}^t(b+y)(x-\mu))$. The ideal generated by these coefficients, for various μ's, contains the elements $1-\exp(4\pi\sqrt{-1}\,{}^t(b+y)\mu)$, which for suitable choices of μ are of the form $1-\zeta$, ζ an n-th root of unity different from 1. It follows that regardless of the choice of (x,y) the coefficients in the q-expansion of $\theta(Z, a+x, b+y)$ generate the unit ideal after we invert n. The same is true if we take the $2m$-th power, and form the product over all (x,y) in an orbit in $(1/2\cdot\mathbb{Z}/\mathbb{Z})^{2g}$. The result is the q-expansion of the norm (for a connected component of $\overline{N}_{g,n}$) of $g(a,b)^{2m}$ over $\overline{\mathsf{A}}_{g,n}$, and as the geometric fibres of $\overline{\mathsf{A}}_{g,n}$ are irreducible it follows that the divisor of this norm contains no vertical components, that is, no irreducible components of fibres. The same must be true for the divisor of $g(a,b)^{2m}$. As it coincides with the divisor of $f(a,b)^{2m}$ in characteristic 0 (because over the complex numbers both divisors are given by the same special value theta series), the assertion follows.

3.3. Corollary. *Suppose $\overline{x}$ is a geometric point of $\overline{N}_{g,n}$ corresponding to an abelian variety A with a level-n structure and a symmetric Θ-divisor $\boldsymbol{\Theta} \subset A$. If $g(a,b)^{2m}$ vanishes at $\overline{x}$ and $a,b \in (n^{-1}\cdot\mathbb{Z}/\mathbb{Z})^g$ are not both zero, then the torsion point of A parametrized by (a,b) lies in $\boldsymbol{\Theta}$.*

Finally we need information about the behavior of $g(a,b)^{2m}$ at the boundary. For the q-expansion of $\theta(Z, a+x, b+y)$ things are easy: If we formally apply the Φ-operator at a standard cusp, from modular forms of genus g to genus $g-r$, we get 0 if $a+x$ is not contained in $((1/n)\cdot\mathbb{Z}/\mathbb{Z})^{g-r} \subset ((1/n)\cdot\mathbb{Z}/\mathbb{Z})^g$ (use the first $(g-r)$ entries) and otherwise the θ-series with the first $g-r$ components of $b+y$ respectively $a+x$ as parameters. Especially θ-series in genus r lie always in the image of the Φ-operator, a fact well known to experts in Siegel modular forms, and "most" of the θ-series vanish on the boundary. To transform these formal calculations into well-defined statements which can be proved, we have to form symmetric functions in powers $\theta(Z, a+x, b+y)^{2m}$, with x and y ranging over all pairs of elements in either orbit of $((1/2)\cdot\mathbb{Z}/\mathbb{Z})^g$. Under the Φ-operator we either get 0, or the corresponding symmetric function in dimension $g-r$, where we symmetrize not just over one set of θ's but over the disjoint union of several copies of it.

We are now ready to prove prop. 2.1 and 2.2. Recall the statements:

Proposition 2.1. *Let $\overline{\mathsf{A}}_g$ be a smooth arithmetic toroidal compactification of A_g. There exists an integer $m > 0$ such that $\omega^{\otimes m}$ is generated by its global sections over $\overline{\mathsf{A}}_g$.*

Proof. We may assume $g \geq 2$. By quasicompactness of $\overline{\mathsf{A}}_g$ it suffices to show that the intersection in $\overline{\mathsf{A}}_g$ of the base loci of all positive powers $\omega^{\otimes m}$ is void. So assume that there is a geometric point x in this intersection. We use the following construction to produce modular forms on $\overline{\mathsf{A}}_g$ from forms on $\overline{\mathsf{A}}_{g,n}$ or $\overline{N}_{g,n}$ (n odd): Denote the bigger space by X, and suppose that $f \in \Gamma(X, \omega^{\otimes m})$ is a modular form of weight m on X. Form the norm (relative to $\overline{\mathsf{A}}_g$) of the polynomial $T - f$. As $\overline{\mathsf{A}}_g$ is normal, its coefficients are modular forms of weight mk on $\overline{\mathsf{A}}_g$, whose values at x are equal to the elementary symmetric functions of the set of values of f on the fibre over x. If they all vanish at x then f vanishes on the whole fibre.

Assume that the fibre G_x of the universal semi-abelian scheme at x is an extension of a principally polarized abelian variety A of dimension $g - r$ by a torus T. On A choose a symmetric Θ-divisor, a level-n structure (n odd and invertible in x) and an n-torsion point $\neq 0$ not on this Θ-divisor (By ch. I, 5.3 this can be done if $n > 1$). For a suitable $m \geq 1$ we obtain a modular form $g'(a', b')^{2m}$ on $N_{g-r,n}$ which does not vanish at the geometric point $\tilde{x}$ defined by A (and the choice of level-n structure and theta divisor above), where (a', b') corresponds to the chosen n-torsion point. Furthermore $g(a', b')$ is the image under Φ (= Φ-operator corresponding to the boundary component containing $\tilde{x}$) of a theta constant $g(a, b)$ on $\overline{N}_{g,n}$. By symmetrizing (relative to $\overline{\mathsf{A}}_g$) the modular form $g(a, b)^{2m}$ we obtain modular forms on $\overline{\mathsf{A}}_g$ which do not all vanish on x: The value of a modular form at the boundary is determined by the constant term of its Fourier-Jacobi expansion. Applying the Φ-operator produces a symmetric function in the $g'(a', b')$'s, which do not all vanish in x by construction. □

2.2. Proposition. *Let C be a proper smooth irreducible curve over an algebraically closed field k, $f : G \to C$ a semi-abelian scheme over C. If $\deg(\omega_{G/C}) \leq 0$, then*

(i) $\omega_{G/C}$ is a torsion bundle (i.e. some positive power of it is trivial), hence $\deg(\omega_{G/C}) = 0$.

(ii) G is an extension of an isotrivial abelian scheme (over C) by a torus (over C).

Here we recall that an abelian scheme $A \to C$ is isotrivial if it becomes constant on a finite étale covering of C. Notice that if $A \times_C C'$ is constant for some proper smooth curve C' sitting over C, then A is isotrivial: For any $n \geq 3$ prime to char.(k), *$S := A[n]$ is finite étale over C and $A \times_C S \to S$ is constant because $A \times_C S \times_C C' \to S \times_C C'$ is.*

Proof. By ch. I, prop. 2.9 the closure of the torus part of the generic fibre of G is a torus over C, so that G is an extension, by a torus over C, of a semi-abelian scheme over C which is generically abelian. This reduces us immediately to the case that G is generically abelian, and we must show that G has good reduction everywhere and is isotrivial. By Zarhin's trick (for every abelian variety A, $A^4 \times (A^t)^4$ admits a principal polarization, see I, 5.5) we may assume that the generic fibre of G carries a principal polarization ($\omega_{G/C}$ and $\omega_{G^t/C}$ are isomorphic, by lemma 3.4 below). Note that we are free to replace C by finite (possibly ramified) coverings. (An abelian variety over a discretely valued field with semi-stable reduction and potentially good reduction actually has good reduction.) Doing this we may assume that there exists a symmetric cubical invertible $\mathcal{O}_G$-module $\mathcal{L}$ on G which defines the principal polarization of the generic fibre. Replacing C by finite coverings we can obtain level structures on the generic fibre, for any n prime to char.(k), and we shall take n odd and ≥ 3. The data $(G, \mathcal{L},$level-n structure) defines a morphism from C into $\overline{N}_{g,n}$ (n odd, ≥ 3), such that G is isomorphic to the pullback of the universal semi-abelian scheme there.

Here is the how the hypothesis is used: Modular forms on $\overline{N}_{g,n}$ pull back to sections of powers of $\omega_{G/C}$. By prop. 2.1 $\omega_{G/C}^{\otimes m}$ is generated by its global sections for some $m > 0$, so in particular $\deg(\omega_{G/C}^{\otimes m}) \geq 0$, while $\deg(\omega_{G/C}^{\otimes m}) \leq 0$ by assumption. Hence global sections of $\omega_{G/C}^{\otimes m}$ either vanish everywhere or nowhere. Therefore $\omega_{G/C}$ is torsion.

Now let us show that G has good reduction everywhere: Consider the global sections of $\omega_{G/C}^{\otimes m}$ of the form $g(a,b)^{2m}$, where $a, b \in (1/n\mathbb{Z}/\mathbb{Z})^g$, n a odd number invertible in k, m divisible enough that the theta constant $g(a,b)^{2m}$ is a modular form. At a point of bad reduction for G they all vanish, except for pairs (a,b) in a proper sublattice $(1/n\mathbb{Z}/\mathbb{Z})^{2g-1} \subset (1/n\mathbb{Z}/\mathbb{Z})^{2g}$. This follows from the q-expansions: Symmetrize $g(a+x, b+y)^{2m}$ over $x, y \in ((1/2)\cdot\mathbb{Z}/\mathbb{Z})^g$ to obtain modular forms on $\overline{\mathsf{A}}_{g,n}$. If the image of the point happens to lie in the standard boundary component this lattice is defined by setting the last (several) components in a equal to 0. Otherwise transform with a suitable element of $\mathrm{Sp}(2g, \mathbb{Z})$. The same must hold in all points of C since $\omega_{G/C}$ is torsion. Hence over the generic fibre of G all $g(a,b)^{2m}$ vanish except for those (a,b) contained in a proper sublattice $(1/n\mathbb{Z}/\mathbb{Z})^{2g-1} \subset (1/n\mathbb{Z}/\mathbb{Z})^{2g}$. In other words the Θ-divisor in the generic fibre of G contains all n-torsion points outside a proper sublattice. But for various n's these points are dense in the generic fibre, hence a contradiction. (Here is a short argument that the n-torsion points on the Θ-divisor are dense: Consider n's which are powers of a fixed odd prime l invertible in k. Let G_η be the generic fibre of G and T_l be its l-adic Tate module. There exists a hyperplane H in $T_l \otimes \mathbb{Q}_l$ such that the subset of l-power torsion points represented by $(T_l \otimes \mathbb{Q}_l) - H$ is contained in the Θ-divisor, since a projective limit indexed by $\mathbb{N}$ of non-empty finite sets is again non-empty. Denote by X the closure of these torsion points on Θ. Since $T_l \otimes \mathbb{Q}_l$ is the union of $(T_l \otimes \mathbb{Q}_l) - H$ and a suitable translate of it, there exists an l-power torsion point x such that $X \cup x + X$ contains all l^∞-torsion points, hence contains the abelian variety G_η. Thus $\dim(X) = \dim(G_\eta)$, so $X = G_\eta$.)

We have now shown that G has good reduction everywhere. We want to show that it is isotrivial. Let us normalize $\mathcal{L}$ so that its restriction to the 0-section is trivial. By the Riemann-Roch computation in I, th. 5.1, $f_*(\mathcal{L}^{\otimes m})$ has degree 0, for any $m > 0$, and its determinant is a torsion invertible sheaf on C. If we assume that G has a level-n structure (n odd and invertible in k) we may evaluate $f_*(\mathcal{L}^{\otimes m})$ at n-torsion points, thus (for big n) injecting this vector bundle into a direct sum of torsion line bundles. Passing over to a finite covering we may assume that these are trivial, so (for a fixed $m \geq 3$) $f_*(\mathcal{L}^{\otimes m})$ is a trivial vector bundle. For any positive multiple am of m, $f_*(\mathcal{L}^{\otimes\, am})$ is globally generated and has degree zero, so it is constant. As a projective scheme over C, G is the projective spectrum of the graded quasicoherent algebra $\oplus_{a\geq 0} f_*(\mathcal{L}^{\otimes\, am})$. We have shown that each graded piece of this algebra is a trivial free $\mathcal{O}_C$-module, therefore the multiplication of this graded algebra must be constant since $H^0(C, \mathcal{O}_C) = k$. Hence G is constant as a scheme, therefore constant as an abelian scheme, by rigidity of abelian schemes. □

3.4. Lemma. *Suppose $f : G \to S$ is a semi-abelian scheme over a normal base S, which is abelian in the generic points, and $G^t \to S$ its dual. Then there is a natural isomorphism $\omega_G \approx \omega_{G^t}$.*

Proof. If G is abelian then by Serre duality ω_G is dual to $R^g f_*(\mathcal{O}_G) = \det(R^1 f_*(\mathcal{O}_G))$, which is dual to ω_{G^t}. Thus we obtain an isomorphism $\rho_G : \omega_G \approx \omega_{G^t}$ over the open set of S where G is abelian, and we have to show that it extends. Let us also note that for an isogeny $\alpha : G \to G_1$, with dual $\alpha^t : G_1^t \to G^t$ we have $\deg(\alpha) \cdot \rho_{G_1} = \alpha^{t*} \cdot \rho_G \cdot \alpha^*$. For the assertion it now suffices to check at points of codimension one, so we may assume that S is the spectrum of a discrete valuation ring R. By descent we are free to replace R by extensions, so we may assume that R is complete, and that G is obtained by applying the Mumford construction to a split pair $(\tilde{G}, \imath : Y \to \tilde{G}(K))$. As usual we denote by A and T the abelian part and the maximal torus of $\tilde{G}$. Choosing polarizations gives an inclusion $Y \subset X$ and maps $c^t : Y \to A(R)$, $c : X \to A^t(R)$ such that $c|Y = \lambda_A \cdot c^t$. We can lift to $\tilde{c}^t : Y \to \tilde{G}(R)$, $\tilde{c} : X \to G(R)$ which define a trivialization of the pullback of the Poincaré sheaf, such that $i = \tilde{c}^t \cdot b$, where $b : Y \to T(K)$ is given by a bilinear form $b(y, \mu) = \pi^{\beta(y,\mu)}$ with a positive definite symmetric $\beta : Y \times X \to \mathbb{Z}$ (π is a uniformizing element of R). Now for n invertible in R, G is isogenous to the quotient of $\tilde{G}$ by $n \cdot \imath(Y)$, and by the isogeny formula above the two ρ's have poles (or zeros) of the same order along $\{\pi = 0\}$. We are finished if we can show that these orders differ by a factor n.

The pairs $(\tilde{c}, \tilde{c}^t)$ as above form a torsor (under a torus) over a group scheme of finite type over R. It follows that for some $n > 1$ invertible in R there exists a normal integral extension $R \subset R_1$ and over R_1 a $\tilde{c}_1^t : Y \to \tilde{G}(R_1)$ which, for suitable sections $R_1 \to R$, induces $\tilde{c}^t$ respectively $n \cdot \tilde{c}^t$. Now apply the Mumford construction to the pair $(\tilde{G}, \imath_1)$ over $R_1[\![T]\!]$, where $\imath_1$ is the product of $\tilde{c}_1^t$ and the map b_1 given by $b_1(y, \mu) = T^{\beta(y,\mu)}$. The corresponding isomorphism ρ_1 has some order r along the divisor $\{T = 0\}$. Under suitable pull-backs ρ_1 induces the isomorphisms ρ either for $G = \tilde{G}/\imath(Y)$ (sending T to π) or for $\tilde{G}/\imath(n \cdot Y)$ (sending T to π^n). But these have the same order of pole, so that $r = r \cdot n$, that is $r = 0$, and we are done.

4. Applications to Diophantine Geometry

The minimal compactification A_g^* together with the ample line bundle $\mathcal{L}$ on it (descended form of $\omega^{\otimes m}$) can be used to define heights for rational points of A_g. For some interesting applications (including the proof of the conjectures of Mordell, Tate and Shafarevich), we refer the reader to [F 2]. (See also [FW], [Sz 3], [ArG] for more detailed exposition and related topics.) There the compactification A_g^* was not known to exist, so some tricks had to be used (thm. 1, loc. cit.), which somewhat obscured the simplicity of the general picture. We use this occasion to simplify the considerations in [F 2].

First of all we need a hermitian metric on the line bundle $\mathcal{L}$ (a descended form of $\omega^{\otimes m}$) over the complex numbers. Over A_g itself there is a very natural hermitian metric on ω, namely square integration of holomorphic differentials.

4.1. Definition. If A is an abelian variety of dimension g over the field of complex numbers $\mathbb{C}$, and $\alpha, \beta \in \omega_A$ are two invariant differential g-forms, define their inner product by $(\alpha|\beta) = \sqrt{-1}^{\,g^2} \int_{A(\mathbb{C})} \alpha \wedge \overline{\beta}$. This is clearly invariant under all automorphisms of A and so defines a hermitian metric for ω on A_g and hence for $\mathcal{L}$ (which corresponds to ω^m) on $[\mathsf{A}_g]$ over $\mathbb{C}$. (In other words, the square of the norm of α is the volume of the abelian variety A w.r.t. the Haar measure determined by α.)

Recall also the notion of logarithmic singularity of a hermitian metric:

4.2. Definition. (formulated for complex analytic varieties) Let X be a complex analytic variety, U an open dense subset which is the complement of a analytic subvariety Z of X (regard Z as the boundary of $X - Z$). Let $\mathcal{L}$ be a complex analytic line bundle on X with a continuous hermitian metric on $\mathcal{L}|_{X-Z}$. We say that this metric has *logarithmic singularities* near Z if for any point $z \in Z$, there exists a neighborhood N of z, a local generator $\alpha \in \Gamma(N, \mathcal{L})$ of $\mathcal{L}$ over N and local generators $f_1, \ldots, f_t \in \Gamma(N, \mathcal{J}_Z)$ of the ideal of regular functions vanishing on Z such that for $x \in N$

$$\operatorname{Max}(\|\alpha(x)\|, \|\alpha(x)\|^{-1}) \leq C \cdot (\operatorname{Min}_i\{|\log|f_i(x)||\})^n$$

for suitable positive constants C and n.

4.3. Remark. This notation is invariant under pull-backs, and can be checked upstairs for proper morphisms, especially birational modifications (exercise).

If $\overline{X}$ is scheme proper over the ring of integers $\mathcal{O}_K$ of a number field K, X an open dense subscheme, and $\mathcal{L}$ is an invertible sheaf on $\overline{X}$ with hermitian metrics on X_v for all archimedian places v of K, then there is a notion of (logarithmic) height w.r.t. $\mathcal{L}$ for all $\overline{K}$-rational points of X, denoted by $h_{\mathcal{L}}$. We refer the reader to [L 2] for an extensive treatment of related basic diophantine geometry.

The interest of the notion of logarithmic singularity lies in the following:

4.4. Proposition. *Notation as above, if $\mathcal{L}$ is ample, $(\ |\ )_v$ is a hermitian metric on $\mathcal{L}|X_v$ for each archimedean prime v of K with logarithmic singularities along $(\overline{X}-X)_v$. Then*

(i) There exists a constant C such that $h_{\mathcal{L}}(x) \geq C\ \forall x \in X(\overline{K})$.

(ii) $\forall t \in \mathbb{R}$, $\forall d \in \mathbb{Q}$, the set of all points $x \in X(\overline{K})$ with $h_{\mathcal{L}}(x) \leq t$ and rational over an overfield of K with relative degree $\leq d$, is finite.

If $\overline{X} = X$, this is well-known. The general case can be reduced to this one; the point is that for any set of continuous hermitian metrics $(\ |\ )'_v$ on $\mathcal{L}|\overline{X}_v$ at each archimedean places v of K, there exist positive constants C_1, C_2 such that $|h_{\mathcal{L}}(x) - h'_{\mathcal{L}}(x)| \leq C_1 + C_2 \cdot \log(h'_{\mathcal{L}}(x))$ for all $x \in X(\overline{K})$, where $h'_{\mathcal{L}}$ denotes the logarithmic height computed using $(\ |\ )'_v$, (c. [F 2]). The reader can also take this as an exercise.

4.5. Proposition. *The hermitian metric on $\mathcal{L}|_{[A_g]/\mathbb{C}}$ defined in 4.1 has logarithmic singularities at the boundary.*

Proof. By remark 4.3 this can be checked on an arithmetic toroidal compactification, which is an easy calculation using the explicit local coordinates of the boundary. It is instructive to look at the totally degenerate case first. Near such a point p over the 0-dimensional cusp, the local coordinates come from $S^2(X) =$ characters of $S^2(X)^* \otimes_{\mathbb{Z}} G_m$. We can choose a $\mathbb{Z}$-basis $x_1, \ldots, x_d$ ($d = g\cdot(g+1)/2$) of $S^2(X)$ so that they give local coordinates at p, and the boundary is defined by their product (some of them may be non-zero at p). The tautological bilinear form is given by $b(\mu,\nu) = \prod_{1\leq j\leq d} x_j^{s_j(\mu,\nu)}$, where the s_j's are the dual basis of x_j's. It is elementary that the covolume of the lattice in $\mathbb{C}^g = X \otimes_{\mathbb{Z}} \mathbb{C}$ generated by $e_1, \ldots, e_g$ and the columns of the period matrix Ω is $|\det(\mathrm{Im}(\Omega))|$. In terms of the local coordinates x_j's, $\mathrm{Im}(\Omega)$ translates into the bilinear form $(2\pi)^{-1} \cdot \sum -\log(|x_j|) \cdot s_j$ on X. Its determinant is clearly bounded below by a positive constant and bounded above up to constant by $(-\log|x_1 \cdots x_d|)^g$ near the point p, which is of the required form since the square of the norm of $(d\chi_1/\chi_1) \wedge \cdots \wedge (d\chi_g/\chi_g)$ is $(2\pi)^{2g}\|dz_1 \wedge \cdots \wedge dz_g\|^2$, where (χ_i) is a basis of X and (z_i) the corresponding linear coordinates.

The general case is not more difficult: near a point p of $\overline{\mathsf{A}}_g$ over a boundary component of genus $g-r$, we find local coordinates $x_1, \ldots, x_a$ ($a = r\cdot(r+1)/2$) and $y_1, \ldots, y_b$ ($b = d-a$) such that the boundary is defined by $\prod x_j = 0$ (again some of them may be non-zero). The universal semi-abelian variety is the quotient of a $\tilde{G}$ sitting in an extension $0 \to T \to \tilde{G} \to A \to 0$, by a subgroup of periods. The period subgroup is the image of $Y \cong \mathbb{Z}^r$ under a map $\imath$, which after integrally trivializing the pull-back of the Poincaré biextension via $c \times c$ is given by a bilinear form b. The bilinear form b is of the form $b(\mu,\nu) = \prod x_j^{s_j(\mu,\nu)}$ up to units, where the s_j are positive semi-definite symmetric bilinear forms on a rank-r quotient X_1 of X, which form a $\mathbb{Z}$-basis of $B(X_1)$.

The square of the norm of a local generator α of ω near p is equal to the volume of G, measured by integration of $\pi^*(\alpha) \wedge \overline{\pi}^*(\overline{\alpha})$, where $\pi^*(\alpha)$ is the pull-back of α to $\tilde{G}$. Up to insignificant positive and bounded factors one obtains the determinant of the matrix of the bilinear form $\sum -\log(|x_j|)\cdot s_j$ (e.g. use C^∞-deformation to the case where $c : X \to A$ is trivial). The proposition follows.

Now we know that $\mathcal{L}$ with its metric can be used to measure heights. What do we get for the rational point of A_g^* defined by a principally polarized abelian variety over a number-field K? By definition we have to compute the degree of the pull-back of $\mathcal{L}$ to $\mathrm{Spec}(\mathcal{O}_K)$, with its metric. Assume that the abelian variety has semistable reduction, so extends to a semi-abelian scheme G over $\mathrm{Spec}(\mathcal{O}_K)$. The projective $\mathcal{O}$-module ω_G has hermitian metrics at the infinite places, by square-integration, and thus a degree. We claim that the height of the corresponding rational point of A_g^* is equal to this degree, multiplied by $m/[K:\mathbb{Q}]$. For G is the pullback of the universal semi-abelian variety by the morphism $\mathrm{Spec}(\mathcal{O}) \to \overline{\mathsf{A}}_g$. It follows that the pullback of $\mathcal{L}$ to $\mathrm{Spec}(\mathcal{O}_K)$ is isomorphic to $\omega_G^{\otimes m}$, as a metrized line-bundle, since this holds over $\overline{\mathsf{A}}_g$. The assertion follows.

For any abelian variety A over a number field F, we define the *modular height* $h_{\mathrm{mod}}(A_{/F})$ as $[K:\mathbb{Q}]^{-1}\deg(\omega_{A/F})$, where $\omega_{A/F}$ is top exterior product of the dual of the relative Lie algebra of the Néron model of $A_{/K}$, the metric at archimedian places being given by square-integration as in 4.1. Note that this does not depend on the choice of a polarization on A. Also the isomorphism in corollary 3.4. is an isometry, so the dual abelian variety A^t has the same height as A. By Zarhin's trick $A^4 \times A^{t^4}$ always has a principal polarization, and if A has semi-stable reduction its modular height is eight times the modular height of A. We have seen that m^{-1} times the height of the K-rationally point of A_g^* w.r.t. $\mathcal{L}$ is equal to $h_{\mathrm{mod}}(A_{/K})$ for any overfield K of F where A acquires semi-stable reduction. We shall call it the geometric height of A and denoted $h_{\mathrm{geom}}(A)$. Since we have a canonical morphism $\omega_{A/K} \to \omega_{A/F} \otimes_{\mathcal{O}_F} \mathcal{O}_K$ which is an isometry at archimedian places for any field extension K of F, the modular height of A will only decrease when passing to bigger number fields and stabilize once semi-stability is reached. From this discussion and 4.5 we easily deduce

4.6. Proposition. *(i) For any positive integer g, there exists a constant C such that $h_{\mathrm{mod}}(A_{/F}) \geq h_{\mathrm{geom}}(A_{/F}) \geq C$ for any principally polarized abelian variety $A_{/F}$ of dimension g defined over a number field F.*

(ii) $\forall g \in \mathbb{N}$, $\forall t \in \mathbb{R}$, $\forall d \in \mathbb{Q}$, the set of isomorphism classed of principally polarized abelian varieties of dimension g defined over a number field K with $[K:\mathbb{Q}] \leq d$, whose modular height is $\leq t$, is finite. The same holds for abelian varieties (forgetting polarizations).

The analogy of global function fields with number fields is by now well-known. In the global function field case height is given by the degree of invertible sheaves over the corresponding algebraic curve. Thus it is purely geometric; hermitian metrics are not needed. Often a statement in the function field case is easier to prove than its number field analogue. Let us illustrate this for a theorem of Zarhin, whose number field analogue is a major result in [F 2]:

4.7. Theorem (Zarhin). *Let C be a smooth geometrically connected curve over a finite field k, $K = k(C)$ its function field, A an abelian variety over K, $\pi = \mathrm{Gal}(\overline{K}/K)$, and l a prime invertible in k. Then $T_l(A) \otimes_{\mathbb{Z}_l} \mathbb{Q}_l$ is a semisimple π-module, and the natural map $\mathrm{End}_K(A) \otimes_{\mathbb{Z}} \mathbb{Z}_l \to \mathrm{End}_\pi(T_l(A))$ is an isomorphism.*

Sketch of Proof. We shall show that any π-invariant subspace W of $T_l(A) \otimes_{\mathbb{Z}_l} \mathbb{Q}_l$ is the image of an idempotent in $\mathrm{End}_K(A) \otimes_{\mathbb{Z}} \mathbb{Q}_l$. For this we may replace A by $A^4 \times A^{t^4}$, which has a principal polarization, and W by the maximal isotropic subspace $W^4 \times W^{\perp 4}$. (The diagonal components of a projector from $U \times V$ to a product subspace $U' \times V'$ are projectors from U to U' and V to V' respectively.) We may also assume that A has a level-n structure for some $n \geq 3$ prime to l and $p = \mathrm{char.}(k)$, so that it extends to a semi-abelian scheme G over C. If we now divide A by the stages of the l-divisible group defined by W, we obtain a sequence A_m of principally polarized abelian varieties over K, which extend to semi-abelian varieties G_m over C. The isogenies $G \to G_m$ are étale, so $\omega_{G_m} \cong \omega_G$, hence the degree of the pullback of the ample invertible sheaf ω on $\mathsf{A}^*_{g,n}$, under the morphism $C \to \mathsf{A}^*_{g,n}$ classifying A_m, is bounded (in fact constant). As these mappings correspond to k-rational points in a k-scheme of finite type (contained in $\mathbf{Morph}(C, \mathsf{A}^*_{g,n})$), there is only a finite number of them, so infinitely many of the A_m's are isomorphic. From the resulting endomorphisms of A one easily constructs an idempotent in $\mathrm{End}_K(A) \otimes_{\mathbb{Z}_l} \mathbb{Q}_l$ with the required properties.

4.8. Corollary. *Suppose C is a smooth complete curve over an algebraically closed field k, $K = k(C)$ its function field, $\pi^0 = \mathrm{Gal}(\overline{K}/K \cdot \overline{k})$, and A an abelian variety over K. Then the action of π^0 in $T_l(A) \otimes_{\mathbb{Z}_l} \mathbb{Q}_l$ (l invertible in k) is semisimple. If it is trivial, then A extends to an abelian scheme over C and is either constant (if k has characteristic 0) or at least isogenous of p-power order (if k has positive characteristic p) to a constant abelian scheme.*

Proof. Let us first show that the π^0-action is semisimple. One way is to use the theory of pure sheaves ([D 5]). We give a more elementary approach. If k is the algebraic closure of a finite field $\mathbb{F}$, then C is defined over some finite extension of $\mathbb{F}$, and A can be descended to the function field $\mathbb{F}'(C)$ of an $\mathbb{F}'$-form of C for some sufficiently large finite extension $\mathbb{F}'$ of $\mathbb{F}$. Since the geometric Galois group π^0 is a normal subgroup of the arithmetic Galois group $\mathrm{Gal}(\overline{F}'(C)/\mathbb{F}'(C))$, π acts semisimply on $T_l(A) \otimes_{\mathbb{Z}_l} \mathbb{Q}_l$ by Zarhin's theorem, and everything follows.

In general, since C (resp. A) is finitely presented over k (resp. K), by [EGA IV] §8, there exists a subalgebra $R \subset k$, of finite type over $\mathbb{Z}$ with quotient field L, an open subscheme U of $\mathrm{Spec}(R)$, a smooth curve $C_U \to U$, a closed subscheme Z of C_U finite étale over U which is the disjoint union of copies of $U : Z = x_{1/U} \amalg \cdots \amalg x_{r/U}$, and an abelian scheme A_T over $T = C_U - Z$, such that if we make base change from U to $\mathrm{Spec}(k)$ via the inclusion $R \subset k$ we recover $C = C_U \times_U \mathrm{Spec}(k)$ and an abelian scheme over $T_k = T \times_U \mathrm{Spec}(k)$

extending $A_k = A$. Furthermore we may assume that A_T has a level-l structure, so that the action of $\pi_1(T)$ on $T_l(A_T)$ factors through the tame fundamental group of $T = C_U$. As formation of this tame fundamental group commutes with base change, we may apply Zarhin's theorem to the geometric fibres of C_U over closed points of U to conclude semisimplicity.

Now assume that π^0 operates trivially on $T_l(A)$. By the criterion of Neron-Ogg-Shafarevich, A extends to an abelian scheme A over C. Choose a symmetric relatively ample line bundle $\mathcal{L}$ on A, trivialized along the unit section. By the theory of Hilbert schemes there exists a projective variety $\mathbf{Sect}^0(A/C)$ which classifies sections $s : S \to A$ such that $\deg(s^*(\mathcal{L})) = 0$. By the theory of Neron-Tate heights $\mathbf{Sect}^0(A/C)$ is a group-scheme. Now $A[l^\infty]$ is a constant l-divisible group over C, so it (or better its constant value) maps into $\mathbf{Sect}^0(A/C)$, and in fact it is isomorphic to the l-torsion in $\mathbf{Sect}^0(A/C)$. Let $B \subset \mathbf{Sect}^0(A/C)$ denote the closure of its image. Then B is geometrically reduced, thus smooth, and one sees easily that it is an abelian variety over k. The evaluation map $B \times_k C \to A$ induces an isomorphism on l-torsion, so it is an isogeny. Once we know this for one l, it follows that for all $l \neq p$ the π-operation on $T_l(A)$ is trivial, and we deduce that $\mathbf{Sect}^0(A/C)$ is an extension of a finite group scheme of p-power order (trivial if k has characteristic 0) by an abelian variety B such that $B \times_k C \to A$ is an isogeny of p-power order.

4.9. Corollary. *Suppose k is a field, K a finitely generated separable extension such that k is algebraically closed in K, G a semi-abelian variety over K such that $\pi^0 = \mathrm{Gal}(\overline{K}/K \cdot \overline{k})$ operates trivially on $T_l(G)$ (l a prime invertible in k). Then either G is defined over k (if k has characteristic 0) or is at least p-isogenous (if k has positive characteristic p) to a semi-abelian variety defined over k.*

For abelian varieties this follows by an easy induction over the transcendence degree of K/k. In general G is an extension $T \to G \to A$, and it follows that T is constant and A is p-isogenous to a constant abelian variety (respectively constant, if k has characteristic 0). We thus may assume that both are constant. It remains to show that the class (in the abelian variety $\mathbf{Ext}^1(A,T)$) of the extension is constant, and for this we may assume that k is algebraically closed. As $T_l(G)$ is constant the class is infinitely l-divisible and the assertion follows, either by the Mordell-Weil theorem or like this:

We may assume that K is the function field of a complete curve C over k. The extension-class gives an infinitely l-divisible map $C \to \mathbf{Ext}^1(A,T)$, so the pullback by it of a symmetric ample line bundle by such a map must have degree 0, so the map must be constant.

Later in this chapter we need the following information about abelian schemes A over $C = \mathbb{P}^1_{/k}$. We know that the π^0-operation on $T_l(A)$ is trivial, hence:

(i) If k has characteristic 0, they are constant:

(ii) in char.$(k) = p > 0$, there are non-constant families. Here is an example due to L. Moret-Baily (c. [Sz 1] exposé 8): Suppose E is a supersingular elliptic curve over an algebraically closed field k of characteristic p, so that E contains a copy of α_p. Over $\mathbb{P}^1_{/k}$, divide the constant abelian scheme $E \times E$ by the (non-constant) subgroup α_p of $\alpha_p \times \alpha_p$ corresponding to $\mathcal{O}(-1) \subset \mathcal{O}^2$. The resulting abelian variety A has $\mathbf{t}_A^* \cong \mathcal{O}(-1) \oplus \mathcal{O}(p)$, so it cannot be constant.

5. Projectivity of $\overline{\mathsf{A}}_{g,n}$

In general, for $n \geq 3$, the arithmetic toroidal compactification $\overline{\mathsf{A}}_{g,n}$ is an algebraic space, but not a scheme. Although in most applications the question whether it is a scheme does not matter, it is still interesting to be able to exhibit an ample invertible sheaf on it (when possible). The same problem for toroidal compactification of general arithmetic quotients of bounded symmetric domains was solved by Y.-S. Tai (c. [AMRT] chap. 4). The point is that if the admissible polyhedral cone decomposition satisfies a certain convexity condition, then the corresponding toroidal compactification is a projective algebraic variety. A projective embedding is given by classical automorphic forms (or, pluricanonical forms) vanishing along the boundary to order at least d for some d, where the order of vanishing is suitably defined. In fact Tai proved a slightly stronger result, to the effect that the toroidal compactification is the normalization of the blowing-up of the Baily-Borel-Satake compactification along the ideal of functions vanishing along the boundary to order at least d. We shall show in this section that the same proof works for the arithmetic toroidal compactification $\overline{\mathsf{A}}_{g,n}$ as well, if the convexity condition is satisfied. In particular $\overline{\mathsf{A}}_{g,n}$ is a scheme in this case. Perhaps the real interest of this result is that we can explicitly reverse the process of blowing down $\overline{\mathsf{A}}_{g,n}$ by finding an ideal of blowing-up.

First we have to explain the crucial convexity condition on an admissible polyhedral cone decomposition. Let us recall definition 2.4 of chapter IV:

5.1. Definition. Let X be a free abelian group of finite rank and $\mathfrak{C} = \{\sigma_\alpha\}_{\alpha \in J}$ a $\mathrm{GL}(X)$-admissible polyhedral cone decomposition of $C(X) \subset B(X) \otimes_{\mathbb{Z}} \mathbb{R}$. An *(invariant) polarization function* on $C(X)$ for the cone decomposition $\{\sigma_\alpha\}$ is a $\mathrm{GL}(X)$-invariant continuous piecewise linear function $\phi : C(X) \to \mathbb{R}_{\geq 0}$ such that

(i) $\phi(x) > 0$ for $x \neq 0$, $\phi(t \cdot x) = t\phi(x)$ $\forall x \in C(X)$ $\forall t \in \mathbb{R}_{\geq 0}$,

(ii) ϕ is convex in the sense that $\phi(x + y) \geq \phi(x) + \phi(y)$ $\forall x, y \in C(X)$. (This is what is usually called a concave function in calculus, but in the torus embedding game this seems to be the accepted terminology.)

(iii) ϕ is linear on each cone σ_α (i.e. ϕ coincides with a linear function on σ_α).

(iv) Any polyhedral cone in $C(X)$ on which ϕ is the restriction of a linear function is contained in some cone σ_α.

(v) ϕ takes integral values on $C(X) \cap B(X)$.

The admissible cone decomposition $\{\sigma_\alpha\}$ is said to be *projective* if such a polarization function exists.

We collect here some relevant information about definition 5.1. The reader can look at [AMRT] chap. 2 for proofs of statements below. (As Looijenga has pointed out, prop. 6, p. 132 of loc. cit. is incorrect, which affects the subsequent contents of chap. 2, but all the main results can be fixed with care.)

(a) By the theory of cores and cocores developed in loc. cit. (refinement of classical reduction theory), there exist GL(X)-admissible polyhedral cone decompositions of $C(X)$. In fact, if ϕ is a GL(X)-invariant strictly convex order function, $\{x \in \overline{C(X)} | \phi(x) \geq 1\}$ is what is called a "co-core" (p.120, loc. cit.), which we shall denote by K_ϕ. It is a convex subset $\overline{C(X)}$ not containing $\{0\}$ such that $\mathbb{R}_{\geq 1}\, K_\phi = K_\phi$, $\mathbb{R}_{\geq 0}\, K_\phi = \overline{C(X)}$. The dual of K_ϕ, defined as $\{\lambda \in S^2(X) \otimes_{\mathbb{Z}} \mathbb{R} | \langle \lambda, x\rangle \geq 1\ \forall x \in K_\phi\}$ and denoted $K_\phi^\vee$, is a convex subset in $C(X^*)^\circ$ (= interior of $C(X^*)$) satisfying $\mathbb{R}_{\geq 1}\, K_\phi^\vee = K_\phi^\vee$, $\mathbb{R}_{>0}\, K_\phi^\vee = C(X^*)^\circ$.

(b) The admissible polyhedral cone decomposition $\{\sigma_\alpha\}$ can be recovered from K_ϕ as the cones over the faces of K_ϕ.

(c) Given a projective GL(X)-admissible polyhedral cone decomposition $\mathfrak{C} = \{\sigma_\alpha\}_{\alpha \in J}$ and a strictly convex GL(X)-invariant polarization function ϕ as above, one can find a subdivision $\mathfrak{C}' = \{\sigma_\nu\}_{\nu \in J'}$ of $\{\sigma_\alpha\}$ which is smooth (w.r.t. the lattice $B(X)$) and projective. In fact, a polarization function for $\mathfrak{C}$ can be obtained by first slightly perturbing ϕ to ϕ' so that conditions (i) – (iv) of 4.1.1 are still satisfied (with $\{\sigma_\alpha\}_{\alpha \in J}$ replaced by $\{\sigma_\nu\}_{\nu \in J'}$), and so that the elements of $\phi'(B(X) \cap C(X)) \subset \mathbb{Q}$ are rational (and hence of bounded denominators). Then we clear the denominators. This statement is an easy corollary of results in chap. 3 of [KKMS].

(d) Let $\mathfrak{C}$, ϕ be as in 5.1. Then each top-dimensional cone σ_α in $\mathfrak{C}$ corresponds to a vertex λ_α of $K_\phi^\vee$, namely the linear form whose restriction to σ coincides with the restriction of ϕ to σ. Every vertex of $K_\phi^\vee$ corresponds to a top-dimensional cone in this way.

(e) Let $\mathfrak{C}$, ϕ be as in 5.1 and X_ξ a free quotient of X. Regarding $C(X_\xi)$ as naturally embedded in $C(X)$ as a rational boundary component, the restriction of ϕ to $C(X_\xi)$ is again a strictly convex support function for the GL(X)-admissible polyhedral cone decomposition of $C(X_\xi)$ induced by $\mathfrak{C}$.

Let $\mathfrak{C}$, ϕ be as in definition 5.1. For any vertex λ_α of $K_\phi^\vee$, let σ_α be the corresponding top-dimensional cone. We shall isolate some easy combinatorial lemmas which are important for our proof of projectivity. To avoid confusion, every cone in $C(X)$ is open in its closure in $B(X) \otimes_{\mathbb{Z}} \mathbb{R}$. The closure of a cone σ will be denoted by $\overline{\sigma}$. The dual cone of σ is defined as $\{\lambda \in S^2(X) \otimes_{\mathbb{Z}} \mathbb{R} | \langle \lambda, x\rangle \geq 0\ \forall x \in \sigma\}$ and denoted $\sigma^\vee$; in particular, it is closed in $S^2(X) \otimes_{\mathbb{Z}} \mathbb{R}$.

5.2. Lemma. *If λ_α, λ_β correspond to different top-dimensional cones $\sigma_\alpha, \sigma_\beta$, then $\langle \lambda_\alpha, x\rangle \leq \langle \lambda_\beta, x\rangle\ \forall x \in \overline{\sigma}_\alpha$, with equality only if $x \in \overline{\sigma}_\alpha \cap \overline{\sigma}_\beta$.*

Proof. Let $x \in \overline{\sigma}_\alpha$, and take $y \in \sigma_\beta$. Then $\exists \varepsilon > 0$ s.t. $y + \varepsilon x \in \sigma_\beta$. So $\lambda_\beta(y) + \varepsilon\lambda_\beta(x) = \lambda_\beta(y+\varepsilon x) = \phi(y+\varepsilon x) \geq \phi(y) + \varepsilon\phi(x) = \lambda_\beta(y) + \varepsilon\lambda_\alpha(x)$, and therefore $\langle \lambda_\alpha, x\rangle \leq \langle \lambda_\beta, x\rangle$. If equality holds, then $\phi(y+\varepsilon x) \geq \phi(y) + \varepsilon\phi(x)$, so ϕ is linear on the cone generated by y and x, which implies that $x \in \overline{\sigma}_\beta$ by strict convexity.

5.3. Lemma. *For any integer $n > 1$ there exists an integer $m > 1$ divisible by some power of n such that*
$\langle \lambda_\alpha, x\rangle < \langle \gamma \cdot \lambda_\alpha, x\rangle \; \forall x \in \overline{\sigma}_\alpha \cap C(X)^\circ,\; \forall \gamma \in \mathrm{GL}(X)$ *s.t.* $\gamma \equiv 1 \pmod m$
for any vertex λ_α of $K_\phi^\vee$.

Proof. We always have $\langle \lambda_\alpha, x\rangle \leq \langle \gamma\cdot\lambda_\alpha, x\rangle \; \forall x \in \overline{\sigma}_\alpha$ by 5.2. Taking $m \geq 3$, we have $\lambda_\alpha \neq \gamma\cdot\lambda_\alpha$ if $\gamma \equiv 1 \pmod m$, therefore if $x \in \sigma_\alpha \cap C(X)^\circ$ and $\langle \lambda_\alpha, x\rangle = \langle \gamma\cdot\lambda_\alpha, x\rangle$, by 5.2.0 we must have $x \in \overline{\sigma}_\beta$, where σ_β corresponds to $\gamma\cdot\lambda_\alpha$. Therefore $x \in \overline{\sigma}_\alpha \cap \gamma\cdot\overline{\sigma}_\beta$.

Now we recall (loc. cit. chap. 2 §4.3 thm. 1, p.113) that the closure of Siegel sets and closure of rational polyhedral cones in $C(X)$ are cofinal, therefore $\tilde{\sigma} = \overline{\sigma} \cap C(X)^\circ$ satisfies the Siegel property for each rational polyhedral cone σ in $C(X)$. In other words there are only finitely many $\gamma \in \mathrm{GL}(X)$ s.t. $\gamma\cdot\tilde{\sigma} \cap \tilde{\sigma} \neq \emptyset$. Since there are only finitely many $\mathrm{GL}(X)$-orbits in $\mathfrak{C}$, there exists $m | n^\infty$, $m > 0$, such that $\gamma\cdot\sigma_\alpha \cap \sigma_\alpha \cap C(X)^\circ = \emptyset$ for any σ_α in $\mathfrak{C}$ if $\gamma \equiv 1 \pmod m$. Combined with the above paragraph, the lemma follows.

5.4. Lemma. *Let σ_0 be a top-dimensional cone in $\mathfrak{C}$, $\sigma_1, \dots, \sigma_k$ the top-dimensional cones in $\mathfrak{C}$ having a common (positive-dimensional) face with σ_0. Let $\lambda_0, \lambda_1, \dots, \lambda_k$ be the vertices of $K_\phi^\vee$ corresponding to $\sigma_0, \sigma_1, \dots, \sigma_k$ (so that $\sigma_1, \dots, \sigma_k$ are the neighboring vertices of σ_0). Then the dual cone $\sigma_0^\vee$ of σ_0 is spanned by $\lambda_1 - \lambda_0, \dots, \lambda_k - \lambda_0$: $\sigma_0^\vee = \mathbb{R}_{\geq 0}(\lambda_1 - \lambda_0) + \cdots + \mathbb{R}_{\geq 0}(\lambda_k - \lambda_0)$.*

Proof. Let $\sigma'^\vee = \mathbb{R}_{\geq 0}(\lambda_1 - \lambda_0) + \cdots + \mathbb{R}_{\geq 0}(\lambda_k - \lambda_0)$ and $\overline{\sigma}'$ be the closed cone in $B(X) \otimes_{\mathbb{Z}} \mathbb{R}$ dual to it. By lemma 5.2, $\sigma'^\vee \subseteq \sigma_0^\vee$, i.e. $\overline{\sigma}_0 \subseteq \overline{\sigma}'$. We must show the opposite inclusion. There are two key facts coming from the geometry of the convex body $K_\phi^\vee$: (i) $K_\phi^\vee \subseteq \mathbb{R}_{\geq 0}(\lambda_1 - \lambda_0) + \ldots + \mathbb{R}_{\geq 0}(\lambda_k - \lambda_0)$, (ii) $\sum_{1 \leq \nu \leq k} \mathbb{R}_{\geq 0}(\lambda_\nu - \lambda_0) = \sum \mathbb{R}_{\geq 0}(\lambda_\alpha - \lambda_0)$, where λ_α runs through all vertices of $K_\phi^\vee$ in the second summation. From (i) it follows that $\sigma'^\vee$ contains the positive cone in $S^2(X) \otimes_{\mathbb{Z}} \mathbb{R}$, hence $\overline{\sigma}_0 \subseteq \overline{\sigma}' \subseteq \overline{C}(X)$. For any $x \in \sigma' =$ interior of $\overline{\sigma}' \subset C(X)^\circ$, we have $\langle \lambda_\nu, x\rangle > \langle \lambda_0, x\rangle$, $\nu = 1, \dots, k$ by definition, hence $\langle \lambda_\alpha, x\rangle > \langle \lambda_0, x\rangle$ for all vertices λ_α of $K_\phi^\vee$ by (ii). Since $x \in C(X)$, we must have $x \in \sigma_0$ by lemma 5.2. This shows the opposite inclusion $\sigma' \subseteq \sigma_0$.

5.5. Lemma (Generalization of 5.4). *Let σ_0 be a (not necessarily top-dimensional) cone in Σ, $\sigma_1, \dots, \sigma_m$ the top-dimensional cones in $\mathfrak{C}$ containing σ_0, and $\sigma_{m+1}, \dots, \sigma_l$ the neighboring top-dimensional cones of σ_1. Let $\lambda_0, \lambda_1, \dots, \lambda_l$ be the vertices of $K_\phi^\vee$ corresponding to $\sigma_0, \sigma_1, \dots, \sigma_l$. Then $\sigma_0^\vee = \mathbb{R}(\lambda_2 - \lambda_1) +$*

$\cdots + \mathbb{R}(\lambda_m - \lambda_1) + \mathbb{R}_{\geq 0}(\lambda_{m+1} - \lambda_1) + \cdots + \mathbb{R}_{\geq 0}(\lambda_l - \lambda_1)$. *(An invariant formulation is* $\sigma_0^\vee = \sum_{1 \leq i,j \leq m} \mathbb{R}(\lambda_i - \lambda_j) + \sum_{1 \leq i \leq m, m+1 \leq \nu \leq l} \mathbb{R}(\lambda_\nu - \lambda_i)$.*)*

Proof. The key observation is that the subspace of $S^2(X) \otimes_{\mathbb{Z}} \mathbb{R}$ consisting of elements vanishing on σ_0 is generated by $\lambda_i - \lambda_j$, $1 \leq i, j \leq m$. 5.5 now follows from 5.4.

Remark. The integral versions of 5.4 and 5.5 are false: in general, we cannot simply replace $\mathbb{R}$ by $\mathbb{Z}$ and get all lattice points of $\sigma_0^\vee$. The $\mathfrak{C}$ for which we can are rather special. For such $\mathfrak{C}$, we do not have to take normalization in the statement of the main theorem, when the level is sufficiently high.

From now on, we assume $\mathfrak{C} = \{\sigma_\alpha\}$ *to be a smooth* GL(X)*-admissible polyhedral cone decomposition of* $C(X)$, rank$(X) = g \geq 2$, ϕ *be a* GL(X)*-invariant polarization function for* Σ. $\overline{\mathsf{A}}_{g,n}$ will denote the associated arithmetic toroidal compactification. For convenience, we shall denote by $K(n)$ the kernel of $\mathrm{GL}(X) \to \mathrm{GL}(X/nX)$.

5.6. Definition. $\forall n \geq 1$, let ${}_n\jmath$ be the invertible coherent sheaf of ideals on $\overline{\mathsf{A}}_{g,n}$ supported on the divisor at infinity $D_\infty = \overline{\mathsf{A}}_{g,n} - \mathsf{A}_{g,n}$ such that for any irreducible component D_α of D_∞ corresponding to a $\mathrm{GL}(g,\mathbb{Z})(n)$-orbit $\mathrm{GL}(g,\mathbb{Z})(n) \cdot \sigma_\alpha$ of 1-dimensional cones (i.e. open half-lines) in $\mathfrak{C}$, the order of ${}_n\jmath$ along D_α is the value of ϕ at the $\mathbb{Z}$-generator of $\sigma_\alpha \cap n \cdot B(X)$. (This value is clearly well-defined because ϕ is GL(X)-invariant.) We shall simply write $\jmath$ if there is no confusion.

5.7. Definition. $\forall d \geq 1$, let ${}_n\mathcal{J}_d = \pi_{n*}({}_n\jmath^d)$, where $\pi_n : \overline{\mathsf{A}}_{g,n} \to \mathsf{A}_{g,n}^*$ is the natural morphism. ${}_n\mathcal{J}_d$ is a coherent sheaf of ideals of $\mathsf{A}_{g,n}^*$. We shall simply write $\mathcal{J}_d$ if there is no confusion.

5.8. Theorem. $\forall n \geq 3$, *there exists an integer* $d \geq 1$ *such that*

(i) $\overline{\mathsf{A}}_{g,n}$ *is (naturally isomorphic to) the normalization of the blowing-up of* $\mathsf{A}_{g,n}^*$ *along the coherent sheaf of ideals* $\mathcal{J}_d$.

(ii) $\jmath^d = \pi_n^*(\mathcal{J}_d)$.

Reduction to the Case of High Level Structure. The proof of theorem 5.8 consists of two parts. When the level structure is high enough in the sense that $\langle \lambda_\alpha, x \rangle < \langle \gamma \cdot \lambda_\alpha, x \rangle$ $\forall x \in \overline{\sigma}_\alpha \cap C(X)^\circ$, $\forall \gamma \in \mathrm{GL}(X)$ s.t. $\gamma \equiv 1 \pmod{n}$ for any vertex λ_α of $K_\phi^\vee$, theorem 5.8 will be proved directly by local calculations, starting with the explicit description of the completed strict local rings of $\mathsf{A}_{g,n}^*$. The general case is reduced to this case by lemma 5.3 and elementary properties of blowing-ups. We shall deal with the reduction part first. We shall need to use the notion of integral structure (or "completion") of an ideal of an integral domain, in the sense of Zariski-Samuel II, appendix 4. (This is very different from formal completion.) We refer the reader to it for a detailed treatment, and only recall that if I is an ideal of an integral domain R, its integral closure consists of all elements x in the quotient field of R satisfying an equation of the form $x^d + a_1\, x^{d-1} + \cdots + a_{d-1}\, x + a_d = 0$, where $a_i \in I^i$ for $i = 1, \ldots, d$. This notion of integral closure (completion) clearly sheafifies for Zariski's topology. In lemmas 5.9–5.13, we collect some elementary facts for the convenience of the reader. It is convenient to introduce a notation: for a locally integral scheme X and a coherent sheaf of ideals $\jmath$, $X_\jmath$ denotes the normalization of the blowing-up $\mathrm{Bl}_\jmath(X)$ along $\jmath$.

5.9. Lemma. *Let X be a locally noetherian and locally integral scheme, $\mathcal{J}$ be a coherent sheaf of ideals of $\mathcal{O}_X$. Let $\mathcal{J}^\vee$ be the integral closure of $\mathcal{J}$. Then*

(a) the blowing-up of X along $\mathcal{J}$ and $\mathcal{J}^m$ $(m \geq 1)$ are canonically isomorphic,

(b) $X_\mathcal{J}$ and $X_{\mathcal{J}^\vee}$ are canonically isomorphic, and the pull-back of $\mathcal{J}$ and $\mathcal{J}^\vee$ to $X_\mathcal{J} = X_{\mathcal{J}^\vee}$ are equal. The same holds for any ideal between $\mathcal{J}$ and $\mathcal{J}^\vee$.

Proof. (a) is standard, as $\forall m \geq 1$ an ideal I in an integral domain is invertible iff I^m is. For (b), let $Y = X_\mathcal{J}$, $Z = X_{\mathcal{J}^\vee}$. Since $\mathcal{J}^\vee \mathcal{O}_Y$ is integral over the principal ideal $\mathcal{J}\mathcal{O}_Y$, which is integrally closed because Z is normal, $\mathcal{J}^\vee \mathcal{O}_Y = \mathcal{J}\mathcal{O}_Y$, thus $\mathcal{J}^\vee \mathcal{O}_Y$ is locally principal, so we obtain a birational morphism $Y \to Z$ over X by the universal property of blowing-up. Now the integral closure of $\mathcal{J}\mathcal{O}_Z$ is clearly the invertible ideal $\mathcal{J}^\vee \mathcal{O}_Z$. Since the integral closure of a proper ideal in a normal domain is a proper ideal, $\mathcal{J}\mathcal{O}_Z = \mathcal{J}^\vee \mathcal{O}_Z$. Hence $\mathcal{J}\mathcal{O}_Z$ is invertible and we get a morphism $Z \to Y$ over X, which must be the inverse of $Y \to Z$. The last part of (b) follows from the first part. □

5.10. Lemma. *Let $f : X \to S$ be a proper birational morphism between integral locally noetherian schemes, $\mathcal{J}$ a coherent sheaf of ideals of $\mathcal{O}_S$. Then $f_* f^* \mathcal{J} \subset \mathcal{J}^\vee =$ the integral closure of $\mathcal{J}$. We have equality if $f^*\mathcal{J}$ is invertible and X is normal or if $\mathcal{J}$ is integrally closed (complete).*

Proof. We may assume $S = \mathrm{Spec}(R)$ to be affine. Then we can check the statement $f_* f^* \mathcal{J} \subset \mathcal{J}^\vee$ by pulling back to (not necessarily discrete) valuation rings R_v of R. Since any finitely generated ideal in a valuation ring is principal, we are reduced to the statement that $f_* \mathcal{O}_X$ is contained in the integral closure of $\mathcal{O}_S$ if f is birational, which is obvious. Since $f_* f^* \mathcal{J}$ is sandwiched between $\mathcal{J}$ and $\mathcal{J}^\vee$, we have equality if $\mathcal{J}$ is integrally closed. If $f^*\mathcal{J}$ is invertible, it is integrally closed, therefore elements of $\mathcal{J}^\vee$ define global sections of $f^*\mathcal{J}$. (Note: The second part is Zariski-Samuel II, appendix 4, §4 lemma, p.354 when $X = S_\mathcal{J}$.) □

5.11. Lemma. *Consider the following commutative diagram*

$$\begin{array}{ccc} X' & \xrightarrow{\ g\ } & X \\ {\scriptstyle f'}\downarrow & & \downarrow{\scriptstyle f} \\ S' & \xrightarrow[\ h\ ]{} & S \end{array}$$

where X', X, S', S are locally noetherian normal integral schemes, f', f are proper. Assume that there is a finite group H acting on X' and S' which is equivariant w.r.t. f' and identifies this diagram as the quotient diagram of f'. Suppose $\jmath$ is a coherent sheaf of ideals of $\mathcal{O}_X$, $\jmath' = g^*\jmath \subset \mathcal{O}_{X'}$, $\mathcal{J}' = f'_*\jmath$, $\mathcal{J} = f_*\jmath$. If $\jmath \xrightarrow{\sim} (g_* g^*\jmath)^H$ (automatic when g is flat), then $\mathcal{J} = (h_*\mathcal{J}')^H =$ the subsheaf of H-invariants in $h_*\mathcal{J}'$.

Proof. $(h_*\mathcal{J}')^H = (h_* f'_*\jmath)^H = (f_* g_* g^*\jmath)^H = f_*((g_* g^*\jmath)^H) = f_*\jmath = \mathcal{J}$. □

5.12. Lemma. *Let X be a normal noetherian scheme on which a finite group H operates, $\pi : X \to Y$ be a morphism identifying Y as the H-quotient of X. Suppose $\jmath$ is an H-invariant coherent sheaf of ideals of $\mathcal{O}_X$. There exists an integer $k \geq 1$ such that if we let $\mathcal{J} = (\pi_*(\jmath^k))^H$, $\mathcal{J}' = \pi_*((\jmath^k)^\vee))^H$, then $X_\jmath/H = Y_\mathcal{J} = Y_{\mathcal{J}'}$, and $\mathcal{J}\mathcal{O}_{\tilde{X}} = \mathcal{J}'\mathcal{O}_{\tilde{X}} = \jmath^k\mathcal{O}_{\tilde{X}}$ where $\tilde{X} = X_\jmath$. ($X_\jmath/H$ exists as a scheme: We may assume X and Y to be affine, in which case $X_\jmath$ is projective over X and there is no problem.)*

Proof. We have a canonical morphism $X_{\pi^*\mathcal{J}} \to Y_\mathcal{J}$ by the universal property of blowing-up, which is clearly finite and H-equivariant, therefore $X_{\pi^*\mathcal{J}} \to Y_\mathcal{J}$ identifies $Y_\mathcal{J}$ with $X_{\pi^*\mathcal{J}}/H$ since $Y_\mathcal{J}$ is normal.

For an integer $k \geq 1$, let $\mathcal{J} = (\pi_*(\jmath^k))^H$, $\mathcal{J}' = (\pi_*((\jmath^k)^\vee))^H$, $\tilde{X} = X_\jmath$. Clearly $\mathcal{J}\mathcal{O}_{\tilde{X}} \subseteq \jmath^k\mathcal{O}_{\tilde{X}}$. We claim that for a suitable integer $k \geq 1$ they are in fact equal. It suffices to show that we can find a suitable $k \geq 1$ such that $\jmath^k\mathcal{O}_{\tilde{X}} \subseteq \mathcal{J}\mathcal{O}_{\tilde{X}}$. We may and do assume that Y and X are both affine. It suffices to show that $\Gamma(X, \jmath^k)^H$ generates $\jmath^k\mathcal{O}_{\tilde{X}}$ for some $k \geq 1$, which follows by using H-norms to produce invariants.

From $\jmath^k\mathcal{O}_{\tilde{X}} = \mathcal{J}\mathcal{O}_{\tilde{X}}$, we get $\jmath^k \subset (\mathcal{J}\mathcal{O}_X)^\vee$ by $(\tilde{X} \to X)_*$ since $\jmath^k$ is integrally closed, which also implies that $\jmath^k = (\mathcal{J}\mathcal{O}_X)^\vee$. Hence $X_\jmath = Y_\mathcal{J}$ from 5.9 and the first paragraph, and everything follows.

5.13. Proposition. *Consider the commutative diagram of 5.11*

$$\begin{array}{ccc} X' & \xrightarrow{g} & X \\ {\scriptstyle f'}\downarrow & & \downarrow{\scriptstyle f} \\ S' & \xrightarrow[h]{} & S \end{array}$$

where X', X, S', S are noetherian normal integral schemes, f', f are proper. Assume that there is a finite group H acting on X' and S' which is equivariant w.r.t. f' and identifies this diagram as the quotient diagram of f'. Suppose $\jmath$ is an invertible sheaf of ideals of $\mathcal{O}_X$, $\jmath' = g^*\jmath \subset \mathcal{O}_{X'}$, $\mathcal{J}' = f'_*\jmath$, $\mathcal{J} = f_*\jmath$ and we assume that $\jmath^m \xrightarrow{\sim} (g_* g^*\jmath^m)^H$ for any $m \geq 1$ (automatic when g is flat). If $X' = S'_{\mathcal{J}'}$ and $\jmath' = f'^*\mathcal{J}$, then for some integer $k \geq 1$, we have $X = S_{\mathcal{J}_k}$ and $f^*\mathcal{J}_k = \jmath^k$, where $\mathcal{J}_k = f_*\jmath^k$.

Proof. By lemma 5.12, there exists $k \geq 1$ such that for $\mathcal{K} = (h_*((\mathcal{J}'^k)^\vee))^H$ we have $X = S_{\mathcal{K}}$ and $\mathcal{K}\mathcal{O}_{X'} = \jmath^k\mathcal{O}_{X'}$. By 5.10, $(\mathcal{J}'^k)^\vee = f'_*(\jmath^k)$, so $\mathcal{K} = (h_* f'_*(\jmath^k))^H = (h_*(\mathcal{J}'_k))^H = \mathcal{J}_k$ by lemma 5.11, where $\mathcal{J}'_k$ denotes $f'_*(\jmath'^k)$. This shows that $X = S_{\mathcal{J}_k}$. It remains to show that $f^*\mathcal{J}_k = \jmath^k$. We know that $\mathcal{J}_k\mathcal{O}_{X'} = \jmath^k\mathcal{O}_{X'}$, i.e., that $g^*f^*\mathcal{J}_k = g^*\jmath^k$. $f^*\mathcal{J}_k$ is an invertible sheaf of ideals contained in $\jmath^k$, and the pull-backs to X' of these two invertible sheaves of ideals are equal, hence they must be equal. □

5.14. Corollary. *If the statement of theorem 5.8 holds for $\overline{\mathsf{A}}_{g,m}$ and an integer $d_1 \geq 1$ where $n|m|n^\infty$, then it also holds for $\overline{\mathsf{A}}_{g,n}$ and a suitable positive multiple d of d_1.*

Proof. Let $\pi = \pi_{n,m} : \overline{\mathsf{A}}_{g,m} \to \mathsf{A}_{g,n}$ be the canonical morphism and $H = H_{n,m} = \Gamma(n)/\Gamma(m)$ ($\Gamma(N)$ = principal congruence subgroup of $\mathrm{Sp}(2g,\mathbb{Z})$). It is clear from the definition and local chart of the arithmetic toroidal compactification that $\pi^*\, {}_n\jmath^k \xrightarrow{\sim} {}_m\jmath^k$, ${}_n\jmath^k \xrightarrow{\sim} (\pi_*\, {}_m\jmath^k)^H$ $\forall k \geq 1$, hence the condition ${}_n\jmath^k \xrightarrow{\sim} (\pi_*\,\pi^*\, {}_n\jmath^k)^H$ $\forall k \geq 1$ is satisfied.

Conclusion of the Reduction Step. In view of corollary 5.14 and lemma 5.3, in order to prove theorem 5.8 we may assume that (*) $\langle \lambda_\alpha, x\rangle < \langle \gamma\cdot\lambda_\alpha, x\rangle$ $\forall x \in \overline{\sigma}_\alpha \cap C(X_\xi)^\circ$, any $\gamma \in \mathrm{GL}(X_\xi)$ s.t. $\gamma \equiv 1 \pmod n$, and any vertex λ_α of $K^\vee_{\xi,\phi}$. Here we consider the $\mathrm{GL}(X_\xi)$-admissible cone decomposition induced on $C(X_\xi)$ and the strictly convex $\mathrm{GL}(X_\xi)$-invariant support function ϕ_ξ induced by ϕ on $C(X_\xi)$ for free quotients X_ξ of X. The cocore $\{\phi_\xi \geq 1\}$ and its dual core in $C(X^*_\xi)^\circ$ will be denoted $K_{\xi,\phi}$ and $K^\vee_{\xi,\phi}$ respectively.

Conclusion of the Proof of Theorem 5.8. From now on we assume the condition (*) above is satisfied, where $n \geq 3$. We want to prove that for a suitable integer $d \geq 1$, $\overline{\mathsf{A}}_{g,n}$ is the normalization of the blowing-up of $\mathsf{A}^*_{g,n}$ along ${}_n\mathcal{J}_d$. Therefore we must show for a suitable $d \geq 1$ (a) the pull-back of ${}_n\mathcal{J}_d$ to $\overline{\mathsf{A}}_{g,n}$ is an invertible sheaf of ideals, in fact equal to ${}_n\jmath^d$, (b) the induced map from $\overline{\mathsf{A}}_{g,n}$ to $(\mathsf{A}^*_{g,n})_{{}_n\mathcal{J}_d}$ is an isomorphism. By Zariski's main theorem, it suffices to prove (a) and a weaker statement (b′): the induced map from $\overline{\mathsf{A}}_{g,n}$ to $(\mathsf{A}^*_{g,n})_{{}_n\mathcal{J}_d}$ induces a finite homomorphism between strict local rings formally completed along the boundary of $\overline{\mathsf{A}}_{g,n}$. We shall see that under condition (*) we can take $d = 1$. Our argument is based on the structure of completed strict local rings of $\mathsf{A}^*_{g,n}$ proved in §2 and the combinatorial lemmas assembled above. In fact (a) follows from the condition (*) and (b′) from 5.5.

Again it is instructive to explain the argument first for points over 0-dimensional cusps (the completely degenerate case), because there is no abelian part to confuse us and we are simply dealing with the classical q-expansion. By fpqc descent, to check (a) and (b′) it suffices to check so after passing to strict local rings, or even their completions along the boundary of $\overline{\mathsf{A}}_{g,n}$.

Thus let $\tilde{x}$ be a geometric point of $\overline{\mathsf{A}}_{g,n}$ and $\overline{x}$ its image in $\mathsf{A}^*_{g,n}$. $\tilde{x}$ lies in a stratum of $\overline{\mathsf{A}}_{g,n}$ corresponding to a $\mathrm{Sp}(2g,\mathbb{Z})(n)$-orbit of cones in $C(X)^\circ$ ($\mathrm{Sp}(2g,\mathbb{Z})(n)$ = the principal congruence subgroup of $\mathrm{Sp}(2g,\mathbb{Z})$ of level-n), and we pick a representative σ_0. The formal completion of $\overline{\mathsf{A}}_{g,n}$ along this stratum can be described as follows: σ_0 corresponds to a torus embedding $\mathrm{Spec}(R_{\sigma_0})$ with $R_{\sigma_0} = \mathbb{Z}[\sigma_0^\vee \cap n^{-1} \cdot S^2(X)]$. The stratum of $\overline{\mathsf{A}}_{g,n}$ corresponding to σ_0 is isomorphic to the subscheme of $\mathrm{Spec}(R_{\sigma_0})$ defined by the ideal I_{σ_0} generated by those elements of $\sigma_0^\vee \cap n^{-1} \cdot S^2(X)$ taking strictly positive values on σ_0. The formal completion of $\overline{\mathsf{A}}_{g,n}$ along this stratum is the formal spectrum of the I_{σ_0}-adic completion $R^\vee_{\sigma_0}$ of R_{σ_0}. What does the support function do? Well, it defines a linear function on σ_0, and the restriction of the ideal $\jmath$ to $\mathrm{Spec}(R^\vee_{\sigma_0})$ is generated by any element of $n^{-1}\cdot S^2(X)$ which coincides with ϕ on σ_0. (Such an element exists by the hypothesis on ϕ.)

We know from theorem 2.7 that the formal completion of the 0-dimensional stratum of $\mathsf{A}^*_{g,n}$ containing x is isomorphic to the formal spectrum of the ring of $\mathrm{GL}(X)(n)$-invariants in the formal power series $\mathbb{Z}[1/n,\zeta_n][\![C(X^*)\cap n^{-1}\cdot S^2(X)]\!]$, where $\mathrm{GL}(X)(n)$ is the principal congruence subgroup of $\mathrm{GL}(X)$ of level-n. What is (the restriction of) the ideal $\mathcal{J}$ to this formal completion? It is just the intersection of $\mathbb{Z}[1/n,\zeta_n][\![C(X^*)\cap n^{-1}\cdot S^2(X)]\!]^{\mathrm{GL}(X)(n)}$ with the ideal of $\mathbb{Z}[1/n,\zeta_n][\![C(X^*)\cap n^{-1}\cdot S^2(X)]\!]$ generated by those elements of $n^{-1}\cdot S^2(X)$ lying in the core $K^\vee_\phi$ dual to the cocore K_ϕ. Now it follows from (*) that the pull-back of $\mathcal{J}$ to $\overline{\mathsf{A}}_{g,n}$ is equal to $\jmath$ at the stratum corresponding to σ_0: after passing to formal completion, for a vertex λ_1 of $K^\vee_\phi$ corresponding to a top-dimensional cone σ_1 (i.e. coinciding with ϕ on σ_0), $\sum_{\gamma\in\mathrm{GL}(X)(n)}\gamma\cdot\lambda_1$ lies in $\mathbb{Z}[1/n,\zeta_n][\![C(X^*)\cap n^{-1}\cdot S^2(X)]\!]^{\mathrm{GL}(X)(n)}$ and gives a local generator. This proves (a). (In short, (*) allows us to form a $\mathrm{GL}(X)(n)$-invariant power series with a preassigned "leading term".) As for (b), lemma 5.5 tells us two things (i) $\mathbb{Z}(\lambda_2-\lambda_1)+\ldots+\mathbb{Z}(\lambda_m-\lambda_1)$ is a subgroup of finite index in the subgroup of elements of $n^{-1}\cdot S^2(X)$ which vanish on σ_0, (ii) $\mathbb{Z}[\mathbb{N}_{\geq 1}(\lambda_{m+1}-\lambda_1)+\ldots+\mathbb{N}_{\geq 1}(\lambda_l-\lambda_1)]$ generates an ideal of definition of $R^\vee_{\sigma_0}$. This implies (b′) according to the definition of blowing-up.

Now we turn to the general case. As usual this is not more difficult than the completely degenerate case. The only (minor) extra points are, in the notation of theorem 2.8, that the invertible sheaves $\mathcal{L}(\lambda)$ are relatively ample on $A_{(\overline{x})}$ for $\lambda\in C(X^*_\xi)^\circ$, and the theorem of Lefschetz that the (≥ 3)-th power of an ample invertible sheaf on an abelian variety is already very ample (c. [Mum 1] chap. 3 §17 theorem, p.163).

We shall use the notation in theorem 2.5 (4). Let $\tilde{x}$ be a geometric point of $\overline{\mathsf{A}}_{g,n}$ and $\overline{x}$ be its image in $\mathsf{A}^*_{g,n}$ which lies in a cusp corresponding to a rank-r quotient X_ξ and a coset of $\mathrm{Sp}(2g,\mathbb{Z}/n\mathbb{Z})/P_{g,r}(\mathbb{Z}/N\mathbb{Z})'$. $\tilde{x}$ lies in a stratum which corresponds to an orbit of the induced cone decomposition of $C(X_\xi)$ under the principal congruence subgroup of $\mathrm{GL}(X_\xi)$ of level-n. Let σ_0 be a representative of this orbit. We shall show that condition (*) implies (a) and lemma 5.5 (with X replaced by X_ξ) implies (b′). Now any vertex λ_α of the core $K^\vee_{\xi,\phi}$ lies in $S^2(X)$, and therefore the corresponding invertible sheaf $\mathcal{L}(\lambda)$ on $A_{(\overline{x})}$ is very ample, since it is isomorphic to the n-th power of the ample invertible sheaf $\mathcal{L}(n^{-1}\cdot\lambda)$ and $n \geq 3$. (The integral structure is given by the lattice $n\cdot B(X)$, or dually $n^{-1}\cdot S^2(X)$!) Now we look at the structure of the completed strict local rings of $\mathsf{A}^*_{g,n}$ as given by 2.7, and the same argument as above gives (a) and (b′). This proves theorem 5.8. □

5.16. Remark. The statement of theorem 5.8. still holds if we replace $\mathbb{Z}$ by a field k: the same argument works.

6. Extension Theorems

The main result of this section is a purity statement about semi-abelian schemes. It says that an abelian scheme over a dense open subscheme of a regular local ring whose complement is a divisor with normal crossings extends to a semi-abelian scheme, provided the l-adic local monodromy is unipotent and the generic characteristic is zero. A consequence of this purity result is that any morphism from the complement of a divisor with normal crossings in a regular scheme S to $\mathsf{A}^*_{g,n}$ automatically extends (uniquely) to S, provided the generic characteristic of S is zero. In characteristic zero, this extension theorem (true for any arithmetic quotient of a bounded symmetric domain) is due to Borel, using curvature estimates and GAGA. In characteristic $p > 0$ the corresponding statements do not hold any more, due to (the existence of continuous families of) purely inseparable isogenies.

We remark that "the method of Lefschetz pencils" used in this section is due to A. Grothendieck, cf. [SGA 2].

Our major tool is a similar purity result for Barsotti-Tate (p-divisible) groups. Recall that the purity theorem of Zariski-Nagata states that for a regular scheme any finite étale covering of the complement of a closed subset of codimension ≥ 2 extends to a finite étale covering of the whole scheme. We need a variant for Barsotti-Tate groups.

It is convenient to assemble some simple facts from commutative algebra here, which are more-or-less well-known. See SGA 2 for much more. We give proofs for the convenience of the reader.

6.1. Lemma. *Let R be an excellent regular local ring of dimension 2, K be the fraction field of R, L be a finite extension of K, $\tilde{R}$ be the integral closure of R in L. Then $\tilde{R}$ is a free R-module.*

Proof. $\tilde{R}$ being normal of Krull dimension 2, its depth is 2. Therefore, as an R-module, it has depth 2. Since R is regular, we have proj. $\dim_{R\text{-mod}}(\tilde{R})$ +

$\operatorname{depth}_{R\text{-mod}}(\tilde{R}) = \dim(R) = 2$. Hence proj. $\dim_{R\text{-mod}}(\tilde{R}) = 0$, i.e. $\tilde{R}$ is a projective R-module.

6.2. Lemma. *Let R be a regular local ring of Krull dimension 2, $\mathfrak{m}$ its maximal ideal, K its fraction field, $s = \operatorname{Spec}(R/\mathfrak{m})$, $\eta = \operatorname{Spec} K$, $U = \operatorname{Spec} R - \{s\}$, $j : U \to \operatorname{Spec} R$ be the inclusion. If $\mathcal{F}$ is a non-zero locally free $\mathcal{O}_U$-module of finite rank, then $j_*\mathcal{F}$ is a (locally) free R-module.*

Proof. Choose and fix an isomorphism $\alpha : \mathcal{F}_\eta \xrightarrow{\sim} K^n$. Then $\mathcal{F}$ is described as follows: The localization of $\mathcal{F}$ at any height-1 prime ideal $\mathfrak{p}$ of R can be identified via α as a free $R_\mathfrak{p}$-submodule $\mathcal{F}_\mathfrak{p}$ of K^n. Moreover $\mathcal{F}_\mathfrak{p} = (R_\mathfrak{p})^n$ for all but finitely many height-1 primes $\mathfrak{p}$, and the R-module M corresponding to $j_*\mathcal{F}$ is the intersection of all $\mathcal{F}_\mathfrak{p}$. It is clear that M is contained in $x^{-1} \cdot R^n$ for some non-zero element $x \in R$, hence M is finitely generated as R-module. As $\operatorname{depth}_{(R,\mathfrak{m})} M = 2$ the assertion follows.

The same argument proves the first part of

6.3. Lemma. *Let R be a regular local ring of dimension $d \geq 2$, $\mathfrak{m} =$ maximal ideal of R, $S = \operatorname{Spec}(R)$, $U = S - \operatorname{Spec}(R/\mathfrak{m})$, $j : U \to S$ the inclusion. Suppose $\mathcal{F}$ is a locally free $\mathcal{O}_U$-module of finite type, then*

(i) $F = \Gamma(j_*\mathcal{F})$ is an R-module of finite type, and $\operatorname{proj.dim}_R(F) \leq d-2$.

(ii) The R-modules $H^i(\mathcal{F})$ are supported in $\operatorname{Spec}(R/\mathfrak{m})$ $\forall i > 0$, *and are of finite type* $\forall 0 < i \leq d-2$.

Proof of (ii). The first part of (ii) is clear from the Leray spectral sequence. The second part of (ii) is an easy consequence of the local duality theorem for coherent sheaves. Let us sketch the proof: we have $H^i(\mathcal{F}) \cong H^{i+1}_\mathfrak{m}(F)$ $\forall i > 0$ from the long exact sequence relating local cohomology to cohomology without support. Local duality gives $H^{i+1}_\mathfrak{m}(F) \cong D(\operatorname{Ext}^{d-i-1}_R(F,R))$ $\forall i$, where D is the local duality functor. Since $\operatorname{Ext}^{d-i-1}_R(F,R)$ is an artinian R-module for $i \leq d-2$ and D transforms artinian R-modules to artinian R-modules, the assertion holds. □

Remark. The reader can also look at (the first few pages of [Ho] for a self-contained proof of lemma 6.3, where it is also proved that $H^i(\mathcal{F}) \cong \operatorname{Ext}^i_R(F^\vee, R)$ $\forall 0 \leq i \leq d-2$, and $\cong \operatorname{Ext}^d_R(\operatorname{Ext}^{d-i-1}_R(F^\vee, R), R)$ if $0 < i \leq d-2$.

The purity result for Barsotti-Tate (p-divisible) groups will be proved under the assumption that p is not a divisor of zero and that the dimension of the p-divisible group is constant:

6.4. Theorem. *Suppose S is a regular scheme, $Z \subset S$ a closed subscheme of codimension ≥ 2, $G = G[p^\infty] = \cup G[p^n]$ a p-divisible group over $U = S - Z$. We assume that p is not a divisor of 0 in $\mathcal{O}_S$ and that the dimension of $G[p^\infty] \times_U \mathbf{Spec}(\mathcal{O}_U/p\mathcal{O}_U)$ is constant. Then $G[p^\infty]$ extends uniquely to a p-divisible group over S.*

6.5. Remark. (1) When p is invertible in $\mathcal{O}_S$, the theorem reduces to the purity theorem of Zariski-Nagata, which says that if U is a dense open subscheme of a regular scheme X of codim. ≥ 2, then any finite étale covering of U extends (uniquely) to a finite étale covering of X, c. [SGA 2] exp. X thm. 3.4. Thus it is of interest only when p is not invertible in $\mathcal{O}_S$.

(2) We recall that the dimension of a p-divisible group is defined when p is nilpotent over the base. Since the dimension is locally constant the assumption that the dimension is constant is automatic if $U \times_{\mathrm{Spec}\,\mathbb{Z}} \mathrm{Spec}(\mathbb{Z}/p\mathbb{Z})$ is connected. Since the statement of theorem 6.4 is obviously Zariski local in S, this assumption is needed only Zariski-locally in S. We do not know whether this condition is implied by the rest.

(3) Theorem 6.4 does not hold if we drop the assumption that p is not a divisor of zero. A counter-example will be given below.

(4) Here is a convenient fact: homomorphisms between p-divisible group over (the same) normal base can be extended from dense open subschemes of the base, if the generic points have characteristic 0. For codimension 1, this follows easily from Tate [T]. In higher codimension the assertion is then obvious, looking at the affine algebras and using the algebraic analogue of Hartog's extension theorem.

The two assumptions that p is not a zero-divisor and the dimension is constant are used in the following way: Each injection $\mathcal{O}_{G[p^n]} \subset \mathcal{O}_{G[p^{n+1}]}$, derived from $p \cdot \mathrm{id} : G[p^{n+1}] \to G[p^n]$ has a canonical splitting (as $\mathcal{O}_S$-modules), given by $p^{-\dim(G)} \cdot \mathrm{trace}_{G[p^{n+1}]/G[p^n]}$. This comes down (by dévissage) to the fact (cf. [T] proof of prop. 2, pp. 164-165) that the trace-form over $\mathcal{O}_S$ on $\mathcal{O}_{G[p]}$ is divisible by $p^{\dim(G)}$, and unimodular after that division. Note that this canonical splitting commutes with any base change (as long as p is not a zero divisor).

Provisionally we call a p-divisible groups G (over *any* base scheme S) such that the injections $\mathcal{O}_{G[p^n]} \subset \mathcal{O}_{G[p^{n+1}]}$ derived from $p \cdot \mathrm{id} : G[p^{n+1}] \to G[p^n]$ all have splitting ($= \mathcal{O}_S$-linear retract) *good.* This notion is obviously stable under pullback. We shall prove

6.4′. Theorem. *Suppose S is a regular scheme, $Z \subset S$ a closed subscheme of codimension ≥ 2, $G = G[p^\infty] = \cup G[p^n]$ a* good *(as above) p-divisible group over $U = S - Z$. Then $G[p^\infty]$ extends uniquely to a p-divisible group over S.*

Proof of theorem 6.4′. We assume $\dim(S) \geq 2$, since otherwise there is nothing to prove. Uniqueness is obvious, as the extension of $G[p^n]$ has to be defined by the direct image (for $U \hookrightarrow S$) of its structure-sheaf. The existence part shall be proved in two steps, first for the case $\dim(S) = 2$ (i.e. for codimension 2 points of S), then in general.

Existence: Case $\dim(S) = 2$**.** By lemma 6.2, any locally free $\mathcal{O}_U$-module of finite rank extends uniquely to a locally free $\mathcal{O}_S$-module (by direct image), and morphisms between locally free $\mathcal{O}_U$-modules of finite rank extend (uniquely). Moreover this extension functor commutes with taking duals and tensor-product. It follows that any finite flat group scheme H over U extends uniquely to a finite flat group over S, which we shall again denote by H. (We can also use lemma 6.1 for this.) Clearly this extension functor commutes with taking Cartier-dual and fibre product over S.

Applying this the finite flat group schemes $G[p^n]$ over U we see that they extend to finite flat group schemes $G[p^n]$ over S. To show that they form a Barsotti-Tate group over S, we must show that the exact sequences $0 \to G[p] \to G[p^{n+1}] \to G[p^n] \to 0$ over U remain exact when extended to S, i.e. over S $G[p^{n+1}] \to G[p^n]$ is faithfully flat, and its kernel is $G[p]$. That $G[p^{n+1}]_{/S} \to G[p^n]_{/S}$ is faithfully flat follows from the existence of splitting of $\mathcal{O}_{G[p^n]} \subset \mathcal{O}_G[p^{n+1}]$ over U, because the extension of a splitting is a splitting. The kernel of $G[p^{n+1}]_{/S} \to G[p^n]_{/S}$, being finite flat, must be the extension of $G[p]$ by the uniqueness of extension. Thus we have found an extension of $G[p^\infty]$ over S.

Reduction Steps. Let $d = \dim(S)$, which we may assume to be ≥ 3. We shall *proceed by induction on* d (we may assume that S is the spectrum of a local ring, so that d is finite). Thus we may assume that the statement of theorem 6.4′ has been proved if the dimension of the base is $\leq d-1$. Clearly the statement is Zariski local in S, so we may assume $S = \operatorname{Spec}(R)$ for R a regular local ring, and U the complement of the closed point.

The next step is to reduce to the case R complete w.r.t. the maximal ideal: The statement of theorem 6.4′ means that the direct images of the structure sheaves $\mathcal{O}_{G[p^n]}$ under $U \to S$ are all locally free $\mathcal{O}_S$-modules, and the resulting maps $G[p^{n+1}]_{/S} \to G[p^n]_{/S}$ are all faithfully flat. Since taking direct image commutes with flat base change, we see by faithfully flat descent that existence of extension over $\operatorname{Spec}(R^\wedge)$ implies the existence of extension over $\operatorname{Spec}(R)$.

The situation now is that R is a complete regular local ring of dimension $d \geq 3$, $\mathfrak{m}$ the maximal ideal, $U = \operatorname{Spec}(R) - \operatorname{Spec}(R/\mathfrak{m})$, and the residue field k has characteristic p. We have made the induction hypothesis that the statement of thm. 6.4′ holds when the base scheme has dimension $\leq d-1$. Choose a regular parameter x in $\mathfrak{m}$. Let $S_0 = \operatorname{Spec}(R/xR)$, $U_0 = U \times_S S_0$, $S_k = \operatorname{Spec}(R/x^{k+1}R)$ for $k \geq 0$, $j : U \to S$ be the inclusion. We have a *good* p-divisible group G over U, which we want to extend to S.

Consider the restriction $G_{/U_k}$ of G to U_k, $k \geq 0$, which clearly are still good. By the induction hypothesis, $G_{/U_0}$ extends uniquely to a p-divisible group over S_0, which we shall write $G_{/S_0}$. We claim that each $G_{/U_k}$ extends uniquely to a p-divisible group $G_{/S_k}$ over S_k, and that they form a compatible tower (i.e. $G_{/U_{k+1}} \times_{U_{k+1}} U_k = G_{/U_k}$). This claim is proved by deformation theory as follows: construct the extension $G_{/S_k}$'s by induction on k. Assume that $G_{/S_k}$ has been constructed. It is always possible to lift $G_{/S_k}$ to a p-divisible group over S_{k+1}, because there is no obstruction (the versal deformation of a p-divisible group is smooth). Choose and fix a lifting, which we use as "base point" to compare with other liftings. Again by deformation theory of p-divisible groups, the set of all liftings of $G_{/S_k}$ to S_{k+1} form a principle homogeneous space under $\Gamma(S_0, \mathbf{t}_G \otimes \mathbf{t}_{G^t} \otimes (x^{k+1}R/x^{k+2}R))$. (Recall that deformations of a Barsotti-Tate group correspond to liftings of the Hodge filtration in the Dieudonné module.) Similarly, the set of all liftings of $G_{/U_k}$ to U_{k+1} form a principal homogeneous space under $\Gamma(U_0, \mathbf{t}_G \otimes \mathbf{t}_{G^t} \otimes (x^{k+1}R/x^{k+2}R))$. Note that $\Gamma(U_0, \mathbf{t}_G \otimes \mathbf{t}_{G^t} \otimes (x^{k+1}R/x^{k+2}R)) = \Gamma(S_0, \mathbf{t}_G \otimes \mathbf{t}_{G^t} \otimes (x^k R/x^{k+1}R))$ because U_0 is the complement of the closed point of the regular scheme $\mathrm{Spec}(R/xR)$, which has dimension $d-1 \geq 2$. Thus the restriction to U_k of the chosen lifting differs from $G_{/U_k}$ by an element of $\Gamma(U_0, \mathbf{t}_G \otimes \mathbf{t}_{G^t} \otimes (x^{k+1}R/x^{k+2}R))$. This element uniquely extends to an element of $\Gamma(S_0, \mathbf{t}_G \otimes \mathbf{t}_{G^t} \otimes (x^{k+1}R/x^{k+2}R))$, which we use to adjust the chosen lifting to get an extension of $G_{/U_{k+1}}$. The claim is proved.

It is now obvious how to construct the extension $G_{/S}$: since R is complete, for each n the finite flat group schemes $G[p^n]_{/S_k}$ defines a finite flat group scheme $G[p^n]_{/S}$ by passing to the limit. Clearly $G[p^{n+1}]_{/S} \to G[p^n]_{/S}$ is flat because it is so over each S_k, say by the "fibre by fibre criterion of flatness", EGA IV 11.3.10. The only (slight) difficulty is to show that the $G_{/S}$ so constructed is indeed an extension of $G_{/U}$.

Let us observe first that it suffices to show that the the structure sheaf of $G[p^n]_{/U}$ extends to a locally free $\mathcal{O}_S$-module (necessarily by j_*) for each n: If so, we have $\mathrm{depth}(R/x^{k+1}R) = d-1 \geq 2\ \forall k \geq 0$, so the reduction modulo x^k of the above extension is the unique locally free extension of the structure sheaf of $G[p^n]_{/U_k}$, therefore coincides with the structure sheaf of $G[p^n]_{/S_k}$. Taking the limit as $k \to \infty$, we see that the finite flat group scheme $G[p^n]_{/S}$ constructed in the previous paragraph using deformation theory is indeed an extension of our original $G[p^n]_{/U}$. The following lemma concludes the proof of theorem 6.4′:

6.6. Lemma. *Let R be a complete regular local ring, $\mathfrak{m}$, $x \in \mathfrak{m}$, S, U, S_k, U_k as above, $j: U \to S$, $\bar{j}: U_k \to S_k$ the inclusion, $\mathcal{F}$ a locally free $\mathcal{O}_U$-module of finite rank, $F = \Gamma(j_* \mathcal{F})$, $\mathcal{F}_k = \mathcal{F} \otimes_{\mathcal{O}_U} \mathcal{O}_{U_k}$, $F_k = \Gamma(\bar{j}_* \mathcal{F}_k)$. Suppose that F_k is a free R/x^{k+1}-module for each $k \geq 0$, and $F_{k+1} \otimes_R (R/x^{k+1}R) \xrightarrow{\sim} F_k$ via the canonical map. Then F is a free R-module and the canonical map from $F \to \varprojlim_k F_k$ is an isomorphism.*

Proof. Consider the exact sequence $0 \to \mathcal{F} \xrightarrow{x^k} \mathcal{F} \to \mathcal{F}_k \to 0$, which gives a short exact sequence $0 \to F/x^kF \to F_k \to \mathrm{Ker}(x^k : H^1(\mathcal{F}) \to H^1(\mathcal{F})) \to 0$. Since $H^1(\mathcal{F})$ is an R-module of finite type supported in $\mathrm{Spec}(R/\mathfrak{m}R)$, it is killed by x^k for k large. Thus we have short exact sequences $0 \to F/x^kF \to F_k \to H^1(\mathcal{F}) \to 0$ for $k \gg 0$. Considering the map from $0 \to F/x^{k+1}F \to F_{k+1} \to H^1(\mathcal{F}) \to 0$ to $0 \to F/x^kF \to F_k \to H^1(\mathcal{F}) \to 0$, we see that $\mathrm{Ker}(F/x^{k+1}F \to F/x^kF) \cong \mathrm{Ker}(F_{k+1} \to F_k)$ for $k \gg 0$. In other words we have $F/xF \xrightarrow{\sim} F_0$ since F is torsion free. By dévissage $F/x^{k+1} \xrightarrow{\sim} F_k$, and we get $F \xrightarrow{\sim} \varprojlim_k F_k$ by passing to the limit, so F is a free R-module. This concludes the proof of the lemma and also the theorem. □

Now we shall apply this to abelian schemes:

6.7. Theorem. *Let S be a regular scheme with maximal points of characteristic 0, $Z \subset S$ a divisor with normal crossings. Let A be an abelian scheme over $S - Z$ which extends to a semi-abelian scheme over an open subscheme V containing $S - Z$ and the maximal points of Z (equivalent to having unipotent local monodromy at the maximal points of Z). Then A extends uniquely to a semi-abelian scheme G over S.*

6.8. Corollary. *If S is a regular scheme with maximal points of characteristic 0, Z a closed subscheme in S of codimension ≥ 2, then any abelian scheme over $U = S - Z$ extends uniquely to an abelian scheme over S.*

6.9. Remark. A similar result for smooth curves of genus ≥ 2 was first proved by Moret-Bailly (cf. [MB 2]) *without* the characteristic zero hypothesis. The proof of theorem 6.4′ follows his method: the codimension 2 case implies the general case by induction on codimension and deformation theory. In the proof of theorem 6.7, a key step is descent w.r.t. a blowing-up of a closed point. A similar thing appeared in [MB 2], but in our case we use the main theorem of chapters II, III. Some of the main ideas about such purity statements already appeared in [GR3].

Proof of Corollary 6.8. We may assume $S = \mathrm{Spec}(R)$ to be strictly local. By induction on codimension of Z, we may assume Z to be the closed point s of S. Then Z is contained in a divisor with normal crossings. By theorem 6.7 the abelian scheme on $S - Z$ has a semi-abelian extension, which is abelian in codimension one and thus everywhere (chapter III, theorem 10.1).

The proof of theorem 6.7 is divided into several steps, which occupy the rest of this section.

Preliminary Reduction Steps

(i) First, we *may assume that $A_{/U}$ carries a principal polarization*, by (the extension theorem of homomorphisms between semi-abelian schemes and) Zarhin's trick (I, 5.5): $A^4 \times A^{t^4}$ always carries a principal polarization. The statement is obviously Zariski local, so we may assume that $S = \mathrm{Spec}(R)$ with R a regular excellent local ring.

(ii) We now want to *reduce to the case R is noetherian complete regular and strictly local.* This is not entirely obvious but not as hard as one might think: Let R' be the completion of the strict henselization of R. Then $\mathrm{Spec}(R') \to \mathrm{Spec}(R)$ is faithfully flat and quasi-compact. Let $S' = \mathrm{Spec}(R')$, $V' = V \times_S S'$, $S'' = \mathrm{Spec}(R' \otimes_R R')$, $V'' = V' \times_V V'$. We know that $A_{/U}$ extends to a semi-abelian scheme $G_{/V}$ over V. Assuming $G_{/V'}$ extends to a semi-abelian scheme G' over S', we must show that $G_{/V}$ extends to a semi-abelian scheme over S. Let $\mathcal{L}_U$ be the pull-back of the Poincaré sheaf via (id, λ), where λ is the principal polarization, so that $\mathcal{L}_U$ is a symmetric relatively ample cubical invertible sheaf on $A_{/U}$. By [Raynaud] cor. XI 1.16, p.173 and the theory of extensions of cubical invertible sheaves, $\mathcal{L}_U$ extends uniquely to a relatively ample cubical invertible sheaf $\mathcal{L}_V$ on $G_{/V}$ and its pull-back extends uniquely to a relatively ample cubical invertible sheaf $\mathcal{L}_{S'}$ on $G'_{/S'}$. We have $\Gamma(G'_{/S'}, \mathcal{L}_{S'}^{\otimes k}) = \Gamma(G'_{/V'}, \mathcal{L}_{V'}^{\otimes k}) = \Gamma(G_{/V}, \mathcal{L}_V^{\otimes k}) \otimes_R R'$ $\forall k \geq 0$ since $\mathrm{codim}(V, S) \geq 2$, and an open immersion $G' \to X' := \mathrm{Proj}(\oplus_{k\geq 0} \Gamma(G'_{/S'}, \mathcal{L}_{S'}^{\otimes k}))$ defined by $\mathcal{L}_{S'}$. We have seen that X' descends to $X := \mathrm{Proj}(\oplus_{k\geq 0} \Gamma(G_{/V}, \mathcal{L}_V^{\otimes k}))$, compatibly with the descent from $G'_{/V'}$ to $G_{/V}$. The problem now is to show that the open subset $G'_{/S'}$ of X' descends. Let p_1, p_2 be the two projections from $X' \times_X X'$ to X'. What we must show is that the the two open subsets $p_1^{-1}(G_{/S'})$, $p_2^{-1}(G_{/S'})$ of $X' \times_X X'$ are in fact equal. Notice that these two open subset both contain the zero section, and that their fibres are connected.

To achieve this we shall use the addition formula: if $\Phi : G' \times_{S'} G' \to G' \times_{S'} G'$ is the isogeny defined by $(x, y) \to (x + y, x - y)$, then $\Phi^*(p_1^*\mathcal{L}_{S'} \otimes p_1^*\mathcal{L}_{S'}) = (p_1^*\mathcal{L}_{S'} \otimes p_1^*\mathcal{L}_{S'})^{\otimes 2}$. The same holds for $G_{/V}$. This defines a rational map $X' \times_{S'} X' \to X' \times_{S'} X'$ containing $G' \times_{S'} G'$ in its domain of definition, and it descends to a rational map $\Phi : X \times_S X \to X \times_S X$ in the obvious sense. Now for any point z of S'', $p_1^{-1}(G^2_{/S'})_z$, $p_2^{-1}(G^2_{/S'})_z$ (p_1 and p_2 the maps to $X' \times_{S'} X'$) are two connected open subsets of $(X'' \times_{S''} X'')_z$ containing the zero section, both irreducible, and both contained in the domain of definition of Φ and stable under it. Hence the two pull-backs of $G'_{/S'}$ are both connected open subgroups, so they coincide, and $G'_{/S'}$ descends to a group scheme $G_{/S}$. Thus we may assume R to be a regular complete local ring with separably closed residue field.

(iii) *Reduction to the case $A_{/U}$ has a level-n structure for some $n \geq 3$ invertible in R.* R is now a noetherian complete regular local ring with algebraically closed residue field, U the complement of a divisor Z with normal crossings in $S = \mathrm{Spec}(R)$. Choose and fix an integer $n \geq 3$ invertible in R. Since $A[n]_{/U}$ is tamely ramified at the maximal points of Z, Abhyankar's lemma tells us that there is a regular local ring R' over R obtained from R by taking roots of regular parameters, such that over $U' = U \times_S S'$ ($S' = \mathrm{Spec}(R')$) the abelian scheme A admits a principal level-n structure. We assume that $A_{/U'}$ extends to a semi-abelian scheme G' over S'. We would like to descend it down to S. Note that R' is a finite flat R-module, so by the theory of finite flat descent, any descent datum for G' w.r.t. $S' \to S$ is effective. It remains to construct a descent datum for G' w.r.t $S' \to S$ which over $V' = V \times_S S' = S' - \{s'\}$ (s' = closed point of S') extends the one defined by $G_{/V}$. By the definition of descent data, it boils down to extending the isomorphism over $V'' = V' \times_S V'$ between the two semi-abelian schemes p_i^*G' over $S'' = S' \times_S S'$ defined by $G_{/V}$. (The 1-cocycle condition results from the uniqueness of extension.) Now we just have to quote a result of Raynaud ([Raynaud] IX 1.4, p.130, which says that if S is a locally noetherian schem, U a dense open subscheme of S containing all points of depth ≤ 1, G a smooth group scheme over S, H a group scheme locally of finite type, then any homomorphism from $G_{/U}$ to $H_{/U}$ extends uniquely to a homomorphism from G to H. So we have a descent datum and G' descends to a semi-abelian scheme G over S extending $A_{/U}$.

A major step of the proof of theorem is to descend a semi-abelian extension of A from the blowing-up of the closed point of S in the situation envisioned above. We interrupt our argument for a moment:

6.10. Proposition. *Let R be a noetherian regular local ring of dimension $d \geq 2$ with generic characteristic 0. Let K be its fraction field, k its residue field, $S = \mathrm{Spec}(R)$, $\eta = \mathrm{Spec}(K)$, $s = Spec(k)$, Z a divisor with normal crossings in S, $U = S - Z$. Let $S^* \to S$ be the blow-up of the closed point of S, E be the exceptional fibre which is isomorphic to $\mathbb{P}^{d-1}_{/k}$. If $A \to U$ is an abelian scheme which extends to a semi-abelian scheme $G^* \to S^*$, then G^* is the pull-back of a unique semi-abelian scheme $G \to S$ over S.*

Proof. Uniqueness is clear by the extension theorem of homomorphism between semi-abelian schemes over a normal base (cf. I. prop. 2.7). Now we turn to the existence part.

(i) As before we may assume that R is complete, k separably closed, $A_{/U}$ principally polarized with a symplectic principal level-n structure. Let us first look at the action of $\mathrm{Gal}(\overline{K}/K) = \pi_1(\eta, \overline{\eta})$ on the l-adic Tate module $T_l(A_{\overline{\eta}})$. It is unramified over U (i.e. factors through $\pi_1(X - Z, \overline{\eta})$) and is tamely ramified at the maximal points of Z by hypothesis. (Notice that this uses only that A extends to a semi-abelian scheme in codimension 2.) Therefore this action factorizes through the tame fundamental group of $X - Z$, which is a product of copies of $\mathbb{Z}^\wedge(1)_{\mathrm{non}-p}$ by Abhyankar's lemma (cf. [SGA 1] exposé XIII). Thus the action of $\mathrm{Gal}(\overline{K}/K)$ on $T_l(A_{\overline{\eta}})$ is commutative and unipotent.

Let $\nu = \mathrm{Spec}(L)$ be the generic point of the exceptional fibre E. We shall identify $\mathrm{Gal}(\overline{L}/L)$ with the quotient of the decomposition group $D_\nu \subset \mathrm{Gal}(\overline{K}/K)$ at ν by the inertia group I_ν at ν. Let $G^*_{/E}$ be the restriction of G^* to E, G^*_ν the fibre over ν. Via the above identification, the action of $\mathrm{Gal}(\overline{L}/L)$ on the l-adic Tate-module $T_l(G^*_{\overline{\nu}})$ comes from the action of D_ν on the I_ν-invariants $T_l(A_{\overline{\eta}})^{I_\nu} \subset T_l(A_{\overline{\eta}})$, hence is commutative and unipotent. On the other hand, the torus part of G^*_ν extends to a subtorus $T_{/E}$ of $G^*_{/E}$ (I. prop. 2.9). Applying the corollary to the theorem of Zarhin (that is 4.8) to generic lines in E together with the Bertini theorem for π_1 ([Ra′]), it follows that the action of $\mathrm{Gal}(\overline{L}/L)$ on the l-adic Tate module of the abelian part of G^*_ν is semisimple. Being both semi-simple and unipotent, it has to be *trivial*. It follows that $G^*_{/E}$ is the extension of an abelian scheme $B \to E$ by a split torus T_E over E. Also as E is simply connected the Tate-module $T_l(G^*_{/E})$ is trivial.

(ii) We would like to argue that $G^*_{/E}$ is constant, i.e., that it is the pull-back of a semi-abelian variety over k to E. (This has to be the case if the proposition is true.) In the trivial case when $\mathrm{char.}(k) = 0$, this follows from corollary 4.9. If $\mathrm{char.}(k) = p > 0$, we come to the same conclusion, as follows: It suffices (again by 4.9) to show that the p-divisible group of G^*_E is constant. We first construct a p-divisible group over the formal completion $S^{*\wedge}$ of S^* along E, by taking p-power-torsion in the formal completion $G^{*\wedge}$ of G^*. This p-divisible group will be temporarily denoted by $H^\wedge = \cup H^\wedge[p^n]$, where each $H^\wedge[p^n]$ is the finite locally free group over $S^{*\wedge}$, defined either as $G^{*\wedge}[p^n]$ or $G^*[p^n]^\wedge$, the equivalence of the two definitions being left to the reader. Remember that $S^{*\wedge}$ is also the $\mathfrak{m}$-adic completion of S^* w.r.t. $S^* \to \mathrm{Spec}(R)$ where $\mathfrak{m}$ is the maximal ideal of R. Since each $H^\wedge[p^n]$ is determined by its structure sheaf, which is a locally free $\mathcal{O}_{G^{*\wedge}}$-module of finite rank, Grothendieck's formal existence theorem ([EGA III] §5) tells us that it is algebraic. In other words $H^\wedge$ comes from a unique p-divisible group $H = \cup H[p^n]$ over S^*. Notice that H has constant dimension over $S^* \times_{\mathrm{Spec}\mathbb{Z}} \mathrm{Spec}(\mathbb{Z}/p\mathbb{Z})$, which is the dimension of G^*. By theorem 6.4, the restriction of H to U extends to a p-divisible group $G_f[p^\infty] = \cup G_f[p^n]$. By the extension theorem of homomorphism between p-divisible groups (6.5.(4)), the pull-back of $G_f[p^\infty]$ to S^* must be (canonically isomorphic to) H. In particular, the pull-back of $G_f[p^\infty] \times_S \mathrm{Spec}(k)$ to E is $H \times_{S^*} E = G^*_{/E}[p^\infty]$, so $G^*_{/E}$ is constant: $G^*_{/E} = G_k \times_{\mathrm{Spec}(k)} E$.

(iii) Now we claim that $G^{*\wedge}$ is the pull-back of a unique formal semi-abelian extension $G^\wedge$ over S (i.e. extension of a formal abelian scheme by a formal torus). The uniqueness of $G^\wedge$ is clear. The existence follows by deformation theory, as in the proof of theorem 6.4′ (the reduction step). The category of deformations of $G^*_{/E}$ to S^*_k (= k-th infinitesimal neighborhood of E in S^*) is equivalent to the category of deformations of G_k over $\mathrm{Spec}(R/\mathfrak{m}^{k+1})$. Use induction over k, and the following fact: If $I \subset \mathcal{O}_{S^*}$ denotes the ideal of E, then $I^k/I^{k+1} \approx \mathcal{O}_E(k)$ has $\mathfrak{m}^k/\mathfrak{m}^{k+1}$ as space of global sections. Note also that the abelian part of $G^\wedge$ is principally polarized, so $G^\wedge$ is algebraic.

(iv) To conclude this proposition, we use the theory of degenerations developed in chapters II, III. Recall that the Raynaud extension $\tilde{G}$ corresponds to a homomorphism from c from X to $B^t_{/S} = B_{/S}$ (via the principal polarization). The homomorphism for $\tilde{G}^*$, denoted $\tilde{c}$, is the pull-back of c to S^*. The last part of the degeneration data for $G^*_{/S^*}$ is a symmetric bimultiplicative formal-meromorphic map from $Y \times X$ to the inverse of the Poincaré sheaf of $B_{/S^*} \times_{S^*} B^t_{/S^*}$ lying over $c^{t*} \times c^*$ satisfying the positivity condition (for a chosen polarization) as specified in chapter II, III. Looking at one element (μ, χ) of $X \times X$ at a time, we have an invertible sheaf $\mathcal{L}$ over S (the fibre over (μ, χ) of the inverse of Poincaré sheaf), its pull-back $\pi^*\mathcal{L}$ to S^*, and a formal meromorphic section of $\pi^*(\mathcal{L})$, and would like to say that this formal meromorphic section comes uniquely from a formal meromorphic section of $\mathcal{L}$. This is true for regular sections (as obviously $H^0(S^{*\wedge}, \pi^*(\mathcal{L})^\wedge) = H^0(S, \mathcal{L})$), and in general choose a non-zero $f \in R$ such that G^* is abelian over R_f. Then multiplication with some power of f makes the meromorphic sections above regular, and we are done.

(v) We are essentially through with the proof of 6.10, but there remains a technical point which has to be clarified. What we have shown is the following: from the semi-abelian scheme G^* over S^* we pass to a semi-abelian scheme over the formal scheme $S^{*\wedge}$, denoted by H. This means that for each affine open $\mathrm{Spf}(B^\wedge)$ of $S^{*\wedge}$, which is the formal completion of an affine open $\mathrm{Spec}(B)$ of S^*, we take the fibre product $G^*_{/\mathrm{Spec}(B)} \times_{\mathrm{Spec}(B)} \mathrm{Spec}(B^\wedge)$. What we have really proved is that H comes from a unique semi-abelian scheme G over S. We still must show that $G \times_S S^*$ is canonically isomorphic to G^*, knowing that they give canonically isomorphic semi-abelian schemes over the formal scheme $S^{*\wedge}$. The key is the fact if two principally polarized abelian varieties with symplectic level-n structure ($n \geq 3$) over a field F are isomorphic (together with polarization and level-n structure) over an over-field of F, then they are in fact isomorphic over F (in fact by a unique isomorphism). This is an easy exercise in descent, or an immediate consequence of the existence of fine moduli spaces. In step (iv) above, we can take the level-n structure into consideration so that the generic fibre of $G_{/S}$ carries a level-n structure, and when pulled back to $S^{*\wedge}$, the canonical isomorphism between $G \times_S S^* \times_{S^*} S^{*\wedge}$ and $G^* \times_{S^*} S^{*\wedge}$ preserves the principal polarization and level-n structure as well. Now we have two semi-abelian schemes $G \times_S S^*$ and G^* over S^*, whose generic fibres are principally polarized abelian varieties with level-n structure, which become isomorphic over $S^{*\wedge}$. Hence the generic fibres of $G \times_S S^*$ and G^* are isomorphic, say using the quotient field of an affine coordinate ring of $S^{*\wedge}$. Extending such an isomorphism between the generic fibres over S (I, prop. 2.7) proves prop. 6.10.

Conclusion of the Proof of Theorem 6.7. (iv) The next step is to prove 6.7 for the *case* $\dim(S) = 2$. Thus we assume as before that $S = \text{Spec}(R)$ where R is the spectrum of a complete regular local ring of dimension 2. The situation is that we have a divisor Z with normal crossings in $S = \text{Spec}(R)$ and a principally polarized abelian scheme $A \to S - Z$ which extends to a semi-abelian scheme over $V = S - \{s\}$ where s is the closed point of S, and we would like to show that it extends to a semi-abelian scheme over S.

By reduction-step (iii) we may (and do) assume that $A \to S - Z$ carries a level-n structure for some $n \geq 3$ invertible in R. Then it defines a morphism f from $S - Z$ to $\mathsf{A}_{g,n}$. Take an arithmetic toroidal compactification $\overline{\mathsf{A}}_{g,n}$ (say a projective one so that we are firmly in the category of schemes). Then after finitely many blow-ups, f extends to a morphism from S to $\overline{\mathsf{A}}_{g,n}$ (elimination of points of indeterminacy for 2-dimensional regular schemes), so A extends to a semi-abelian scheme. By proposition 6.10 we are done. We remark that elimination of points of indeterminacy for 2-dimensional regular schemes can be done by basic commutative algebra without resolution of singularities, cf. [Zari] thm. 10, p. 681, [Z-S]. For a proof by overkill, see [Li].

(v) Proceeding by induction on the dimension of the base, we assume theorem 6.7 holds if $\dim(S) \leq d - 1$ (where $d \geq 3$), and would like to show that the theorem is true for $\dim(S) = d$ as well. As shown before we may assume $S = \text{Spec}(R)$ where R is a complete regular local ring with separably closed residue field k, $\dim(R) = d$. By induction, A extends to a semi-abelian scheme $G_{/V}$ over V = complement of the closed point s of S. Our strategy is to show that $G_{/V}$ extends to a semi-abelian scheme over S^* = blowing-up of the closed point s in S. By 6.10, this suffices. The advantage is that $G_{/V}$ extends to a semi-abelian scheme over the complement of a finite set Q of the exceptional fibre E, by induction. This gets us started.

(vi) Let $G^*_{/S^*-Q}$ be the semi-abelian scheme over $S^* - Q$ extending $G_{/V}$. As $E - Q$ is simply connected the l-adic monodromy of the Tate module of $G^*_{/E-Q}$ is trivial for l invertible in R, therefore over $E - Q$ the semi-abelian scheme $G^*_{/E-Q}$ is the extension of an abelian scheme $A_{/E-Q}$ over $E - Q$ by a split torus T over $E - Q$. (Notice that we have used the letter A again for the abelian part of $G^*_{/E-Q}$ and we can forget about the original abelian scheme over U.) First we want to show that $G^*_{/E-Q}$ extends to a semi-abelian scheme over E. If char.$(k) = 0$, this is the case because the monodromy is trivial (4.9). If char.$(k) = p > 0$ $G^*_{/E-Q}$ is already constant up to p-isogeny. Let us look at the p-divisible group $G^*_{/E-Q}[p^\infty]$ attached to $G^*_{/E-Q}$. We claim that it is "good" in the sense of 6.4: for any point x of $E - Q$, choose an affine open Spec(B) in $S^* - Q$ containing x with $E \cap \text{Spec}(B)$ a closed subscheme Spec(B/I). Then the restriction of $G^*_{/E-Q}[p^\infty]$ to Spec(B/I) is the reduction (mod I) of the p-power torsion of the I-adic completion of $G^*_{/\text{Spec}(B)}$, thought of as a p-divisible group over Spec$(B^\wedge)$, where $B^\wedge$ is the I-adic completion of B. The trace gives the required splitting as in 6.1 and the claim is proved. It follows from theorem 6.4′ that $G^*_{/E-Q}[p^\infty]$ extends to a p-divisible group over E, which we denote by $G^*_{/E}[p^\infty]$, and thus also $G^*_{/E-Q}$ extends to a unique semi-abelian scheme $G^*_{/E}$ over E, which is an extension of an abelian scheme $A_{/E}$ extending $A_{/E-Q}$ (necessarily unique) by the split torus T over E extending the previous $T_{/E-Q}$.

(vii) Now that we have $G^*_{/E}$, we want to deform it along the infinitesimal neighborhoods of E in S^*. In fact, we claim the category of deformations of $G^*_{/E}$ to the k-th infinitesimal neighborhood of E inside S^* is equivalent to the category of deformations of $G^*_{/E-Q}$ to the k-th infinitesimal neighborhood of $E - Q$ inside $S^* - Q$, $\forall k \geq 0$. We shall give a local proof to the effect that we can change S^* to any affine open subscheme of S^* and the statement still holds. Thus we can Zariski localize and take the local ring B at a point q of Q. Provisionally we let $X = \text{Spec}(B)$, $V = \text{Spec}(B) - \{q\}$, $E \cap X = \text{Spec}(B/I)$ where I is generated by a regular parameter of the regular local ring B. We would like to show that deformations of G^* to the k-th order neighborhood for X and V are the same. We prove this by induction on k. Suppose this has been proved for $k = n - 1$; let us prove it for $k = n$. Let $X_k = \text{Spec}(B/I^{k+1})$, $V_k = X_k - \{q\}$, and $G^*_{/X_n}$ a deformation of $G^*_{/E}$ to X_k. Since the versal deformation space of abelian varieties is formally smooth, $G^*_{/X_{n-1}}$ can be lifted to X_n. Choose and fix such a lifting as a "base point", the set of all liftings to X_n is identified with $H^0(X_0, \mathbf{F})$, where $\mathbf{F} = \mathbf{t}_H \otimes_{B/I} \mathbf{t}_{A^t}$, where H stands for $G^*_{/E}$ and A^t for the dual of its abelian part. Similarly, liftings of (the restriction to V_{n-1} of) $G^*_{/X_{n-1}}$ to V_n are identified with $H^0(V_0, \mathbf{F})$. Restriction of liftings over X_n to V_n corresponds to the restriction map from $H^0(X_0, \mathbf{F})$ to $H^0(V_0, \mathbf{F})$, which is an isomorphism because $\dim(B) = d-1 \geq 2$. The claim is now proved. Therefore $G^*_{/S^*-Q}{}^\wedge$ extends to $G^*_{/S^*}{}^\wedge$ which is an extension of a formal abelian scheme A by a torus. Again A is principally polarized and thus algebraic. Also the formal p-divisible group group $G^*_{/S^*-Q}{}^\wedge[p^\infty]$ can be uniquely extended to a formal p-divisible group $G^*_{/S^*}{}^\wedge[p^\infty]$ over $S^{*\wedge}$.

(viii) We can apply GFGA (Grothendieck's formal existence theorem) to conclude that $G^*_{/S^*}{}^\wedge[p^\infty]$ is the formal completion of a unique p-divisible group $G^*_f[p^\infty]_{/S^*}$ over S^*. Restrict to $S^* - E = S - \{s\}$, extend to S^* (by 6.1), then pull back and compare. We see that $G^*_f[p^\infty]_{/S^*}$ is the pull-back of a unique p-divisible group $G^*[p^\infty]_{/S}$ over S. This implies in particular that $G^*_{/E}[p^\infty]$ is constant along E, therefore $G^*_{/E}$ is constant (by 4.9), i.e. $G^*_{/E}$ is the pull-back of a semi-abelian variety $G_{/k}$ over k. By the usual computations of deformation spaces (as in step iii) of the proof of 6.10) liftings of $G_{/k}$ to the k-th infinitesimal neighborhood of $\mathfrak{m}$ in S correspond bijectively to liftings of $G^*_{/E}$ to the k-th infinitesimal neighborhood of E in S^*. Thus $G^*_{/S^*}{}^\wedge$ is the pull-back of a unique $G^\wedge_{/S}$. This is comforting, but we still have to get a semi-abelian scheme over S^*. We do this by the theory of degeneration (of course). The semi-abelian scheme $G^*_{/S^*-Q}$ gives a semi-abelian scheme over $(S^* - Q)^\wedge$. We would like to extend it to a semi-abelian scheme over $S^{*\wedge}$. Let us first show that this would conclude the proof of theorem 6.7: The same argument used in part (iv) of the proof of 6.10 shows that this semi-abelian scheme over $S^{*\wedge}$ descends uniquely to a semi-abelian scheme G over S. By the argument of 6.10 (v), $G \times_S S^*$ extends $G^*_{/S^*-Q}$, so G is the required extension.

(ix) It remains to extend the semi-abelian scheme $G^*_{/(S^*-Q)^\wedge}$ to $S^{*\wedge}$. We translate this problem using the theory of degeneration in chapters II, III: We have shown that the formal completion of G^* extends to $S^{*\wedge}$, and therefore we have an abelian scheme A over $S^{*\wedge}$, the character group X of the split torus T, and a homomorphism $c : X \to A$ over $S^{*\wedge}$. (We can dispense with the Y and A^t because we assume that the generic fibre is principally polarized.) The last part of the degeneration data for $G^*_{/(S^*-Q)^\wedge}$ is a symmetric bimultiplicative formal-meromorphic map ψ from $X \times X$ to the inverse of the Poincaré sheaf lying over $c \times c$, which satisfies a certain transitivity condition. We want to extend this degeneration data to $S^{*\wedge}$, which comes down to extending meromorphic sections of line bundles $\mathcal{L}$. As in the proof of 6.10 everything follows from the fact that $H^0(S^{*\wedge}, \mathcal{L}) \xrightarrow{\sim} H^0(S^{*\wedge} - Q, \mathcal{L})$, which holds because all S^*_k are Cohen-Macaulay and Q has codimension ≥ 2.

6.11. Corollary. *If U is the complement of a divisor with normal crossings in a regular scheme S whose maximal points are all of characteristic zero, then any morphism $f : U \to \mathsf{A}_{g,n}$ with $n \geq 3$ extends to a morphism $S \to \mathsf{A}^*_{g,n}$.*

Proof. On U we have a principally polarized abelian variety with level-n structure. It extends to a semi-abelian scheme G in codimension one, thus everywhere by theorem 6.7. Let X and Y denote the closures of the graph of f in $S \times \overline{\mathsf{A}}_{g,n}$ respectively $S \times \mathsf{A}^*_{g,n}$, so $X \to Y \to S$. The projection from Y to $\mathsf{A}^*_{g,n}$ maps $y \in Y$ to a point classifying (see theorem 2.5) the abelian part of G, so it is locally constant on the fibres of $Y \to S$. By Zariski's main theorem Y is isomorphic to S, so f extends.

A Counter-Example. Let us give an example that the assertions of theorem 6.4, theorem 6.7, proposition 6.10 and corollary 6.11 all fail in characteristic p. It is a variation of L. Moret-Bailly's example in section 4:

Let $R = k[\![x,y]\!]$, k algebraically closed of characteristic p, $S = \mathrm{Spec}(R)$, $V = X - \{\text{closed point}\}$, E a supersingular elliptic curve over k, so that E contains a copy of α_p. Over V we denote by A the quotient of $E \times E$ by the subgroup of $\alpha_p \times \alpha_p$ determined by the kernel of $(x,y) : R^2 \to R$. A extends to an abelian scheme A^* over S^* = the blow-up of the closed point of S, and the isogeny $E \times E \to A$ over U extends to an isogeny $E \times E \to A^*$ over S^*. The restriction of A^* to the exceptional locus is given by L. Moret-Bailly's example, hence is non-constant, so A^* does not extend to S itself. Also $V \to \mathsf{A}_g$ does not extend to S, nor does the p-divisible group $A[p^\infty]$ over V: Otherwise the homomorphism $E[p^\infty] \times E[p^\infty] \to A[p^\infty]$ over U would extend to an isogeny over S (look at the finite flat algebras) and its kernel to a flat subgroup of $\alpha_p \times \alpha_p$ over S. This is a contradiction, because the pull-back of this extended isogeny must coincide with the non-constant isogeny $E[p^\infty] \times E[p^\infty] \to A^*[p^\infty]$ attached to $E \times E \to A^*$.

7. p-adic Monodromy

In this section we give another application of the minimal compactification, to the p-adic monodromy of the "physical" p-adic Tate module of a generic principally polarized abelian variety in characteristic p. Let us explain this problem more precisely. We choose an algebraically closed field k of characteristic $p > 0$. To fix ideas let us fix an integer $N \geq 3$, $(N,p) = 1$ and another integer $g \geq 1$. Consider the moduli scheme $\mathsf{A}_{g,N}$ and the universal abelian scheme $A \to \mathsf{A}_{g,N} = \mathsf{A}$ over k. Let $\mathsf{A}^\circ \subset \mathsf{A}$ be the ordinary locus, so that A° is open dense in A and all geometric points of A° correspond to ordinary abelian varieties, and A° is maximal w.r.t. this property. (Recall that an abelian variety A over k is ordinary iff $A[p](k) \cong (\mathbb{Z}/p\mathbb{Z})^{\dim(A)}$.) Let $A[p^\infty] = \varprojlim_n A[p^n]$ be the Barsotti-Tate group (or p-divisible group) associated to A. Over A°, we have an exact sequence

$$0 \to A[p^\infty]_{\mathrm{tor}} \to A[p^\infty] \to A[p^\infty]_{\text{ét}} \to 0$$

where $A[p^\infty]_{\mathrm{tor}}$ (resp. $A[p^\infty]_{\text{ét}}$) are the toric part (resp. étale quotient) of $A[p^\infty]$. Over A°, $A[p^\infty]_{\text{ét}}$ can be identified with an étale sheaf of torsion free $\mathbb{Z}_p$-modules of rand g, therefore corresponds to a representation $\rho : \pi_1(\mathsf{A}^\circ) \to \mathrm{GL}_g(\mathbb{Z}_p)$ well-defined up to conjugation, called the p-adic monodromy representation of $A \to \mathsf{A}^\circ$. A natural guess is that ρ is actually surjective (i.e. the p-adic monodromy is as big as possible), a question which was raised explicitly in [Ka-La].

Historical Remarks. In the elliptic modular case this was first proved by Igusa [I 1] by analysing ramification at supersingular points, thereby showing that the local monodromy at any supersingular point is already $\mathbb{Z}_p^\times$. In the Hilbert modular case the analogous statement was first proved by Ribet [Ri], [De-Ri] by arithmetic methods, using Deligne's result on canonical liftings of abelian varieties over finite fields. Unfortunately it is not easy to use Ribet's method in our present case, because the target group of ρ is not commutative.

The goal of this section is to deduce the surjectivity as an easy consequence of the existence of the minimal compactification $\mathsf{A}_{g,N}^*$ (over k):

7.1. Proposition. *The p-adic monodromy representation* $\rho : \pi_1(\mathsf{A}^\circ) \to \mathrm{GL}_g(\mathbb{Z}_p)$ *is surjective.*

What we actually prove is that the p-adic local monodromy at the 0-dimensional cusps of $\mathsf{A}_{g,N}^*$ is already very big:

7.2. Proposition. *Let c be a 0-dimensional cusp of* $\mathsf{A}^* = \mathsf{A}_{g,N}^*$ *(over k), R be the completion of the local ring of* A^* *at c. Consider the local p-adic monodromy representation* $\rho_c : \pi_1(\mathrm{Spec}(R) - \{\infty\}) \to \mathrm{GL}_g(\mathbb{Z}_p)$*, defined as the composition of* ρ *with* $\pi_1(\mathrm{Spec}(R) - \{\infty\}) \to \pi_1(\mathsf{A}^\circ)$*. Then the image of* ρ_c *is equal to the subgroup* $\mathrm{SL}_g(\mathbb{Z}_p)$*.*

Remarks. (i) Igusa's theorem together with proposition 7.2 clearly implies proposition 7.1. But our method does not yield anything for the case $g = 1$. In particular, it does not give a new proof of Igusa's theorem.

(ii) As we remarked before, our method of constructing arithmetic toroidal compactifications actually generalizes to Shimura varieties of PEL-type. Similarly, the method of proposition 7.2 can be applied to show that the local p-adic monodromy of Shimura varieties of PEL-type at the boundary of their minimal compactification is big. For instance, this method gives a geometric proof of Ribet's theorem.

Proof of Proposition 7.2. First of all, over A°, the Cartier dual of $A[p^\infty]_{\mathrm{tor}}$ is $A[p^\infty]_{\text{ét}}$, so it suffices to show that the local monodromy of (the Cartier dual of) $A[p^n]_{\mathrm{tor}}$ over A° near the 0-dimensional cusp c of A^* already contains $\mathrm{SL}(g, \mathbb{Z}/p^n\mathbb{Z})$. Let $\overline{\mathsf{A}}$ be an arithmetic toroidal compactification of A, $\pi : \overline{\mathsf{A}} \to \mathsf{A}^*$ be the natural morphism. Over $\mathrm{Spec}(R)$ the Barsotti-Tate group $A[p^\infty]_{\mathrm{tor}} \to \mathsf{A}^\circ$ extends to a neighborhood of $\pi^{-1}(c)$, because G is a torus over $\pi^{-1}(c)$. By chap. IV theorem 5.4.(4), the formal completion of $G \to \overline{\mathsf{A}}$ along $\pi^{-1}(c)$ is isomorphic to $({}^{\heartsuit}G \to \overline{E}^\wedge)/\mathrm{GL}(X)(N)$, using the notation of chap. IV 3.1, where $\mathrm{GL}(X)(N) = \mathrm{Ker}\{\mathrm{GL}(X) \to \mathrm{GL}(X/NX)\}$, and $\overline{E}^\wedge$ is connected. Also $A[p^\infty]_{\text{ét}}$ is given by the representation of $\mathrm{GL}(X)(N)$ on X. Now we can simply read off the local monodromy of $A[p^n]$ over $\mathrm{Spec}(R) - \{\infty\}$ by looking at the $\mathrm{GL}(X)$-action on the cone decomposition: it is just $\mathrm{GL}(X)(N)/\mathrm{GL}(X)(N \cdot p^n)$, which is canonically isomorphic to $\mathrm{SL}(X/p^nX) \cong \mathrm{SL}(g, \mathbb{Z}/p^n\mathbb{Z})$. □

Chapter VI. Eichler Integrals in Several Variables

In this chapter we study the cohomology of A_g. In sections 3, 4, 5 we study the Betti cohomology of $\mathsf{A}_g(\mathbb{C})$ and its Hodge structure, and show the degeneration of various spectral sequences. Our method, developed in [F 1], is based on the Bernstein-Gelfand-Gelfand (abbreviated as BGG) resolution (cf. [BGG]), Mumford's extension of equivariant vector bundles to toroidal compactifications (cf. [Mum 6]) and Deligne's Hodge theory (cf. [D 2], [D 3]). Making use of geometric information available in our present case, we obtain results not contained in [F 1]. In some sense the major work is contained in section 2, where we compute the formal cohomology of toroidal compactifications of fibre products of the universal abelian scheme. In section 6 we describe (but do not present) the p-adic analogue, namely the p-adic étale cohomology groups, which turn out to be crystalline (hence Hodge-Tate) Galois representations and closely related to the crystalline cohomology groups. (Since the crystalline cohomology is closely tied to the de Rham cohomology as is well-known, this analogy is a very good one.) The first section furnishes geometric information needed for studying cohomology, namely explicit compactification of the fibre products of the universal abelian scheme.

1. Compactification of the Universal Abelian Variety

A good way of getting information about cohomology of open varieties (resp. maps), first conceived by Grothendieck, is to use "good" compactifications of the varieties (resp. maps) in question. It is often helpful to have an explicit compactification which can be controlled. In this section we compactify the universal abelian variety, which gives geometric information needed later in the chapter. For some applications it suffices to construct compactifications locally in the étale topology of $\overline{\mathsf{A}}_g$ (we choose once and for all an admissible cone decomposition), so the reader only interested in these need not consider global gluing.

We shall need to compactify not only the universal abelian variety, but its fibre products.Thus we shall use the following notation: For some $s \geq 1$, consider $Y = A^s$, the s-fold product of the universal abelian variety A over $\overline{\mathsf{A}}_g$ or $\overline{\mathsf{A}}_{g,n}$. The reader should note that our compactification will *not* contain the universal semi-abelian scheme G^s over $\overline{\mathsf{A}}_{g,n}$ as an open subset. First we work only locally in $\overline{\mathsf{A}}_{g,n}$ so we can use schemes everywhere, and we do not need stacks or algebraic spaces. These however make their appearance in the global statement. Before formulating the main result let us fix some notation: If $\overline{Y}$ is a smooth S-scheme, and $Y \subset \overline{Y}$ an open subscheme whose complement is a relative divisor with normal crossings, we abbreviate $\Omega^b_{\overline{Y}/S}[d\log\infty]$ by $\overline{\Omega}^b_{Y/S}$. When $S = \mathrm{Spec}(\mathbb{Z})$, the subscript will be omitted.

Let us now formulate the main result of this section and the next:

1.1. Theorem. *Let $X = \mathsf{A}_{g,n} \subset \overline{X} = \overline{\mathsf{A}}_{g,n}$, and $f : Y = A^s \to X$ denote the s-fold fibre product of the universal abelian scheme over X. Then there exists an open embedding $Y \subset \overline{Y}$, and a proper extension $\overline{f} : \overline{Y} \to \overline{X}$ such that*

(i) $\overline{Y}$ is smooth over $\mathbb{Z}$, and $\overline{Y} - Y$ is a relative divisor with normal crossings.

(ii) The translation-action of Y on itself extends to an action of G^s on $\overline{Y}$.

(iii) $\overline{\Omega}^1_{Y/X} = \overline{\Omega}^1_Y / \overline{f}^*(\overline{\Omega}^1_X)$ is locally free, and isomorphic to $\overline{f}^*(\mathcal{T}^{*s}_G)$.

(iv) $R^a\,\overline{f}_*(\Lambda^b\,\overline{\Omega}^1_{Y/X}) \cong \Lambda^a\,\mathcal{T}^s_G \otimes \Lambda^b\,\mathcal{T}^{*s}_G$.

Here G denotes the universal semi-abelian scheme over $\overline{X}$, $\mathcal{T}_G = \mathbf{Lie}(G/\overline{X})$, and the isomorphisms above extend the canonical isomorphisms over X (which determine them uniquely).

The proof is quite lengthy and is spread out over two sections, but the principle is pretty easy: In the construction of G we have already used compactifications, via the notion of relative complete models. We use a variant of this notion (called weak relatively complete models, obtained by relaxing the condition that $\tilde{G}$ be contained in $\tilde{P}$ in the notation of chapter 3), to get smooth compactifications, for which we make use of toroidal methods. Parts (iii) and (iv) are proved by computing formal cohomology of $\overline{Y}$ at the boundary, which will be done in §2.

Bibliographical note: Yukihiko Namikawa constructed a toroidal compactification of the universal family of abelian varieties in the category of complex analytic spaces using the Voronoi-Delaunay decomposition, c. [Na 1]. In [Na 2], he axiomatized his construction, which is very similar to our notion of weak relatively models. He also announced results about the projectivity of his compactification.

We shall first introduce the combinatorial data to be used to produce $\overline{Y} \to \overline{X}$. Instead of the cone $C = C(X)$ of positive semi-definite symmetric bilinear forms on $X = \mathbb{Z}^g$ with rational radicals, we use a more complicated cone $\tilde{C} = \tilde{C}_s(X)$. Let $\tilde{B}_s(X) = B(X) \times (X^*)^s$, a typical element of which will be denoted by $(b; l_1, \ldots, l_s)$. Usually we shall think of an element $(b; l_1, \ldots, l_s)$ of $\tilde{B}_s(X)$ as an s-tuple $(a_1, \ldots, a_s)$ of quadratic functions up to constants $a_j : X \to \mathbb{Z}$ with $a_j(\mu) = b(\mu, \mu) + 2 \cdot l_j(\mu)\ \forall \mu \in X$. In $\tilde{B}_s(X)_{\mathbb{R}} = \tilde{B}_s(X) \otimes_{\mathbb{Z}} \mathbb{R}$ we have the cone $\tilde{C}_s(X)$ consisting of all $q = (b; l_1, \ldots, l_s) \in \tilde{B}_s(X)_{\mathbb{R}} = B(X)_{\mathbb{R}} \times (X^*_{\mathbb{R}})^s$ such that $b \in C(X)$ and each l_j vanishes on the radical of b (in other words each a_j is bounded below and their common quadratic part has rational radical). Clearly the surjection $\tilde{B}_s(X) \to B(X)$ induces a surjection $\tilde{C} \to C$, and the fibre of $\tilde{C}$ over an element $b \in C$ is isomorphic to the image of the map $X_{\mathbb{R}}^s \to (X^*)_{\mathbb{R}}^s$ induced by b. Instead of the symmetry group $\mathrm{GL}(X)$ we now have the semidirect product $\mathrm{GL}(X) \ltimes X^s$ operating, the second factor by translation in the arguments of the a_j. More precisely, for $\gamma \in \mathrm{GL}(X)$, $q = (b; l_1, \dots, l_s) \in Q_s(X)_{\mathbb{R}}, \mu = (\mu_1, \dots, \mu_s) \in X^s, (\gamma \cdot q) := (b \circ \gamma^{-1}; l_1 \circ \gamma^{-1}, \dots, l_s \circ \gamma^{-1}) = (b \circ \gamma^{-1}; {}^t\gamma^{-1} \cdot l_1, \dots, {}^t\gamma^{-1} \cdot l_s), (\mu \cdot q) = T_\mu q := (b; l_1', \dots, l_s')$ with $l_i'(\nu) = l_i(\nu) + b(\mu_i, \nu)$, so $\gamma \cdot \mu \cdot \gamma^{-1} = (\gamma \cdot \mu_1, \dots, \gamma \cdot \mu_s)$. This operation respects $\tilde{C}$, and the projection onto $B(X)$ is equivariant with respect to the canonical morphism $\mathrm{GL}(X) \ltimes X^s \to \mathrm{GL}(X)$. Note that $\tilde{C}_s(X)$ is the s-fold fibre product of $\tilde{C}_1(X)$ over $C(X)$, and the inverse image in $\tilde{C}_s(X)$ of a rational boundary component $C(X_1)$ of $C(X)$ determined by a surjection of lattices $X \to X_1$ is naturally isomorphic to $\tilde{C}_s(X_1)$.

1.2. Remark. It is of some interest to consider the lattice $Q_s(X)$ consisting of s-tuples $(a_1, \dots, a_s)$ of quadratic functions $a_j : X \to \mathbb{Z}$, up to constants, such that the associated symmetric bilinear forms of the a_j are all equal. There exists a surjection $Q_s(X) \to B(X)$, sending $(a_1, \dots, a_s)$ to the associated bilinear form b, and the kernel of this surjection is isomorphic to $(X^*)^s$ (as we consider only quadratic functions up to constants). Thus we have a short exact sequence $0 \to (X^*)^s \to Q_s(X) \to B(X) \to 0$, but there is no natural splitting: $Q_s(X)$ is only non-canonically isomorphic to $\tilde{B}_s(X) = B(X) \times (X^*)^s$. On the other hand $Q_s(X) \otimes \mathbb{Z}[1/2] \to B(X) \otimes \mathbb{Z}[1/2]$ is canonically isomorphic to $\tilde{B}_s(X) \otimes \mathbb{Z}[1/2] \to B(X) \otimes \mathbb{Z}[1/2]$. Thus $Q_s(X)_{\mathbb{R}} = \tilde{B}_s(X)_{\mathbb{R}}$, inside of which we have the cone $\tilde{C}_s(X)$. The action of $\mathrm{GL}(X) \ltimes X^s$ is given as follows: for $\gamma \in \mathrm{GL}(X)$, $q = (a_1, \dots, a_s) \in Q_s(X)_{\mathbb{R}}$, $\mu = (\mu_1, \dots, \mu_s) \in X_s$, $(\gamma \cdot q) = (a_1 \circ \gamma^{-1}, \dots, a_s \circ \gamma^{-1})$, $(\mu \cdot q) = T_\mu q = (a_1', \dots, a_s')$ with $a_i'(\nu) = a_i(\nu + \mu_i) - a_i(\mu_i)$, so $\gamma \cdot \mu \cdot \gamma^{-1} = (\gamma \cdot \mu_1, \dots, \gamma \cdot \mu_s)$. If we use the integral structure given by $Q_s(X)$ instead of $\tilde{B}_s(X)$ and make the necessary modifications in what follows, we get a toroidal compactification of the s-fold fibre product of the component of the Picard scheme of the universal abelian scheme corresponding to the principal polarization.

1.3. Definition. Let $\{\sigma_\alpha\}$ be a $\mathrm{GL}(X)$-admissible polyhedral cone decomposition of C. A $\mathrm{GL}(X) \ltimes X^s$*-admissible polyhedral cone decomposition of* $\tilde{C}$ *relative to* $\{\sigma_\alpha\}$ is collection of simplices τ such that

(i) Each τ is a non-degenerate rational polyhedral cone which is open in the smallest $\mathbb{R}$-subspace containing it.

(ii) Any face of a τ in the collection also belongs to this collection.

(iii) $\tilde{C}$ is the disjoint union of the τ's.

(iv) The collection of τ's is invariant under $\mathrm{GL}(X) \ltimes X^s$, and there are only finitely many orbits under this group.

(v) Any τ maps under the natural projection $\tilde{C} \to C$ into a σ_α. Such a cone decomposition of $\tilde{C}$ is called *smooth* if each τ is generated by part of a $\mathbb{Z}$-basis of $\tilde{B}_s(X)$. It is said to be *equidimensional* if any τ maps onto a σ_α under the projection $\tilde{C} \to C$.

1.4. Remarks. Reduction theory implies that admissible polyhedral cone decompositions of $\tilde{C}$ exist, that any two of them have a common refinement, and that any one has a smooth refinement ([KKMS], Ch. I, §2, Th. 11). Let us also remark that for any quotient lattice X_1 of X, we have an induced $\mathrm{GL}(X_1) \ltimes X_1^s$-admissible decomposition of $\tilde{C}_s(X_1)$. In general the natural section of $\tilde{B}_s(X) \to B(X)$ may not respect the cone decompositions. However if we refine the cone decomposition of C and $\tilde{C}$, it can be arranged that each $\sigma_\alpha \times \{0\}$ is a τ, in which case $\overline{Y}$ would contain G as a dense open subset. The compactification $\overline{Y} \to \overline{X}$ will turn out to be equidimensional (i.e. fibres have the same dimension) if the cone decomposition of $\tilde{C}$ is equidimensional in the above sense. However we shall not make use of this property and therefore refrain from further combinatorial analysis of this notion.

1.5. Definition. Let $\{\tau_\beta\}$ be a $\mathrm{GL}(X) \ltimes X^s$-admissible polyhedral cone decomposition of $\tilde{C}_s(X)$. An *admissible homogeneous principal polarization function* of $\{\tau_\beta\} \to \{\sigma_\alpha\}$ is a piecewise linear $\mathrm{Gl}(X)$-invariant continuous function $\tilde{\phi}$ on $\tilde{C}_s(X)$ such that

(i) $\tilde{\phi}$ takes rational values on $\tilde{B}_s(X) \cap \tilde{C}$ with bounded denominators.

(ii) $\tilde{\phi}(t \cdot q) = t \cdot \tilde{\phi}(q) \forall t \in \mathbb{R}_{\geq 0}$, $\forall q \in \tilde{C}$.

(iii) $\tilde{\phi}$ is linear on each τ_β.

(iv) $\tilde{\phi}$ is convex in the sense that $\{q \in \tilde{C}_s(X) | \tilde{\phi}(q) \geq a\}$ is convex $\forall a \in \mathbb{R}$; equivalently, $\tilde{\phi}(t \cdot q + (1-t) \cdot q') \geq t \cdot \tilde{\phi}(q) + (1-t) \cdot \tilde{\phi}(q') \forall 0 \leq t \leq 1, \forall q, q' \in \tilde{C}_s(X)$.

(v) $\tilde{\phi}$ is strictly convex over each cone σ_α. More precisely, for each cone σ_α in $C(X)$ and each τ_β lying over σ_α, there exist a finite number of linear functionals $l_i : \tilde{B}_s(X) \to \mathbb{R}, 1 \leq i \leq m$ with $l_i \geq \tilde{\phi}$ on the preimage of σ_α for each i, and $\tau_\beta = \{q \in Q_s(X) | q \text{ lies over } \sigma_\alpha \text{ and } \tilde{\phi}(q) = l_i(q) \text{ for each } i\}$.

(vi) There exists a rational number $r > 0$ such that for each $\mu = (\mu_1, ..., \mu_s) \in X^s$, $\tilde{\phi} - \tilde{\phi} \circ T_\mu : q \mapsto f(q) - f(\mu \cdot q)$ is equal to r times (restriction to $\tilde{C}$ of) the linear functional $\tilde{\chi}_\mu$ on $\tilde{B}_s(X)$, where for $q = (b; l_1, \ldots, l_s) \in \tilde{C}$, $\tilde{\chi}_\mu(q) = \sum_{1 \leq i \leq 1} a_i(\mu_i) = \sum_{1 \leq i \leq s} (b(\mu_i, \mu_i) + 2 \cdot l_i(\mu_i))$.

1.6. Remarks. (a) The conditions (i)–(vi) above constitute a kind of convexity condition on $\{\tau_\beta\} \to \{\sigma_\alpha\}$. We shall see that they imply that the morphism $\overline{Y} \to \overline{X}$ attached to $\{\tau_\beta\} \to \{\sigma_\alpha\}$ is projective. In fact even if we are only interested in constructing local compactifications, we need a local version of this for algebraization. (Of course, if one works in the category of complex analytic spaces this is not needed.)

(b) The basic example of such a polarization function $\tilde{\phi}$ is $\tilde{C} \ni q = (a_1, \ldots, a_s) \mapsto \sum_{1 \leq i \leq s} \mathrm{Min}_{\mu \in X}\{a_i(\mu)\})$. This function determines a cone decomposition of $\tilde{C}_s(X)$ equidimensional over a cone decomposition of $C(X)$, which is a mild generalization of the Voronoi-Delaunay decomposition. Taking subdivisions and slightly modifying this polarization function, we obtain other cone decompositions admitting polarization functions, which can be chosen to be smooth.

(c) The reason for using the linear functionals $\tilde{\chi}_\mu$ in the above definition is that such a polarization function will induce an ample invertible sheaf on $\overline{Y}$

extending a power of $\mathcal{O}(2)$ on Y, where $\mathcal{O}(2)$ denotes the canonical symmetric ample invertible sheaf on the (s-fold fibre product of) the universal abelian scheme giving twice the principal polarization.

We need a mild generalization of the notion of relatively complete model; we shall see that Mumford's construction works in this (generalized) setting. As in chapter III, $S = \mathrm{Spec}(R)$ for a noetherian normal domain R complete w.r.t. an ideal I with $\mathrm{rad}(I) = I, K =$ fraction field of R, $\eta =$ generic point of S, $R_k = R/I^{k+1}$ for $k \geq 0$, $S_k = R_k$.

1.7. Definition. A *weak relatively complete model* of a split object $(\tilde{G}, X, Y, c, c^t, \lambda, \phi, \tau, \imath, \tilde{\mathcal{L}}, \psi, \mathcal{M})$ in DD_{ample} consists of the following 5 pieces of data satisfying 3 conditions to be specified later:

(a) an integral scheme $\tilde{P}$ locally of finite type over A (= the abelian part of $\tilde{G}$) whose generic fibre is (canonically isomorphic to) $\tilde{G}_\eta$. The structural morphism $\tilde{P} \to A$ is denoted by $\tilde{\pi}$.

(b) an action of $\tilde{G}$ on $\tilde{P}$ extending the translation action on the generic fibre. This action will be denoted by T_g for any functorial point g of $\tilde{G}$.

(c) a rigidified invertible sheaf $\tilde{\mathcal{L}}$ on $\tilde{P}$ extending $\tilde{\pi}^*\mathcal{M}|_{\tilde{\pi}^{-1}(V)}$ where V is the largest open subscheme of S over which $\tilde{P}$ is isomorphic to $\tilde{G}$ (extending the isomorphism between generic fibres).

(d) an action of Y on $(\tilde{P}, \tilde{\mathcal{L}})$ extending the action of Y on $(\tilde{G}_\eta, \tilde{\pi}^*\mathcal{M}_\eta)$ defined by $\imath, \phi, \psi$. This action will be denoted by $S_y : \tilde{P} \to \tilde{P}$ and $\tilde{S}_y : S_y^*\tilde{\mathcal{L}} \to \tilde{\mathcal{L}}$ for $y \in Y$.

(e) An action of $\tilde{G}$ on the rigidified invertible sheaf $\mathcal{N} := \tilde{\mathcal{L}} \otimes \tilde{\pi}^*\mathcal{M}^{\otimes -1}$ extending the translation action of $\tilde{G}_\eta$ on its structural sheaf. This action will be denoted by $\tilde{T}_g : T_g^*\mathcal{N} \to \mathcal{N}$ for any functorial point g of $\tilde{G}$.

The above data are required to satisfy the following three conditions:

(i) There exists an open $\tilde{G}$-invariant subscheme $U \subset \tilde{P}$ of finite type over S such that $\tilde{P} = \cup_{y \in Y} S_y(U)$.

(ii) $\tilde{\mathcal{L}}$ is ample on $\tilde{P}$ in the sense that complements of zero-sets of global sections in $\Gamma(\tilde{P}, \tilde{\mathcal{L}}^{\otimes n}), n \geq 1$, give a basis of the Zariski topology of $\tilde{P}$.

(iii) (completeness condition) For any valuation (not necessarily of rank one) v of the function field $K(\tilde{G})$ of $\tilde{G}$ with $v(R) \geq 0, v$ has a center on $\tilde{P}$ if and only if $\forall \mu \in X, \exists y \in Y$ such that $v(I_{y,\mu} \cdot (\mathcal{O}_\mu)_{s_v}) \geq 0$, where s_v is the center of v on S, $(\mathcal{O}_\mu)_{s_v}$ is the subsheaf of functions over $\mathcal{O}_{S,s_v}$ transforming with character μ of the torus T, and $I_{y,\mu}$ is the principal fractional ideal of R defined by $\tau(y, \mu)$ w.r.t. the integral structure given by $(c^t \times c)^*\mathcal{P}_A^{\otimes -1}|_{(y,\mu)}$.

1.8. Remarks. (a) The above definition differs from that of a relatively complete model only in that we do not require $\tilde{G}$ be contained in $\tilde{P}$.

(b) Our major examples of weak relatively complete models come from admissible polyhedral cone decompositions of $\tilde{C}$ which admit homogeneous principal polarization functions, as we shall see below.

Mumford's Construction Revisited

Given a weak relatively complete model $(\tilde{P}, \tilde{\mathcal{L}})$ as above, we would like to construct a Y-quotient $(P, \mathcal{L})$ of it. Of course we just follow the steps of chapter III, and check that all the constructions for relatively complete models go through for weak models: We form the formal (I-adic) completion of $(\tilde{P}, \tilde{\mathcal{L}})$, divide by Y, and algebraize it to a pair $(P, \mathcal{L})$ consisting of a proper scheme P over S and an ample invertible sheaf $\mathcal{L}$ on it. The generic fibre of P is smooth over K just as before. Let us formulate the basic properties of the quotient construction in the present setting:

(i) The semi-abelian scheme $G \to S$ obtained from the given ample degeneration data operates on $P \to S$.

(ii) The generic fibre P_η is canonically isomorphic to G_η, respecting the zero section and the translation action. In particular the generic fibre P_η is independent of the choice of the weak relatively complete models. The zero section of G_η extends to a section of $P \to S$.

(iii) If U is an open subset of S over which $\tilde{P}$ contains $\tilde{G}$, then $P \times_S U$ contains $G \times_S U$.

(iv) P is irreducible with connected geometric fibres. G_η is schematically dense in P.

We shall only sketch the proofs here, as they are nearly identical to those in chapter III. First one checks that the theory of integrable semi-abelian subschemes carries over in the present context. This gives the action of G on P and also shows that P_η is independent of the choice of weak relatively complete models. In fact the same argument shows (iii) if one is careful about open subsets of S over which $\tilde{P}$ contains $\tilde{G}$. As in chapter III, irreducibility of P follows by basechange to complete discrete valuation rings. Notice that we do not need to assume S to be excellent (cf. chap. III), since normalization is unnecessary. That P has connected geometric fibres follows from the completeness condition as in [Mum 5].

1.9. Remark. It is possible to generalize the notion of weak relatively complete models even further so that P_η is only a G_η-torsor. We shall not need this generality, and leave this to the diligent reader.

1.10. Construction of Good Compact Formal Models at the Boundary: The Totally Degenerate Case. Let us choose and fix a $\mathrm{GL}(X) \ltimes X^s$-admissible polyhedral cone decomposition $\{\tau_\beta\}$ of $\tilde{C}$ relative to a $\mathrm{GL}(X)$-admissible polyhedral cone decomposition $\{\sigma_\alpha\}$, which admits a principal homogeneous polarization function $\tilde{\phi}$. Clearing the denominators, we can *assume that $\tilde{\phi}$ takes integral values on integral elements of $\tilde{C}$ and that $\tilde{\phi} - \tilde{\phi} \circ T_\mu = k \cdot \tilde{\chi}_\mu$ for some integer* $k > 0$. For simplicity we shall also *assume that both $\{\tau_\beta\}$ and $\{\sigma_\alpha\}$ are smooth,* although it not logically necessary.

By the theory of torus embeddings, to the map $(\tilde{C}, \{\tau_\beta\}) \to (C, \{\sigma_\alpha\})$ corresponds a map of torus embeddings $\overline{F} \to \overline{E}$, equivariant for the morphism of tori $F \to E$ with cocharacter groups $\tilde{B}_s(X)$ and $B(X)$, and for the map of symmetry groups $\mathrm{GL}(X) \ltimes X^s \to \mathrm{GL}(X)$. Note that the kernel of $F \to E$ is the torus $\tilde{G}^s := (X^*)^s \otimes G_m$ = the torus with character-group X^s. F is canonically isomorphic to $E \times ((X^*)^s \otimes G_m)$. Also there is a proper birational modification $\overline{E}^* \to \overline{E}$, which is isomorphic over E, such that $\overline{E}^*$ has an $\overline{E}$-map into $\overline{F}$ extending the natural inclusion $E \to F$. $\overline{E}^*$ corresponds to a subdivision of $\{\sigma_\alpha\}$.

Finally we need a relatively ample invertible sheaf $\mathcal{L}$ on $\overline{F}$. For this we make use of the polarization function $\tilde{\phi} : \tilde{C} \to \mathbb{R}$: Define a fractional ideal $\mathcal{I} = \mathcal{I}_{\tilde{\phi}}$ of $\overline{F}$, by requiring that over the open affine set defined by a simplex τ, $\mathcal{I}_{\tilde{\phi}}$ is generated by the character χ of F corresponding to the element of $Q_s(X)^*$ such that $\chi|\tau = \tilde{\phi}|\tau$. As ϕ is strictly convex over each simplex σ_α in C, [KKMS], Ch. I, §3, Th. 13 shows that $\mathcal{I}_{\tilde{\phi}}$ is a relatively ample invertible sheaf for $\overline{F} \to \overline{E}$. By construction, $\mathcal{I}$ is equivariant under the torus F and also its subtorus $\tilde{G}^s$. We can extend the operation of X^s on $\overline{F}$ to $\mathcal{I}$, by twisting the standard operation on rational functions with multiplication by the rational functions $\tilde{\chi}_\mu^k$ for $\mu \in X^s$ (i.e. the map $T_\mu^* \mathcal{I}_{\tilde{\phi}} \to \mathcal{I}_{\tilde{\phi}}$ given by the action is "multiplication by $\tilde{\chi}_\mu^k$"). That this is indeed an action of X^s on $\mathcal{I}$, follows from the transformation rule for $\tilde{\phi} : \tilde{\phi} - \tilde{\phi} \circ T_\mu = k \cdot \tilde{\chi}_\mu$. However, this operation does not commute with that of $\tilde{G}^s$, and we obtain as commutator between a $\mu \in X^s$ and $\tilde{G}^s$ the character μ^{2k} on $\tilde{G}^s$, since $\tilde{\chi}_\mu | T^s = \mu^2$ as character of $\tilde{G}^s = (X^*)^s \otimes G_m$. More precisely $\tilde{T}_a^* \tilde{T}_\mu^* = \mu^{2k}(a) \tilde{T}_\mu^* \tilde{T}_a^*$ for each functorial point a of $\tilde{G}^s$. This should remind us of the definition of relatively complete models in [Mum 5] (in which such a commutator is specified as part of the definition), in the sense that $\overline{F}$ looks like such a model over $\overline{E}$. Some other similarities:

(i) Over the 0-stratum of $\overline{E}$, the action of X^s on $\overline{F}$ is derived from the T^s-action (translation) via the bilinear form $(\mu, \nu) \to \sum_{1 \le j \le s} \mu_j \cdot \nu_j \in S^2(X)$, on $X^s \times X^s$, where $\mu = (\mu_1, \dots, \mu_s)$, $\nu = (\nu_1, \dots, \nu_s)$ and $\sum_{1 \le j \le s} \mu_j \cdot \nu_j$ is regarded as a regular function on E. This is clear from the definition.

(ii) The 0-stratum in $\overline{F}$ is the product of the 0-stratum in $\overline{E}$ and the split torus T^s.

(iii) If V is a valuation ring (not necessarily discrete), with a dominant map $\mathrm{Spec}(V) \to \overline{E}$, and field of quotients K, an $\overline{E}$-morphism $\mathrm{Spec}(K) \to \overline{F}$ extends to $\mathrm{Spec}(V)$ if and only if $\forall \mu = (\mu_1, \dots, \mu_s) \in X^s$ the valuation of the function $\tilde{\chi}_\mu$ on $\overline{F}$ is bounded above and below by suitable integral multiples of the valuation of $\sum \mu_j \otimes \mu_j \in S^2(X)$. This follows from the definition of $\tilde{C}_s(X)$: To make this clear we first explain the case of a discrete valuation. It corresponds to an element of $q = (b, l_1, \dots, l_s) \in \tilde{B}_s(X)$ such that the corresponding bilinear form b is positive semi-definite. It lies in $\tilde{C}$ if and only if the l_i are bounded by multiples of b.

In general if Γ denotes the value group of V, the valuations of the characters of E define a homomorphism $b_\Gamma : S^2(X) \to \Gamma$, and the valuations of the characters of F define a homomorphism $q_\Gamma : \tilde{B}_s(X)^* \to \Gamma$ extending b_Γ. In general if we have a homomorphism $h : M \to \Gamma$ from a lattice M to an ordered group Γ, $h^{-1}(\Gamma_{\ge 0})$ is a sub-semigroup of M which is either equal to M or is the intersection of M with a closed half-space of $M \otimes \mathbb{R}$. In either case we obtain a (possibly degenerate) rational half-line $\mathbb{R}_{\geq 0}h_1$ in $M^* \otimes \mathbb{R}$. If $\mathbb{R}_{\geq 0}h_1 \neq \{0\}$, restrict h to the $\ker(h_1) \cap M$ and repeat the above procedure until we get a degenerate half-line. This way we obtain a sequence of half-lines $\mathbb{R}_{\geq 0}h_1 \subseteq M^* \otimes \mathbb{R}$, $\mathbb{R}_{\geq 0}h_2 \subseteq M^* \otimes \mathbb{R}/\mathbb{R}h_1, \cdots, \mathbb{R}_{\geq 0}h_k \subseteq M^* \otimes \mathbb{R}/\mathbb{R}h_1 + \ldots + \mathbb{R}h_{k-1}$, $\mathbb{R}_{\geq 0}h_{k+1} = \{0\}$, $k \leq \operatorname{rank}(M)$. For an element $m \in M$ we have $h(m) \geq 0$ if and only if $h_i(m) > 0$ for the first i with $h_i(m) \neq 0$. This whole package of half-lines in $M^* \otimes \mathbb{R}$ determined by h will be denoted $[h]$, which is reminiscent of the notion of infinitely near points in algebraic geometry. Coming back to our previous situation, b_Γ (resp. q_Γ) determines a package half-lines $[b_\Gamma]$ (resp. $[q_\Gamma]$) in $B(X) \otimes \mathbb{R}$ (resp. $\tilde{B}_s(X) \otimes \mathbb{R}$), and $[q_\Gamma]$ lies over $[b_\Gamma]$ in the obvious sense (that is half-lines map to half-lines). Thus all the usual properties for discrete valuations still hold with obvious modifications. For instance, that b_Γ comes from a morphism $\operatorname{Spec}(V) \to \overline{E}$ means that $[b_\Gamma]$ lies in the closure of a simplex σ in (the cone decomposition of) $C(X)$, that is, that we can find representatives for the b_i which lie in the closure of σ. To say that $\overline{E}$-morphism $\operatorname{Spec}(K) \to \overline{F}$ extends to $\operatorname{Spec}(V)$ is equivalent to saying that $[q_\Gamma]$ lies in the closure of a simplex τ in (the cone decomposition of) $\tilde{C}_s(X)$. By the definition of $\tilde{C}_s(X)$ it is easy to see that this is the case if and only if $\forall \mu = (\mu_1, \ldots, \mu_s) \in X^s$, $q_\Gamma(\tilde{\chi}_\mu)$ is bounded above and below by suitable integral multiples of $b_\Gamma(\mu_j \otimes \mu_j)$.

(iv) Finally for each affine open subscheme $E(\sigma)$ of $\overline{E}$ attached to a simplex σ there exists an T^s-invariant open subset U of finite type such that the inverse image of $E(\sigma)$ in $\overline{F}$ is the union of the X^s-translates of U. (Because the inverse image in $\tilde{C}$ of the closure of σ is covered by finitely many X^s-orbits of τ_β's.)

We derive that for a simplex $\sigma \subset C(X)^\circ$, if we replace $\overline{E}$ by its formal completion along the σ-stratum, we obtain by base change a scheme $\tilde{P}$ over a noetherian complete normal domain $R_\sigma^\wedge$, an ample invertible sheaf $\mathcal{I}$, a split torus $\tilde{G}^s$ with character group X^s operating on $\tilde{P}$, and an action of X^s on $(\tilde{P}, \mathcal{I})$ which satisfy all axioms for a relatively complete model, except that $\tilde{P}$ does not contain $\tilde{G}^s$ as an open subset, but only its restriction to the 0-stratum in the base; i.e. $(\tilde{P}, \mathcal{I})$ is a weak relatively complete model. Indeed this example is the motivation behind the definition of weak relatively complete models. Now by our previous result on quotients of weak relatively complete models, we obtain for each simplex $\sigma \subset C(X)^\circ$ a scheme P_σ proper over $\operatorname{Spec}(R_\sigma^\wedge)$ extending the s-fold fibre product A_σ^s of the universal abelian scheme over the 0-stratum of $\operatorname{Spec}(R_\sigma^\wedge)$, an ample invertible sheaf $\mathcal{L}_\sigma$ on P_σ extending the symmetric ample invertible sheaf $\mathcal{O}(2k)$ over A_σ^s and an action of the s-fold fibre product G_σ^s of the semi-abelian scheme on P_σ extending the translation action of A_σ^s.

As in chapter IV we can do better: Let $\overline{E}^\wedge$ be the formal completion of $\overline{E}$ along the (reduced) locally closed subscheme whose support is the union of strata corresponding to cones σ_α in $C(X)^\circ$. Then the above construction gives us a proper scheme $P \to \overline{E}^\wedge$ extending A^s over the 0-stratum, a symmetric relatively ample invertible sheaf $\mathcal{L}$ on $P \to \overline{E}^\wedge$ extending $\mathcal{O}(2k)$ on A^s, an action of G^s on P extending the translation action, and a $\mathrm{GL}(X)$-action on all these. So everything descends to $\overline{E}^\wedge/\mathrm{GL}(X)$ (as a formal algebraic stack).

It is also clear from the construction that $P \to \overline{E}^\wedge/\mathrm{GL}(X)$ has a natural stratification coming from the cone decompositions.

Finally we can introduce level-n structures just as before. In the present situation, introducing the principal level-n structure just means that we work over $\mathbb{Z}[\zeta_n, 1/n]$, pull back the weak relatively complete model to $\overline{E}_n^\wedge$ (defined by the same cone decomposition with integral structure given by the lattice $n \cdot B(X)$), divide by X^s, and pass to quotient by the principal congruence subgroup of $\mathrm{GL}(X)$ of level-n. The details are left to the readers.

1.11. Construction of Good Compact Formal Models at the Boundary: The Mixed Case. In general, of course, we need mixed versions of weak relatively complete models (and their X^s-quotients), attached to the rational boundary components of $C(X)$. Now $\tilde{G}$ does not denote a torus any more, but always a semi-abelian scheme sitting in a short exact sequence $0 \to T \to \tilde{G} \to A \to 0$, where T is a (split) torus and A an abelian scheme.

Let X_ξ be a quotient lattice of X, $r = \mathrm{rank}(X)$. The cone decomposition $\{\tau_\beta\}$ of $\tilde{C}_s(X)$ induces a $\mathrm{GL}(X_\xi) \ltimes X_\xi^s$-admissible polyhedral cone decomposition of $\tilde{C}_s(X_\xi)$ as we remarked before. As in chapter IV we have a tautological (or universal) extension $0 \to T_\xi \to \tilde{G}_\xi \to A \to 0$ lying over $\mathbf{Hom}(X_\xi, A^t) = \mathbf{Hom}(X_\xi, A)$ where $A \to \mathsf{A}_{g-r}$ is (the pull-back to $\mathbf{Hom}(X, A)$ of) the universal abelian scheme, with principal polarization λ. Let $E_\xi = B(X_\xi) \otimes G_m$, $F_\xi = \tilde{B}_s(X_\xi) \otimes G_m$, so that E_ξ has character group $S^2(X_\xi)$, and F_ξ has character-group $S^2(X_\xi) \times (X_\xi)^s$. We have a E_ξ-torsor Ξ_ξ over $\mathbf{Hom}(X, A)$ which, if we push it forward by $[\mu \times \nu] : E_\xi \to G_m$, gives the rigidified G_m-torsor $(c(\mu) \times c(\nu))^* \mathcal{P}_A^{\otimes -1}$, where $c : X_\xi \to A$ is the characteristic map over $\mathbf{Hom}(X, A)$, and $\mathcal{P}_A$ is the Poincaré G_m-torsor. Let $\tilde{\Xi}_\xi = \tilde{\Xi}_{\xi,s} := \Xi_\xi \times_{\mathbf{Hom}(X_\xi, A)} (\tilde{G}_\xi)^s$, an F_ξ-torsor Ξ_ξ over $\mathbf{Hom}(X, A)$ lying over Ξ_ξ and equivariant w.r.t. $F_\xi \to E_\xi$. By the theory of torus embeddings, we have a morphism of torus embeddings $\overline{F}_\xi \to \overline{E}_\xi$ attached to the map between cone decompositions underlying $\tilde{C}_s(X) \to C(X)$, just as in the previous section. Now we use the contraction product to define $\overline{\Xi}_\xi := \Xi_\xi \times^{E_\xi} \overline{E}_\xi$, $\overline{\overline{\Xi}}_\xi := \tilde{\Xi}_\xi \times^{F_\xi} \overline{F}_\xi$, and define $\mathcal{S}_\xi$ to be the formal completion along the (reduced) locally closed subscheme of $\overline{\Xi}_\xi$ supported in the union of strata corresponding to cones in $C(X_\xi)^\circ$. We shall see that $\overline{\overline{\Xi}}_\xi \to \mathcal{S}_\xi$ underlies a weak relatively complete model.

It is easy to see that the translation action of $(\tilde{G}_\xi)^s$ extends to an action of $(\tilde{G}_\xi)^s$ on $\overline{\overline{\Xi}}_\xi$. We need a relatively ample invertible sheaf on $\overline{\overline{\Xi}}_\xi \to \overline{\Xi}_\xi$ with an X^s-action, for which we start with the locally principal sheaf of fractional ideals $\mathcal{I}_{\tilde{\phi}}$ given by the homogeneous principal polarization function $\tilde{\phi}$ just as in the totally degenerate case. More precisely, $\tilde{\phi}$ defines an F_ξ-equivariant locally principal sheaf $\mathcal{I}_{\tilde{\phi}}$ of fractional ideals on $\overline{F}_\xi$, generated on each affine open corresponding to a simplex τ in the cone decomposition of $\tilde{C}_s(X_\xi)$ by the character $l_\tau \in \tilde{B}_s(X_\xi)^*$ determined by $\tilde{\phi}|\tau$. As this fractional ideal is F_ξ-equivariant, it passes to the contraction product and gives a locally principal sheaf of fractional ideals there, which we still denote by $\mathcal{I}_{\tilde{\phi}}$. Notice that $(\tilde{G}_\xi)^s$ operates on $(\overline{\overline{\Xi}}_\xi, \mathcal{I}_{\tilde{\phi}})$ extending the translation. Now we define $\tilde{\mathcal{L}}$ to be $\mathcal{I}_{\tilde{\phi}} \otimes \tilde{\pi}^*\mathcal{M}$, where $\tilde{\pi}$ is the morphism from $\overline{\overline{\Xi}}_\xi$ to $(A_\xi)^s$ and $\mathcal{M} = \mathcal{O}(2k)$, the canonical symmetric relatively ample invertible sheaf on $(A_\xi)^s$ giving $2k\cdot$(the principal polarization on $(A_\xi)^s$), k being the positive integer with $\tilde{\phi} - \tilde{\phi} \circ T_\mu = k \cdot \tilde{\chi}_\mu \, \forall \mu \in (X_\xi)^s$.

It remains to check that the conditions for weak relatively complete models are satisfied, when we restrict to an affine open formal subscheme $\mathcal{S}_\sigma$ of $\mathcal{S}_\xi$ corresponding to a cone σ in $C(X_\xi)^\circ$. It is clear from the construction that the translation action of the period-subgroup $(X_\xi)^s$ on $(\tilde{G}_\xi)^s_\eta$ extends to $\overline{\overline{\Xi}}_\xi$. The existence of a $(\tilde{G}_\xi)^s$-invariant open subset of finite type whose $(X_\xi)^s$-translates cover $\overline{\overline{\Xi}}_\xi$ is immediate from the fact that modulo $(X_\xi)^s$, there are only finitely many cones in $\tilde{C}_s(X_\xi)$ over a cone σ in $C(X_\xi)^\circ$. The completeness condition is verified exactly as before. In fact after restricting a valuation to (the function field of) $\mathbf{Hom}(X_\xi, A)$, the proof for the totally degenerate case works in the present case. The only part that still requires some explanation is the action of $Y = (X_\xi)^s$ on $\tilde{\mathcal{L}}$: Let $\tilde{\pi} : \overline{\overline{\Xi}}_\xi \to (A_\xi)^s_{/\overline{\overline{\Xi}}_\xi}$ be the natural map from $\overline{\overline{\Xi}}_\xi$ to $(A_\xi)^s$, $\mathcal{M} = \mathcal{O}(2k)$ as before, and write $\tilde{\pi}_*\tilde{\pi}^*\mathcal{M}_\eta = \oplus_\nu \mathcal{M}_\eta \otimes \mathcal{O}_\nu$, where ν runs over $(X_\xi)^s$ and $\mathcal{O}_\nu$ denotes the subsheaf of functions on $(\tilde{G})^s_\eta$ transforming according to ν w.r.t. the action of $(T_\xi)^s$. Let $\psi : 1_Y \to c^*\mathcal{O}(2)^{\otimes -k}_\eta$ be the symmetric cubical trivialization obtained from $\tau^{\otimes k}$ by diagonal pull-back, where τ is part of the the (s-fold fibre product of) the degeneration data over $\overline{\overline{\Xi}}_\xi$ as explained in chapter IV. For $\mu \in Y = (X_\xi)^s$, the map $\tilde{S}^*_\mu : S^*_\mu\tilde{\mathcal{L}} \to \tilde{\mathcal{L}}$ describing the action of $(X_\xi)^s$ on $\tilde{\mathcal{L}}$ can finally be given in terms of its restriction to the $\tilde{G}_\eta$ as follows (remember that now $\phi : Y = (X_\xi)^s \to (X_\xi)^s$ is multiplication by $2k$):

$$\tilde{S}^*_\mu : S^*_\mu(\mathcal{M}_\eta \otimes \mathcal{O}_\nu) = (\mathcal{M}_\eta \otimes \mathcal{O}_{2k\mu+\nu}) \otimes \mathcal{M}_\eta(c(\mu)) \otimes \mathcal{O}_\nu(c(\mu)) \to \mathcal{M}_\eta \otimes \mathcal{O}_{2k\mu+\nu}$$

is defined by "multiplication by $\psi(\mu)\cdot\tau(c(\nu),\mu)$". This indeed defines an action of $(X_\xi)^s$ on $\tilde{\mathcal{L}}$, which was precisely the reason for requiring that $\tilde{\phi} - \tilde{\phi} \circ T_\mu = k \cdot \tilde{\chi}_\mu \forall \mu \in (X_\xi)^s$, in the definition of homogeneous principal polarization function. We leave the fun of checking this, which is after all a twisted version of the corresponding fact in the totally degenerate case, to the reader.

We have seen that $(\overline{\overline{\Xi}}_\xi, \tilde{\mathcal{L}})$ over $\mathcal{S}_\xi$ is indeed a weak relatively complete model. Complete formally, divide by $(X_\xi)^s$, algebraize and finally pass to the $\mathrm{GL}(X_\xi)$-quotient, to obtain a pair $(P, \mathcal{L})$ consisting of a proper scheme P and a relatively ample invertible sheaf $\mathcal{L}$ over the formal algebraic stack $\mathcal{S}_\xi/\mathrm{GL}(X_\xi)$, with an action of G^s on P over $\mathrm{GL}(X_\xi)$, such that $(P, \mathcal{L})$ extends $(A^s, \mathcal{O}(2k))$ over the 0-stratum, and the action of G^s on P extends to the translation action of A^s. We shall call this pair (or just P) a *good compact formal model.*

We remark that there are obvious variants of the above construction if add level-n structures and work over $\mathbb{Z}[\zeta_n, 1/n]$. Since we have discussed level-n structures in chapter IV §6, the adaption to the present situation is left to the reader as an exercise, as usual.

Approximation, Algebraization, and Gluing

We have discussed the approximation process in chapter IV §4 fairly extensively, and everything carries over to our present situation. This already gives us compactification of A^s locally in the étale topology of the boundary of A_g. In IV §4, the approximation result led directly to algebraization and gluing by the theory of degeneration, as explained in IV §§4, 5. The point there is that we can control the approximation so that (after recompleting the base) the relevant cones ("of period matrices") do not change. This is also the key to algebraization in our present situation. But unlike the situation for semi-abelian schemes, which are pretty rigid and which admit a good theory of degeneration, a theory of degeneration "with compactified fibres" is lacking (and seems likely to remain so). Now comes the only extra point: we can characterize the toroidal compactification of A^s as constructed above by its (relative) projective coordinate ring, which depends only on the cone decomposition, which does not change under approximation. This gives algebraization and gluing, just as in IV §5.

Let us first review the method and results in IV §4, and establish some notation along the way. Let $\sigma \subset C(X_\xi)^\circ$ be a cone, R be the strict local ring of a geometric point of the σ-stratum in $\overline{\Xi}_\xi$, and $R^\wedge$ the I-adic completion of R, where I is the ideal for the σ-stratum of $\mathrm{Spec}(R)$. We shall slightly change notation and denote by $({}^\heartsuit P, {}^\heartsuit \mathcal{L})$ the good compact formal model over $\mathrm{Spec}(R^\wedge)$ constructed before using $\{\tau_\beta\} \to \{\sigma_\alpha\}$ and $\tilde{\phi}$, and by ${}^\heartsuit G$ the semi-abelian scheme attached to the degeneration data. What we did in IV §4 is this: Write ${}^\heartsuit R$ as an inductive limit of algebras R_γ of finite type over $\mathbb{Z}$. All the objects of interest, e.g. ${}^\heartsuit P, {}^\heartsuit \mathcal{L}, {}^\heartsuit G$, the group law, are finitely presented over ${}^\heartsuit R$, and hence defined over a certain R_γ of finite type over $\mathbb{Z}$, i.e. they come from a base change $R_\gamma \to R^\wedge$. Thus we can apply M. Artin's approximation theorem to approximate this ring homomorphism $h^\wedge : R_\delta \to R^\wedge$ by a ring homomorphism $h : R_\delta \to R^\wedge$ in the I-adic topology as close as we like. If we first base change ${}^\heartsuit G \to \mathrm{Spec}(R_\delta)$ to $G \to \mathrm{Spec}(R)$ via h, I-adically recomplete it, then compare the resulting semi-abelian scheme with the original ${}^\heartsuit G \to \mathrm{Spec}(R^\wedge)$, we find that they do not differ very much if h is close enough to $h^\wedge$. This was proved by analyzing sections of ${}^\heartsuit \mathcal{L}$, which gives the degeneration data. More precisely, they are the same over $\mathrm{Spec}(R/I)$ (or, $\mathrm{Spec}(R/I^{n_0})$, their X's and Y's are the same, their b's differ only by a unit, their cones σ are the same, the Kodaira-Spencer map for $G \to \mathrm{Spec}(R)$ induces an isomorphism $S^2(\Omega^1_{G/R}) \cong \Omega^1_{R/\mathbb{Z}}[d\log\infty]$, and the I-adic completion of $G \to \mathrm{Spec}(R)$ is the pull-back of the original ${}^\heartsuit G \to \mathrm{Spec}(R^\wedge)$ by an automorphism $\zeta : R^\wedge \to R^\wedge$ congruent to the identity modulo I (or, mod I^{n_0}), which is given by the action of an element of $E(R^\wedge)$ I-adically close to the identity. In particular, if we change the canonical embedding $R \to R^\wedge$ by a suitable automorphism of $R^\wedge$ congruent to the identity modulo I, we get an embedding $\hbar : R \to R^\wedge$ and a semi-abelian scheme $G \to \mathrm{Spec}(R)$, which gives back ${}^\heartsuit G \to \mathrm{Spec}(R^\wedge)$ if we base-change to $R^\wedge$ via $\hbar$. This is an indirect way of algebraizing ${}^\heartsuit G \to \mathrm{Spec}(R^\wedge)$.

Obviously we can carry out the approximation step in the same way: just choose R_δ large enough so that ${}^\heartsuit P, {}^\heartsuit\mathcal{L}, {}^\heartsuit G$, the group law and the operation ${}^\heartsuit G^s \times_{\mathrm{Spec}(R^\wedge)} {}^\heartsuit P \to {}^\heartsuit P$ are all defined over R_δ. Approximating $h^\wedge : R_\delta \to R^\wedge$ I-adically by $h : R_\delta \to R$, we obtain P, $\mathcal{L}$, G over $\mathrm{Spec}(R)$, and all the properties stated above are satisfied. Moreover, we can achieve $P \otimes_R (R/I^{n_0}) = {}^\heartsuit P \otimes_{R^\wedge} (R^\wedge/I^{n_0}R^\wedge)$ for any positive integer n_0 we care to pick. As above, if we change the canonical embedding $R \to R^\wedge$ by a suitable automorphism of $R^\wedge$ congruent to the identity modulo I, we get an embedding $\hbar : R \to R^\wedge$, a semi-abelian scheme $G \to \mathrm{Spec}(R)$, a proper scheme $P \to \mathrm{Spec}(R)$, a relatively ample invertible sheaf $\mathcal{L}$ on P, a action of G^s on P, such that the over the 0-stratum P is canonically identified with A^s ($= G^s$ over the 0-stratum) such that $\mathcal{L}$ extends $\mathcal{O}(2k)$ on A^s and the operation of G on P extends the translation action. We know that if we I-adically complete $G \to \mathrm{Spec}(R)$ we get back the original ${}^\heartsuit G \to \mathrm{Spec}(R^\wedge)$; in particular ${}^\heartsuit P$ and P are birationally the same, and $P \to \mathrm{Spec}(R^\wedge)$ can be made I-adically close to the original ${}^\heartsuit P \to \mathrm{Spec}(R^\wedge)$. Of course this is not good enough: we want to know that $P \to \mathrm{Spec}(R^\wedge)$ is exactly ${}^\heartsuit P \to \mathrm{Spec}(R^\wedge)$. This is actually a pretty easy matter: in brief, we use the projective coordinate ring of ${}^\heartsuit P \to \mathrm{Spec}(R^\wedge)$. More precisely, ${}^\heartsuit P$ is characterized as $\mathrm{Proj}_{R^\wedge}(\oplus_{m\geq 0}\Gamma({}^\heartsuit P, {}^\heartsuit\mathcal{L}^{\otimes m}))$. The point now is that $\Gamma({}^\heartsuit P, {}^\heartsuit\mathcal{L}^{\otimes m})$ can be described explicitly in terms of the map $\{\tau_\beta\} \to \{\sigma_\alpha\}$ between admissible cone decompositions: Write $R^\wedge$ (non-canonically) as $R_1[\![\lambda^S]\!]$, where S runs over $\sigma^\vee \cap S^2(X)$, and R_1 is a strict regular local ring formally étale over $\mathbb{Z}[x_1, ..., x_N]$ with $N = g\cdot(g+1)/2 - r\cdot(r+1)/2$ (R_1 is the strict localization at a geometric point of $\mathbf{Hom}(X_\xi, A_\xi)$. Then via the Fourier-Jacobi expansion, $\Gamma({}^\heartsuit P, {}^\heartsuit\mathcal{L}^{\otimes m})$ is canonically identified with the $R^\wedge$-module of all $(X_\xi)^s$-invariant infinite series of the form

$$\sum_{(S,\mu)} a_{(S,\mu)} \cdot \lambda_S \cdot s_\mu$$

where $a_{(S,\mu)} \in R_1$, $s_\mu \in \Gamma((A_\xi)^s, (\mathcal{M}\otimes\mathcal{O}_\mu)^{\otimes m})$, and $(S,\mu) \in S^2(X_\xi)\times(X_\xi)^s = \tilde{B}_s(X_\xi)^*$ runs through all elements of $\tilde{B}_s(X_\xi)^*$ with $(S,\mu) \geq m\cdot\tilde{\phi}$ on the inverse image of σ in $\tilde{C}_s(X_\xi)$. This is invariant under ζ, so everything we said holds for $P \to \mathrm{Spec}(R)$ as well, which shares the same semi-abelian scheme G with ${}^\heartsuit P$. Thus ${}^\heartsuit P$ and P have the same projective coordinate ring over R, so they are the same.

After the algebraization step we might try to glue as in IV §5. Unfortunately "openness of versality" poses difficulties. However in our case we can proceed as follows: It suffices to exhibit a quasicoherent graded algebra on $\overline{X}$ which extends the direct sum of the direct images of $\mathcal{L}^{\otimes m}$, $m \geq 0$. These direct images form vector bundles over X, and we have already exhibited (by the existence of local compactifications P) an extension to the formal completion of $\overline{X}$ along any stratum. We have to show that these formal completions can be algebraized. By routine calculations one checks that this is equivalent to the following statement:

Suppose in $\overline{X} = \overline{\mathsf{A}}_g$ we have points x in the σ-stratum and y in the τ-stratum ($\tau \subset \overline{\sigma}$) such that y specializes to x. Then our data give modules over the completions R_1 and R_2 of the strict local rings of $\overline{X}$ in x and y respectively. These have to be compatible in the sense that for any point $z \in \mathrm{Spec}(R_1)$ lying over y, they induce the same module over the completion of R_1 in z.

We are now in the situation of the 2-step constructions of chapter IV, §10. Translated into the geometry of convex bodies, the necessary compatibility amounts to the following two statements, each of which corresponds to one inclusion between the two induced modules:

(i) If $(S,\mu) \geq m \cdot \tilde{\phi}$ on the inverse image of σ, then also $(S,\mu) \geq m \cdot \tilde{\phi}$ on the inverse image of τ.

(ii) Any pair (S,μ) such that $(S,\mu) \geq m \cdot \tilde{\phi}$ on the inverse image of τ can be extended to such a pair (with the same μ) such that $(S,\mu) \geq m \cdot \tilde{\phi}$ on the inverse image of σ.

The first assertion is totally trivial, and the proof of the second one is only slightly more involved.

1.12. Remark. The above arguments allows us to construct other extensions of the direct images of $\mathcal{L}^{\otimes m}$. For example in the above we might use series $\sum_{(S,\mu)} a_{(S,\mu)} \cdot \lambda_S \cdot s_\mu$, where we sum over all pairs (S,μ) such that $m \cdot S \geq [\mu \otimes \mu]$ on σ. This is as much a canonical extension as any other, and it corresponds somehow (that is up to some minor technical points) to the polarization function $\tilde{\phi}$ defined by $\tilde{C} \ni q = (a_1, ..., a_s) \mapsto \sum_{1 \leq i \leq s} (\mathrm{Min}_{\mu \in X}\{a_i(\mu)\})$. If m divides n this actually defines a vector bundle on $\overline{\mathsf{A}}_{g,n}$.

We summarize the results proved so far:

1.13. Theorem. *Let X be a free abelian group of rank g. Given a smooth* $\mathrm{GL}(X)$*-admissible polyhedral cone decomposition $\{\sigma_\alpha\}$ of $C(X)$, a smooth* $\mathrm{GL}(X) \ltimes X_s$*-admissible polyhedral cone decomposition $\{\tau_\beta\}$ of $\tilde{C}_s(X)$ over it, and a principal polarization function $\tilde{\phi}$ with $\tilde{\phi} - \tilde{\phi} \circ T_\mu = k \cdot \chi_\mu \, \forall \mu \in X^s$, there exists an associated (relative) scheme $P = P(\{\tau_\beta\})$ proper over the arithmetic toroidal compactification $\overline{A}_g(\{\sigma_\alpha\})$ and a relatively ample invertible sheaf $\mathcal{L}$ on $P(\{\tau_\beta\}) \to \overline{A}_g(\{\sigma_\alpha\})$ with the following properties:*

(1) $P_{/A_g}$ is canonically isomorphic to the s-fold fibre product A^s of the universal abelian scheme A over A_g. $\mathcal{L}_{/A_g}$ is canonically isomorphic to $\mathcal{O}(2)^{\otimes k}$, where $\mathcal{O}(2)$ is the canonical symmetric ample invertible sheaf on A^s which gives twice the principal polarization of A^s.

(2) The semi-abelian scheme $G^s \to \overline{A}_g(\{\sigma_\alpha\})$ extending A^s operates on P extending the translation action of A^s. G^s is contained in P as a dense open subscheme if each $\sigma_\alpha \times \{0\}$ is a τ_β.

(3) The formal completions of $P(\{\tau_\beta\}) \to \overline{A}_g(\{\sigma_\alpha\})$ along strata of $\overline{A}_g$ are canonically isomorphic to the good compact formal models constructed before. Especially, P is smooth over $\mathrm{Spec}(\mathbb{Z})$ *and the boundary of P is a divisor with normal crossings relative to* $\mathrm{Spec}(\mathbb{Z})$.

(4) P has a natural stratification induced by the isomorphism in (3). The strata are parametrized by $\{\tau_\beta\}/\mathrm{GL}(X) \ltimes X^s$ (via the dual complex construction).

Finally, we remark that there is an obvious variant if we throw in principal level-n structures, say for $n \geq 3$. In that case P itself is an algebraic space proper and smooth over $\mathrm{Spec}(\mathbb{Z}[\zeta_n, 1/n])$. Moreover if $\{\sigma_\alpha\}$ is projective (i.e. admits a $\mathrm{GL}(X)$-invariant polarization function), $P(\{\tau_\beta\})$ and $\overline{A}_g(\{\sigma_\alpha\})$ are both schemes projective over $\mathrm{Spec}(\mathbb{Z})$.

2. Computation of Formal Cohomology

We have proved only half of theorem 1.1 of the previous section. The purpose of this section is to prove statements (iii) and (iv) of this theorem. (iv) is important in that it is the key to the degeneration of spectral sequences to be studied in §5.

We have constructed compactifications $P = \overline{Y}$, such that their formal completions (along certain strata) are explicitly known. To check their properties we pass to these completions, and use descent and the fact that nice properties tend to be open; that is, if they hold at a certain point they hold nearby (we may have to refine the covering of $\overline{X}$). For example, it follows that the compactifications are formally smooth over $\mathrm{Spec}(\mathbb{Z})$, with infinity a divisor with normal crossings, which follows from the corresponding fact for $\overline{F}$. The same technique applies to the assertions about differentials in (iii): There exists a natural isomorphism from $j^*\overline{\Omega}^1_{Y/X}$ to $j^*\overline{f}^*(\mathcal{T}^*_G)^s$, where $j : A^s \hookrightarrow P$ is the natural inclusion of the 0-stratum, because of the A^s-action. We want to see that this isomorphism extends to an isomorphism between $\overline{\Omega}_{Y/X}$ and $\overline{f}^*(\mathcal{T}^*_G)^s$. In any case the G^s-action on P induces a map from $\overline{\Omega}_{Y/X}$ to $\overline{f}^*(\mathcal{T}^*_G)^s$. To see that this is an isomorphism it suffices to check after passing to the formal completion. By construction G^s operates on P extending the translation action of A^s, and its formal completion is canonically the formal completion of an extension of an abelian scheme by a torus. The assertion that $\Omega^1_{Y/X}$ is naturally isomorphic to $\overline{f}^*(\mathcal{T}^*_G)^s$ now follows from the (well-known and easy) fact that $\Omega^1_{\overline{F}}[d\log\infty]$ (resp. $\Omega^1_{\overline{E}}[d\log\infty]$) is canonically isomorphic to $\mathcal{O}_{\overline{F}} \otimes_{\mathbb{Z}} \tilde{B}_s(X)^*$ (resp. $\mathcal{O}_{\overline{E}} \otimes_{\mathbb{Z}} B(X)^*$), where a character χ goes to the attached invariant differential $d\chi/\chi$. Thus (iii) is proved.

To prove statement (iv) of theorem 1.1, it suffices to show that $R^1\overline{f}^*(\mathcal{O}_{\overline{Y}})$ is locally free and naturally isomorphic to $\mathcal{T}^s_G$, and that $R^a\overline{f}_*(\mathcal{O}_{\overline{Y}}) \cong \Lambda^a(R^1\overline{f}_*(\mathcal{O}_{\overline{Y}}))$. We shall first reduce these to corresponding statement for the formal completions along strata, and via GFGA to statements about formal cohomology. The formal cohomology can be computed directly via the quotient construction, and we shall see that it reduces to the cohomology of the group X^s.

First of all we have natural maps $\Lambda^a(R^1\overline{f}_*(\mathcal{O}_{\overline{Y}})) \to R^a\overline{f}_*(\mathcal{O}_{\overline{Y}})$ given by the cup-product. To check that they are isomorphisms we can localize in the Zariski or étale topology, and pass to the formal completion. Next the principal polarization on A induces over X an isomorphism from $(\mathcal{T}_G)^s$ to $R^1\overline{f}_*(\mathcal{O}_{\overline{Y}})$. To check that it extends to an isomorphism over $\overline{X}$ we can again localize in the Zariski or étale topology, or pass to the formal completion. We should mention that when one works over the formal completions of the local rings in $\overline{X}$, the differentials are to be taken as "universally finite modules of differentials".

We have completed our reduction step. It remains only to check that $R^1\overline{f}^\wedge_*(\mathcal{O}_{\overline{Y}^\wedge})$ is naturally isomorphic to $\mathcal{T}_{G/\overline{X}^\wedge}$, and that $R^a\overline{f}_*(\mathcal{O}_{\overline{Y}^\wedge}) \cong \Lambda^a(R^1f_*(\mathcal{O}_{\overline{Y}^\wedge}))$, over the completed strict local ring at a geometric point of the boundary of $\overline{X}$.

Another tool in our hands is the G^s-action on the direct images $R^af_*(\mathcal{O}_{\overline{Y}})$. Since G^s is an abelian scheme over X it is at least plausible that it operates trivially on the cohomology sheaves $R^a\overline{f}_*(\mathcal{O}_{\overline{Y}})$, and this is indeed a consequence of the fact that $R^a\overline{f}_*(\mathcal{O}_{\overline{Y}})$ is locally free. To use this idea efficiently, let us pass to the formal completion $R^\wedge$ of a strict local ring at a geometric point $\overline{x}$ of the boundary of $\overline{X}$, say in the σ-stratum for a cone σ in $C(X_\xi)^\circ$. Then the formal completion of G is canonically the formal completion of a $\tilde{G}_\xi$ sitting in an exact sequence $0 \to T_\xi \to \tilde{G}_\xi \to A_\xi \to 0$ in our previous notation. Moreover, we have an explicit construction of $P^\wedge$ (or $\overline{Y}^\wedge$) over Spf(R) as the X^s_ξ-quotient of $\tilde{P}^\wedge$. The operation of G^s on $R^a\overline{f}_*(\mathcal{O}_{\overline{Y}})$ passes to the formal completion, and we obtain an action of $(T^s_\xi)^\wedge$ on $R^a\overline{f}_*(\mathcal{O}_{\overline{Y}^\wedge})$ over Spf$(R^\wedge)$. So over $R^\wedge$, $R^a\overline{f}^\wedge_*(\mathcal{O}_{\overline{Y}^\wedge})$ decomposes into a topological direct sum of $(T^s_\xi)^\wedge$-eigenspaces. We shall show that for any non-trivial character of $(T^s_\xi)^\wedge$ the corresponding eigenspace in the cohomology vanishes, while the eigenspaces for the trivial character of $(T^s_\xi)^\wedge$ are locally free and behave as they should. So instead of computing the whole cohomology of the formal scheme we only have to consider eigenspaces for $(T^s_\xi)^\wedge$. This will simplify things considerably.

Finally we can start with the real work. We have to compute the $(T^s_\xi)^\wedge$-eigenspaces for the cohomology of the formal scheme associated to P. For simplicity we first do it for the totally degenerate case, then indicate what to do in general.

We first consider the situation that $R^\wedge$ is the completed strict local ring of a geometric point $\overline{x}$ in the σ-stratum for a cone σ in $C(X)^\circ$, so $T_\sigma = \tilde{G}_\sigma$. In this special case we consider the formal completion of $\overline{F}$ along the pre-image of the σ-stratum in $\overline{E}$, and divide by the action of the group X^s. It suffices to show that the $\tilde{G}^s$-eigenspaces in the cohomology of the formal completion of $\overline{F}$ are all trivial, except for the trivial eigenspace and cohomology in degree 0, where we obtain the constants. For then the cohomology of the quotient is equal to the group cohomology of X^s with values in the constants, and all assertions follow: $R^a\overline{f}^\wedge_*(\mathcal{O}_{\overline{Y}^\wedge}) = \Lambda^a(R^1\overline{f}^\wedge_*(\mathcal{O}_{\overline{Y}^\wedge}))$, and $R^1\overline{f}^\wedge_*(\mathcal{O}_{\overline{Y}^\wedge}) = \mathrm{Hom}(X^s, R^\wedge)$ is canonically isomorphic to the relative Lie algebra of $\tilde{G}$.

The cohomology of torus embeddings can be computed via the canonical affine coverings corresponding to the cone decomposition. It is known that the result is independent of the particular cone decomposition, since subdividing a given decomposition induces a proper map between the toroidal embeddings with vanishing higher direct images, see [KKMS], Ch. I, §3, Th. 12, p.42, and its corollaries. The same is true for formal completions, as direct images and formal completions commute for proper maps. We apply this as follows: The canonical affine covering of the formal completion of $\overline{F}$ is derived from the induced cone decomposition of the pre-image of the closure of σ in $\tilde{C}_s(X)$. By subdividing this decomposition we may assume that the pre-image of the closure of σ in $\tilde{C}_s(X)$ is an increasing union $\cup\tau_m$, where the τ_m are convex subsets which are finite unions of polyhedra occurring in the decomposition. It follows that the formal scheme $\tilde{P}^\wedge$ is the union of the increasing sequence of open sets corresponding to the τ_m's, and that the cohomology of each such open set is the same as the cohomology of the formal completion of the affine torus-embedding defined by the single polyhedron τ_m.

To go on we need some notation: Let $R = \mathbb{Z}[S^2(X) \cap \sigma^\vee]$ denote the ring of the affine subset of $\overline{E}$ corresponding to σ, $I \subset R$ the ideal which defines the σ-stratum. So I is generated by elements in $S^2(X) \cap \sigma^\vee$ which are strictly positive in the interior of σ. We change our notation slightly so that now $R^\wedge$ stands for the I-adic completion of the affine coordinate ring R. By $S_m = \mathbb{Z}[\tilde{B}_s(X)^* \cap \tau_m^\vee]$ we denote the affine coordinate ring of the torus embedding given by τ_m. It contains R, and the affine formal scheme corresponding to τ_m is defined by the I-adic completion of S_m. We have to compute $\tilde{G}^s$-eigenspaces of cohomology. of $\tilde{P}^\vee$. If suitable Mittag-Leffler conditions hold the cohomology we are after is equal to the projective limit (over m) of those eigenspaces. We get all the information we need from the following result:

2.1. Lemma. *Notation as above.*

(a) If μ is a non-trivial character of $\tilde{G}^s$, then for any $k \geq 0$ the projective system of μ-eigenspaces $\{S_m/I^k \cdot S_m\}_{m \geq 0}$ is essentially 0.

(b) If μ is the trivial character of $\tilde{G}^s$, for m big enough and $k \geq 0$ the map $R/I^k \to S_m/I^k \cdot S_m$ induces an isomorphism between μ-eigenspaces.

Proof. If $\mu = 0$ the μ-eigenspace in S_m consists of elements of $S^2(X)$ which are ≥ 0 on the image of τ_m in $C(X)$. If m is big enough this image is equal to σ, and so the eigenspace is R. This proves (b).

Assume that $\mu \neq 0$. It suffices to show that for any m we can find $l \geq m$ such that the mapping $S_l \to S_m$ sends the μ-eigenspace in S_l into $I \cdot S_m$. We choose l so big that τ_l contains the translate of τ_m under the action of $-\mu \in X^s$ on $\tilde{C}_s(X)$. For any $\chi \in \tilde{B}_s(X)^*$ positive on τ_l such that the restriction of χ to $(X^*)^s$ is equal to μ, the $-\mu$-translate of χ is equal to the difference of χ and $\sum \mu_j \otimes \mu_j \in S^2(X)$, where $\mu = (\mu_1, \ldots, \mu_s)$. As the restriction of this translate to τ_m is ≥ 0, in S_m, χ is divisible by $\sum \mu_j \otimes \mu_j$ (regarded as a regular function in S_m), which is an element of I since σ is assumed to be contained in $C(X)^\circ$ (i.e. to consist of positive definite quadratic forms on X), and $\mu \neq 0$. The lemma is proved.

It follows that the higher derived functors $\lim\limits_{\leftarrow}{}^{(\nu)}$ (limit over m and k) of the μ-eigenspaces in $S_m/I^k \cdot S_m$ vanish, so the μ-eigenspaces in the cohomology of the formal completion of $\overline{F}$ equal the projective limits over k and m of the $S_m/I^k \cdot S_m$-eigenspaces, which either vanish or equal the I-adic completion of R. This finishes the computation of the eigenspaces in the formal cohomology of $\overline{F}$, and also of its quotient under X^s as explained before. After some routine checking of compatibilities of isomorphisms we have shown statement (iv) of theorem 1.1 in the case of total degeneration.

Finally we come to the mixed case: Here the formal completion of $\tilde{P}$ is a fibration over the abelian variety $A^s = A^s_\xi$, and we can apply the previous theory (in smaller dimension) "to the fibres". It follows that the T^s_σ-eigenspaces on the formal cohomology are either trivial, or equal to the cohomology of A^s. Dividing out by X^s_σ amounts to considering a spectral sequence converging to the cohomology of the formal completion of P, with E_2-term given by $H^a(X^s_\sigma, H^b(A^s, \mathcal{O})) \cong \Lambda^a(X^*_\sigma)^s \otimes \Lambda^b H^1(A^s, \mathcal{O})$. This spectral sequence must degenerate, because we know that the ranks of the E_∞-terms add up to the rank of the cohomology of the generic fibre of G^s over X, which is 2^{gs} and equals the sum of the ranks of the E_2-terms. It follows that the cohomology groups are locally free, and careful study of cup-products in the spectral sequence reveal that they are equal to the exterior powers of the H^1. That $R^1\overline{f}^*(\mathcal{O}_{\overline{Y}})$ is naturally isomorphic to $\mathcal{T}^s_G$ follows from the five-lemma once we have shown that the isomorphism over X (defined by the principal polarization) extends to a regular map. However, the the composition of the isomorphism with multiplication by 2 extends, as the line bundle $\mathcal{L}$ on P defines twice the polarization. As the strata of $\overline{X}$ are smooth over Spec($\mathbb{Z}$), the map is regular in codimension one, so regular overall. So after much work theorem 1.1 has been shown completely.

3. $\Omega^\bullet_X$-Modules and $\mathcal{D}_X$-Modules

We have constructed toroidal compactifications $(f : Y \to X) \subset (\overline{f} : \overline{Y} \to \overline{X})$ of the s-fold fibre product of the universal abelian scheme over the toroidal compactification $X \subset \overline{X}$ of the moduli space X. In this section we shall construct a filtered complex $\mathcal{G}^\bullet$ over $\overline{X}$ with logarithmic integrable connection, which extends the relative de Rham cohomology complex $\mathbb{R}f_{*DR}(Y/X)$ over X (with its Hodge filtration and Gauss-Manin connection) and whose associated logarithmic de Rham complex is (canonically quasi-isomorphic to) $\mathbb{R}\overline{f}_*\overline{\Omega}^\bullet_Y$. In the next section $\mathcal{G}^\bullet$ and its logarithmic de Rham complex will be explicitly described. Since the cohomology of $\mathbb{R}\overline{f}_*\overline{\Omega}^\bullet$ is just the cohomology of Y, we shall be able to deduce degeneration of various spectral sequences from Deligne's Hodge theory. Our main tool is a correspondence between (filtered) complexes of $\Omega^\bullet_X$-

modules (resp. $\overline{\Omega}^\bullet_X$-modules) and (filtered) complexes of $\mathcal{D}_X$-modules (resp. $\overline{\mathcal{D}}_X$-modules). Although we are ultimately interested in the $\Omega^\bullet_X$-modules, for one key calculation we need to pass to the $\mathcal{D}_X$-module side. This is because there we obtain sheaves instead of objects in a derived category, so that our local information can be glued to obtain the global picture. We recall that $\overline{\Omega}^\bullet_X$ stands for $\Omega^\bullet_{\overline{X}}[d\log\infty]$, and likewise for $\overline{\Omega}^\bullet_Y$, $\overline{\Omega}^\bullet_{Y/X}$ and $\overline{\mathcal{D}}_X$.

As a preview of things to come let us recall a correspondence between graded complexes modules over the symmetric algebras and and graded complexes of modules over the exterior algebras, which seems to be well-known, but for which we could not find a good reference. They are very much inspired by the book [GHV]. This correspondence can be presented for ringed topoi, but it is essentially a "pointwise" result, and we shall do it in the context of schemes.

Suppose X is a scheme, $\mathcal{E}$ a locally free $\mathcal{O}_X$-module of finite rank with dual $\mathcal{E}^\vee$. We shall consider the symmetric algebra $S^\bullet(\mathcal{E}^\vee)$ and the exterior algebra $\Lambda^\bullet(\mathcal{E})$ as graded algebras: elements of $\mathcal{E}$ are homogeneous of degree 1, elements of $\mathcal{E}^\vee$ have degree -1, and elements of $\mathcal{O}_X$ degree zero. Note that the gradings are bounded above in the sense that for $d \gg 0$ sufficiently large there are no non-zero homogeneous elements of degree d. We require this property for all graded modules which we encounter in the sequel, without explicitly mentioning it. More precisely let $\mathrm{CG}(S^\bullet(\mathcal{E}^\vee))$(resp. $\mathrm{CG}(\Lambda^\bullet(\mathcal{E}))$) be the abelian category of quasi-coherent graded complexes of modules over $S^\bullet(\mathcal{E}^\vee)$ (resp. $\Lambda^\bullet(\mathcal{E})$) such that multiplication is compatible with gradation (in the sense to be specified in a moment) and the gradings are bounded above, and let $\mathrm{DG}(S^\bullet(\mathcal{E}^\vee))$ (resp. $\mathrm{DG}(\Lambda^\bullet(\mathcal{E}))$) be its derived category (inverting quasi-isomorphisms). Any complex we consider in this subsection belongs to one of the above categories. Note that $\Lambda^\bullet(\mathcal{E})$ is a commutative algebra in the graded sense, hence we can regard left $\Lambda^\bullet(\mathcal{E})$-modules as right $\Lambda^\bullet(\mathcal{E})$-modules following the usual sign convention, or even better, as $\Lambda^\bullet(\mathcal{E})$-bi-modules.

Definitions for $\mathrm{CG}(S^\bullet(\mathcal{E}^\vee))$ and $\mathrm{CG}(\Lambda^\bullet(\mathcal{E}))$. We shall systematically use $\mathcal{F}$ to denote an object in $\mathrm{CG}(\Lambda^\bullet(\mathcal{E}))$, $\mathcal{G}$ an object in $\mathrm{CG}(S^\bullet(\mathcal{E}^\vee))$. An object $\mathcal{G}$ in $\mathrm{CG}(S^\bullet(\mathcal{E}^\vee))$ consists of a complex of graded quasi-coherent $S^\bullet(\mathcal{E}^\vee)$-modules $\mathcal{G}^n$ with differentials $d_\mathcal{G} : \mathcal{G}^n \to \mathcal{G}^{n+1}$, $n \in \mathbb{Z}$. Each $\mathcal{G}^n$ is a graded module over $S^\bullet(\mathcal{E}^\vee) : \mathcal{G}^n = \oplus_i k^i(\mathcal{G}^n)$, $k^i(\mathcal{G}) = \oplus_n k^i(\mathcal{G}^n) = (0)$ for $i \gg 0$, and the differentials $d_\mathcal{G} : \mathcal{G}^n \to \mathcal{G}^{n+1}$ preserve the grading. The requirement for $\mathcal{F}$ is similar but slightly different in that the grading of $\Lambda^\bullet(\mathcal{E})$ enters in the multiplication: an object $\mathcal{F}$ in $\mathrm{CG}(\Lambda^\bullet(\mathcal{E}))$ consists of a complex of graded quasi-coherent $\mathcal{O}_X$-modules $\mathcal{F}^n (n \in \mathbb{Z})$, $\mathcal{F}^n = \oplus_i k^i(\mathcal{F}^n)$, $k^i(\mathcal{F}) = \oplus_n k^i(\mathcal{F}^n) = (0)$ for $i \gg 0$, with differentials $d_\mathcal{F} : \mathcal{F}^n \to \mathcal{F}^{n+1}$ together with graded $\mathcal{O}_X$-bilinear pairings $\Lambda : \mathcal{F}^n \times \Lambda^m(\mathcal{E}) \to \mathcal{F}^{n+m}$, such that the following conditions hold:

(i) each differential $d_\mathcal{F} : \mathcal{F}^n \to \mathcal{F}^{n+1}$ preserves the grading

(ii) $(f \wedge \omega_1) \wedge \omega_2 = f \wedge (\omega_1 \wedge \omega_2)\, \forall f \in \mathcal{F}^n, \forall \omega_1 \in \Lambda^{m_1}(\mathcal{E}), \omega_2 \in \Lambda^{m_2}(\mathcal{E})$

(iii) $k^i(\mathcal{F}^n) \wedge \Lambda^m(\mathcal{E}) \subset k^{i+m}(\mathcal{F}^{n+m})$

(iv) $d_\mathcal{F}(f \wedge \omega) = d_\mathcal{F} f \wedge \omega\, \forall f \in \mathcal{F}^n, \forall \omega \in \Lambda^m(\mathcal{E})$. Morphisms between objects of $\mathrm{CG}(S^\bullet(\mathcal{E}^\vee))$ (resp. $\mathrm{CG}(\Lambda^\bullet(\mathcal{E}))$) are defined in the obvious way: they are $S^\bullet(\mathcal{E}^\vee)$-linear (resp. $\Lambda^\bullet(\mathcal{E})$-linear) cochain maps (of degree 0) preserving gradation. A morphism between objects of $\mathrm{CG}(S^\bullet(\mathcal{E}^\vee))$ (resp. $\mathrm{CG}(\Lambda^\bullet(\mathcal{E}))$) is a quasi-isomorphism if it induces isomorphisms between all cohomology sheaves, regarded as $\mathcal{O}_X$-modules. Two morphisms $u,\ v : \mathcal{G} \to \mathcal{G}'$ between objects in $\mathrm{CG}(S^\bullet(\mathcal{E}^\vee))$ are homotopic if there exist graded $S^\bullet(\mathcal{E}^\vee)$-linear maps $H^n : \mathcal{G}^n \to \mathcal{G}'^{\,n-1}$ such that $d_{\mathcal{G}}'^{n-1} \circ H^n + H^{n+1} \circ d_{\mathcal{G}}^{n-1} = u^n - v^n\ \forall n \in \mathbb{Z}$. (In brief, $\exists H : \mathcal{G}[1] \to \mathcal{G}'$ s.t. $dH = u - v$.) Similarly, two morphisms $u,\ v : \mathcal{F} \to \mathcal{F}'$ between objects in $\mathrm{CG}(\Lambda^\bullet(\mathcal{E}))$ are homotopic if there exist graded $\mathcal{O}_X$-linear maps $H^n : \mathcal{F}^n \to \mathcal{F}'^{\,n-1}$ such that $H^n(f) \wedge \omega = H^{n+m}(f \wedge \omega)\ \forall f \in \mathcal{F}^n$, $\forall \omega \in \Lambda^m(\mathcal{E})$, and $d_{\mathcal{F}'}^{n-1} \circ H^n + H^{n+1} \circ d_{\mathcal{F}}^{n-1} = u^n - v^n\ \forall n \in \mathbb{Z}$. We also have shift functors $\mathcal{F} \Rightarrow \mathcal{F}[n]$, $\mathcal{G} \Rightarrow \mathcal{G}[n]$ shifting the degree of the complexes and twist functors $\mathcal{F} \Rightarrow \mathcal{F}\{n\}$, $\mathcal{G} \Rightarrow \mathcal{G}\{n\}$ shifting the gradation: $(\mathcal{F}[n])^m = \mathcal{F}^{m+n}$, $k^i(\mathcal{F}\{n\}) = k^{i+n}(\mathcal{F})$, and similarly for $\mathcal{G}$.

We shall define two (additive) exact functors $\Phi : \mathrm{CG}(\Lambda^\bullet(\mathcal{E})) \to \mathrm{CG}(S^\bullet(\mathcal{E}^\vee))$, $\Psi : \mathrm{CG}(S^\bullet(\mathcal{E}^\vee) \to \mathrm{CG}(\Lambda^\bullet(\mathcal{E}))$ such that Φ is the left adjoint of Ψ. Φ and Ψ respect homotopies and quasi-isomorphisms (in the graded sense) and consequently pass to the derived categories. We shall see that $\Phi : \mathrm{DG}(\Lambda^\bullet(\mathcal{E})) \to \mathrm{DG}(S^\bullet(\mathcal{E}^\vee))$, $\Psi : \mathrm{DG}(S^\bullet(\mathcal{E}^\vee) \to \mathrm{DG}(\Lambda^\bullet(\mathcal{E}))$ are equivalences of categories, inverse to each other. We shall also see that these equivalences of categories preserve complexes with coherent cohomology. There are also variants for DG^+, DG^-, DG^b, DG^b_c, $\mathrm{DG}_{\mathrm{perf}}$, which we leave to the readers to formulate and prove.

Definition of Φ. Let $\mathcal{F}$ be an object of $\mathrm{CG}(\Lambda^\bullet(\mathcal{E}))$, define a graded $S^\bullet(\mathcal{E}^\vee)$-complex $\mathcal{G} = \Phi(\mathcal{F})$ by the rule $\mathcal{G}^n = S^\bullet(\mathcal{E}^\vee) \otimes_{\mathcal{O}_X} \mathcal{F}^n$, with differential $d_{\mathcal{G}} : \mathcal{G}^n \to \mathcal{G}^{n+1}$ given locally by $d_{\mathcal{G}}(\phi \otimes f) = \phi \otimes d_{\mathcal{F}}(f) - (-1)^n \sum_j (\partial_j \cdot \phi) \otimes (f \wedge \omega_j)$. where $\{\partial_j\}$ and $\{\omega_j\}$ denote local dual bases of $\mathcal{E}^\vee$ and $\mathcal{E}$ respectively. The grading of $\mathcal{G} = \Phi(\mathcal{F})$ is defined in the usual way: $k^i(\mathcal{G}^n) = \sum_{j+j'=i} S^{-j}(\mathcal{E}^\vee) \otimes_{\mathcal{O}_X} k^{j'}(\mathcal{F}^n)$. We check by straight-forward computation that the definition for differentials does not depend on this specific choice of dual basis, that $\mathcal{G} = \Phi(\mathcal{F})$ thus defined is indeed an object of $\mathrm{CG}(S^\bullet(\mathcal{E}^\vee))$, and that Φ defines an exact functor from $\mathrm{CG}(\Lambda^\bullet(\mathcal{E}))$ to $\mathrm{CG}(S^\bullet(\mathcal{E}^\vee))$ which respects homotopies. It is also obvious that Φ commutes with the shift and twist functors.

Definition of Ψ. Let $\mathcal{G}$ be an object of $\mathrm{CG}(S^\bullet(\mathcal{E}^\vee))$, define $\mathcal{F} = \Psi(\mathcal{G})$ by the rule $\mathcal{F}^n = \prod_q \mathrm{Hom}(\Lambda^q(\mathcal{E}), \mathcal{G}^{n+q})$, with differential and exterior multiplication defined for an element ψ (= a family of maps $\Lambda^q(\mathcal{E}) \to \mathcal{G}^{n+q}$) by $d_{\mathcal{F}}(\psi)(\omega) = d_{\mathcal{G}}(\psi(\omega)) + (-1)^{n+m} \sum \partial_j \cdot \psi(\omega \wedge \omega_j)$, where $\{\partial_j\}$ and $\{\omega_j\}$ are local dual bases of $\mathcal{E}^\vee$ and $\mathcal{E}$; $(\psi \wedge \tilde{\omega})(\omega) = \psi(\tilde{\omega} \wedge \omega)\, \forall \tilde{\omega} \in \Lambda^i(\mathcal{E})$, $\forall \omega \in \Lambda^m(\mathcal{E})$. The gradation of $\mathcal{F} = \Psi(\mathcal{G})$ is defined in the obvious way: $k^i(\mathcal{F})$ consists of ψ's such that $\psi(\Lambda^q(\mathcal{E})) \subset k^{q+i}(\mathcal{G})$. Simple computation shows that the differentials are well-defined, and that Ψ defines an exact functor from $\mathrm{CG}(S^\bullet(\mathcal{E}^\vee))$ to $\mathrm{CG}(\Lambda^\bullet(\mathcal{E}))$ which respects homotopies. Clearly Ψ commutes with the shift and twist functors.

It is easy to check that the functor Ψ is the right adjoint of Φ: if $\mathcal{F}$ is object of $\mathrm{CG}(\Lambda^\bullet(\mathcal{E}))$, $\mathcal{G}$ is an object of $\mathrm{CG}(S^\bullet(\mathcal{E}^\vee))$, then $\mathrm{Hom}_{\mathrm{CG}(S^\bullet(\mathcal{E}^\vee))}(\Phi(\mathcal{F}), \mathcal{G})$ and $\mathrm{Hom}_{\mathrm{CG}(\Lambda^\bullet(\mathcal{E}))}(\mathcal{F}, \Psi(\mathcal{G}))$ are both canonically identified with graded $\mathcal{O}_X$-

linear cochain maps from the graded complex of $\mathcal{O}_X$-modules underlying $\mathcal{F}$ to the graded complex of $\mathcal{O}_X$-modules underlying $\mathcal{G}$. The identifications are given as follows: a morphism μ from $\Phi(\mathcal{F})$ to $\mathcal{G}$ corresponds to the map $\mathcal{F} \ni f \mapsto \mu(1 \otimes f) \in \mathcal{G}$, while a morphism ν from $\mathcal{F}$ to $\Psi(\mathcal{G})$ corresponds to the map $\mathcal{F} \ni f \mapsto \nu(f)(1) \in \mathcal{G}$. Details are left to the reader. By adjointness we obtain natural transformation of functors $\mathrm{id}_{\mathrm{CG}(\Lambda^\bullet(\mathcal{E}))} \to \Psi\Phi$ and $\Phi\Psi \to \mathrm{id}_{\mathrm{CG}(S^\bullet(\mathcal{E}^\vee))}$. We shall see later that $\mathcal{F} \to \Psi\Phi(\mathcal{F})$ and $\Phi\Psi(\mathcal{G}) \to \mathcal{G}$ are all quasi-isomorphisms.

A Variant of Ψ. Instead of Ψ we might consider the more familiar functor "Koszul-complex" from $\mathrm{CG}(S^\bullet(\mathcal{E}^\vee)$ to $\mathrm{CG}(\Lambda^\bullet(\mathcal{E}))$, which associates to an object $\mathcal{G}$ of $\mathrm{CG}(S^\bullet(\mathcal{E}^\vee))$, the tensor product $\mathrm{Kos}(\mathcal{G}) = \mathcal{G} \otimes \Lambda^\bullet(\mathcal{E})$, with $\mathrm{Kos}(\mathcal{G})^n = \sum_{i+j=n} \mathcal{G}^i \otimes \Lambda^j(\mathcal{E})$, gradation $k^i(\mathrm{Kos}(\mathcal{G})) = \sum_{j+j'=i} k^j(\mathcal{G}) \otimes \Lambda^{j'}(\mathcal{E})$, and differential $d(g \otimes \omega) = d_{\mathcal{G}} g \otimes \omega + (-1)^n \sum \partial_j \cdot g \otimes (\omega_j \wedge \omega)$ for $g \in \mathcal{G}^n$, $\omega \in \Lambda^m(\mathcal{E})$), where $\{\partial_j\}$ and $\{\omega_j\}$ are local dual bases of $\mathcal{E}^\vee$ and $\mathcal{E}$. $\mathrm{Kos}(\mathcal{G})$ is canonically isomorphic to $\Psi(\mathcal{G} \otimes \det(\mathcal{E})[-\mathrm{rk}(\mathcal{E})]) = \Psi(\mathcal{G}) \otimes \det(\mathcal{E})[-\mathrm{rk}(\mathcal{E})]$, where $\det(\mathcal{E})[-\mathrm{rk}(\mathcal{E})])$ is considered as a complex concentrated at degree $= \mathrm{rk}(\mathcal{E})$ and also graded by $\mathrm{rk}(\mathcal{E})$. Thus properties of Ψ can be derived from those of Kos and vice versa.

3.1. Proposition. *(1) Φ and Ψ transform quasi-isomorphisms into quasi-isomorphisms. Consequently they pass to the derived categories and define a pair of adjoint functors, Φ being the left adjoint of Ψ.*

(2) $\Phi : \mathrm{DG}(\Lambda^\bullet(\mathcal{E})) \to \mathrm{DG}(S^\bullet(\mathcal{E}^\vee))$, $\Psi : \mathrm{DG}(S^\bullet(\mathcal{E}^\vee) \to \mathrm{DG}(\Lambda^\bullet(\mathcal{E}))$ are equivalences of categories, inverse to each other via the adjointness transformations.

(a) *Φ preserves quasi-isomorphisms.* The question is local in X, so we may assume that $\mathcal{E}$ is free. Furthermore if $\mathcal{E}$ is the direct sum of two locally free sheaves $\mathcal{E}_1$ and $\mathcal{E}_2$ of finite rank, we may first apply the Φ-operator for $\mathcal{E}_1$, to get a $\Lambda^\bullet(\mathcal{E}_2) - S^\bullet(\mathcal{E}_1^\vee)$-complex, and then apply the Φ-operator for $\mathcal{E}_2$, the result of which is canonically isomorphic to applying the Φ-operator for $\mathcal{E}$. We thus reduce to the case $\mathcal{E} = \mathcal{O}_X$, so that a $\Lambda^\bullet(\mathcal{E})$-complex is just a graded complex $\mathcal{F}$ with a cochain map λ, graded of weight 1 (i.e. λ is a morphism from $\mathcal{F}$ to $\mathcal{F}\{1\}$), such that $\lambda^2 = 0$. Using mapping-cones we must show that if $\mathcal{F}$ is acyclic, so is $\Phi(\mathcal{F})$. Note that if $0 \to \mathcal{F}_1 \to \mathcal{F}_2 \to \mathcal{F}_3 \to 0$ is an exact sequence such that the assertion is true for two of the $\mathcal{F}_i$, it also holds for the third.

The situation now is that $\mathcal{E} = \mathcal{O}_X$, $\mathcal{F}$ is acyclic, and we must show that $\Phi(\mathcal{F})$ is acyclic. Let us first show this for bounded complexes, for which there are only finitely many non-zero terms. For an arbitrary bounded acyclic complex $\mathcal{F}$, choose n maximal with $\mathcal{F}^n \neq 0$. We find an acyclic graded subcomplex $\mathcal{F}'$ in $\mathcal{F}$ by replacing $\mathcal{F}^{n-2}$ by the kernel of $d_{\mathcal{F}} \circ \lambda = \lambda \circ d_{\mathcal{F}} : \mathcal{F}^{n-2} \to \mathcal{F}^n$, $\mathcal{F}^{n-1}$ by $\mathrm{Ker}(\lambda) \cap \mathrm{Ker}(d_{\mathcal{F}})$, $\mathcal{F}^n$ by 0, and not changing the other $\mathcal{F}^m$'s. It is easy to see that $\mathcal{F}'$ is still acyclic. By induction on the number of non-zero terms we may assume that $\Phi(\mathcal{F}')$ is acyclic, so we divide out and assume that the subcomplex defined above vanishes, by a previous remark. Again we find an acyclic graded subcomplex, this time with only non-zero terms $\mathrm{Ker}(\lambda)$ in degree $n-1$ and

$d_{\mathcal{F}}(\mathrm{Ker}(\lambda))$ in degree n. Dividing again we get a complex with λ vanishing on $\mathcal{F}^{n-2}$, so we may divide out by a new acyclic subcomplex with terms $\mathcal{F}^{n-2}$ and $d_{\mathcal{F}}(\mathcal{F}^{n-2})$. It follows that we have to check the assertion only for Φ applied to a graded complex of the form $0 \to \mathcal{F} \cong \mathcal{F} \to 0$, with λ given by a endomorphism of $\mathcal{F}$ of grade 1, and $\lambda^2 = 0$. Consequently $(1 \pm T \cdot \lambda) : \mathcal{F}[T] \to \mathcal{F}[T]$ is an isomorphism, and this is exactly the assertion we want to check.

Let us still assume that $\mathcal{E} = \mathcal{O}_X$, $\mathcal{F}$ acyclic, and λ as before, but $\mathcal{F}$ not necessarily bounded. For any pair of integers $m \leq n$, first define a sub-object $\mathcal{F}'$ of $\mathcal{F}$ as follows: $(\mathcal{F}')^j = \mathcal{F}^j$ if $j \leq n$, $(\mathcal{F}')^{n+1} = \mathrm{Ker}(d \circ \lambda = \lambda \circ d : \mathcal{F}^{n+1} \to \mathcal{F}^{n+3})$, $(\mathcal{F}')^{n+2} = \mathrm{Ker}(d) \cap \mathrm{Ker}(\lambda) \cap \mathcal{F}^{n+2}$, $(\mathcal{F}')^j = (0)$ if $j \geq n+3$. Then define a quotient object $\mathcal{F}''$ of $\mathcal{F}'$ by: $(\mathcal{F}'')^j = (0)$ if $j \leq m-3$, $(f'')^{m-2} = (\mathcal{F}')^{n+2}/\mathrm{Ker}(d) + \mathrm{Im}(\lambda)$, $(\mathcal{F}'')^{m-1} = (\mathcal{F}')^{m-1}/\mathrm{Im}(d \circ \lambda)$, $(\mathcal{F}'')^j = (\mathcal{F}')^j$ if $j \geq n+3$. It is easy to see that $\mathcal{H}^j(\Phi(\mathcal{F})) = \mathcal{H}^j(\Phi(\mathcal{F}''))$ if $m \leq j \leq n$. Hence the general case follows from the case for bounded complexes.

(b) *Ψ preserves quasi-isomorphisms.* All the reduction steps in (a) still hold, the proofs are similar and only easier: the $S^\bullet(\mathcal{E}^\vee)$-module structure of $\mathcal{G}$ preserves the degree of the complex, so we can use the usual truncation process for complexes. By the same steps as above it suffices to show that Ψ transforms an $S^\bullet(\mathcal{E}^\vee)$-complex $0 \to \mathcal{G} \cong \mathcal{G} \to 0$ into an acyclic complex, which is immediate: Filter $\Psi(\mathcal{G}^\bullet)$ by the subcomplexes defined by taking the sum $\oplus_{0 \leq q \leq r-p} \mathrm{Hom}(\Lambda^q(\mathcal{E}), \mathcal{G}^{n+q})$ of the first $r-p+1$ terms in $\oplus_q \mathrm{Hom}(\Lambda^q(\mathcal{E}), \mathcal{G}^{n+q})$, for $0 \leq p \leq r = \mathrm{rank}(\mathcal{E})$. (More canonically, $\mathrm{Ker}\{\prod_q \mathrm{Hom}(\Lambda^q(\mathcal{E}), \mathcal{G}^{n+q}) \to \prod_{r-p+1 \leq p \leq r} \mathrm{Hom}(\Lambda^q(\mathcal{E}), \mathcal{G}^{n+q})\}$.) The successive quotients are obviously acyclic.

(c) *$\mathcal{F} \to \Psi\Phi(\mathcal{F})$ and $\Phi\Psi(\mathcal{G}) \to \mathcal{G}$ are quasi-isomorphisms.* By dévissage we reduce to the case that complexes $\mathcal{F}^\bullet$ and $\mathcal{G}^\bullet$ have only one non-zero term, denoted by $\mathcal{F}$ respectively $\mathcal{G}$. (Note the slight change of notation.) Then $\Psi\Phi(\mathcal{F}^\bullet)$ reduces to $\mathrm{Kos}(S^\bullet(\mathcal{E}^\vee) \otimes \mathcal{F}) \otimes \det(\mathcal{E}^\vee)[\mathrm{rk}(\mathcal{E})]$. Similarly computing $\Phi\Psi(\mathcal{G}^\bullet)$ reduces to the $\mathrm{Kos}(S^\bullet(\mathcal{E}^\vee) \otimes \mathcal{G}) \otimes \det(\mathcal{E}^\vee)[\mathrm{rk}(\mathcal{E})]$. Now it is known that for any quasi-coherent $\mathcal{O}_X$-module $\mathcal{M}$, for the Koszul complex $\Lambda^\bullet(\mathcal{E}) \otimes S^\bullet(\mathcal{E}^\vee) \otimes \mathcal{M}$, we have canonically defined maps of complexes $\mathcal{M} \otimes \det(\mathcal{E})[-\mathrm{rk}(\mathcal{E})] \to \Lambda^\bullet(\mathcal{E}) \otimes S^\bullet(\mathcal{E}^\vee) \otimes \mathcal{M}$, $\Lambda^\bullet(\mathcal{E}) \otimes S^\bullet(\mathcal{E}^\vee) \otimes \mathcal{M} \to \mathcal{M} \otimes \det(\mathcal{E})[-\mathrm{rk}(\mathcal{E})]$, which are quasi-isomorphisms. These maps preserve the gradation and degrees (of complexes), hence the assertion. It follows that Φ and Ψ define equivalences of derived categories, and that the same is true for the functor "Koszul-complex". Moreover Φ and Ψ are adjoint pairs as functors on the derived categories.

3.2. Remarks. (a) The Koszul-complex has the advantage that it is easier to describe products: As $\Lambda^\bullet(\mathcal{E})$ is commutative in the graded sense, any right $\Lambda^\bullet(\mathcal{E})$-complex $\mathcal{F}$ becomes a bicomplex, by the rule $\omega \wedge f = (-1)^{nm} f \wedge \omega (f \in \mathcal{F}^n, \omega \in \Lambda^m(\mathcal{E}))$. We thus can define the notion of a $\Lambda^\bullet(\mathcal{E})$-bilinear pairing $\mathcal{F}_1 \times \mathcal{F}_2 \to \mathcal{F}_3$ in the obvious way. If the $\mathcal{F}_i$ are Koszul complexes of $S^\bullet(\mathcal{E}^\vee)$-complexes $\mathcal{G}_i$, this corresponds to an $S^\bullet(\mathcal{E}^\vee)$-linear pairing $\mathcal{G}_1 \otimes_{S^\bullet(\mathcal{E}^\vee)} \mathcal{G}_2 \to \mathcal{G}_3$, after replacing $\mathcal{G}_3$ by a quasi-isomorphic complex (Just compute the Koszul-complex of $\mathcal{G}_1 \otimes_{S^\bullet(\mathcal{E}^\vee)} \mathcal{G}_2$).

(b) If X is noetherian, there is a correspondence between coherent objects: Complexes $\mathcal{F}$ with $\mathcal{O}_X$-coherent cohomology correspond to $\mathcal{G}$ with $S^\bullet(\mathcal{E}^\vee)$-coherent cohomology (use the same dévissage as before). Other variants/generalizations are left to the reader.

We need a non-commutative version of these considerations: Consider a scheme X which is smooth over some base scheme S, and the vector-bundle $\mathcal{E} = \Omega^1_{X/S}$, abbreviated by Ω. The exterior algebra $\Omega^\bullet = \Lambda^\bullet\Omega$ has a differential d of degree 1 (exterior differentiation), and we have to replace the symmetric algebra by $\mathcal{D} = \mathcal{D}_{X/S}$, the sheaf of relative HPD-differential operators (of finite order). (The reason for using HPD-differential operators is that, if $x_1, \dots, x_d$ are local coordinates, the $\partial^I/\partial x^I$'s, with I ranging over all multi-indices, form an $\mathcal{O}_X$-basis of $\mathcal{D}_{X/S}$, and $\partial/\partial x_1, \dots, \partial/x_d$ generate $\mathcal{D}_{X/S}$ over $\mathcal{O}_X$ as algebra. If we used all differential operators, we would need terms $(I!)^{-1} \cdot \partial^I/\partial x^I$, and $\partial/\partial x_1, \dots, \partial/\partial x_d$ would no longer generate $\mathcal{D}_{X/S}$. See [Be 1], [BO 1] for more discussion of HPD-differential operators.) Both $\Omega^\bullet$ and $\mathcal{D}$ have decreasing filtrations $F^\bullet$, with $F^p(\Omega)$ = (relative) differential forms of degree $\geq p$, and $F^p(\mathcal{D})$ = differential operators of degree $\leq -p$. The corresponding graded objects are (canonically isomorphic to) $\Lambda^\bullet(\Omega)$ (with trivial differential) and $S^\bullet(\Omega^\vee)$.

We shall define two additive categories $\mathrm{CF}(\Omega^\bullet)$ and $\mathrm{CF}(\mathcal{D})$ consisting of filtered complexes of $\Omega^\bullet$-modules and $\mathcal{D}$-modules, and two exact functors $\Phi : \mathrm{CF}(\Omega^\bullet) \to \mathrm{CF}(\mathcal{D})$, $\Psi : \mathrm{CF}(\mathcal{D}) \to \mathrm{CF}(\Omega^\bullet)$ (abusing notation). As before, we shall show that Φ is the left adjoint of Ψ, and Φ and Ψ respect quasi-isomorphisms and therefore pass to derived categories. Furthermore the adjointness maps $\mathcal{F} \to \Psi\Phi(\mathcal{F})$ and $\Phi\Psi(\mathcal{G}) \to \mathcal{G}$ are quasi-isomorphisms, so Φ, Ψ form an adjoint pair of equivalence of categories on the derived categories.

Definition for $\mathrm{CF}(\Omega^\bullet)$**.** We consider filtered complexes $\mathcal{F}$ of (right) $\Omega^\bullet$-modules, which form an additive category $\mathrm{CF}(\Omega^\bullet)$. An object $\mathcal{F}$ of $\mathrm{CF}(\Omega^\bullet)$ is a complex of quasicoherent $\mathcal{O}_X$-modules, with a right multiplication $\wedge$ by $\Omega^\bullet$ (as before) satisfying $d_\mathcal{F}(f \wedge \omega) = d_\mathcal{F}(f) \wedge \omega + (-1)^n f \wedge d(\omega)(f \in \mathcal{F}^n)$. Furthermore each $\mathcal{F}^n$ is endowed with a decreasing filtration by $\mathcal{O}_X$-quasicoherent submodules $F^p(\mathcal{F}^n)(p \in \mathbb{Z})$, with $F^p(\mathcal{F}^n) = 0$ for $p \gg 0$. The multiplication by $\Omega^\bullet$ should respect filtrations $(F^p(\mathcal{F}^n) \wedge \Omega^q \subset F^{p+q}(\mathcal{F}^{n+q}))$, as should $d_\mathcal{F}(d(F^p(\mathcal{F}^n)) \subset F^p(\mathcal{F}^{n+1})$. A morphism between two such complexes should respect the structures, and it is a quasi-isomorphism if it induces a quasi-isomorphisms on the associated graded objects $\mathrm{gr}_F(-)$, which are graded $\Omega^\bullet$-complexes in the previous sense (i.e. in $\mathrm{CG}(\Lambda^\bullet(\Omega))$. Again we have shift and twist functors $\mathcal{F} \Rightarrow \mathcal{F}[n]$ and $\mathcal{F} \Rightarrow \mathcal{F}\{n\} : (\mathcal{F}[n])^m = \mathcal{F}^{m+n}$, $F^i(\mathcal{F}\{n\}) = F^{i+n}(\mathcal{F})$. The filtered derived category $\mathrm{DF}(\Omega^\bullet)$ is obtained from $\mathrm{CF}(\Omega^\bullet)$ by inverting quasi-isomorphisms.

Definition for $\mathrm{CF}(\mathcal{D})$**.** Similarly we consider filtered $\mathcal{D}$-complexes $\mathcal{G}$, which form an additive category $\mathrm{CF}(\mathcal{D})$. An object $\mathcal{G}$ in $\mathrm{CF}(\mathcal{D})$ is a complex of quasicoherent (left) $\mathcal{D}$-modules, with $\mathcal{D}$-linear differentials and a decreasing filtration $F^\bullet$

by sub-complexes $F^p(\mathcal{G})$ of quasi-coherent $\mathcal{O}_X$-modules, such that $F^p(\mathcal{G}) = 0$ for $p \gg 0$, and $F^q(\mathcal{D}) \cdot F^p(\mathcal{G}) \subset F^{p+q}(\mathcal{G}) \forall p, q \in \mathbb{Z}$. The definition of morphisms is the obvious one. The associated graded $\mathrm{gr}_F(\mathcal{G})$ is an object in $\mathrm{CG}(S^\bullet(\Omega^\vee))$. A morphism in $\mathrm{CF}(\mathcal{D})$ is a quasi-isomorphism if it induces an isomorphism on $\mathrm{gr}_F(-)$. The filtered derived category $\mathrm{DF}(\mathcal{D})$ is obtained from $\mathrm{CF}(\mathcal{D})$ by inverting quasi-isomorphism. Again we have the shift functor and the twist functor $\mathcal{G} \Rightarrow \mathcal{G}[n]$ and $\mathcal{G} \Rightarrow \mathcal{G}\{n\}$, which shift the degree and filtration respectively. Note that a quasicoherent $\mathcal{D}$-module $\mathcal{M}$ can also be described as a quasicoherent $\mathcal{O}_X$-module $\mathcal{M}$ with an integrable connection $\nabla : \mathcal{M} \to \mathcal{M} \otimes \Omega$. In this description the filtration on $\mathcal{M}$ has to satisfy Griffiths transversality: $\nabla(F^p(\mathcal{M})) \subset F^{p-1}(\mathcal{M}) \otimes \Omega$, and $\mathcal{D}$-linear maps become parallel maps.

Definition of Φ. We define the functor $\Phi : \mathrm{CF}(\Omega^\bullet) \to \mathrm{CF}(\mathcal{D})$ from filtered $\Omega^\bullet$-complexes to filtered $\mathcal{D}$-complexes as follows: for $\mathcal{F}$ in $\mathrm{CF}(\Omega^\bullet)$, set the n-th term of $\Phi(\mathcal{F})$ to be $\Phi(\mathcal{F})^n = \mathcal{D} \otimes_{\mathcal{O}_X} \mathcal{F}^n$ (as a left $\mathcal{D}$-module) with filtration $F^p(\Phi(\mathcal{F})^n) = \sum_{q+r=p} F^q(\mathcal{D}) \otimes F^r(\mathcal{F}^n)$ and differential $d(\phi \otimes f) = \phi \otimes d_{\mathcal{F}}(f) - (-1)^n \sum (\phi \cdot \partial_j) \otimes (f \wedge \omega_j)$, $(\phi \in \mathcal{D}, f \in \mathcal{F}^n)$, where ∂_j and ω_j again denote local dual bases of $\Omega^\vee$ and Ω. One checks by local calculation that the right side in the formula for $d(\phi \otimes f)$ is independent of the choice of dual basis,and is $\mathcal{O}_X$-bilinear in ϕ and f. Moreover $d(F^p(\Phi(\mathcal{F})^n)) \subset F^p(\Phi(\mathcal{F})^{n+1})$, $d^2 = 0$. So $\Phi(\mathcal{F})$ thus defined is indeed an object of $\mathrm{CF}(\mathcal{D})$. Clearly Φ is an exact functor.

Definition of Ψ. For the other direction, define $\Psi(\mathcal{G})$ for a filtered $\mathcal{D}$-complex $\mathcal{G}$ in $C(\mathcal{D})$, by $\Psi(\mathcal{G})^n = \prod_q \mathrm{Hom}(\Omega^q, \mathcal{G}^{n+q})$, with right $\Omega^\bullet$-module structure and differentiation given by $(\psi \wedge \tilde{\omega})(\omega) = \psi(\tilde{\omega} \wedge \omega)$, $d(\psi)(\omega) = d_{\mathcal{G}}(\psi(\omega)) - (-1)^n \psi(d\omega) + (-1)^{n+m} \sum \partial_j \cdot \psi(\omega \wedge \omega_j) (\psi \in \Psi(\mathcal{G}^\bullet)^n$, $\omega \in \Omega^m)$. The filtration is the natural filtration on Hom's: $F^p(\Psi(\mathcal{G})^n)$ consists of all ψ's such that $\psi(F^r(\Omega^\bullet)) \subset F^{r+p}(\mathcal{G}) \forall r$. Of course one has to do some checking: $d(F^p(\Psi(\mathcal{G})) \subset F^p(\Psi(\mathcal{G})$, $F^p(\Psi(\mathcal{G}) \wedge F^q(\Omega^\bullet) \subset F^{p+q}(\Psi(\mathcal{G})$, $d^2 = 0$. The first two are immediate, the last is a straightforward calculation. Clearly this defines an exact functor $\Psi : \mathrm{CF}(\mathcal{D}) \to \mathrm{CF}(\Omega^\bullet)$.

A Variant of Ψ. Instead of Ψ we also could have used the functor "de Rham complex", which sends an object $\mathcal{G}$ in $\mathrm{CF}(\mathcal{D})$ to $\mathrm{DR}(\mathcal{G}) = \mathcal{G} \otimes_{\mathcal{O}_X} \Omega^\bullet$ in $\mathrm{CF}(\Omega^\bullet)$, with $\mathrm{DR}(\mathcal{G})^n = \sum_{i+j=n} \mathcal{G}^i \otimes \Omega^j$, $F^p(\mathrm{DR}(\mathcal{G})) = \sum_{q+r=p} F^q(\mathcal{G}^i) \otimes F^r(\Omega)$, and the usual differential: $d(g \otimes \omega) = d_{\mathcal{G}}(g) \otimes \omega + (-1)^n g \otimes d\omega$ for $g \in \mathcal{G}^n$, $\omega \in \Omega^m$. $\mathrm{DR}(\mathcal{G})$ is canonically isomorphic to $\Psi(\mathcal{G}) \otimes \det(\Omega)[-\dim(X/S)]$ with product filtration, where $\det(\Omega)$ has filtration degree $\dim(X/S)$ as usual. It has the advantage that $\Omega^\bullet$-products on $\mathcal{F}$'s correspond to parallel products of $\mathcal{G}$'s.

Results. From the definitions one derives that $\mathrm{Hom}_{\mathrm{CF}(\mathcal{D})}(\Phi(\mathcal{F}), \mathcal{G}) = \mathrm{Hom}_{\mathrm{CF}(\Omega^\bullet)}(\mathcal{F}, \Psi(\mathcal{G}))$, because both are canonically identified with $\mathcal{O}_X$-linear maps of complexes from $\mathcal{F}$ to $\mathcal{G}$ respecting filtrations. In other words Ψ is the right adjoint of Φ. The way we defined $\Phi : \mathrm{CF}(\Omega^\bullet) \to \mathrm{CF}(\mathcal{D})$, $\Psi : \mathrm{CF}(\mathcal{D}) \to \mathrm{CF}(\Omega^\bullet)$ makes it clear that they pass to associated graded categories and give the old $\Phi : \mathrm{CG}(\Lambda^\bullet(\Omega)) \to \mathrm{CG}(S^\bullet(\Omega^\vee))$, $\Psi : \mathrm{CG}(S^\bullet(\Omega^\vee)) \to \mathrm{CG}(\Lambda^\bullet(\Omega))$ defined before. Therefore it is clear that our new Φ and Ψ respect quasi-isomorphisms, and that the natural maps $\mathcal{F} \to \Psi\Phi(\mathcal{F})$ and $\Phi\Psi(\mathcal{G}) \to \mathcal{G}$ are quasi-isomorphisms.

Finally we need the logarithmic variant of the above constructions in the context of compactifications. Our situation now is that X is smooth over S, and contained in a smooth S-scheme $\overline{X}$ such that $\overline{X} - X$ is a relative divisor with normal crossings over S. Denote by $\overline{\Omega}$ the sheaf of relative 1-differentials with logarithmic poles along $\overline{X}-X$, so that $\overline{\Omega}^\bullet = \Lambda^\bullet(\overline{\Omega})$ is a differential graded algebra. Instead of $\mathcal{D}_{\overline{X}}$ we use $\overline{\mathcal{D}} \subset \mathcal{D}_{\overline{X}}$ which locally is generated over $\mathcal{O}_{\overline{X}}$ (as subalgebra of $\mathcal{D}_{\overline{X}}$) by $\partial/\partial x_1, \dots, \partial/\partial x_a, y_1 \cdot \partial/\partial y_1, \dots, y_b \cdot \partial/\partial y_b$, if $x_1, \dots, x_a$, $y_1, \dots, y_b$ are local parameters such that the boundary is defined by $\prod y_j = 0$. Using $\overline{\Omega}^\bullet$, $\overline{\mathcal{D}}$ instead of $\Omega^\bullet_{\overline{X}}$, $\mathcal{D}_X$, we define an additive category $\mathrm{CF}(\overline{\Omega}^\bullet)$ of filtered $\overline{\Omega}^\bullet$-complexes and an additive category $\mathrm{CF}(\overline{\mathcal{D}})$ as before. Again we can define exact functors $\Phi : \mathrm{CF}(\overline{\Omega}^\bullet) \to \mathrm{CF}(\overline{\mathcal{D}})$, $\Psi : \mathrm{CF}(\overline{\mathcal{D}}) \to \mathrm{CF}(\overline{\Omega}^\bullet)$ and a variant $\mathrm{DR} : \mathrm{CF}(\overline{\mathcal{D}}) \to \mathrm{CF}(\overline{\Omega}^\bullet)$ of Ψ. By exactly the same method, we derive that Φ is left adjoint to Ψ, that Φ, Ψ, and DR all respects quasi-isomorphisms, and that the natural maps $\mathcal{F} \to \Psi\Phi(\mathcal{F})$ and $\Phi\Psi(\mathcal{G}) \to \mathcal{G}$ for $\mathcal{F}$ in $\mathrm{CF}(\overline{\Omega}^\bullet)$, $\mathcal{G}$ in $\mathrm{CF}(\overline{\mathcal{D}})$ are quasi-isomorphisms. In particular Φ, Ψ, DR are all equivalence of categories on the level of (filtered) derived categories.

Let us summarize our results in a form to be used later:

3.3. Theorem. *Notation as above.*

(a) The functors $\Phi : \mathrm{DF}(\overline{\Omega}^\bullet) \to \mathrm{DF}(\overline{\mathcal{D}})$, $\Psi : \mathrm{DF}(\overline{\mathcal{D}}) \to \mathrm{DF}(\overline{\Omega}^\bullet)$ *are equivalences of categories, and* Φ *is the left adjoint of* Ψ.

(b) The functor "de Rham complex" $\mathrm{DR} : \mathrm{CF}(\overline{\mathcal{D}}) \to \mathrm{CF}(\overline{\Omega}^\bullet)$, *defines an equivalence of categories from* $\mathrm{DF}(\overline{\mathcal{D}})$ *to* $\mathrm{DF}(\overline{\Omega}^\bullet)$. $\mathrm{DR}(-)$ *is canonically isomorphic to* $\Psi(-) \otimes \det(\overline{\Omega})[-\dim(X/S)]$ *(already on the level of filtered complexes.) Under* DR, $\overline{\Omega}^\bullet$*-linear pairings correspond to parallel (w.r.t. the connection) pairings of* $\overline{\mathcal{D}}$*-modules.*

We want to apply this theory in the following context: Suppose X is a smooth $\mathbb{C}$-scheme, and $X \subset \overline{X}$ a smooth compactification such that $\overline{X} - X$ is a divisor with normal crossings. Suppose furthermore that $f : Y \to X$ is a proper smooth map. Then we can find a smooth compactification $Y \subset \overline{Y}$, with infinity a divisor with normal crossings, such that f extends to $\overline{f} : \overline{Y} \to \overline{X}$. The direct image $\mathbb{R}\overline{f}_*(\Omega^\bullet_{\overline{Y}})$ of the complex of differential forms with logarithmic poles is a filtered $\overline{\Omega}^\bullet_X$-complex, via multiplication of differential forms on Y with pull-backs of forms on X. Furthermore, up to canonical quasi-isomorphism it is independent of the choice of the compactification $\overline{Y}$, by resolution of singularities [Hi] and the following result:

3.4. Lemma. *Suppose* $Y \subset \overline{Y}_1$ *and* $Y \subset \overline{Y}_2$ *are two smooth compactifications with normal crossings at infinity,* $g : \overline{Y}_1 \to \overline{Y}_2$ *a morphism which is the identity on* Y. *Then the maps* $\overline{\Omega}^q_Y \to \mathbb{R}g_*(\overline{\Omega}^q_Y)$ *are quasi-isomorphisms.*

Proof. As $\overline{\Omega}^{\bullet}_{Y_2} \cong \mathbb{R}g_*(g^*\overline{\Omega}^{\bullet}_{Y_2})$ (by [Hi]) we must check that the complex $\mathcal{K}^{\bullet} = \mathbb{R}g_*(\overline{\Omega}^{q}_{Y_1}/g^*\overline{\Omega}^{q}_{Y_2}))$ is acyclic, for each q. This problem is local in $\overline{Y}_2$, so we may assume that $\overline{Y}_2$ is projective. (Use resolution of singularities to extend locally defined compactifications to global ones.) Furthermore Hodge theory tells us that the $\overline{\Omega}^{q}_{Y}$'s on $\overline{Y}_1$ and $\overline{Y}_2$ have the same cohomology, so $\mathbb{H}^{\bullet}(\overline{Y}_2, \mathcal{K}^{\bullet}) = 0$. Now use induction over the dimension of the support of the cohomology of $\mathcal{K}^{\bullet}$: If the dimension is 0, the cohomology must vanish, and we are done. If the dimension is positive, choose a generic hyperplane $H \subset \overline{Y}_2$. Then $H \cap Y \subset H$ and $\overline{f}^{-1}(H \cap Y) \subset \overline{f}^{-1}(H)$ are both smooth compactifications, with infinity divisors with normal crossings, and the formation of $\mathcal{K}^{\bullet}$ commutes with restriction to H, as this holds for $\overline{\Omega}^{q}_{Y_1}/g^*\overline{\Omega}^{q}_{Y_2}$. By induction $\mathcal{K}^{\bullet}$ is acyclic when restricted to H, so the support of its cohomology has dimension 0, and the lemma is proved.

As $\mathbb{R}\overline{f}_*(\overline{\Omega}^{\bullet}_{Y})$ is canonical, endomorphisms of Y/X act on it.

Now we work over $\mathbb{C}$, and suppose $X \subset \overline{X}$ is a smooth toroidal compactification of $X = A_{g,n}/\mathrm{Spec}(\mathbb{C})$, $n \geq 3$, and $f : Y = \mathsf{A}^s \to X$ the s-fold fibre product of the universal abelian variety over X. Extend f to some good compactification $\overline{f} : \overline{Y} \to \overline{X}$ (for instance a toroidal one constructed in §1), and collect all the information we have accumulated so far, which almost determines $\mathbb{R}\overline{f}_*(\overline{\Omega}^{\bullet}_{Y})$:

(1) By the Φ-Ψ-formalism (theorem 3) we get from $\mathbb{R}\overline{f}_*(\overline{\Omega}^{\bullet}_{Y})$ a filtered $\overline{\mathcal{D}}$-complex $\mathcal{G}^{\bullet}$, determined up to quasi-isomorphism, whose de Rham complex is $\mathbb{R}\overline{f}_*(\overline{\Omega}^{\bullet}_{Y})$ (with its usual Hodge filtration). The restriction of $\mathcal{G}^{\bullet}$ to X is nothing else than (or, canonically quasi-isomorphic to) the relative de Rham complex $\mathcal{H}^{\bullet}_{\mathrm{DR}}(Y/X)$ for $\mathsf{A}^s \to X$, with $\mathcal{D}$-module structure via the Gauss-Manin connection. The cup-product on $\mathbb{R}\overline{f}_*(\overline{\Omega}^{\bullet}_{Y})$ induces a product structure on $\mathcal{G}^{\bullet}$.

(2) We may use the toroidal compactifications constructed in §1 to compute $\mathbb{R}\overline{f}_*(\overline{\Omega}^{\bullet}_{Y})$. It follows that there is a increasing filtration $G_{\bullet}$ of $\mathrm{gr}_F(\overline{\Omega}^{\bullet}_{Y})$, with $G_p(\overline{\Omega}^{q}_{Y}) = \overline{f}^*\overline{\Omega}^{q-p}_{X} \wedge \overline{\Omega}^{p}_{Y}$, so $\mathrm{gr}^G_p \mathrm{gr}_F(\overline{\Omega}^{\bullet}_{Y}) = \overline{f}^*\overline{\Omega}^{\bullet}_{X}[-p] \otimes \overline{\Omega}^{p}_{Y/X}$. It induces filtrations $G_{\bullet}$ on $\mathrm{gr}_F(\mathbb{R}\overline{f}_*(\overline{\Omega}^{\bullet}_{Y}))$ and $\mathrm{gr}_F(\mathcal{G}^{\bullet})$.

(3) Thus $\mathrm{gr}_F\mathbb{R}\overline{f}_*(\overline{\Omega}^{\bullet}_{Y}) = \mathbb{R}\overline{f}_*(\mathrm{gr}_F(\overline{\Omega}^{\bullet}_{Y}))$ has an increasing filtration such that the successive quotients are isomorphic to the Koszul complexes associated to $\mathbb{R}\overline{f}_*(\overline{\Omega}^{p}_{Y/X}) \cong \Lambda^p((\mathcal{T}^*_G)^s) \otimes \Lambda^{\bullet}((\mathcal{T}_G)^s)[-q]$, with the trivial ($= 0$) multiplication by $\overline{\Omega}_X$.

(4) The spectral sequence associated to the filtration $G_{\bullet}$ degenerates: This is true over X, and the E_2-terms are locally free.

(5) Again by the Φ-Ψ-formalism, $\mathrm{gr}_F(\mathcal{G}^{\bullet})$ has an increasing filtration $G_{\bullet}$ such that the successive quotients are quasi-isomorphic to $\mathbb{R}\overline{f}_*(\overline{\Omega}^{p}_{Y/X}) \cong \Lambda^p((\mathcal{T}^*_G)^s) \otimes \Lambda^{\bullet}((\mathcal{T}_G)^s)[-p]$.

(6) For any m the cohomology $\mathcal{H}^m(\mathrm{gr}_F(\mathcal{G}^{\bullet}))$ is locally free, and $\mathcal{H}^m(\mathrm{gr}^q_F(\mathcal{G})))$ is isomorphic to $\Lambda^q((\mathcal{T}^*_G)^s) \otimes \Lambda^{m-q}((\mathcal{T}_G)^s)$ (there is a canonical isomorphism on X, and it extends by the local calculations near the boundary of X: Only one gr_G enters.)

(7) (Liebermann's trick) For any $l \in \mathbb{Z}$, multiplication by l defines an endomorphism $[l]$ of $\mathsf{A}^s_{/X}$, and $[l]^*$ operates as an endomorphism on $\mathcal{G}$ and $\mathrm{gr}_F(\mathcal{G})$. On $\mathcal{H}^m((\mathrm{gr}_F(\mathcal{G})))$ the action of $[l]^*$ is multiplication with l^m, *so* $\mathcal{G}$ *must be split as filtered* $\overline{\mathcal{D}}$*-complex* (w.r.t. the F-filtration): We may represent $\mathcal{G}$ by a sequence of $\overline{\mathcal{D}}$-modules $\mathcal{G}^m$, which form a complex with trivial differentials. Furthermore, $\mathcal{G}^m$ has a decreasing filtration F, such that $\mathrm{gr}^q_F(\mathcal{G}^m) \cong \Lambda^q((\mathcal{T}^*_G)^s) \otimes \Lambda^{m-q}((\mathcal{T}_G)^s)$ is locally free over $\mathcal{O}_{\overline{X}}$, so $\mathcal{G}^m$ is a vector bundle with an integrable logarithmic connection $\nabla : \mathcal{G}^m \to \mathcal{G}^m \otimes \overline{\Omega}_X$. On X itself $\mathcal{G}^m$, ∇, and F are given by the m-th relative de Rham cohomology $\mathcal{H}^m_{\mathrm{DR}}(Y/X)$ of $\mathsf{A}^s \to X$, its Gauss-Manin connection, and its Hodge filtration.

(8) Finally the cup-product gives bilinear parallel pairings $\mathcal{G}^m \times \mathcal{G}^n \to \mathcal{G}^{m+n}$, which extend the cup-product on the relative de Rham cohomology of $\mathsf{A}^s \to X$. They are commutative in the graded sense, by continuity, and so induce transformations $\Lambda^m(\mathcal{G}^1) \to \mathcal{G}^m$. These give isomorphisms on the gr_F-level, hence are isomorphisms.

We shall see that the accumulated data are sufficient to determine the $\mathcal{G}^m$'s.

4. Extending Vector Bundles

In this section and the next, the field of definition is $\mathbb{C}$ unless otherwise specified. For convenience of notation we shall again denote $\mathsf{A}_{g,n}$ by X and $X \subset \overline{X}$ a smooth toroidal compactification of $X = \mathsf{A}_{g,n}$. Here $n \geq 1$ is some integer, which the reader may assume to be ≥ 3 if he prefers. Of course, in general $\overline{X}$ is just an algebraic stack even if X is a scheme, so strictly speaking we shall have to do things in this context. But this causes no problem as we are dealing mostly with local questions. Thus we shall ignore considerations of this sort and pretend $\overline{X}$ is a scheme.

We shall first review and establish some notation to be used in this section and the next. Most of what we shall say belongs the the general theory of bounded symmetric domains, but we try to be explicit and not to be too heavily Lie-theoretic. References for bounded symmetric domains are [AMRT], [Sa 3], [He].

In this section and the next, $\mathbf{G}$ denotes the Chevalley group scheme $\mathrm{GSp}(2g)$ over $\mathbb{Z}$ of symplectic similitudes on the symplectic space $\mathbb{Z}^{2g}$ with its standard alternating form. More explicitly, for any (commutative ring with identity) R,

$$\mathbf{G}(R) = \left\{ \gamma = \begin{pmatrix} A & B \\ C & D \end{pmatrix} \middle| \begin{array}{c} A \cdot {}^tB - B \cdot {}^tA = C \cdot {}^tD - D \cdot {}^tC = 0, \\ A \cdot {}^tD - B \cdot {}^tC = \eta \cdot I_g \text{ for some } \eta \in R^* \end{array} \right\}$$

The invertible element η is clearly determined by γ, so we call η the multiplier of γ. η defines a homomorphism $\eta : \mathbf{G} \to G_m$ called the multiplier representation of $\mathbf{G}$, which sends an element γ of $\mathbf{G}$ to its multiplier. η is faithfully flat and surjective in the Zariski topology.

$\mathbf{G}^\circ := \mathrm{Ker}(\eta)$ is by definition the Chevalley group scheme $\mathrm{Sp}(2g)$ of all symplectic transformations on the symplectic space $\mathbb{Z}^{2g}$ with its standard alternating form. Notice that $\det = \eta^g$ as homomorphisms from $\mathbf{G}$ to G_m.

The conditions defining $\mathbf{G}$ above are equivalent to

$$^tD \cdot B - {}^tB \cdot D = {}^tC \cdot A - {}^tA \cdot C = 0, \quad {}^tD \cdot A - {}^tB \cdot C = \eta \cdot I_g$$

(for the same $\eta \in R^*$). These relations can be easily presented in matrix form:

$$\gamma \cdot \gamma^\vee = \gamma^\vee \cdot \gamma = \lambda(\gamma) \cdot 1_{2g}\ , \quad \text{where}$$

$$\gamma^\vee = \begin{pmatrix} {}^tD & -{}^tB \\ -{}^tC & {}^tA \end{pmatrix} = \eta(\gamma) \cdot \gamma^{-1}.$$

Notice that $\eta(\gamma) = \eta(\gamma^\vee)$, and $\gamma \mapsto \gamma^\vee$ is an involution on $\mathbf{G}$.

$\mathbf{G}$ has an involution (i.e. an automorphism of order 2) σ defined by $\sigma(\gamma) = \eta(\gamma) \cdot {}^t\gamma^{-1} = J \cdot \gamma \cdot J^{-1}$, where

$$J = \begin{pmatrix} 0 & I_g \\ -I_g & 0 \end{pmatrix}, \quad \text{so } \sigma(\gamma) = \begin{pmatrix} D & -C \\ -B & A \end{pmatrix} \quad \text{if } \gamma = \begin{pmatrix} A & B \\ C & D \end{pmatrix}.$$

$\mathbf{G}$ operates on $\mathbb{Z}^{2g}$ (thought of as column vectors) on the left via matrix multiplication; we call this the standard representation. Unfortunately the classical formulas for the complex uniformization of A_g requires a different action: $\mathbf{G}$ must operate on $\mathbb{Z}^{2g}$ on the left via the composition of σ and the standard representation. We shall come back to this point later.

There are two ways to embed G_m into $\mathbf{G}$:

$$d : t \mapsto \begin{pmatrix} t & 0 \\ 0 & t \end{pmatrix} \qquad \alpha : t \mapsto \begin{pmatrix} t & 0 \\ 0 & 1 \end{pmatrix}$$

$d(G_m)$ is central in $\mathbf{G}$, while $\mathbf{G}^\circ \cdot \alpha(G_m) = \mathbf{G}$.

In $\mathbf{G}$ we have Chevalley subgroup schemes $\mathbf{T} \subset \mathbf{M} \subset \mathbf{Q} = \mathbf{M} \ltimes \mathbf{U}$, where

(a) $\mathbf{T} \cong \mathbf{G}_m^{g+1}$ is the group of diagonal matrices. An element of $\mathbf{T}$ has entries $a_1, \ldots, a_g, d_1, \ldots, d_g$, with $a_1 \cdot d_1 = a_2 \cdot d_2 = \ldots = a_g d_g$ ($= \eta$, the multiplier for the element). Such an element will be written as $\mathrm{diag}(a_1, \ldots, a_g, d_1, \ldots, d_g)$. $\mathbf{T}^0 := \mathbf{T} \cap \mathbf{G}^\circ$.

(b) $\mathbf{M} \cong \mathrm{GL}(g) \times G_m$ is the subgroup of $\mathbf{G}$ which respects the standard-decomposition $\mathbb{Z}^{2g} = \mathbb{Z}^g \oplus \mathbb{Z}^g$. Elements $m \in \mathbf{M}$ have two $g \times g$-matrices A and D on the two diagonal blocks, with $A \cdot {}^tD = \eta \cdot I_g$ a scalar which is a unit, so that η is the multiplier of m as an element of $\mathbf{G}$. Such an element will be written as $\mathrm{diag}(A, D)$. $\mathbf{M}^0 := \mathbf{M} \cap \mathbf{G}^\circ$.

(c) $\mathbf{U}$ is determined by $\mathbf{U}(\mathbb{Z}) \cong \Gamma^2(\mathbb{Z}^g) \cong \mathrm{Hom}_{\mathbb{Z}}(S^2(X), \mathbb{Z})$ = the space of all $\mathbb{Z}$-valued bilinear symmetric forms. In other words $\mathbf{U}(R)$ is the subgroup of $\mathbf{G}(R)$ which fixes the first direct summand $R^g \subset R^{2g}$. Elements $u \in \mathbf{U}$ are upper triangular matrices with I_g's on the diagonal blocks, and symmetric

$g \times g$-matrices B in the upper right corner. $\mathbf{U}$ is the unipotent radical of $\mathbf{Q}$, where

(d) $\mathbf{Q} = \mathbf{M} \ltimes \mathbf{U}$ is the parabolic subgroup of $\mathbf{G}$ which stabilizes the first direct summand $\mathbb{Z}^g$ of $\mathbb{Z}^{2g}$ (also a maximal isotropic sublattice). $\mathbf{Q}$ is actually a maximal parabolic subgroup. $\mathbf{Q} = \mathbf{M} \ltimes \mathbf{U}$ is a Levi decomposition, with $\mathbf{M}$ identified with the subgroup of elements stabilizing both the first and the second direct summand $\mathbb{Z}^g \subset \mathbb{Z}^{2g}$. It consists of elements of $\mathbf{G}$ with zero in the lower left corner when written in block-form. $\mathbf{Q}^0 := \mathbf{Q} \cap \mathbf{G}^\circ = \mathbf{M}^\circ \ltimes \mathbf{U}$.

Along the way we also need the subgroups $\mathbf{N}$ and $\mathbf{B} = \mathbf{T} \ltimes \mathbf{N}$, where

(e) $\mathbf{B} \subset \mathbf{G}$ is the Borel subgroup of matrices in $\mathbf{G}$ with entries $A, B, 0, D$ in block-form with A upper-triangular and D lower-triangular, and

(f) $\mathbf{N}$ is the unipotent radical of $\mathbf{B}$. $\mathbf{N}$ consists of the elements of $\mathbf{B}$ with 1's along the diagonal. $\mathbf{B}^0 := \mathbf{B} \cap \mathbf{G}^\circ = \mathbf{T}^\circ \ltimes \mathbf{N}$.

The adjoint action of $\mathbf{M}$ on the Lie algebra $\mathfrak{u}$ of $\mathbf{U}$ has as determinant a character $e^{2\rho_n} : \mathbf{M} \to \mathbf{G}_m$, which sends an element $m = (A, D) \in \mathbf{M}$ into $\det(A)^{g+1} \cdot \eta^{-g(g+1)/2} = \det(D)^{-g-1} \cdot \eta^{g(g+1)/2} (A \cdot {}^tD = \eta \cdot I_g)$. The notation is explained by the fact that on the level of Lie algebras the derivative is given by $2\rho_n$, that is the sum of the roots in $\mathfrak{u}$.

Similarly the adjoint action of $\mathbf{T}$ on the Lie algebra of $\mathbf{N} \cap \mathbf{M}$ defines a character $e^{2\rho_M} : \mathbf{T} \to G_m$, which sends an element $t = \operatorname{diag}(a_1, \ldots, a_g, d_1, \ldots, d_g)$ to $\prod_{1 \le i \le g} a_i^{g-2i+1} = \prod_{1 \le i \le g} d_i^{2i-g-1}$. The adjoint action of $\mathbf{T}$ on the Lie algebra of $\mathbf{N}$ defines a character $e^{2\rho} : \mathbf{T} \to G_m$, which is equal to the product of $e^{2\rho_n}$ and $e^{2\rho_M}$.

$W_G \cong \mathcal{S}_g \ltimes (\pm 1)^g$ and $W_M \cong \mathcal{S}_g$ denote the Weyl groups of $\mathbf{G}$ and $\mathbf{M}$ respectively w.r.t. the maximal torus $\mathbf{T}$. They act on elements $t = \operatorname{diag}(a_1, \ldots, a_g, d_1, \ldots, d_g)$ of $\mathbf{T}$ by permutation or exchange of the a's and d's.

Let $\mathbf{K}$ be the maximal compact subgroup of $\mathbf{G}^\circ(\mathbb{R})$, whose elements have block-form $(A, -B, B, A)$ with B in the lower left corner. Associating to such a typical element $A + \sqrt{-1} \cdot B$ establishes an isomorphism from $\mathbf{K}$ to $\mathbf{U}(g) =$ the unitary group of rank g. $\mathbf{M}_\mathbb{C}$ is conjugate to the complexification of $\mathbf{K}$. Let $\mathbf{G}(\mathbb{R})^+ =$ the connected component of $\mathbf{G}(\mathbb{R}) =$ subgroup of elements of $\mathbf{G}(\mathbb{R})$ whose multipliers are positive. $\mathbb{D} = \mathbf{G}^\circ(\mathbb{R})/\mathbf{K} = \mathbf{G}(\mathbb{R})^+/\mathbf{K} \cdot \mathbb{R}_{>0}$ is a bounded symmetric domain, which is embedded in its compact dual $\mathbb{D}^\vee = (\mathbf{G}/\mathbf{Q})(\mathbb{C})$ as a $\mathbf{G}(\mathbb{R})^+$-invariant open subset of the $\mathbf{G}(\mathbb{C})$-homogeneous space $\mathbb{D}^\vee$. If we identify $\mathbb{D}^\vee$ with the space of all Lagrangian (i.e. maximal isotropic) subspaces of $\mathbb{C}^{2g}$ w.r.t. the standard symplectic pairing $\langle\ ,\ \rangle$, then $\mathbb{D}$ consists of all Lagrangian subspaces W such that $-\sqrt{-1}\langle v, \overline{v} \rangle > 0\ \forall v \in W$, clearly stable under $\mathbf{G}(\mathbb{R})^+$. Each W is spanned by the columns of the $2g \times g$ matrix ${}^t(-I_g, \Omega)$ with a unique $\Omega \in \mathbb{H}_g (=$ Siegel's upper half space). This gives the unbounded realization of $\mathbb{D}$ as a Siegel domain. $\mathbf{G}(\mathbb{C})$ operates naturally on $\mathbb{D}^\vee$, and the subgroup $\mathbf{G}(\mathbb{R})^+$ leaves $\mathbb{D}$ invariant with the transported action on $\mathbb{H}_g$ given by the usual formula. Some people may prefer to consider $\mathbb{D}^\pm$ and $\mathbb{H}_g^\pm$, but we refrain from doing so.

Remark about base points: the base point $\mathbf{Q}(\mathbb{C})$ of $\mathbb{D}^\vee = (\mathbf{G}/\mathbf{Q})(\mathbb{C}) = (\mathbf{G}^\circ/\mathbf{Q}^\circ)(\mathbb{C})$ does not lie in $\mathbb{D}$, but only in the boundary of $\mathbb{D}$. Explicitly, the base point of $\mathbb{D}^\vee$ is the Lagrangian subspace spanned by the columns of ${}^t(-I_g, 0)$, while the usual choice of base point of $\mathbb{D}$ is the Lagrangian subspace spanned by the columns of ${}^t(-I_g, \sqrt{-1} \cdot I_g)$. These two base points are related by a Cayley transform.

The abelian variety A_Ω determined by $\Omega \in \mathbb{H}_g$ can be described as follows: we identify $H_1(A_\Omega(\mathbb{C}), \mathbb{Z})$ with $\mathbb{Z}^{2g}$ (column vectors), $\mathrm{Lie}(A_\Omega(\mathbb{C}))$ with $\mathbb{C}^g$ (column vectors), and the projection $\mathbb{Z}^{2g} \otimes \mathbb{C} = H_1(A_\Omega(\mathbb{C}), \mathbb{Z}) \otimes \mathbb{C} \to \mathrm{Lie}(A_\Omega(\mathbb{C})) = \mathbb{C}^g$ is defined to be left multiplication by (Ω, I_g), i.e. the period subgroup is $\Omega \cdot \mathbb{Z}^g + \mathbb{Z}^g$. The standard symplectic form defines a polarization on $H_1 = \mathbb{Z}^{2g}$, equal to the imaginary part of the Hermitian form $(z, w) \mapsto {}^t z \cdot (\mathrm{Im}\Omega) \cdot \overline{w}$ on $\mathrm{Lie}(A_\Omega(\mathbb{C})) = \mathbb{C}^g$. For $\gamma \in \mathbf{G}^\circ(\mathbb{Z})$ with entries A, B, C, D as usual, the formula

$$ {}^t(C\Omega + D)^{-1} \cdot (\Omega, I_g) = (\gamma \cdot \Omega, I_g) \cdot \gamma^\sigma, \quad \text{where } \gamma \cdot \Omega = (A\Omega + B) \cdot (C\Omega + D)^{-1} $$

tells us that left multiplication by ${}^t(C\Omega + D)^{-1}$ defines an isomorphism from A_Ω to $A_{\gamma \cdot \Omega}$; $\mathbf{G}^\circ(\mathbb{Z})$ operates on $\mathbf{H}_1$ via the composition of σ with the standard representation, on **Lie** by the 1-cocycle ${}^t(C\Omega + D)^{-1}$, on **Lie*** by the 1-cocycle $(C\Omega + D)$.

σ being an inner automorphism fixing the center G_m of $\mathbf{G}$, this twisting by σ does not cause any mathematical problem, but it does tend to create psychological problems. Fortunately, since we shall not use any explicit formula of the action of Γ on $\mathbb{H}_g$ in this chapter, we can (and will) precompose with σ to undo this unpleasant involution.

Let Γ be the arithmetic subgroup (torsion free for those who do not like stacks) of $\mathbf{G}^\circ(\mathbb{Q})$ such that $X = \Gamma \backslash \mathbb{D}$. As X is a locally homogeneous space, we can construct vector bundles in the usual manner: For any finite dimensional complex representation (τ, E_τ) of $\mathbf{Q}^\circ(\mathbb{C})$, we can attach a $\mathbf{G}^\circ(\mathbb{C})$-equivariant bundle $\mathbf{G}^\circ(\mathbb{C}) \times_{\mathbf{Q}^\circ(\mathbb{C}),\tau} E_\tau$ on $\mathbb{D}^\vee$, where the contraction product $\mathbf{G}^\circ(\mathbb{C}) \times_{\mathbf{Q}^\circ(\mathbb{C}),\tau} E_\tau$ is by definition the quotient of $\mathbf{G}^\circ(\mathbb{C}) \times E_\tau$ under the equivalence relation $(gq, \tau(q)^{-1}v) \sim (g, v)\ \forall g \in \mathbf{G}^\circ(\mathbb{C}), \forall q \in \mathbf{Q}^\circ(\mathbb{C})$. Restricting to $\mathbb{D}$ and dividing by Γ, we obtain a vector bundle $\mathcal{E}(\tau, \Gamma)$ on X, which we shall sometimes abbreviate as $\mathcal{E}(\tau)$ or $\mathcal{E}$. In brief: as $\mathbf{G}^\circ(\mathbb{C})$-equivariant bundles on $\mathbb{D}^\vee$ correspond to complex representations E of $\mathbf{Q}^\circ$ (E = fibre of $\mathcal{E}$ at the origin), we obtain a functor $E \Rightarrow \mathcal{E}$ from $\mathbf{Q}^\circ$-representations to vector-bundles on X. Notice that we can replace $\mathbf{G}^\circ$ by $\mathbf{G}$ and $\mathbf{Q}^\circ$ by $\mathbf{Q}$ everywhere in the above discussion. Whether one uses $\mathbf{G}, \mathbf{Q}, \mathbf{M}$ or $\mathbf{G}^\circ, \mathbf{Q}^\circ, \mathbf{M}^\circ$ will have no substantial consequence in this section: in any case, the only difference is the central G_m-action. The center G_m of $\mathbf{G}$ plays a role only in the subsequent sections, when we consider weights of cohomology. (Powers of η correspond to Tate twists.)

For a (finite dimensional complex) representation of $\mathbf{K}$, we can first extend to $\mathbf{M}^\circ(\mathbb{C})$ by analytic continuation and then to $\mathbf{Q}^\circ(\mathbb{C})$ so that $\mathbf{U}(\mathbb{C})$ operates trivially. Such representations of $\mathbf{Q}^\circ(\mathbb{C})$ will be called $\mathbf{K}$-representations or $\mathbf{M}^\circ$-representations by abuse of terminology. Notice that although every holomorphic $\mathbf{G}(\mathbb{R})^+$-equivariant vector bundle is C^∞-isomorphic to an equivariant holomorphic vector bundle coming from a $\mathbf{K}$-representation, this isomorphism is not necessarily holomorphic: there are $\mathcal{E}(\tau, \Gamma)$'s which do not come from any $\mathbf{K}$-representations (as holomorphic vector bundles).

When τ is the restriction to $\mathbf{Q}^\circ(\mathbb{C})$ of a (finite dimensional complex) representation ρ of $\mathbf{G}^\circ(\mathbb{C})$, $\mathcal{E}(\tau,\Gamma)$ carries a natural integrable connection, as follows: $v \mapsto (g \mapsto \rho(g)v)$ for $v \in E_\rho$, $g \in \mathbf{G}^\circ(\mathbb{C})$ defines a trivialization of $\mathbf{G}^\circ(\mathbb{C}) \times_{\mathbf{Q}^\circ(\mathbb{C}),\tau} E_\tau$, hence an integrable connection on $\mathbf{G}^\circ(\mathbb{C}) \times_{\mathbf{Q}^\circ(\mathbb{C}),\tau} E_\tau$, which descends to $\mathcal{E}(\tau,\Gamma)$. Similar statements hold if we use $\mathbf{G}, \mathbf{Q}$ instead of $\mathbf{G}^\circ, \mathbf{Q}^\circ$.

4.1. Examples. (a) When τ is the $\mathbf{K}$-representation coming from the standard representation of $\mathbf{M}^\circ = \mathrm{GL}(g)$ (sending (A,D) to A), $\mathcal{E}(\tau,\Gamma)$ is (canonically isomorphic to) $\mathcal{T}_G^* = \mathbf{Lie}(G/X)^*$.

(b) The line bundle ω is associated to the determinant of the standard representation of GL_g.

(c) Ω_X is associated to the second symmetric power of the standard representation of GL_g.

(d) The canonical line bundle $\mathcal{K}_X$ is associated to the character $(\det)^{g+1}$ of GL_g, thus $\mathcal{K}_X \cong \omega^{\otimes g+1}$.

(e) $\mathbf{H}^1_{\mathrm{DR}}(G/X)$ with the Gauss-Manin connection is associated to the standard representation of $\mathbf{G}^\circ$.

In the sequel we shall often identify a group with its $\mathbb{C}$-points and similarly for Lie algebras if there is no possible confusion. We have seen that $\mathbf{G}$-equivariant (resp. $\mathbf{G}^\circ$-equivariant) vector bundles on $\mathbb{D}^\vee$ correspond to finite-dimensional $\mathbf{Q}$-modules (resp. $\mathbf{Q}^\circ$-modules). Similarly $\mathbf{G}(\mathbb{C})$-equivariant (resp. $\mathbf{G}^\circ$-equivariant) quasicoherent $\mathcal{D}$-modules on $\mathbb{D}^\vee$ correspond to $\mathfrak{g}$-$\mathbf{Q}$-modules (resp. $\mathfrak{g}^\circ$-$\mathbf{Q}^\circ$-modules). These are $\mathbf{Q}$-modules (resp. $\mathbf{Q}^\circ$-modules) with an operation of the Lie-algebra $\mathfrak{g}$ (resp. $\mathfrak{g}^\circ$) of $\mathbf{G}$ (resp. $\mathbf{G}^\circ$), whose restriction to the Lie-algebra $\mathfrak{q}$ (resp. $\mathfrak{q}^\circ$) of $\mathbf{Q}$ (resp. $\mathbf{Q}^\circ$) coincides with the differential of the $\mathbf{Q}$-action (resp. $\mathbf{Q}^\circ$-action). Also recall that a quasicoherent equivariant $\mathcal{D}$-module on $\mathbb{D}^\vee$ is a quasicoherent $\mathcal{D}$-module on $\mathbb{D}^\vee$ with a compatible $\mathbf{G}(\mathbb{C})$ or $\mathbf{G}^\circ(\mathbb{C})$-action, i.e. such that suitable diagrams involving differential operators and group actions commute. Under this correspondence, coherent equivariant $\mathcal{D}$-modules on $\mathbb{D}^\vee$ correspond to finite dimensional $\mathfrak{g}$-$\mathbf{Q}$ (or $\mathfrak{g}^\circ$-$\mathbf{Q}^\circ$)-modules, or equivalently finite dimensional representations of $\mathbf{G}$ or $\mathbf{G}^\circ$, which brings us back to the previous discussion. Restricting $\mathcal{D}$-modules on $\mathbb{D}^\vee$ obtained as above to $\mathbb{D}$ and then descending to X, we have a functor from the category of $\mathfrak{g}^\circ$-$\mathbf{Q}^\circ$-modules to the category of $\mathcal{D}$-modules on X. Again the action of the center will be felt later only.

Any $\mathbf{G}^\circ(\mathbb{C})$-homogeneous differential operator between $\mathcal{E}$'s on $\mathbb{D}^\vee$ or $\mathbb{D}$ induces a differential operator between the $\mathcal{E}$'s on X. Later on we shall determine those operators in terms of the representations E.

To convince the reader that the central G_m does something, let us at least provide some concrete examples. To the $\mathbf{Q}$-representation $\mathbb{C}^g$ where an element of $\mathbf{Q}$ with blocks $A, 0, 0, D$ operates on $\mathbb{C}^g$ via left-multiplication by D, we attach the $\mathbf{G}(\mathbb{R})^\circ$-equivariant vector bundle $\mathbf{Lie}(A/\mathbb{D})$ on $\mathbb{D}$. This is justified by the action of $\mathbf{G}(\mathbb{Q}) \cap \mathbf{G}(\mathbb{R})^\circ$ (Hecke operators) on $A/\mathbb{D}$, given by the formula

$$\eta(\gamma) \cdot {}^t(C\Omega + D)^{-1} \cdot (\Omega, I_g) = (\gamma \cdot \Omega, I_g) \cdot \gamma^\sigma \ ,$$

where $\gamma \in \mathbf{G}(\mathbb{Q}) \cap \mathbf{G}(\mathbb{R})^\circ$ has blocks A, B, C, D and $\gamma \cdot \Omega = (A\Omega + B) \cdot (C\Omega + D)^{-1}$. Consequently $\Omega^1(A/\mathbb{D})$ is attached to the $\mathbf{M}$-representation $x = (A, B, 0, D) \mapsto$ left multiplication by $\eta(x)^{-1} \cdot A$ on $\mathbb{C}^g$ of $\mathbf{Q}$. Also $\Omega^1_{\mathbb{D}/\mathbb{C}}$ corresponds to the representation on symmetric $g \times g$-matrices which sends a symmetric S to $A \cdot S \cdot D^{-1}$, and differs from the second symmetric power of $\Omega^1(A/\mathbb{D})$ by a twist with η. Similarly to the standard representation of $\mathbf{G}(\mathbb{C})$ we attach the $\mathbf{H}_1(A/\mathbb{D})$ (thought of as either the relative de Rham homology or Betti homology). Therefore $\mathbf{H}^1(A/\mathbb{D})$ is attached to the contragradient of the standard representation of $\mathbf{G}$ (isomorphic to $\eta^{\otimes -1} \otimes$(standard)). The justification is the Hecke action and consideration of weights.

We have seen that the description of X as a locally homogeneous space gives us some naturally defined vector bundles on it, which we want to extend to $\overline{X}$. More precisely we intend to define for each $\mathcal{E} = \mathcal{E}(\tau)$ a *canonical extension* $\overline{\mathcal{E}} = \overline{\mathcal{E}}(\tau)$, a vector bundle on $\overline{X}$, such that the functor $\mathcal{E} \Rightarrow \overline{\mathcal{E}}$ commutes with direct sums, tensor products and duals. This is the major purpose of this section.

For $\mathbf{M}^\circ$-representations (i.e. $\mathbf{U}$ operates trivially) this has been done by Mumford ([Mum 6]), but in our case we can simplify his arguments. Let us recall his construction: Locally near the boundary the inclusion $X \subset \overline{X}$ looks like a bundle over some smooth base, with fibres a torus embedding $E_1 \subset \overline{E}_1$ (see [AMRT], or our construction of $\overline{A}_g$). Here E_1 is a torus with character group $n^{-1} \cdot S^2(X_1)$, X_1 a quotient free $\mathbb{Z}$-module of $X = \mathbb{Z}^g$. The operation of E_1 can be described in the homogeneous space picture, as follows: $\mathbf{U}_1(?) = \mathrm{Hom}_{\mathbb{Z}}(S^2(X_1), ?)$ is contained in the unipotent radical of a maximal parabolic subgroup of $\mathbf{G}$ (in fact the stabilizer of X_1^* if we define $\mathbf{G}$ using the standard symplectic pairing on $X^* \oplus X$). We have $E_1(\mathbb{C}) \cong \mathbf{U}_1(\mathbb{C})/\mathbf{U}_1(\mathbb{Z})$, and the (locally defined) E_1-operation on X is derived from the $\mathbf{U}_1$-operation on $\mathbb{D}^\vee$. It follows that locally $\mathcal{E}$ has a basis consisting of $\mathbf{U}_1(\mathbb{C})$-invariant or equivalently E_1-invariant sections, and we locally define $\overline{\mathcal{E}}$ as the submodule of $j_*(\mathcal{E})(j : X \subset \overline{X}$ the inclusion) generated by $E_1(\mathbb{C})$-invariant sections. Locally the formation of $\overline{\mathcal{E}}$ clearly commutes with direct sums, tensor-products and duals, and the canonical extension of a subobject (resp. quotient) is again a subobject (resp. quotient). Moreover, locally any $E_1(\mathbb{C})$-homogeneous differential operator on $\mathcal{E}$ extends to $\overline{\mathcal{E}}$: E_1-homogeneous differential operators on E_1 are polynomials with constant coefficients in $x_j \cdot \partial/\partial x_j$, x_j running through a basis for the characters of E_1, thus extend to $\overline{E}_1$. The same argument proves similar assertions for $\overline{E}_1$-bundles and the claim follows.

It remains to show that these local constructions glue to give a global $\overline{\mathcal{E}}$. Let us call a representation E of $\mathbf{Q}^\circ$ *good* if this is the case. It is easy to see that the set of good representations is closed under direct sums, tensor products, duals and subquotients. Thus if we know that a representation τ_0 of $\mathbf{Q}^\circ$ is good, all representations of $\mathbf{Q}$ generated by τ_0 under the above four operations are also good. For example the standard representation of $\mathbf{M}^\circ$ on $\mathbb{C}^g$ is good, because one checks easily (by looking at local coordinates) that the local extensions coincide with $\mathcal{T}_G^* = \Omega^1(G/\overline{X})$, so they glue (to form $\mathcal{T}_G^*$). It follows that any $\mathbf{M}^\circ$-representation of $\mathbf{Q}^\circ$ is good. This proves Mumford's theorem in our present case. This proof is simpler than the original one because the geometry (i.e. the existence of the universal semi-abelian variety G) already furnishes the canonical extension for a faithful representation of $\mathbf{M}^\circ$ (namely the standard representation).

Notice that we can identify the canonical extensions for some vector bundles on X of particular interest:

(a) the canonical extension of $\Omega^1(A/X)$ is $\Omega^1(G/\overline{X})$

(b) the canonical extension of $\omega(A/X)$ is $\omega(G/\overline{X})$

(c) the canonical extension of Ω^1_X is given by $\overline{\Omega}^1_X := \Omega^1_{\overline{X}}[d\log\infty]$, the differentials of degree one with logarithmic poles, which is isomorphic to $S^2(\mathcal{T}^*_G)$

(d) the canonical extension of $\mathcal{K}_X$ is $\mathcal{K}_{\overline{X}}(\infty)$, the top exterior differentials with logarithmic poles, which is isomorphic to $\omega(G/\overline{X})^{\otimes g+1}$.

(a) has been shown, (b) follows from (a) and (d) from (c). For (c), that the canonical extension Ω^1_X is $\Omega^1_{\overline{X}}[d\log\infty]$ is an easy local calculation, or follows from the Kodaira-Spencer isomorphism, and the rest follows because the operation of taking canonical extension is compatible with tensor products and coinvariants under flipping.

This is an appropriate place to state the main result of this section:

4.2. Theorem. *(i) The procedure "canonical extension" above defines an exact, faithful functor $E \Rightarrow \overline{\mathcal{E}}$ from the category of (finite dimensional) $\mathbf{Q}^\circ(\mathbb{C})$-modules to the category of vector-bundles on $\overline{X}$.*

(ii) The above functor commutes with tensor-products and duals. Homogeneous differential operators between the $\mathcal{E}$'s extend to the corresponding $\overline{\mathcal{E}}$'s. For $E =$ the standard representation of $\mathbf{M}^\circ(\mathbb{C})$ on $\mathbb{C}^g$ ($\mathbf{U}$ operates trivially), we have $\overline{\mathcal{E}} = \mathcal{T}^*_G$.

(iii) The properties listed in (ii) determine this functor uniquely. Moreover those $\overline{\mathcal{E}}$'s coming from (the restriction to $\mathbf{Q}^\circ$ of finite dimensional) $\mathbf{G}^\circ(\mathbb{C})$-modules are naturally modules under $\overline{\mathcal{D}} = \overline{\mathcal{D}}_{\overline{X}}$.

Proof of Theorem 4.2. The statement of theorem 4.2 already gives the hint of the proof: We just have to prove the theorem (i.e. that the local extensions glue globally) for a single faithful representation of $\mathbf{Q}^\circ$, a natural one to use is the restriction to $\mathbf{Q}^\circ$ of the standard representation of $\mathbf{G}^\circ$. The corresponding vector bundle $\mathcal{E}$ on X is $\mathcal{H}^1_{\mathrm{DR}}(A/X)$, which has an integrable connection $\nabla : \mathcal{E} \to \mathcal{E} \otimes \Omega^1_X$ (the Gauss-Manin connection); that is, it is a $\mathcal{D}_X$-module. The key to the proof of theorem 4.2 is to use ∇.

Clearly ∇ is (defined by) a homogeneous differential operator from $\mathcal{E}$ to $\mathcal{E} \otimes \Omega^1_X$, so it extends locally to $\nabla : \overline{\mathcal{E}} \to \overline{\mathcal{E}} \otimes \overline{\Omega}^1_X$. It follows that locally $\overline{\mathcal{E}}$ is a $\overline{\mathcal{D}}$-module. Furthermore it sits in an extension $0 \to \mathcal{T}^*_G \to \overline{\mathcal{E}} \to \mathcal{T}_G \to 0$, and the map $\mathcal{T}^*_G \to \mathcal{T}_G \otimes \overline{\Omega}^1_X$ induced by ∇ is derived from the Kodaira-Spencer isomorphism $\overline{\Omega}^1_X \cong S^2(\mathcal{T}^*_G)$. It follows that locally $\overline{\mathcal{E}}$ is generated as a $\overline{\mathcal{D}}$-module by $\mathcal{T}^*_G$. This characterizes $\overline{\mathcal{E}}$ as a subsheaf of $(X \to \overline{X})_*\mathcal{E}$, and so the local extensions must glue. This proves the theorem for the standard representation of $\mathbf{Q}^\circ$. As any finite dimensional $\mathbf{Q}$-representation is a subquotient of some tensor product of a direct sum of copies of the standard representation $\mathbb{C}^{2g}$ (which is self-dual), every (finite dimensional) $\mathbf{Q}^\circ$-module is good and statements (i) and (ii) of theorem 3 follow. (iii) follows similarly because it holds for the standard representation of $\mathbf{G}^\circ$. □

4.3. Remarks. Mumford showed ([Mum 6]) that the canonical metrics (i.e. the Bergman metrics, well-defined defined up to multiplicative constants) on $\mathbf{M}^\circ$-representations define hermitian metrics with logarithmic singularities on the canonical extensions $\overline{\mathcal{E}}$'s associated to these representations. (This implies gluing of local extensions.) Here we can check that directly: It suffices to show it for $\mathcal{T}_G^*$, which is easily done using local coordinates.

If $f : \overline{X}_1 \to \overline{X}_2$ is a map of toroidal compactifications of X induced by a subdivision of the cone decomposition used in the construction of $\overline{X}_2$, the two $\overline{\mathcal{E}}$'s are related by $\overline{\mathcal{E}}_2 = g^*(\overline{\mathcal{E}}_1)$: check locally. Especially they have the same coherent cohomology: $H^*(\overline{X}_1, \overline{\mathcal{E}}_1) \cong H^*(\overline{X}_2, \overline{\mathcal{E}}_2)$, since the fibres of f are rational varieties.

If $D \subset \overline{X}$ is any divisor with support in the boundary, the extensions $\overline{\mathcal{E}} \otimes \mathcal{O}(D)$ are again stable under all homogeneous differential operators between the $\mathcal{E}$'s: any homogeneous differential operator from $\mathcal{E}_1$ to $\mathcal{E}_2$ extends to a differential operator from $\overline{\mathcal{E}}_1 \otimes \mathcal{O}(D)$ to $\overline{\mathcal{E}}_2 \otimes \mathcal{O}(D)$. This follows again from the structure of invariant differential operators on a torus embedding $E_1 \subset \overline{E}_1$, as these operators respect divisors in the boundary, and similarly for $\overline{E}_1$-bundles. However these extensions do not behave very well with respect to subdivisions, and the formation of $\overline{\mathcal{E}} \otimes \mathcal{O}(D)$ does not commute with duals or tensor products.

The spaces $X, \overline{X}$ usually are naturally defined over some number field $K \subset \mathbb{C}$ (for example $K = \mathbb{Q}(e^{2\pi i/n})$ for $A_{g,n}$), and $\mathbf{G}$ and $\mathbf{Q}$ have natural $\mathbb{Q}$-structures. (The number field K here is not to be confused with the maximal compact subgroup of $\mathbf{G}^\circ(\mathbb{R})$.) One might ask whether the bundles $\mathcal{E}$ and $\overline{\mathcal{E}}$ are also naturally defined over K, provided E is. This holds indeed: It suffices to check that the principal $\mathbf{Q}$-homogeneous space over X derived from $\mathbf{G}(\mathbb{C}) \to \mathbb{D}^\vee$, as well as its extension to $\overline{X}$, have canonical K-structures. This follows because $\mathcal{T}_G^*$ as well as the relative de Rham cohomology sheaves of A over X have canonical K-structures, and the K-structures on the relative de Rham cohomology extend to $\overline{X}$ (by uniqueness). The principally homogeneous $\mathbf{Q}$-space in question can be derived from the exact sequence $0 \to \Omega^1_{A/X} \to \mathcal{H}^1_{\mathrm{DR}}(A/X) \to \mathbf{Lie}(A^t/X) \to 0$, hence the result. Of course K-rational maps between E's define K-rational maps between $\mathcal{E}$'s. However these K-structures on X or $\overline{X}$ should not be confused with the natural $\mathbb{Q}$-structure on $\mathbb{D}^\vee = \mathbf{G}/\mathbf{Q}$: For example, the bundle $\mathbf{V}$ associated to the standard representation of $\mathbf{Q}$ on $\mathbb{C}^{2g}$ has a K-structure by its interpretation as relative de Rham cohomology of A, which does not coincide with the rational structure derived from its interpretation as the sheaf associated to the topological first derived image of the constant sheaf $\mathbb{Q}$ on A. (The latter corresponds to the natural $\mathbb{Q}$-structure on $\mathbb{D}^\vee = \mathbf{G}/\mathbf{Q}$.) Also $\mathbb{Q}$-rational points in $\mathbb{D} \subset \mathbb{D}^\vee$ usually define transcendental points in X: periods of abelian varieties over $\overline{\mathbb{Q}}$ tend to be transcendental.

We may apply this to situation in §1 where $f : A^s = Y \to X$ is compactified to $\overline{f} : \overline{Y} \to \overline{X}$. We construct (in §3) vector bundles $\mathcal{G}^m = \Lambda^m(\mathcal{G}^1)$, with a filtration and a logarithmic connection, which extend the relative de Rham cohomology of A^s. The filtration on $\mathcal{G}^1$ is given by the exact sequence $0 \to (\mathcal{T}_G^*)^s \to \mathcal{G}^1 \to (\mathcal{T}_G)^s \to 0$, and on X, $\mathcal{G}^1$ is equal to $\mathcal{V}^s$, $\mathcal{V}$ the vector bundle associated to the standard representation of $\mathbf{Q}$ on $\mathbb{C}^{2g}$. The above arguments show that $\mathcal{G}^1 = \overline{\mathcal{V}}^s$, and so $\mathcal{G}^m = \Lambda^m(\overline{\mathcal{V}}^s)$. Now we have fulfilled our promise to determine the $\mathcal{G}^m$'s. Let us close with the remark that these isomorphisms are defined over any number field K over which X, $\overline{X}$ and $\overline{Y}$ are naturally defined.

5. The BGG-Complex and Eichler Integrals

This section contains the main results of this chapter. We shall use ideas of Bernstein-Gelfand-Gelfand to obtain resolutions of the locally constant systems attached to finite-dimensional representations of $\mathbf{G}$. Then we show that the spectral sequences computing the cohomology of these locally constant systems degenerate, and we obtain mixed Hodge structures as well as Lie-theoretic extra structures coming from BGG resolutions.

Now things become more Lie-theoretic: We shall construct resolutions of locally constant systems using induced modules. Let $\mathfrak{g}$ and $\mathfrak{q}$ denote the complex Lie algebras of $\mathbf{G}$ and $\mathbf{Q}$ respectively, and $U(\mathfrak{g})$ respectively $U(\mathfrak{q})$ the universal enveloping algebras. For a (left) $\mathfrak{q}$-module E, we can induce from $\mathfrak{q}$ to $\mathfrak{g}$ to obtain the induced (left) $\mathfrak{g}$-module $U(\mathfrak{g}) \otimes_{U(\mathfrak{q})} E$. These induced modules are useful for determining homogeneous differential operators between equivariant vector bundles on $\mathbb{D}^\vee = \mathbf{G}(\mathbb{C})/\mathbf{Q}(\mathbb{C})$:

5.1. Proposition. *Suppose E_1 and E_2 are complex $\mathbf{Q}$-representations, giving rise to equivariant vector bundles $\mathcal{E}_1$ and $\mathcal{E}_2$ on $\mathbb{D}^\vee$. Then there is a (contravariant) functorial isomorphism between the space of all homogeneous differential operators from $\mathcal{E}_1$ to $\mathcal{E}_2$, and $\mathrm{Hom}_{\mathfrak{g}}(U(\mathfrak{g}) \otimes_{U(\mathfrak{q})} E_2^\vee, U(\mathfrak{g}) \otimes_{U(\mathfrak{q})} E_1^\vee)$, where $E_i^\vee$ is the contragradient of E_i.*

Proof. If the homogeneous bundle $\mathcal{E}$ is associated to the $\mathbf{Q}$-module E, the homogeneous $\mathcal{D}$-module $\mathcal{D} \otimes \mathcal{E}$ is associated to the $\mathfrak{g}$-$\mathbf{Q}$-module $U(\mathfrak{g}) \otimes_{U(\mathfrak{q})} E$, as follows from comparing the universal mapping properties of both objects. One also checks that differential operators (of finite order) from $\mathcal{E}_1$ to $\mathcal{E}_2$ correspond to $\mathcal{O}$-linear maps $\mathcal{E}_2^\vee \to \mathcal{D} \otimes \mathcal{E}_1^\vee$, hence homogeneous differential operators correspond to elements of
$\mathrm{Hom}_{\mathfrak{q}}(E_2^\vee, U(\mathfrak{g}) \otimes_{U(\mathfrak{q})} E_1^\vee) = \mathrm{Hom}_{\mathfrak{g}}(U(\mathfrak{g}) \otimes_{U(\mathfrak{q})} E_2^\vee, U(\mathfrak{g}) \otimes_{U(\mathfrak{q})} E_1^\vee)$.
This bijective correspondence is evidently functorial.

5.2. Remark. Again there is a question of K-rationality, for $K \subset \mathbb{C}$ a natural field of definition of X. One shows that K-rational elements of $\mathrm{Hom}_{\mathfrak{g}}(U(\mathfrak{g}) \otimes_{U(\mathfrak{q})} E_2^\vee, U(\mathfrak{g}) \otimes_{U(\mathfrak{q})} E_1^\vee)$ induce K-rational differential operators on X, from $\mathcal{E}_1$ to $\mathcal{E}_2$: The K-rational structure on X induces such a structure on Ω_X, which coincides with the K-rational structure on Ω_X derived from the fact that it is associated to a $\mathbb{Q}$-rational $\mathbf{Q}$-representation. Also the differential $d: \mathcal{O}_X \to \Omega_X$ is the differential operator associated to the obvious $\mathbb{Q}$-rational map $\mathfrak{g}/\mathfrak{q} \to U(\mathfrak{g}) \otimes_{U(\mathfrak{q})} \mathbb{C}$. As $\mathcal{D}$ can be constructed from Ω_X and d, it follows that for any K-rational $\mathbf{Q}$-representation E, with associated K-rational vector bundle $\mathcal{E}$ on X, the K-rational $\mathcal{D}$-module $\mathcal{D} \otimes \mathcal{E}$ is defined by the K-rational representation $U(\mathfrak{g}) \otimes_{U(\mathfrak{q})} E$. The assertion follows.

The effect of proposition 5.1 is that elements of $\mathrm{Hom}_{\mathfrak{g}}(U(\mathfrak{g}) \otimes_{U(\mathfrak{q})} E_2^\vee, U(\mathfrak{g}) \otimes_{U(\mathfrak{q})} E_1^\vee)$ induce differential operators from $\mathcal{E}_1(\Gamma)$ to $\mathcal{E}_2(\Gamma)$ on X, which extend to the canonical extensions $\overline{\mathcal{E}}_1(\Gamma)$ and $\overline{\mathcal{E}}_2(\Gamma)$ on $\overline{X}$. In other words we can translate $\mathfrak{g}$-maps between induced modules into differential operators. For example in the computation of the relative Lie algebra homology $H_*(\mathfrak{g}, \mathfrak{q}; \mathbb{C})$ one uses a complex with q-th term $U(\mathbf{G}) \otimes_{U(\mathfrak{q})} \Lambda^q(\mathfrak{g}/\mathfrak{q})$ (the bar resolution). This translates into a complex of equivariant bundles on $\mathbb{D}^\vee$ (differentials in this complex are not $\mathcal{O}$-linear!) with q-th term associated to $\Lambda^q(\mathfrak{g}/\mathfrak{q})^*$, namely Ω^q. Of course this is just the de Rham complex on $\mathbb{D}^\vee$. More generally for any $\mathbf{G}$-module V, the tensor product of the above complex of $U(\mathfrak{g})$-modules with V corresponds to the de Rham complex of the locally constant system associated to the dual $V^\vee$: we have $V \otimes_{\mathbb{C}} (U(\mathfrak{g}) \otimes_{U(\mathfrak{q})} \Lambda^q(\mathfrak{g}/\mathfrak{q})) \xleftarrow{\sim} U(\mathfrak{g}) \otimes_{U(\mathfrak{q})} (V \otimes_{\mathbb{C}} \Lambda^q(\mathfrak{g}/\mathfrak{q}))$, where the isomorphism is given in terms of the "diagonal map" $\Delta : U(\mathfrak{g}) \to U(\mathfrak{g}) \otimes U(\mathfrak{g})$, by $\sum x_i v \otimes y_i \otimes w \leftarrow x \otimes v \otimes w$, with $\Delta(x) = \sum x_i \otimes y_i$. Note that this isomorphism holds for all $\mathfrak{q}$-modules and not just for the $\Lambda^q(\mathfrak{g}/\mathfrak{q})$'s.

Before we go on we need some notation and basic facts about roots, weights and Weyl groups. Recall that $\mathbf{T}(\subset \mathbf{M} \subset \mathbf{G})$ is the diagonal maximal torus of $\mathbf{G}$, with complex Lie algebra $\mathfrak{t}(\subset \mathfrak{m} \subset \mathfrak{g})$. The conjugation operation of $\mathbf{T}$ on $\mathfrak{m}$ and $\mathfrak{g}$ defines root systems $\Phi_{\mathbf{M}}$ and $\Phi_{\mathbf{G}}$ in $\mathfrak{t}^\vee$. Let $\Phi^{\mathbf{M}} = \Phi_{\mathbf{G}} - \Phi_{\mathbf{M}}$. (Via the Cayley transform, $\Phi_{\mathbf{M}}$ goes to compact roots and $\Phi^{\mathbf{M}}$ goes to non-compact roots.) We choose compatible systems of positive roots $\Phi_{\mathbf{M}}^+$ and $\Phi_{\mathbf{G}}^+$ such that $\Phi_{\mathbf{G}}^+$ is the union of $\Phi_{\mathbf{M}}^+$ and the roots $\Phi^{\mathbf{M}+}$ of $\mathbf{T}$ appearing in the Lie algebra $\mathfrak{u}$ of $\mathbf{U}$, the unipotent radical of $\mathbf{Q}$. To fix ideas, we choose $\Phi_{\mathbf{G}}^+$ as those roots appearing in the Lie algebra $\mathfrak{n}$ of $\mathbf{N}$. This defines systems of simple roots for $\Phi_{\mathbf{G}}$ and $\Phi_{\mathbf{M}}$. Let $\rho = \rho_{\mathbf{G}}$ and $\rho_{\mathbf{M}}$ denote respectively half the sum of the positive roots in $\Phi_{\mathbf{G}}^+$ and in $\Phi_{\mathbf{M}}^+$, and let $\rho^{\mathbf{M}} = \rho_{\mathbf{G}} - \rho_{\mathbf{M}}$. (Our choice of positive roots is such that $2\rho^{\mathbf{M}}$ corresponds to $\mathcal{K}_{\mathbb{D}^\vee} = \Omega_{\mathbb{D}^\vee}^{\mathrm{top}}$.)

A weight of $\mathfrak{t}$ is called $\mathbf{G}$-*dominant* if it has non-negative scalar product with elements of $\Phi_{\mathbf{G}}^+$. The irreducible finite dimensional complex representations of $\mathbf{G}$ are parametrized by $\mathbf{G}$-dominant integral weights $\lambda \in \mathfrak{t}^\vee$ which integrate to characters of $\mathbf{T}$, with λ corresponding to the irreducible representation $V(\lambda)$, which has highest weight λ. Similarly for $\mathbf{M}$: Any $\mu \in \mathfrak{t}^\vee$ which integrates to a character of $\mathbf{T}$ and is $\mathbf{M}$-dominant (having non-negative scalar products with elements of $\Phi_{\mathbf{M}}^+$) parametrizes an irreducible $\mathbf{M}$-module $W(\mu)$, with highest weight μ.

The Weyl groups $W_{\mathbf{G}}$ and $W_{\mathbf{M}}$ are defined as $N_G(\mathbf{T})/\mathbf{T}$ and $N_M(\mathbf{T})/\mathbf{T}$, where $N_{\mathbf{G}}(\mathbf{T})$ is the normalizer of $\mathbf{T}$ in $\mathbf{G}$ and $N_{\mathbf{M}}(\mathbf{T})$ is the normalizer of $\mathbf{T}$ in $\mathbf{M}$. They operate on $\mathfrak{t}$ and $\mathfrak{t}^\vee$, and are generated by reflections given by simple roots. The minimal number $l(w)$ of factors needed in a representation of an element $w \in W_{\mathbf{G}}$ as a product of such reflections is called the length of w. Similarly for $W_{\mathbf{M}}$, but as the length-functions for $W_{\mathbf{G}}$ and $W_{\mathbf{M}}$ coincide on $W_{\mathbf{M}}$ we obtain nothing new. $l(w)$ also can be characterized as the $\mathrm{Card}(\Phi_{\mathbf{G}}^- \cap w(\Phi_{\mathbf{G}}^+)) = \mathrm{Card}(\Phi_{\mathbf{G}}^+ \cap (w(\Phi_{\mathbf{G}}^-))$, which explains why the length-functions coincide on $W_{\mathbf{M}}$. The element $w_0 \in W_{\mathbf{G}}$ of maximal length in $W_{\mathbf{G}}$ is induced by $\mathrm{Ad}(J) = \sigma$, hence operates as $-\mathrm{id}$ on $\mathfrak{t}^\circ$ = complex Lie algebra of $\mathbf{T} \cap \mathbf{G}^\circ$ and leaves the center of $\mathbf{G}$ fixed. Similarly there is the element $w_1 \in W_{\mathbf{M}}$ of maximal length in $W_{\mathbf{M}}$, characterized by $w_1(\Phi_{\mathbf{M}}^+) = \Phi_{\mathbf{M}}^-$. These elements describe duality for finite-dimensional irreducible representations, in the sense that $V(\lambda)^\vee = V(-w_0(\lambda)) = V(\lambda) \otimes \eta^k$ for suitable $k \in \mathbb{Z}$, and $W(\mu)^\vee = W(-w_1(\mu))$. Define a subset $W^{\mathbf{M}} \subset W_{\mathbf{G}}$ by $W^{\mathbf{M}} = \{w \in W_{\mathbf{G}} | \Phi_{\mathbf{M}}^+ \subset w(\Phi_{\mathbf{G}}^+)\}$. Equivalently $W^{\mathbf{M}} = \{w \in W_{\mathbf{G}} | w(\lambda)$ is $\mathbf{M}$-dominant for any $\mathbf{G}$-dominant weight $\lambda \in \mathfrak{t}^\vee\} = \{w \in W_{\mathbf{G}} | \Phi_{\mathbf{G}}^+ \cap w(\Phi_{\mathbf{G}}^-) \subset \Phi^{\mathbf{M}+}\} = \{w \in W_{\mathbf{G}} | w(\rho) - \rho$ is $\mathbf{M}$-dominant$\}$, the last equality following from the structure of roots of $\mathbf{G}$. $W^{\mathbf{M}}$ is a set of representatives for the quotient $W_{\mathbf{M}} \backslash W_{\mathbf{G}}$, in other words any $w \in W_{\mathbf{G}}$ can be written as $w = w' \cdot w''$ with $w' \in W_{\mathbf{M}}$, $w'' \in W^{\mathbf{M}}$. $W^{\mathbf{M}}$ has an involution which maps $w \in W^{\mathbf{M}}$ into $w_1 w w_0$, and $l(w_1 w w_0) = d - l(w)$ with $d = g \cdot (g+1)/2$, for any $w \in W^{\mathbf{M}}$. In more concrete terms $W_{\mathbf{M}}$ is isomorphic to $\mathcal{S}_g$, the permutation group on g letters, and $W_{\mathbf{G}}$ to the semi-direct product $W_{\mathbf{M}} \ltimes \{\pm 1\}^g$. Elements of $\mathbf{T}$ are diagonal matrices with entries of the form $(a_1, \ldots, a_g, d_1, \ldots, d_g)$, with $a_i \cdot d_i = \eta \, \forall i$. $W_{\mathbf{M}}$ acts on them by permuting the t_j's, while elements of the second factor $\{\pm 1\}^g$ exchange a_j and d_j for some j. Thinking of a_i, d_i as characters of $\mathbf{T}$,

$$\Phi_{\mathbf{M}}^+ = \{a_i \cdot a_j^{-1} | 1 \le i < j \le g\} = \{d_j \cdot d_i^{-1} | 1 \le i < j \le g\},$$
$$\Phi_{\mathbf{M}}^- = \{a_i^{-1} \cdot a_j | 1 \le i < j \le g\} = \{d_j^{-1} \cdot d_i | 1 \le i < j \le g\},$$
$$\Phi^{\mathbf{M}+} = \{a_i \cdot d_j^{-1} | 1 \le i, j \le g\},$$
$$\Phi^{\mathbf{M}-} = \{a_i^{-1} \cdot d_j^{-1} | 1 \le i, j \le g\}.$$

We recall the theorem of Harish-Chandra which says the center of the enveloping algebra $Z(\mathfrak{g}) \subset U(\mathfrak{g})$ is isomorphic to the ring of $W_{\mathbf{G}}$-invariants in the symmetric algebra $S^\bullet(\mathfrak{t})$, where $W_{\mathbf{G}}$ acts by a the standard action. More precisely, the (twisted) H.-C. homomorphism $\gamma' : Z(\mathfrak{g}) \to S^\bullet(\mathfrak{t})$ is defined as follows: $U(\mathfrak{g}) = U(\mathfrak{n}^-)U(\mathfrak{t})U(\mathfrak{n}) = U(\mathfrak{t}) \oplus \mathfrak{n}^- U(\mathfrak{n}^-)U(\mathfrak{t}) \oplus U(\mathfrak{g})\mathfrak{n}$ by the Poincaré-Birkhoff-Witt theorem, where $\mathfrak{n}^-$ is the sum of negative root spaces. Any $z \in U(\mathfrak{g})$, being $\mathbf{T}$-invariant, can be written as a sum $z = z' + z''$, with $z' \in U(\mathfrak{t})$, $z'' \in U(\mathfrak{g})\mathfrak{n}$, and $\gamma'(z)$ is defined to be $z' \in U(\mathfrak{t}) = S^\bullet(\mathfrak{t})$. γ' defines the isomorphism mentioned above. If we compose γ' with the map $T_{-\rho}$ (= translation by $-\rho$) : $S^\bullet(\mathfrak{t}) \to S^\bullet(\mathfrak{t})$ induced by $x \mapsto x - \langle \rho, x \rangle \cdot 1$ for $x \in \mathfrak{t}$, we obtain an isomorphism γ from $Z(\mathfrak{g})$ to $S^\bullet(\mathfrak{t})^{W_{\mathbf{G}}}$, the H.-C. isomorphism. Regarding $S^\bullet(\mathfrak{t})$ as the algebra of polynomial functions on $\mathfrak{t}^\vee$, characters $\chi : Z(\mathfrak{g}) \to \mathbb{C}$ are parametrized by elements $\lambda \in \mathfrak{t}^\vee$ via γ: each λ defines a character $\chi_\lambda : z \mapsto \gamma(z)(\lambda)$, and two such elements λ_1 and λ_2 define the same χ_λ

if and only if λ_1 and λ_2 are in the same W_G-orbit in $\mathfrak{t}^\vee$. Also $Z(\mathfrak{g})$ acts on the highest weight module $V(\lambda)$ via $\chi_{\lambda+\rho} : Z(\mathfrak{g})$ acts on a highest weight vector via the character $\chi_{\lambda+\rho}$ by the definition of γ'. The assertion follows since $V(\lambda)$ is irreducible. Notice that this argument shows that $Z(\mathfrak{g})$ operates on any $U(\mathfrak{g})$-module generated by a highest weight vector of weight λ by $\chi_{\lambda+\rho}$.

In [BGG] I.N. Bernstein, I.M. Gelfand and S.I. Gelfand defined a canonical resolution of $V(\lambda)$. They dealt with Borel subgroups, but their arguments carry over to our case (the parabolic subgroup $\mathbf{Q} \subset \mathbf{G}$), and even simplify a little bit. The resulting complex $C_{\lambda\bullet}$ is the generalized $\chi_{\lambda+\rho}$-eigenspace in the standard resolution $V(\lambda) \otimes_{\mathbb{C}} \Lambda^\bullet(\mathfrak{g}/\mathfrak{q}) \to V(\lambda)$, which we call the B-G-G resolution of $V(\lambda)$. It is a direct summand in the standard-resolution, and its degree-q term is given by $C_{\lambda,q} = \oplus_{w\in W^{\mathbf{M}},l(w)=q} U(\mathfrak{g}) \otimes_{U(\mathfrak{q})} W(w(\lambda+\rho)-\rho)$. To see this assertion, let us first handle the case $\lambda = 0$, i.e. $V(\lambda) = \mathbb{C}$ with trivial $\mathfrak{g}$-action. One can check (as $\mathfrak{u}$ is abelian) that for any irreducible constituent $W(\mu)$ of the $\mathfrak{q}$-module $\Lambda^\bullet(\mathfrak{g}/\mathfrak{q})$ $Z(\mathfrak{g})$ operates on $U(\mathfrak{g}) \otimes_{U(\mathfrak{q})} W(\mu)$ by χ_ρ. So there must be some $w \in W_G$ with $w(\rho) = \mu + \rho$. Moreover $\mu = w(\rho) - \rho$, being a weight of $\Lambda^\bullet(\mathfrak{g}/\mathfrak{q})$, is the sum of elements in a subset of $\Phi^{\mathbf{M}-}$. Now we recall (cf. lemma 9.8, p.47 of [BGG]) that for any $w \in W_{\mathbf{G}}$, $\rho - w(\rho)$ is the sum of the elements of $\Phi^+_{\mathbf{G}} \cap w(\Phi^-_{\mathbf{G}})$, and it can be written as a sum of elements in a subset of $\Phi^+_{\mathbf{G}}$ only in this way. This tells us that

(a) $\Phi^+_{\mathbf{G}} \cap w(\Phi^-_{\mathbf{G}}) \subset \Phi^{\mathbf{M}+}$, so $w \in W^{\mathbf{M}}$

(b) $\mu = w(\rho) - \rho = \sum_{\alpha\in\Phi^-_{\mathbf{G}}\cap w(\Phi^+_{\mathbf{G}})} \alpha = w(\rho) - \rho$ (c) $W(\mu)$ occurs only in $\Lambda^q(\mathfrak{g}/\mathfrak{q})$ with $q = l(w)$. In fact one sees that any such $W(w(\rho) - \rho)$ occurs.

We conclude that we have a resolution $C_\bullet \to \mathbb{C}$ (as $U(\mathfrak{g})$-module) with $C_q = \oplus_{w\in W^M,l(w)=q} U(\mathfrak{g}) \otimes_{U(\mathfrak{q})} W(w(\rho)-\rho)$. Tensoring the above resolution $C_\bullet \to \mathbb{C}$ with $V(\lambda)$ over $\mathbb{C}$, we proceed to compute its generalized $\chi_{\lambda+\rho}$-eigenspace. Thus for a given $w \in W^{\mathbf{M}}$ we have to determine those $W(\mu)$'s occurring in the $\mathfrak{q}$-module $V(\lambda) \otimes_{\mathbb{C}} W(w(\rho) - \rho)$ with infinitesimal character $\chi_{\lambda+\rho}$. For this to happen, a there must be some $w' \in W_G$ such that $\mu + \rho = w'(\lambda + \rho)$, and there has to be a weight ν of $V(\lambda)$ with $\mu = \nu + w(\rho) - \rho$. But then $w'(\lambda+\rho) = \mu+\rho = \nu + w(\rho)$, or $w'^{-1}\nu = \lambda + \rho - w'^{-1}w(\rho)$. Since λ is the highest weight of $V(\lambda)$ and $\rho - w'^{-1}w(\rho)$ is a sum of positive roots, we must have $w' = w$ and $\mu = w(\lambda + \rho) - \rho$, and $W(\mu)$ occurs with multiplicity one. This proves that $\oplus_{w\in W^{\mathbf{M}},l(w)=q} U(\mathfrak{g}) \otimes_{U(\mathfrak{q})} W(w(\lambda+\rho)-\rho)$ is the generalized $\chi_{\lambda+\rho}$-eigenspace of $V(\lambda) \otimes_{\mathbb{C}} \Lambda^q(\mathfrak{g}/\mathfrak{q})$. Note that this complex is already defined over $\mathbb{Q}$, as $\mathbf{G}, \mathbf{Q}, \mathbf{M}$ and $\mathbf{T}$ are.

Finally we shall define filtrations on the BGG complexes constructed above. In some sense these filtrations correspond to Hodge filtrations on locally constant systems on $\mathbb{D}$ coming from $\mathbf{G}$-representations. There exists an element $H \in \mathfrak{t}$ characterized by the fact that $\alpha(H) = 0$ for $\alpha \in \Phi_{\mathbf{M}}$, $\alpha(H) = 1$ for $\alpha \in \Phi^{\mathbf{M}+}$, and $d\eta(H) = -1$. Explicitly H is (identified with) the diagonal matrix with entries $(0, \dots, 0, -1, \dots, -1)$, where there are g 0's and g 1's. For each finite dimensional $\mathbf{Q}$-module E, define a decreasing filtration $F^\bullet$ of E by $\mathbf{Q}$-submodules defined as $F^p(E)$ = sum of generalized H-eigenspaces for eigenvalues $\geq p$, so that the F-weights are just the H-weights. (Here we see the advantage of $\mathbf{G}$ over $\mathbf{G}^\circ$: using $\mathbf{G}^\circ$ would lead to the element $(1/2,\dots,1/2,-1/2,\dots,-1/2) \in \mathfrak{t}^\circ$ and the slight inconvenience of having half-integral F-weights, which deviates from the Hodge filtration.) For example, the representation $\mathbb{C}^g$ on which $\mathbf{Q}$ operates by $(A,B,0,D) \mapsto D$ has F-weight -1. The standard representation of $\mathbf{G}$ on $\mathbb{C}^{2g}$ has F-weights 0 and -1, with filtration $0 \subset \mathbb{C}^g \subset \mathbb{C}^{2g}$, $\mathbb{C}^g = F^0$ being the subspace of $\mathbb{C}^{2g}$ whose last g-entries are zero. The F-filtration defined above passes over to the induced modules: for any $\mathbf{Q}$-module E, $U(\mathfrak{g}) \otimes_{U(\mathfrak{q})} E$ gets a decreasing filtration $U(\mathfrak{g}) \otimes_{U(\mathfrak{q})} F^\bullet(E)$. Any $\mathfrak{g}$-linear map between such induced modules respects the $F^\bullet$-filtration: We must show that if E_1 has H-weights $> p$, and E_2 has H-weights $\leq p$, there exists only the trivial $\mathfrak{q}$-linear map from E_1 to $U(\mathfrak{g}) \otimes_{U(\mathfrak{q})} E_2$. But this is obvious as the second module has only H-weights $\leq p$. The same argument shows that the maps derived on the $\mathrm{gr}_F(\mathcal{E})$'s from the corresponding differential operators are actually $\mathcal{O}$-linear (i.e. maps of vector bundles in the usual sense): if E_1, E_2 both have H-weight p, then any $\mathfrak{q}$-linear map from E_1 to $U(\mathfrak{g}) \otimes_{U(\mathfrak{q})} E_2$ factorizes through $E_2 \subset U(\mathfrak{g}) \otimes_{U(\mathfrak{q})} E_2$.

The inclusion from the BGG-complex into $U(\mathfrak{g}) \otimes_{U(\mathfrak{q})} V(\lambda) \otimes_{\mathbb{C}} \Lambda^\bullet(\mathfrak{g}/\mathfrak{q})$ is a filtered quasi-isomorphism: It is a direct summand, and also a direct summand in the filtered sense since $Z(\mathfrak{g})$ commutes with H. Furthermore the bar resolution $U(\mathfrak{g}) \otimes_{U(\mathfrak{q})} V(\lambda) \otimes \Lambda^\bullet(\mathfrak{g}/\mathfrak{q})$ is filtered-quasi-isomorphic to $V(\lambda)$, and the assertion follows. Note also that gr_F of the BGG-resolution has trivial differentials: We have to check that for elements $w_1, w_2 \in W^{\mathbf{M}}$ such that $l(w_1) = q, l(w_2) = q+1$, and such that $\mu_1 = w_1(\lambda+\rho)-\rho$ and $\mu_2 = w_2(\lambda+\rho)-\rho$ satisfy $\mu_1(H) = \mu_2(H)$, there exists only the trivial $\mathfrak{q}$-morphism from $W(\mu_2)$ to $U(\mathfrak{g}) \otimes_{U(\mathfrak{q})} W(\mu_1)$. Again this is rather obvious, since the morphism cannot send $W(\mu_2)$ into $W(\mu_1)$, and all other irreducible $\mathfrak{m}$-submodules of $U(\mathfrak{g}) \otimes_{U(\mathfrak{q})} W(\mu_1)$ have H-weights $> \mu_1(H)$. Let us summarize the above results:

5.3. Proposition. *(1) For any irreducible finite dimensional* $\mathbf{G}$*-module* $V(\lambda)$ *of highest weight* λ*, the bar resolution*

$$U(\mathfrak{g}) \otimes_{U(\mathfrak{q})} V(\lambda) \otimes_{\mathbb{C}} \Lambda^\bullet(\mathfrak{g}/\mathfrak{q}) \twoheadrightarrow V(\lambda)$$

has a direct summand $C_{\lambda\bullet}$*, called the BGG resolution of* $V(\lambda)$*, defined as the generalized* $\chi_{\lambda+\rho}$*-eigenspace of the bar resolution, so that the inclusion* j_λ *of* $C_{\lambda\bullet}$ *into the bar resolution is a quasi-isomorphism.* $C_{\lambda\bullet}$ *is given explicitly by*

$$C_{\lambda q} = \oplus_{w \in W^{\mathbf{M}}, l(w)=q} U(\mathfrak{g}) \otimes_{U(\mathfrak{q})} W(w(\lambda+\rho)-\rho) .$$

(2) Define filtrations on the standard resolution and the BGG resolution by:

$$F^p(U(\mathfrak{g}) \otimes_{U(\mathfrak{q})} V(\lambda) \otimes_{\mathbb{C}} \Lambda^\bullet(\mathfrak{g}/\mathfrak{q})) = U(\mathfrak{g}) \otimes_{U(\mathfrak{q})} F^p(V(\lambda) \otimes_{\mathbb{C}} \Lambda^\bullet(\mathfrak{g}/\mathfrak{q})) ,$$

$$F^p(C_\lambda)_q = \oplus_{w \in W^{\mathbf{M}}, l(w)=q} U(\mathfrak{g}) \otimes_{U(\mathfrak{q})} F^p(W(w(\lambda+\rho)-\rho)) ,$$

where for each $U(\mathfrak{q})$*-module* E*,* $F^p(E)$ *is defined to be sum of generalized* H*-eigenspaces for eigenvalues* $\geq p$*, and* $H = (0,\dots,0,-1,\dots,-1) \in \mathfrak{t}$*. Another expression is*

$F^p(C_{\lambda\bullet}) = \oplus_{w\in W^{\mathbf{M}}, w(\lambda+\rho)(H)\geq\rho(H)+p}\, U(\mathfrak{g}) \otimes_{U(\mathfrak{q})} (W(w(\lambda+\rho)-\rho))$.
Then the inclusion j_λ *above defines a filtered quasi-isomorphism from the BGG resolution to the standard resolution of* $V(\lambda)$. *Moreover the differentials of the graded complex* $\mathrm{gr}_F(C_{\lambda\bullet})$ *are all trivial, and the graded complex* $\mathrm{gr}_F(U(\mathfrak{g})\otimes_{U(\mathfrak{q})} V(\lambda)\otimes_{\mathbb{C}} \Lambda^\bullet(\mathfrak{g}/\mathfrak{q}))$ *is induced from the graded complex of* $\mathfrak{q}$*-modules* $\mathrm{gr}_F(V(\lambda)\otimes_{\mathbb{C}} \Lambda^\bullet(\mathfrak{g}/\mathfrak{q}))$.

For later purposes, we define an element $H_w \in \mathfrak{t}$ in the center of $\mathfrak{g}$ characterized by $\alpha(H_w) = 0$ for $\alpha \in \Phi_{\mathbf{G}}$, and $d\eta(H_w) = -2$. Explicitly, $H_w = (-1,\dots,-1,-1,\dots,-1) = -I_{2g} \in \mathfrak{t}$. Using H_w, we can define an increasing filtration $W_\bullet$ on finite dimensionl $\mathbf{G}$-modules: W_n is the sum of generalized H_w-eigenspaces of weights $\leq n$. For example the standard representation of $\mathbf{G}$ has W-weight -1, and the multiplier representation η has W-weight -2. The filtration $W_\bullet$ will correspond to the weight filtration, while η corresponds to $\mathbb{C}(1)$ (or $\mathbb{Q}(1)$ if we use $\mathbb{Q}$-coefficients), as we shall see later.

Now let us translate the above discussions into geometry. Since we would like to have the F-filtrations as defined before, we shall use $\mathbf{G}$-modules and $\mathbf{Q}$-modules, although it is possible to use $\mathbf{G}^\circ$ and $\mathbf{Q}^\circ$ at the expense of introducing half-integral F-weights.

Construction of the Dual BGG Complex. Let $\mathbb{V}(\lambda)$ denote the locally constant sheaf on $\mathbb{D}^\vee$ (or $\mathbb{D}$, or X) associated to the finite-dimensional $\mathbf{G}$-module $V(\lambda)$ with highest weight λ, and $\mathcal{V}(\lambda)$ the corresponding $\mathcal{D}$-module. The locally constant sheaf and the corresponding $\mathcal{D}$-module attached to the multiplier representation $\eta^{\otimes k}$ will both be denoted by $\mathbb{C}(k)$. $\mathcal{E}\otimes\mathbb{C}(k)$ will be denoted $\mathcal{E}(k)$, following usual convention. The effect of such a "Tate twist" will not be felt until we consider the weight filtration. Similarly, let $\mathcal{W}(\mu)$ denote the vector bundle attached to the finite dimensional $\mathbf{M}$-module $W(\mu)$ with highest weight μ. The BGG-complex $C_{\lambda\bullet}$ defines a complex $\mathcal{K}_\lambda^\bullet$ of vector bundles, called the dual BGG-complex, with

$$\mathcal{K}_\lambda^q = \oplus_{w\in W^{\mathbf{M}}, l(w)=q}\, \mathcal{W}(w(\lambda+\rho)-\rho)^\vee ,$$

and the differentials of $\mathcal{K}_\lambda^\bullet$ are differential operators between vector bundles. $\mathcal{K}_\lambda^\bullet$ is a direct summand in the de Rham complex of $\mathbb{V}(\lambda)^\vee$, and is a filtered resolution of this locally constant sheaf. The differentials in $\mathcal{K}_\lambda^\bullet$ are given by homogeneous differential operators on $\mathbb{D}^\vee$, so the complex $\mathcal{K}_\lambda^\bullet$ on X extends to a complex $\overline{\mathcal{K}}_\lambda^\bullet$ on $\overline{X}$. Similarly if $D\subset\overline{X}$ is the reduced divisor of the boundary of $\overline{X}$, the differentials of $\mathcal{K}_\lambda^\bullet$ extend to give a complex $\overline{\mathcal{K}}_\lambda^\bullet\otimes\mathcal{O}(-D)$ on $\overline{X}$ as we have seen in the previous section. Notice that both complexes $\overline{\mathcal{K}}_\lambda^\bullet$ and $\overline{\mathcal{K}}_\lambda^\bullet\otimes\mathcal{O}(-D)$ have (decreasing) $F^\bullet$-filtrations, induced from the filtration of the BGG-complex $C_{\lambda\bullet}$, such that associated graded complexes have trivial differentials. Explicitly

$$F^p(\overline{\mathcal{K}}_\lambda^\bullet) = \oplus_{w\in W^{\mathbf{M}}, w(\lambda+\rho)(H)+p\leq\rho(H)}\, \overline{\mathcal{W}}(w(\lambda+\rho)-\rho)^\vee ,$$

and similarly for $\overline{\mathcal{K}}_\lambda^\bullet\otimes\mathcal{O}(-D)$.

5.4. Proposition. *(1) The dual BGG-complex* $\overline{\mathcal{K}}_\lambda^\bullet$ *on* $\overline{X}$ *is quasi-isomorphic to* $\mathbb{R}j_*(\mathbb{V}(\lambda)^\vee)$, *where* $j : X \to \overline{X}$ *is the inclusion.*

(2) The complex $\overline{\mathcal{K}}_\lambda^\bullet \otimes \mathcal{O}(-D)$ *on* $\overline{X}$ *represents the direct image with compact support* $\mathbb{R}j_!(\mathbb{V}(\lambda))$. *The inclusion* $\overline{\mathcal{K}}_\lambda^\bullet \otimes \mathcal{O}(-D) \subset \overline{\mathcal{K}}_\lambda^\bullet$ *corresponds to the canonical map* $\mathbb{R}j_!(\mathbb{V}(\lambda)^\vee) \to \mathbb{R}j_*(\mathbb{V}(\lambda)^\vee)$.

Proof. (1) As the map j is Stein, $\mathbb{R}j_*(\mathbb{V}(\lambda)^\vee)$ is quasi-isomorphic to $j_*(\mathcal{K}_\lambda^\bullet)$, and we have to check that the natural inclusion $\overline{\mathcal{K}}_\lambda^\bullet \to j_*(\mathcal{K}_\lambda^\bullet)$ is a quasi-isomorphism. Instead of $\mathcal{K}_\lambda^\bullet$ we may consider the de Rham complex for $\mathbb{V}(\lambda)^\vee$, which contains $\mathcal{K}_\lambda^\bullet$ as a direct summand. In other words it suffices to show that the natural inclusion $(\mathcal{V}(\lambda)^\vee \otimes_\mathbb{C} \Omega_X^\bullet)^{\mathrm{can}} \to j_*(\mathcal{V}(\lambda)^\vee \otimes_\mathbb{C} \Omega_X^\bullet)$ is an isomorphism, where $(\mathcal{V}(\lambda)^\vee \otimes_\mathbb{C} \Omega_X^\bullet)^{\mathrm{can}}$ denotes the canonical extension of $\mathcal{V}(\lambda)^\vee \otimes_\mathbb{C} \Omega_X^\bullet$. The question is local around the boundary, and locally only the action of a certain unipotent subgroup $\mathbf{U}_1 \subset \mathbf{G}$ on $V(\lambda)$ matters ($\mathbf{U}_1$ depends on the stratum of the boundary). We may filter $V(\lambda)$ by $\mathbf{U}_1$-subspaces and thus reduce to the case of the trivial representation: $V(\lambda) = \mathbb{C}$. But there the canonical extension of $\Omega_X^\bullet$ is the logarithmic de Rham complex $\Omega_{\overline{X}}^\bullet(d\log\infty)$, which is well-known to represent $\mathbb{R}j_*(\mathbb{C})$, see e.g. Deligne [D 1] II 6.9, p.102, or make a local calculation, which can be reduced to the one-variable case using Künneth.

(2) $\mathbb{R}j_!(\mathbb{V}(\lambda)^\vee)$ is quasi-isomorphic to $j_!(\mathbb{V}(\lambda)^\vee)$, and we have to check that the natural map $j_!(\mathbb{V}(\lambda)^\vee) \to \overline{\mathcal{K}}_\lambda^\bullet \otimes \mathcal{O}(-D)$ is a quasi-isomorphism. As in (1) we are reduced to the case $V(\lambda)^\vee = \mathbb{C}$, and the result follows from an easy local calculation, which can be reduced to the one-variable case by the Künneth formula. It is obvious that the inclusion $\overline{\mathcal{K}}_\lambda^\bullet \otimes \mathcal{O}(-D) \subset \overline{\mathcal{K}}_\lambda^\bullet$ corresponds to the canonical map $\mathbb{R}j_!(\mathbb{V}(\lambda)^\vee) \to \mathbb{R}j_*(\mathbb{V}(\lambda)^\vee)$.

Remark about rationality: If $K \subset \mathbb{C}$ is a natural field of definition for X, the dual BGG-complexes $\mathcal{K}_\lambda^\bullet$, $\overline{\mathcal{K}}_\lambda^\bullet$ are defined over K.

Finally we can formulate the main result of this chapter:

5.5. Theorem. *(i) The spectral sequence associated to the "filtration bête"*

$$\begin{aligned} E_1^{p,q} &= H^q(\overline{X}, \overline{\mathcal{K}}_\lambda^p) \\ &= \oplus_{w \in W^{\mathbf{M}}, l(w)=p} H^q(\overline{X}, \overline{\mathcal{W}}(w(\lambda+\rho)-\rho)^\vee) \\ &\Rightarrow H^{p+q}(X, \mathbb{V}(\lambda)^\vee) \end{aligned}$$

and the spectral sequence for the F-filtration

$$\begin{aligned} E_1^{p,q} &= H^{p+q}(\overline{X}, \mathrm{gr}_F^p(\overline{\mathcal{K}}_\lambda^\bullet)) \\ &= \oplus_{w \in W^{\mathbf{M}}, w(\lambda+\rho)(H)+p=\rho(H)} H^{p+q-l(w)}(\overline{X}, \overline{\mathcal{W}}(w(\lambda+\rho)-\rho)^\vee) \\ &\Rightarrow H^{p+q}(X, \mathbb{V}(\lambda)^\vee) \end{aligned}$$

both degenerate at E_1. *The filtration on* $H^m(X, \mathbb{V}(\lambda)^\vee)$ *defined by the second spectral sequence is the F-filtration (i.e. the Hodge filtration) of a mixed Hodge structure on* $H^m(X, \mathbb{V}(\lambda)^\vee)$, *of weights* $\geq m + \lambda(H_w)$ *where* H_w *is the central element defined before.*

(ii) The spectral sequences

$$E_1^{p,q} = H^q(\overline{X}, \overline{\mathcal{K}}^p_\lambda(-D)) \Rightarrow H_!^{p+q}(X, \mathbb{V}(\lambda)^\vee) \text{ and}$$
$$E_1^{p,q} = H^{p+q}(\overline{X}, \mathrm{gr}^p_F(\overline{\mathcal{K}}^\bullet_\lambda(-D))) \Rightarrow H_!^{p+q}(X, \mathbb{V}(\lambda)^\vee)$$

both degenerate on E_1. The filtration on $H_!^m(X, \mathbb{V}(\lambda)^\vee)$ defined by the second spectral sequence is the F-filtration of a mixed Hodge structure on $H_!^m(X, \mathbb{V}(\lambda)^\vee)$, of weights $\leq m + \lambda(H_w)$.

Proof of Theorem 5.5. The main ingredients of theorem are all here; we just have to put them together and deduce the theorem from Deligne's Hodge theory. We remark that the major work is the computation of formal cohomology of the toroidal compactification of fibre products of the universal abelian scheme at the boundary of A_g.

Degeneration of Spectral Sequences. We may assume that $X = \mathsf{A}_{g,n}(\mathbb{C})$ for some $n \geq 3$, and the the general case follows by taking invariants under finite groups, as every arithmetic subgroup of $\mathbf{G}(\mathbb{Z})$ is a congruence subgroup. Also (i) and (ii) are equivalent, by duality. It also suffices to consider the spectral sequence associated to the F-filtration: its degeneration implies the degeneration of the first spectral sequence by dimension count, as $\mathrm{gr}_F(\overline{\mathcal{K}}^\bullet_\lambda)$ has trivial differentials. Clearly we may twist $V(\lambda)$ by powers of the multiplier representation η, so we may assume that there is an integer $s \geq 0$ such that $V(\lambda)$ is contained in the s-fold tensor product of the standard representation of $\mathbf{G}$ on $\mathbb{C}^{2g}$. So if $Y = A^s$ is the s-fold product of the universal abelian variety over $X, \mathbb{V}(\lambda)^\vee$ is a direct summand in $R^s f_*(\mathbb{C})(f : Y \to X$ is the projection). This holds in the filtered sense w.r.t. the F-filtration. The de Rham complex associated to $\overline{\mathcal{V}}(\lambda)$ is a direct summand in $\mathbb{R}\overline{f}_*(\overline{\Omega}^\bullet_Y)$, in the filtered sense (w.r.t. the F-filtration), so that the associated spectral sequence for $H^m(X, \mathbb{V}(\lambda))$ is a direct summand in the spectral sequence for $H^{m+s}(\overline{Y}, \overline{\Omega}^\bullet_Y)$, which degenerates by Deligne's Hodge theory. Finally $\overline{\mathcal{K}}^\bullet_\lambda$ is filtered quasi-isomorphic to the extended de Rham complex, so degeneration also holds for the spectral sequence associated to its F-filtration.

Weights and Mixed Hodge Structure. To get a mixed Hodge structure we need a good W-filtration (the weight filtration), and the rest follows from Deligne's results for $Y \subset \overline{Y}$. For this we could appeal to the general theory of weights, but in fact there is a better way, using the idea that we can derive the mixed Hodge structure on $H^m(X, \mathbb{V}(\lambda)^\vee)$ from that of $H^{m+s}(Y, \mathbb{C})$ as a direct summand (respecting the mixed Hodge structure).

Our strategy is to characterize the subspace $H^m(X, \mathbb{V}(\lambda)^\vee) \subset H^{m+s}(Y, \mathbb{C})$ by the action of some algebraic correspondences. For this let us recall how correspondences operate on the cohomology $H^*(Y, \mathbb{C})$: If $Z \subset Y \times_X Y$ is an algebraic cycle of dimension $d + r$ ($d = \dim(Y)$), it defines a morphism from $H^l(Y, \mathbb{C})$ into $H^{l-2r}(Y, \mathbb{C})$ for each i, by pulling a class back by pr_1 to Z and then pushing forward via pr_2. These mappings generate a graded algebra of endomorphisms of $H^*(Y,\mathbb{C})$, whose elements respect the Hodge and weight filtrations. The same algebra also operates on $\mathbb{R}f_*(\mathbb{C})$.

Here are some examples of such algebraic correspondences:

(1) Graphs of endomorphisms of A^s operate in the obvious manner on $\mathbb{R}f_*(\mathbb{C})$. In particular we obtain a ring of endomorphisms of $\mathbb{R}f_*(\mathbb{C})$ which is isomorphic to $M_{s\times s}(\mathbb{Z})$ using the endomorphisms $[n]_A$.

(2) For any two indices $1 \leq i < j \leq s$, the Poincaré bundle $\mathcal{P}$ on $A \times A$ This gives us a cycle of codimension 1 on $A \times A$. Pull this divisor back to Y via the projection to factors i and j and identify Y as the diagonal in $Y \times_X Y$. We get a cycle of codimension one on the diagonal in $Y \times_X Y$, and thus an endomorphism of $\mathbb{R}f_*(\mathbb{C})$ of degree 2, which is just multiplication with the first Chern class of $\mathcal{P}$.

(3) For any two indices $1 \leq i < j \leq s$, the Poincaré bundle $\mathcal{P}$ on $A \times A$ defines a cycle of codimension one in $A \times A$. Take its $(2g-1)$th-power, and push forward to Y via the projection to factors i and j and pull back again to $Y \times_X Y$ via the first projection.Then intersect with the pull-back to $Y \times_X Y$ of the diagonal in the double-product of the remaining $(s-2)$-factors in $Y = A^s$. This gives a cycle of dimension one more than the dimension of of Y, and hence an endomorphism of $\mathbb{R}f_*(\mathbb{C})$ of degree -2.

We claim that from those three types of algebraic correspondences we can build an idempotent in the algebra of endomorphisms of $\mathbb{R}f_*(\mathbb{C})$ with image $\mathbb{V}(\lambda)$. (If we want to be truly canonical, we should throw in Tate twists. But we shall ignore Tate twists, which are not essential, for the rest of the argument.) First, if $\mathbb{V}_1$ denotes the relative cohomology of A, so that $\mathbb{R}f_*(\mathbb{C}) = \Lambda^\bullet(\mathbb{V}_1^s)$, the subspace $\mathbb{V}_1 \otimes \mathbb{V}_1 \otimes \cdots \otimes \mathbb{V}_1 \subset \mathbb{R}f_*(\mathbb{C})$ (s-factors) is characterized by the fact that for $m_1, \ldots, m_s \in \mathbb{Z}$ the endomorphism $[m_1] \times \cdots \times [m_s]$ of A^s acts on it by $m_1 \cdot \ldots \cdot m_s$. So one easily constructs an idempotent with image $\mathbb{V}_1 \otimes \mathbb{V}_1 \otimes \cdots \otimes \mathbb{V}_1 \subset \mathbb{R}f_*(\mathbb{C})$. Furthermore, by classical invariant theory ([Weyl], ch. VI, §1) there exists an idempotent endomorphism of $\mathbb{V}_1 \otimes \mathbb{V}_1 \otimes \cdots \otimes \mathbb{V}_1$ with image $\mathbb{V}(\lambda)$, which is in the algebra of endomorphism generated by permutations of the factors, and endomorphisms of the following special type: Extension to more factors by tensoring with the identity on the rest of the map $\mathbb{V}_1 \otimes \mathbb{V}_1 \to \mathbb{C} \to \mathbb{V}_1 \otimes \mathbb{V}_1$, defined by first contracting via the symplectic form, then multiplying with the Chern class of the Poincaré bundle (dual to the symplectic form).

All these endomorphism can be generated by algebraic correspondences: Use type (1) for permutations, and types (2) and (3) for contractions. So we find an element in our algebra of endomorphisms of $\mathbb{R}f_*(\mathbb{C})$ which respects the subspace $\mathbb{V}_1 \otimes \mathbb{V}_1 \otimes \cdots \otimes \mathbb{V}_1 \subset \mathbb{R}f_*(\mathbb{C})$, and acts on it as an idempotent with image $\mathbb{V}(\lambda)$. Since the action of algebraic correspondences respects the mixed Hodge structure, this finishes the proof. □

Remarks. (1) If we used $\mathbf{G}^\circ$ instead of $\mathbf{G}$, we would get half integral F-weights and W-weights, which differ from the mixed Hodge structure coming from that of $\mathbb{R}f_*(\mathbb{C})$ by a (possibly half-integral) Tate twist.

(2) We could have done the whole theory of correspondences with coefficients $\mathbb{Q}$ instead of $\mathbb{C}$, which gives us rational mixed Hodge structure. No additional effort is needed.

(3) The result for cohomology with compact support can be proved either by duality or the same method as above.

(4) The action of our idempotent on $H^q(\overline{Y}, \overline{\Omega}^p)$ cuts out a subspace isomorphic to $H^q(\overline{X}, \mathrm{gr}_F^p(\overline{\mathcal{K}}_\lambda^\bullet))$, as is clear from the proof. Also $\mathrm{gr}_F^p(\overline{\mathcal{K}}_\lambda^\bullet)$ consists of those $\mathcal{W}(\mu)^\vee$ for which $\mu(H) = -p$.

(5) If X and $\overline{X}$ are naturally defined over a number-field $K \subset \mathbb{C}$, we may assume that Y and $\overline{Y}$ are also defined over K, and the action of the correspondences above on $H^q(\overline{Y}, \overline{\Omega}^p)$ is K-rational. It follows that the subspace $H^q(\overline{X}, \mathrm{gr}_F \overline{\mathcal{K}}_\lambda^\bullet)$ cut out by the algebraic correspondences constructed in the proof is also defined over K, since these algebraic correspondences are defined over K. Its K-structure coincides with its natural one: The maps $\mathrm{gr}_F(\overline{\mathcal{K}}_\lambda^\bullet) \to \mathrm{gr}_F(\overline{\mathcal{V}}(\lambda)^\vee \otimes \overline{\Omega}^\bullet) \to R\overline{f}_*(\mathrm{gr}_F(\overline{\Omega}^\bullet))$ are all K-rational.

(6) Again we have to be careful not to confuse the two rational structures on $H^*(X, \mathbb{V}(\lambda))$: If we use the topological interpretation, this space gets a $\mathbb{Q}$-structure, from the $\mathbb{Q}$-structure on $\mathbb{V}(\lambda)$. On the other hand the K-structure on the associated bundle $\mathcal{V}(\lambda)$, and its connection, define a K-structure on the de Rham cohomology $H^*(X, \mathbb{V}(\lambda))$, which is related to the previous one by a (probably highly transcendental) period matrix. The F-filtration is rational for the K-structure, while the W-filtration is rational as well in the $\mathbb{Q}$- as in the K-structure, just as in the usual formalism of Hodge theory.

Let us study *duality*: The pairing $V(\lambda) \times V(\lambda)^\vee \to \mathbb{C}$ defines an $\Omega^\bullet$-pairing from the corresponding de Rham complexes into $\Omega^\bullet$, hence also into the $\Omega^d[-d]$ ($d = \dim(X) = g \cdot (g+1)/2$) via the natural map $\Omega^\bullet \to \Omega^d[-d]$. We have $V(\lambda)^\vee = V(\lambda) \otimes \eta^{\otimes k}$ with $k = \lambda(H_w)$, where H_w is the generator of the center defined above. Restricting to the dual BGG-complexes $\mathcal{K}_\lambda^\bullet$ and extending to $\overline{X}$ gives a pairing $\overline{\mathcal{K}}_\lambda^\bullet \times \overline{\mathcal{K}}_\lambda^\bullet \to \overline{\Omega}^d(k)[-d]$ with $k = \lambda(H_w)$. Hence for each q we have pairings

$$\oplus_{w\in W^M, l(w)=q} \overline{\mathcal{W}}(w(\lambda+\rho)-\rho)^\vee \times \oplus_{w\in W^M, l(w)=d-q} \overline{\mathcal{W}}(w(\lambda+\rho)-\rho)^\vee \to \overline{\Omega}^d(\lambda(H_w))$$

Using $W(\mu)^\vee = W(-w_1(\mu))$ and the involution $w \to w_1 w w_0$ on $W^M (l(w_1 w w_0) = d - l(w)$ with $d = g \cdot (g+1)/2)$, we have

$$\begin{aligned} W(w_1 w w_0(\lambda+\rho)-\rho)^\vee &= W(-w w_0(\lambda+\rho) + w_1(\rho)) \\ &= W(w(\lambda+\rho) + w_1(\rho)) \otimes \eta^{\otimes k} (k = \lambda(H_w)) \\ &= W(w(\lambda+\rho)-\rho) \otimes W(2\rho^M) \otimes \eta^{\otimes k} , \end{aligned}$$

since $w_1(\rho) = -\rho + 2\rho^M$. Since $W(2\rho^M)$ corresponds to $\Omega^d_{\mathbb{D}^\vee}$, the second term in the pairing above can be written as $\oplus_{w\in W^M, l(w)=q} \overline{\mathcal{W}}(w(\lambda+\rho)-\rho) \otimes \overline{\Omega}^d(\lambda(H_w))$. As these pairings are derived from equivariant pairings on $\mathbb{D}^\vee$, it follows that the above pairings are derived from the obvious pairings

$$\overline{\mathcal{W}}(w(\lambda+\rho)-\rho)^\vee \times \overline{\mathcal{W}}(w(\lambda+\rho)-\rho) \otimes \overline{\Omega}^d(\lambda(H_w)) \to \overline{\Omega}^d(\lambda(H_w)) .$$

Similarly, the derived pairing $\overline{\mathcal{K}}^\bullet_\lambda(-D) \times \overline{\mathcal{K}}^\bullet_\lambda \to \overline{\Omega}^d_{\overline{X}}(k)$ $(k = \lambda(H_w))$ is given by pairings as above between the individual constituents of the $\overline{\mathcal{K}}^q_\lambda$. On the level of cohomology it induces pairings $\mathbb{H}^l(\overline{X}, \overline{\mathcal{K}}^\bullet_\lambda(-D)) \times \mathbb{H}^{2d-l}(\overline{X}, \overline{\mathcal{K}}^\bullet_\lambda) \to \mathbb{H}^d(\overline{X}, \Omega^d_{\overline{X}}(k)) = \mathbb{C} = \mathbb{C}_{\mathrm{DR}}(2d+k)(k = \lambda(H_w))$, identical to Poincaré duality between cohomology with compact support and ordinary cohomology: $H^l_!(X, \mathbb{V}(\lambda)) \to H^{2d-l}(X, \mathbb{V}(\lambda)) \to \mathbb{C}(2d+k)(k = \lambda(H_w))$. For the spectral sequences $E_1^{p,q} = H^q(\overline{X}, \mathcal{K}^p_\lambda(-D)) \Rightarrow H^{p+q}_!(X, \mathbb{V}(\lambda))$ and $E_1^{p,q} = H^q(\overline{X}, \mathcal{K}^p_\lambda) \Rightarrow H^{p+q}(X, \mathbb{V}(\lambda))$ Serre duality defines a duality between the E_1-terms, which is compatible with Poincaré duality on E_∞.

6. Other Cohomology Theories

There are analogues of the theory for various other kinds of cohomology. We let $X = \mathsf{A}_{g,n}$, considered as a scheme over $\mathrm{Spec}(\mathbb{Z}[1/n])$, and $\overline{X}$ a smooth toroidal compactification of X. That is, we consider level-n-structures up to similitude, and always assume that the level n is invertible.

Coherent Cohomology and Automorphic Forms. Over $\mathbb{C}$ we obtain a pure Hodge structure of weight m on the image of the natural map $H^m_!(X, \mathbb{V}(\lambda)^\vee) \to H^m(X, \mathbb{V}(\lambda)^\vee)$. Its (p,q)-term is isomorphic to

$$\sum_{w \in W^{\mathbf{M}}, w(\lambda+\rho)(H) = \rho(H) - p} \mathrm{Image}\{H^{m-l(w)+k}(\overline{X}, \overline{\mathcal{W}}(w(\lambda+\rho)-\rho)^\vee(-D)) \to H^{m-l(w)+k}(\overline{X}, \overline{\mathcal{W}}(w(\lambda+\rho)-\rho)^\vee)\}\ ,$$

where $k = \lambda(H_w)$. This is just the part of the cohomology which can be represented by harmonic cusp-forms, where the 2^g choices for $w \in W^M$ correspond to the 2^g discrete series representations of $\mathbf{G}(\mathbb{R})$ with central character χ_λ.

We have obtained injections from the coherent cohomology of certain $\overline{\mathcal{W}}(\mu)^\vee$ into singular cohomology. However this applies only to μ's for which $\mu + \rho$ is non-singular (i.e. not perpendicular to any element of Φ), since this is the condition for a $\Phi^+_{\mathbf{M}}$-dominant weight $\mu \in \mathbf{t}^*$ to be of the form $w(\lambda+\rho)-\rho$ for some $\Phi^+_{\mathbf{G}}$-dominant weight λ and some $w \in W^{\mathbf{M}}$. For example for $\overline{\mathcal{W}}(\mu)^\vee = \omega^k$, $k \in \mathbb{N}$, this condition means $k \geq g+1$, the border-case $k = g+1$ corresponding to the differentials on X. Let us check this using roots of $\mathbf{G}^\circ$: $\Phi^+_{\mathbf{G}}$ consists of roots $a_i \pm a_j$, $i \leq j$, $\rho = g \cdot a_1 + (g-1) \cdot a_2 + \ldots + a_1$, $\mu = -k \cdot (a_1 + \ldots + a_g)$, so $\rho + \mu = (g-k) \cdot a_1 + (g-k-1) \cdot a_2 + \ldots + (1-k) a_1$. This is non-singular iff all coefficients are negative, that is, iff $k > g$. For these k we obtain an injection of $\Gamma(\overline{X}, \omega^k)$ into the middle cohomology of X, since we have to take $w \in W^{\mathbf{M}}$ to be $a_i \mapsto -a_{g-i+1}$, which has length $g \cdot (g+1)/2$ (so $\lambda = (k-2) \cdot a_1 + (k-4) \cdot a_2 + \ldots + (k-2g) \cdot a_g$ as dominant weight of $\mathbf{G}^\circ$). Cusp-forms end up in the image of the cohomology with compact support.

l-adic Sheaves. We associate to any $\mathbf{G}$-representation V a $\mathbb{Q}_l$-adic smooth étale sheaf $\mathbb{V}_l$ on $X \otimes \mathbb{Z}[1/l]$, as follows: To the standard representation of $\mathbf{G}$ corresponds the $\mathbb{Q}_l$-sheaf $R^1\pi_*\mathbb{Q}_l(1)$, and η corresponds to $\mathbb{Q}_l(1)$. (This is the only plausible candidate, as the determinant of the standard representation is η^g, while $\det(R^1\pi_*\mathbb{Q}_l) = \mathbb{Q}_l(-g)$ by Poincaré duality.) A general representation of $\mathbf{G}$ is a subquotient of tensor products of copies of the standard representations and its dual; its associated smooth $\mathbb{Q}_l$-sheaf on A_g is cut out by algebraic cycles in Tate-twisted tensor products of copies of $R^1\pi_*\mathbb{Q}_l$.

An equivalent way to say the same thing is the following: smooth ℓ-adic sheaves on A_g correspond (after choosing base point) to continuous ℓ-adic representations of the (arithmetic) fundamental group $\pi_1(\mathsf{A}_g)$ of A_g. The smooth sheaf $R^1\pi_*\mathbb{Q}_l(1)$ corresponds to the representation std: $\pi_1(\mathsf{A}_g) \to \mathbf{G}(\mathbb{Q}_l)$. For any given representation of $\mathbf{G}$, we get by composition a continuous representation of $\pi_1(\mathsf{A}_g)$, which "is" the ℓ-adic sheaf attached to the given representation of $\mathbf{G}$.

An easy local calculation shows that the direct images of $\mathbb{V}_l$ under the inclusion $X \subset \overline{X}$ are smooth when restricted to a stratum of $\overline{X}$. Thus:

6.1. Theorem. *The representation of* $\mathrm{Gal}(\overline{\mathbb{Q}}/\mathbb{Q})$ *on the étale cohomology* $H^*(\mathsf{A}_{g,n} \otimes \overline{\mathbb{Q}}, \mathbb{V}_l)$ *is unramified outside* ln.

De Rham Cohomology and Crystalline Cohomology. The theory applies to the logarithmic de Rham cohomology (over $\mathbb{Q}$) of the logarithmic crystals given by the canonical extensions $\overline{\mathcal{V}}$ of the locally constant systems associated to $\mathbf{G}$-representations V. That is the de Rham cohomology of $\overline{\mathcal{V}}(\lambda)^\vee$ (associated to an irreducible $\mathbf{G}$-representation $V(\lambda)^\vee$) is equal to the hypercohomology of $\overline{\mathcal{K}}^\bullet_\lambda$, and the corresponding spectral sequence degenerates.

Moreover if X has good reduction in characteristic p (that is, if the level n is not divisible by p) the $\mathbb{Q}_p$-valued de Rham cohomology is equal to the crystalline cohomology of X modulo p. It thus admits a Frobenius endomorphism. More precisely we consider $Y = s$-fold product of the abelian variety over $X = \mathsf{A}_g$ (or $\mathsf{A}_{g,n}$). In section 1, we constructed a good compactification $\overline{Y}$. If $f : Y \to X$ denotes the projection and $\overline{f} : \overline{Y} \to \overline{X}$ its extension (which is smooth in the logarithmic sense), we decompose the crystalline direct image under the action of the algebraic cycles, as before. We obtain the canonical extensions $\overline{\mathcal{V}}(\lambda)^\vee$ of the crystals on X associated to the irreducible representations $V(\lambda)^\vee$ of $\mathbf{G}$, which are filtered by the F-filtrations, and which admit Frobenius endomorphisms. That is, the crystals are invariant under local Frobenius lifts on the p-adic completion of $\overline{X}$. Such objects are called filtered F-isocrystals (see [Ogus] for more details).

Furthermore the crystalline cohomology of Y is the direct sum of the crystalline cohomology groups of X with values in these logarithmic crystals $\overline{\mathcal{V}}(\lambda)$.

There is also a $\mathbb{Z}_p$-version, where we do not neglect p-torsion. However we can only use $V(\lambda)$ where λ is not too big (compared with p), as the various direct sum decompositions used in our constructions have to be defined over $\mathbb{Z}_p$. We only indicate the fact that for the crystal given by the dual of standard representation of $\mathbf{G}$, or equivalently by the first de Rham cohomology of A, the Hodge filtration has degrees 0 and 1, and on F^1 Frobenius is divisible by p (modulo p, Frobenius annihilates differentials). Furthermore, the determinant of Frobenius is (up to units) equal to p^g, which means that this first direct image is an object of $\mathcal{MF}^\nabla(X)$ (in the notation of [F 6]).

Relations Between Crystalline and p-adic Étale Cohomology

The result states the p-adic Galois representations coming from the p-adic étale cohomology tend to be crystalline, consequently de Rham, hence also Hodge-Tate. Moreover if p is big compared with the degree of the cohomology and everything has good reduction, an integral version (i.e. $\mathbb{Z}_p$ instead of $\mathbb{Q}_p$) holds. This is proved in [F 6] theorem 5.3 and 5.6 (see also [F 7]), which is based on work of many people: [Fo 1] [Fo 2] [Fo 3] [Fo-La] [Fo-Me] [BO 2] [Og] [Be 2] [F 4] [F 5].

For the reader's convenience, we review some relevant definitions and refer to the sources above for more information.

Let K be a mixed characteristic local field, i.e. the quotient field of a complete discrete valuation ring $\mathcal{O} = \mathcal{O}_K$ with perfect residue field k, with $\mathrm{char}(K) = 0$ and $\mathrm{char}(k) = p > 0$. Denote by $\overline{K}$ a (fixed) algebraic closure of K and $\mathrm{Gal} = \mathrm{Gal}(\overline{K}/K)$ the Galois group of K. $\mathbb{Q}_p$ is canonically embedded in K. The natural action of Gal on $\overline{K}$ extends by continuity to the p-adic completion C (or, $\mathbb{C}_p$) of $\overline{K}$.

Let $B_{\mathrm{HT}} = C[\mathbb{Q}_p(1), \mathbb{Q}_p(-1)] = \oplus_{n\in\mathbb{Z}} C(n)$. This is a graded C-algebra with natural semi-linear Gal-action. For any p-adic representation V of Gal (i.e. a continuous representation of Gal on a finite-dimensional vector space over $\mathbb{Q}_p$), define $\mathbf{D}_{\mathrm{HT}}(V)$ to be $(B_{\mathrm{HT}} \otimes_{\mathbb{Q}_p} V)^{\mathrm{Gal}}$, considered as a graded K-vector space. A fundamental theorem of Tate (cf. [T]) says that $\dim_K(\mathbf{D}_{\mathrm{HT}}(V)) \le \dim_{\mathbb{Q}_p}(V)$ always. The p-adic representation V is said to be *Hodge-Tate* iff equality holds, in other words $C \otimes_{\mathbb{Q}_p} V \cong \oplus_k \mathrm{gr}^k(\mathbf{D}_{\mathrm{HT}}(V))(-k)$ as Gal-modules. Hodge-Tate representations are stable under direct sums, duals and tensor products and form a Tannakian category.

The p-adic étale cohomology of algebraic varieties furnishes the most interesting examples of Hodge-Tate p-adic representations: The main theorem of [F 5] states that for any smooth variety X over K contained in a proper smooth variety $\overline{X}$ with infinity a divisor with normal crossings (or empty), there exists natural $\mathrm{Gal}(\overline{K}/K)$-linear isomorphisms $H^m_{\text{ét}}(X \otimes_K \overline{K}, \mathbb{Q}_p) \otimes_{\mathbb{Q}_p} C \cong \sum_{a+b=m} H^a(\overline{X}, \overline{\Omega}^b_X) \otimes_K C(-b)$ for all m, where "$(-b)$" indicates Tate-twist. In fact the p-adic étale cohomology is crystalline, as we explain next.

Let us assume for simplicity that K is the fraction field of $W(k)$, k a perfect field of characteristic p. The notion of crystalline representations depends heavily on an important ring $B^+_{\mathrm{cris}}(\mathcal{O}_{\overline{K}})$. This is a wonderful ring, introduced first by Fontaine in a slightly different form in [Fo 2]. We shall first list some of its properties and come back to its (somewhat complicated) definition later:

(1) $B^+_{\mathrm{cris}}(\mathcal{O}_{\overline{K}})$ has a natural (decreasing) filtration $\mathrm{Fil}^i = \mathrm{Fil}^i(B^+_{\mathrm{cris}}(\mathcal{O}_{\overline{K}}))$ with $\mathrm{Fil}^0 = B^+_{\mathrm{cris}}(\mathcal{O}_{\overline{K}})$, $\cap_i \mathrm{Fil}^i = (0)$, $\mathrm{Fil}^i \cdot \mathrm{Fil}^j \subset \mathrm{Fil}^{i+j}$.

(2) $W(\overline{k})$ = the ring of classical Witt-vectors is naturally contained in $B^+_{\text{cris}}(\mathcal{O}_{\overline{K}})$, and there is a natural σ-linear action Φ on $B^+_{\text{cris}}(\mathcal{O}_{\overline{K}})$ extending the Frobenius action σ on $W(\overline{k})$.

(3) Gal $= \text{Gal}(\overline{K}/K)$ operates naturally on $B^+_{\text{cris}}(\mathcal{O}_{\overline{K}})$. This action is $W(\overline{k})$-semi-linear and continuous on every Gal-stable finite dimensional $W(\overline{k})$-subspace. Moreover it commutes with Φ and preserves the filtration $\{\text{Fil}^i\}$.

(4) There exists a canonical embedding $\alpha : \mathbb{Q}_p(1) \hookrightarrow (B^+_{\text{cris}}(\mathcal{O}_{\overline{K}})^*, \times)$ which induces a canonical embedding $\beta : \mathbb{Z}_p(1) \hookrightarrow (B^+_{\text{cris}}(\mathcal{O}_{\overline{K}}), +)$. α, β are both Galois-linear, $\Phi(\alpha(x)) = \alpha(x)^p \, \forall x \in \mathbb{Q}_p(1)$, $\Phi(\beta(x)) = p \cdot \beta(x) \, \forall x \in \mathbb{Z}_p(1)$.

From the ring $B^+_{\text{cris}}(\mathcal{O}_{\overline{K}})$ we define another ring $B_{\text{cris}}(\overline{K})$ by $B_{\text{cris}}(\overline{K}) := B^+_{\text{cris}}(\mathcal{O}_{\overline{K}})[p^{-1}, \beta(x)^{-1}]$, where x denotes any generator of $\mathbb{Z}_p(1)$. $B_{\text{cris}}(\overline{K})$ contains the fraction field of $W(\overline{k})$, has an induced filtration, and the action Φ on $B^+_{\text{cris}}(\mathcal{O}_{\overline{K}})$ extends naturally to $B_{\text{cris}}(\overline{K})$. Moreover $B_{\text{cris}}(\overline{K})^{\text{Gal}} = K$, and $F^0(B_{\text{cris}}(\overline{K})) \cap B_{\text{cris}}(\overline{K})^{\Phi} = \mathbb{Q}_p$.

For a p-adic representation V of Gal $= \text{Gal}(\overline{K}/K)$ (i.e. a continuous representation of G on a finite-dimensional vector space over $\mathbb{Q}_p$), we always have $\dim_K((V \otimes B_{\text{cris}}(\overline{K}))^{\text{Gal}}) \leq \dim_K((V \otimes B_{\text{HT}})^{\text{Gal}}) \leq \dim_{\mathbb{Q}_p}(V)$. V is said to be *crystalline* if $\dim_K((V \otimes B_{\text{cris}}(\overline{K}))^{\text{Gal}}) = \dim_{\mathbb{Q}_p}(V)$, which especially implies that V is Hodge-Tate. In this case $E = (V \otimes B_{\text{cris}}(\overline{K}))^{\text{Gal}}$ is a finite dimensional filtered K-vector space with a semi-linear Frobenius automorphism, and this correspondence between crystalline Galois representations and certain vector spaces of this kind is fully faithful, and preserves duals, direct sums and tensor products. In fact $V \otimes B_{\text{cris}}(\overline{K}) \cong E \otimes_K B_{\text{cris}}(\overline{K})$, and we can recover V from this as the subspace of elements in F^0 which are invariant under Frobenius. One of the main results of [F 6] is that the p-adic étale cohomology of a proper smooth algebraic K-variety with good reduction is crystalline, and is associated to the crystalline cohomology. In fact, this is even true of a non-proper variety with a good compactification, that is, of the complement of a divisor with normal crossings of a smooth proper scheme, such that both scheme and divisor have good reduction. For instance if X is a scheme proper and smooth over $\mathcal{O}_K$, then there is a canonical isomorphism $(H^*_{\text{ét}}(X \otimes_K \overline{K}, \mathbb{Q}_p) \otimes B_{\text{cris}}(\overline{K}))^{\text{Gal}} \cong H^*_{\text{cris}}(X \otimes k/K)$ $(= H^*_{\text{DR}}(X/K)$ in this case), preserving Frobenius and the Hodge filtration. There are, in fact, integral versions of the above result, and variants with non-trivial coefficient systems and open varieties. We refer the interested reader to [F 6] and the related papers cited above.

We may apply the general result to Y = s-fold product of the abelian variety over $X = \mathsf{A}_g$(or $\mathsf{A}_{g,n}$), where we have constructed (in section 1) good compactifications $\overline{Y}$. If $f : Y \to X$ denotes the projection and $\overline{f} : \overline{Y} \to \overline{X}$ its extension (which is smooth in the logarithmic sense), we decompose as before the crystalline direct image under the action of the algebraic cycles exhibited in section 5, to obtain the canonical extensions $\overline{\mathcal{V}}$ of the crystals on X associated to the representations V of $\mathbf{G}$. They are filtered by the F-filtration and admit a Frobenius endomorphism, and the crystalline cohomology of Y is the direct sum of the crystalline cohomology groups of X with values in these logarithmic crystals $\overline{\mathcal{V}}$.

Similarly we can do everything on the étale side, thus decomposing the p-adic étale cohomology of (the geometric fibre of) Y into the direct sum of the p-adic étale cohomology groups of X with values in the p-adic étale sheaves (on $X \otimes \mathbb{Q}_p$) $\mathbb{V}_p$ associated to the V. Moreover the correspondence between crystalline and étale cohomology respects these decompositions, as it respects algebraic cycles. It follows that the p-adic Galois representations associated to the Siegel-spaces $\mathsf{A}_{g,n}$ are Hodge-Tate at p, and crystalline if p does not divide n. There is also a statement in the $\mathbb{Z}_p$-theory, which applies if the dimension of Y is $\leq p-2$. However as the dimension of Y tends to be quite big this condition is very restrictive, so we leave the details to the reader.

We admit that there is one problem which we cannot settle to our total satisfaction: There exists a relative version of the comparison theorem, and it should follow that the filtered F-crystals $\overline{\mathcal{V}}$ are associated to the p-adic étale sheaves $\mathbb{V}_p$. We cannot show this if $p = 2$. The problem is that although the maps $\overline{Y} \to \overline{X}$ are smooth in the logarithmic sense, for the relative comparison we need a stronger statement than this. For $p > 2$ however the assertion can be shown, using the $\mathbb{Z}_p$-theory, as follows: The first crystalline direct image of $\overline{f}$ has a natural $\mathbb{Z}_p$-structure which defines an object of $\mathcal{MF}^{\nabla}(X)$ (in the terminology of [F 6]). It follows that there exists an associated étale sheaf on (the generic fibre of) X, and we have to show that this is equal to the first direct image of $\mathbb{Z}_p$ in the étale cohomology. There is no problem for $g = 1$, so assume that $g > 1$. We get an equality of étale sheaves on "the generic fibre" of the formal p-adic completion of $\mathsf{A}_{g,n}$. That is for each affine ring R of $\mathsf{A}_{g,n}$ we get an equality of étale sheaves on $\mathrm{Spec}(R^\wedge \otimes K)$. Use the minimal compactification A_g^* (and the fact that its boundary has codimension ≥ 2 in each fibre) to conclude that this equality is algebraic. This proves the assertion for this first direct image, which suffices, since up to Tate-twist all $\overline{\mathcal{V}}$ are contained in it (for a suitable choice of Y).

We conclude that theorem 5.5 holds for p-adic étale cohomology as well: Hence for $H^m_{\text{ét}}(X \otimes \overline{K}, \mathbb{V}(\lambda)^\vee \otimes \mathbb{Q}_p)$, we have

6.2. Theorem. *(i) For all X (even with bad reduction at p) there is a $\mathrm{Gal}(\overline{\mathbb{Q}}_p/\mathbb{Q}_p)$-isomorphism*

$$\begin{aligned} &H^m_{\text{ét}}(X \otimes \overline{K}, \mathbb{V}(\lambda)^\vee \otimes \mathbb{Q}_p) \otimes_{\mathbb{Q}_p} \overline{\mathbb{Q}}_p^\wedge \\ &\quad \cong \oplus_{w \in W^{\mathrm{M}}} H^{m-l(w)}(\overline{X}, \overline{\mathcal{W}}(w(\lambda+\rho)-\rho)^\vee) \otimes_K \overline{\mathbb{Q}}_p^\wedge((w(\lambda+\rho)-\rho)(H))) \end{aligned}$$

where $k = \lambda(H_w)$.

(ii) If X has good reduction at p, there is an isomorphism

$$(H^*_{\text{ét}}(X \otimes_K \overline{K}, \mathbb{V}(\lambda)_p^\vee \otimes \mathbb{Q}_p) \otimes B_{\text{cris}}(\overline{\mathbb{Q}}_p))^{\text{Gal}} \cong H^*(\overline{X}, \overline{\mathcal{K}}^\bullet_\lambda) \ ,$$

preserving Frobenius and the Hodge filtration.

(iii) If $p > 2$ and X has good reduction at p, $\overline{\mathcal{V}}(\lambda)$ is associated to $\mathbb{V}_p(\lambda)$, for all irreducible $\mathbf{G}$-representations $V(\lambda)$.

We also could have used p-adic methods (instead of Hodge theory) to conclude that the spectral-sequence associated to the F-filtration on $\mathcal{K}_\lambda^\bullet$ degenerates. This is in accordance with the general principle that all places of $\mathbb{Q}$ should be treated on an equal footing.

For completeness we give a definition of $B_{\text{cris}}^+(\mathcal{O}_{\overline{K}})$. The shortest one is that it is inv. $\lim_n H_{\text{cris}}^0(\text{Spec}(\mathcal{O}_{\overline{K}}/p^n\mathcal{O}_{\overline{K}})/W_n(k))$. Here is a more explicit and useful definition. Let $\mathcal{O}_C = \mathcal{O}_{\overline{K}}^\wedge$ be the p-adic completion of $\mathcal{O}_{\overline{K}}$, R be the inverse limit of $\mathcal{O}_{\overline{K}}/p\mathcal{O}_{\overline{K}} \leftarrow \mathcal{O}_{\overline{K}}/p\mathcal{O}_{\overline{K}} \leftarrow \mathcal{O}_{\overline{K}}/p\mathcal{O}_{\overline{K}} \leftarrow \dots$, where the transition maps are all the $a \mapsto a^p$. To an element $u = (x_0, x_1, x_2, \dots)$ of R we associate a sequence $u^{(n)}$ in $\mathcal{O}_C$, $n \in \mathbb{Z}$, with $u^{((n+1)p)} = u^{(n)}$, by taking $u^{(n)}$ to be the limit of $(x_{n+m}^\wedge)^{p^m}$ as $m \uparrow \infty$, where $x_{n+m}^\wedge$ is a representative of x_{n+m} in $\mathcal{O}_{\overline{K}}$. Let $W(R)$ be the ring of all Witt-vectors with entries in R, with the addition and multiplication given by the usual formulas. The Frobenius on R induces an automorphism Φ on $W(R)$. Define a homomorphism $\theta : W(R) \to \overline{R}^\wedge$ from $W(R)$ to the p-adic completion $\overline{R}^\wedge$ of $\overline{R}$, by sending a Witt vector $(u_0, u_1, u_2, \dots)$ to $\sum_{n\geq 0} p^n \cdot u_n^{(n)}$. The kernel of θ is a principal ideal, generated for instance by $\xi = [u_0] + p$, where $u_0 \in R$ has $u_0^{(0)} = -p$, and $[u_0] = (u_0, 0, 0, \dots) \in W(R)$. Then $B_{\text{cris}}^+(\mathcal{O}_{\overline{K}})$ is defined to be the p-adic completion of the divided power envelope $D_I(W(R))$ of $W(R)$ with respect to $I = \xi W(R)$. One can also use the completion of $D_I(W(R))$ with respect to the filtration F^n on $D_I(W(R))$, where $F^n(D_I(W(R)))$ consists of all elements of the n-th divided power of I whose image under Φ is divisible by p^n, as done in [Fa 6], which means using inv. $\lim_n H_{N\text{cris}}^0(\text{Spec}(\mathcal{O}_{\overline{K}}/p^n\mathcal{O}_{\overline{K}})/W_n(k))$ instead of inv. $\lim_n H_{\text{cris}}^0(\text{Spec}(\mathcal{O}_{\overline{K}}/p^n\mathcal{O}_{\overline{K}})/W_n(k))$, where N_{cris} stands for the nilpotent crystalline site.

Chapter VII. Hecke Operators and Frobenii

This chapter contains an elementary discussion of Hecke operators and Frobenii operating on locally constant systems on A_g, and we do not pretend to have proved any serious theorem here. Difficulties arise on two sides: in geometry, with the Lefschetz trace formula for Hecke correspondences and in the harmonic analysis, with the Selberg trace formula for automorphic representations of the symplectic group. Both call for further work.

1. The Local Unramified Hecke Algebra and Its Structure

Notations are as in VI, §4.

Definition of Hecke Algebras

From now on we *fix a prime p*. (In applications, p will be an unramified prime for the given level structure.) In this subsection and the next we will define the relevant unramified local Hecke algebras (or, spherical functions) and review their basic properties. (In applications, p will usually be an unramified prime.) The reader can consult [Sa 2] or the review article [Cartier, in Corvallis I] for a treatment of Hecke algebras of general reductive p-adic groups.

Denote by $G = \mathbf{G}(\mathbb{Q}_p)$, $Q = \mathbf{Q}(\mathbb{Q}_p)$, $M = \mathbf{M}(\mathbb{Q}_p)$, $T = \mathbf{T}(\mathbb{Q}_p)$, $U = \mathbf{U}(\mathbb{Q}_p)$ the p-adic Lie groups corresponding to G, Q, M, T, and U (VI, §4). *By abuse of notation*, we obtain characters $e^{2\rho_n} : M \to p^{\mathbb{Z}} \subset \mathbb{Q}^*$ and $e^{2\rho} : T \to p^{\mathbb{Z}} \subset Q^*$, by composing the algebraic characters with the modulus-homomorphism $|\ |$: $\mathbb{Q}_p^* \to p^{\mathbb{Z}}$ (normalized by $|p| = p^{-1}$). Their positive square roots are denoted by e^{ρ_n} and e^{ρ} respectively.

We need the (hyperspecial) maximal compact subgroups $K = \mathbf{G}(\mathbb{Z}_p) \subset G$ and $L = \mathbf{M}(\mathbb{Z}_p) = M \cap K \subset M$. Fix (left) Haar measures on all the groups by requiring that the open and compact subgroups of integral (i.e. $\mathbb{Z}_p$-) points have measure 1. Note that G, M, U, N are all unimodular.

The following integration formulas are easily checked, as the right sides have the correct invariance properties under left translation with K, and right translation with M and U or T and N: Suppose $f : G \to \mathbb{R}$ is a continuous function with compact support. Then

$$\int_G f(g)\,dg = \iiint_{K\times U\times M} f(kum)\,dk\,du\,dm = \iiint_{K\times M\times U} f(kmu)e^{2\rho_n}(m)\,dk\,dm\,du$$
$$= \iiint_{K\times N\times T} f(knt)\,dk\,dn\,dt = \iiint_{K\times T\times N} f(ktn)e^{2\rho}(t)\,dk\,dt\,dn\ .$$

1.1. Definition. (a) The Hecke algebra $\mathcal{H}(G,\mathbb{Q})$ is defined as follows: its underlying $\mathbb{Q}$-vector space is $C_0(K\backslash G/K)$ = The $\mathbb{Q}$-vector space consisting of all $\mathbb{Q}$-valued K-bi-invariant continuous (i.e. locally constant) functions $\phi : G \to \mathbb{Q}$, with compact support (or more intrinsically, all K-bi-invariant measures on G with compact support). The algebra structure of $\mathcal{H}(G,\mathbb{Q})$ is given by convolution product:

$$(f * g)(x) = \int_{y\in G} f(y)g(y^{-1}x)\,dy\ .$$

(b) As a $\mathbb{Q}$-vector space $\mathcal{H}(G,\mathbb{Q})$ has a basis consisting of the characteristic functions of double classes KgK, and the characteristic function of K is the unit in $\mathcal{H}(G,\mathbb{Q})$. A basic fact is that $\mathcal{H}(G,\mathbb{Q})$ is *commutative.* The structure of $\mathcal{H}(G,\mathbb{Q})$ will be reviewed later.

(c) We define $\mathcal{H}(M,\mathbb{Q})$ and $\mathcal{H}(T,\mathbb{Q})$ in exact analogy. We remark that instead of $\mathbb{Q}$ we could have chosen $\mathbb{R}$ or $\mathbb{C}$ as coefficients. Clearly $\mathcal{H}(G,\mathbb{R}) = \mathcal{H}(G,\mathbb{Q}) \otimes_{\mathbb{Q}} \mathbb{R}$, $\mathcal{H}(G,\mathbb{C}) = \mathcal{H}(G,\mathbb{Q}) \otimes_{\mathbb{Q}} \mathbb{C}$, and similarly for M and T.

Structure of Hecke Algebras

The structure of $\mathcal{H}(T,\mathbb{Q})$ is easily determined: If $Y = \mathrm{Hom}(G_m,\mathbf{T})$ is the cocharacter group of $\mathbf{T}$, we obtain $\mathcal{H}(T,\mathbb{Q}) \cong \mathbb{Q}[Y]$. The isomorphism is defined by sending an element of Y to the characteristic function of the double class which contains the image of $p \in \mathbb{Q}_p^* = G_m(\mathbb{Q}_p)$ under the corresponding homomorphism.

Observe that Y can be identified with the sublattice $Y \subset \mathbb{Z}^{2g}$ consisting of all sequences $(\lambda_1,...,\lambda_g,\mu_1,...,\mu_g)$ with $\lambda_i+\mu_i$ = constant: $(\lambda_1,...,\lambda_g,\mu_1,...,\mu_g)$ corresponds to $t \mapsto \mathrm{diag}(t^{\lambda_1},\dots,t^{\lambda_g},t^{\mu_1},\dots,t^{\mu_g})$, $t \in G_m$. Via this isomorphism, $\mathcal{H}(T,\mathbb{Q})$ is identified with the subalgebra of $\mathbb{Q}[U_1^{\pm1},\dots,U_g^{\pm1},V_1^{\pm1},\dots,V_g^{\pm1}]$ generated by $(U_iV_i^{-1})^{\pm1}$, $1 \le i \le g$, and $(U_1\cdot\ldots\cdot U_g)^{\pm1}$: $\mathcal{H}(T,\mathbb{Q}) \cong \mathbb{Q}[U_1V_1^{-1},V_1U_1^{-1},\dots,U_gV_g^{-1},V_gU_g^{-1},(U_1\cdot\ldots\cdot U_g)^{\pm1}]$. The Weyl group $W_{\mathbf{G}}$ operates on this in the obvious manner: the permutations in $W_{\mathbf{G}} \cong \mathcal{S}_g \ltimes \{\pm1\}^g$ act on $\mathcal{H}(T,\mathbb{Q}) \cong \mathbb{Q}[U_1V_1^{-1},V_1U_1^{-1},\dots,U_gV_g^{-1},V_gU_g^{-1},(U_1\cdot\ldots\cdot U_g)^{\pm1}] \subset \mathbb{Q}[U_1^{\pm1},\dots,U_g^{\pm1},V_1^{\pm1},\dots,V_g^{\pm1}]$ by permuting the U's and V's in the same way. The subgroup $\{\pm1\}^g$ acts by exchanging U's and V's. Obvious analogues hold for $\mathcal{H}(T,\mathbb{R})$ and $\mathcal{H}(T,\mathbb{C})$.

In [S 2] Satake has shown how to extend the above results to $\mathcal{H}(G,\mathbb{R})$ and $\mathcal{H}(M,\mathbb{R})$ and obtain structure theorems for these Hecke algebras as well: Define a homomorphism (called the Satake homomorphism)

$$S_T^G : \mathcal{H}(G,\mathbb{R}) \to \mathcal{H}(T,\mathbb{R}) \text{ by the rule } S_T^G(\phi)(t) = e^{\rho}(t) \cdot \int_N \phi(tn)\, dn\ .$$

His basic result is that S_T^G defines an isomorphism of $\mathcal{H}(G,\mathbb{R})$ onto the $W_{\mathbf{G}}$-invariant elements in $\mathcal{H}(Y,\mathbb{R}) : \mathcal{H}(G,\mathbb{R}) \cong \mathcal{H}(T,\mathbb{R})^{W_{\mathbf{G}}}$.

Similarly, define

$$S_M^G : \mathcal{H}(G,\mathbb{R}) \to \mathcal{H}(M,\mathbb{R}) \text{ by } S_M^G(\phi)(m) = e^{\rho_n}(m) \cdot \int_U \phi(mu)\, du$$

and

$$S_T^M : \mathcal{H}(M,\mathbb{R}) \to \mathcal{H}(T,\mathbb{R}) \text{ by } S_T^M(\phi)(t) = e^{\rho}M(t) \cdot \int_{M\cap N} \phi(tn)\, dn\ .$$

Similar results hold: $\mathcal{H}(G,\mathbb{R})$ is identified with a subalgebra of $\mathcal{H}(M,\mathbb{R})$ under S_M^G and $\mathcal{H}(M,\mathbb{R})$ is identified with $\mathcal{H}(T,\mathbb{R})^{W_{\mathbf{M}}}$ under S_T^M. Moreover we have $S_T^G = S_T^M \circ S_M^G$, so we have $\mathcal{H}(G,\mathbb{R}) \subset \mathcal{H}(M,\mathbb{R}) \subset \mathcal{H}(T,\mathbb{R})$ in a canonical manner. We remark that we obviously can replace $\mathbb{R}$ by $\mathbb{C}$ in all results above. Heuristically, these Satake homomorphisms are similar to the Harish-Chandra homomorphisms used in chapter VI §5.

Recall that we have defined an involution " $^\vee$ " on $\mathbf{G}$, which sends the element with blocks (A,B,C,D) to the element with blocks $({}^tD, -{}^tB, -{}^tC, {}^tA)$. This involution clearly preserves K, M, L, T and therefore induces involutions on the Hecke algebras $\mathcal{H}(G)$, $\mathcal{H}(M)$, $\mathcal{H}(T)$ via $\phi \mapsto \phi^\vee$ with $\phi^\vee(x) = \phi(x^\vee)$.

We note first that the involution on $\mathcal{H}(G)$ is actually the identity. This follows from the following facts: every K-double-class contains an element in T, and

$$J \cdot \operatorname{diag}(a_1, \ldots, a_g, d_1, \ldots, d_g) \cdot J^{-1} = \operatorname{diag}(d_1, \ldots, d_g, a_1, \ldots, a_g)\ ,$$

where $J \in K$ has blocks $(0, 1_g, -1_g, 0)$.

We claim that all three Satake isomorphisms commute with this involution. Let us check this claim for S_M^G. The other two cases are similar. Let $\phi \in \mathcal{H}(G,\mathbb{R})$, we compute:

$$\begin{aligned} S_M^G(\phi)^\vee(m) &= e^{\rho_n}(m^\vee) \cdot \int_U \phi(m^\vee \cdot u)\, du \\ &= e^{\rho_n}(m^\vee) \cdot \int_U \phi(m^\vee \cdot u^\vee)\, du \quad \text{because } {}^\vee \text{ preserves Haar measure} \\ &= e^{\rho_n}(m^\vee) \cdot \int_U \phi(um)\, du \quad \text{because } {}^\vee \text{ acts trivially on } \mathcal{H}(G,\mathbb{R}) \\ &= e^{\rho_n}(m^\vee) \cdot e^{2\rho_n}(m) \int_U \phi(mu)\, du \quad \text{write } um = m \cdot m^{-1}um \text{ in above} \\ &= e^{\rho_n}(m) \cdot \int_U \phi(mu)\, du = S_M^G(\phi)(m)\ . \end{aligned}$$

Notice that we cannot use $\mathbb{Q}$-coefficients in the above because of the use of e^ρ (the best we can do is to use $\mathbb{Q}(\sqrt{p})$-coefficients. However if we suitably twist the $W_{\mathbf{G}}$-action and the involution on $\mathcal{H}(T,\mathbb{Q})$ "by a translation with ρ", and correspondingly twist the Satake homomorphisms, we can indeed use $\mathbb{Q}$-coefficients. More precisely:

Define modified Satake homomorphisms

$$\tilde{S}_M^G : \mathcal{H}(G,\mathbb{Q}) \to \mathcal{H}(M,\mathbb{Q}) \text{ by the rule}$$
$$\tilde{S}_M^G(\phi)(m) = \int_U \phi(um)\,du = e^{2\rho_n}(m)\int_U \phi(mu)\,du\ ,$$
$$\tilde{S}_T^G : \mathcal{H}(G,\mathbb{Q}) \to \mathcal{H}(T,\mathbb{Q}) \text{ by the rule}$$
$$\tilde{S}_T^G(\phi)(t) = \int_N \phi(nt)\,dn = e^{2\rho}(t)\int_N \phi(tn)\,dn\ , \text{ and}$$
$$\tilde{S}_T^M : \mathcal{H}(M,\mathbb{Q}) \to \mathcal{H}(T,\mathbb{Q}) \text{ by the rule}$$
$$\tilde{S}_T^M(\phi)(t) = \int_{N\cap M} \phi(nt)\,dn = e^{2\rho_{\mathbf{M}}}(t)\int_{N\cap M} \phi(tn)\,dn\ .$$

The relation between twisted and untwisted Satake homomorphisms can be summarized by the following commutative diagram of homomorphisms of algebras:

$$\begin{array}{ccccc}
\mathcal{H}(G,\mathbb{R}) & \xrightarrow{S_M^G} & \mathcal{H}(M,\mathbb{R}) & \xrightarrow{S_T^M} & \mathcal{H}(T,\mathbb{R}) \\
\big\downarrow{=} & & \big\downarrow{\cdot e^{\rho_n}} & & \big\downarrow{\cdot e^{\rho}} \\
\mathcal{H}(G,\mathbb{Q})\otimes\mathbb{R} & \xrightarrow[\tilde{S}_M^G\otimes\mathbb{R}]{} & \mathcal{H}(M,\mathbb{Q})\otimes\mathbb{R} & \xrightarrow[\tilde{S}_T^M\otimes\mathbb{R}]{} & \mathcal{H}(T,\mathbb{Q})\otimes\mathbb{R}
\end{array}$$

As before $\tilde{S}_T^G = \tilde{S}_T^G \circ \tilde{S}_M^G$, and all three maps are injective finite homomorphisms of $\mathbb{Q}$-algebras. In fact, $\mathcal{H}(T,\mathbb{Q})$ is finite (but not flat if $g \geq 3$) over (the image of) $\mathcal{H}(M,\mathbb{Q})$ of degree $\mathrm{Card}(W_{\mathbf{M}}) = g!$, $\mathcal{H}(T,\mathbb{Q})$ is finite over (the image of) $\mathcal{H}(G,\mathbb{Q})$ of degree $\mathrm{Card}(W_{\mathbf{G}}) = 2^g \cdot g!$, and therefore $\mathcal{H}(M,\mathbb{Q})$ is finite over (the image of) $\mathcal{H}(G,\mathbb{Q})$ of rank $[W_{\mathbf{G}} : W_{\mathbf{M}}] = 2^g$.

We define twisted involutions on $\mathcal{H}(M,\mathbb{Q})$ and $\mathcal{H}(T,\mathbb{Q})$ by

$$\phi^\tau(m) = e^{2\rho_n}(m)\cdot\phi(m^\vee)\ \forall m \in M,\ \forall\phi \in \mathcal{H}(M,\mathbb{Q})$$
$$\phi^\tau(t) = e^{2\rho}(t)\cdot\phi(t^\vee)\ \forall t \in T,\ \forall\phi \in \mathcal{H}(T,\mathbb{Q})\ .$$

Then all the twisted Satake homomorphisms preserve involutions, and so do the two vertical maps in the diagram above.

We define a new (left) action of $W_{\mathbf{G}} \cong \mathcal{S}_g \ltimes \{\pm 1\}^{2g}$ on $\mathcal{H}(T,\mathbb{Q})$ by

$$(w\cdot\phi)(t) = e^\rho(w^{-1}\cdot t)e^{-\rho}(t)\phi(t)$$

$\forall w \in W_{\mathbf{G}}$, $\forall\phi \in \mathcal{H}(T,\mathbb{Q})$. It is immediate that this indeed defines a representation of $W_{\mathbf{G}}$. Observe that $e^\rho(w^{-1}\cdot t)e^{-\rho}(t) \in \mathbb{Q}^*$ $\forall w \in W_{\mathbf{G}}$, $\forall t \in T$, because we have the (trivial) congruences

$$\sum_{1\le i<j\le g} \pm\lambda_i \pm \lambda_j \equiv \sum_{1\le i<j\le g} \pm\lambda_{\sigma(i)} \pm \lambda_{\sigma(j)} \pmod 2 \ \forall\sigma \in \mathcal{S}_g$$

and

$$\sum_{1\le i\le j\le g} \pm\lambda_i \pm \lambda_j \equiv \sum_{1\le i\le j\le g} \pm\lambda_{\sigma(i)} \pm \lambda_{\sigma(j)} \pmod 2 \ \forall\sigma \in \mathcal{S}_g \ .$$

Using the corresponding property of the Satake isomorphism, it is easy to see that $\tilde{S}_T^G(\mathcal{H}(G,\mathbb{Q}))$ is the subring of $W_{\mathbf{G}}$-invariants in $\mathcal{H}(T,\mathbb{Q})$ under the new (twisted) action. Similarly $\tilde{S}_T^M(\mathcal{H}(M,\mathbb{Q}))$ is the subring of $W_{\mathbf{M}}$-invariants in $\mathcal{H}(T,\mathbb{Q})$ under the twisted action.

So far we have given a fairly explicit description of the structure of and relations between various Hecke algebras (if we recall the explicit description of the characters e^{ρ_n} and e^{ρ} given before).

Frobenius

From now on we will stick to the twisted Satake homomorphisms and the twisted Weyl group actions.

Of special interest for us is the element Frob $\in \mathcal{H}(M,\mathbb{Q})$ defined as the characteristic function of the double class Lm_0L, for the matrix $m_0 = \mathrm{diag}(p\cdot I_g, I_g)$. We claim that it generates the fraction field of $\mathcal{H}(M,\mathbb{Q})$ over that of $\mathcal{H}(G,\mathbb{Q})$.

Before proving the claim, we remark that Frob does not generate $\mathcal{H}(M,\mathbb{R})$ over $\mathcal{H}(G,\mathbb{R})$, as can be seen by explicit computation using the untwisted Satake homomorphisms and the diagram above. Therefore Frob does not generate $\mathcal{H}(M,\mathbb{Q})$ over $\mathcal{H}(G,\mathbb{Q})$.

To prove the claim, it suffices (and is also necessary) to prove that the isotropy subgroup of $\tilde{S}_T^M(\mathrm{Frob})$ in $W_{\mathbf{G}}$, which always contains $W_{\mathbf{M}}$, actually equals $W_{\mathbf{M}}$. First we have to compute $\tilde{S}_T^M(\mathrm{Frob})$. By definition,

$$\begin{aligned}\tilde{S}_T^M(\mathrm{Frob})(t) &= \int_{N\cap M} \mathrm{Frob}(nt)\,dn \\ &= \text{the Haar measure of the open compact subset} \\ &\quad\ N\cap M\cap Lm_0Lt^{-1} \text{ in } N\cap M \ \forall t\in T\ ,\end{aligned}$$

where m_0 is the element $\mathrm{diag}(p\cdot I_g, I_g)$ in M. Since m_0 lies in the center of M, $Lm_0Lt^{-1} = Lm_0t^{-1}$, and therefore $N\cap M\cap Lm_0Lt^{-1}$ is empty unless $m_0t^{-1}\in L$. When $m_0t^{-1}\in L$, $N\cap M\cap Lm_0Lt^{-1} = N\cap M\cap L$ is the maximal compact subgroup with which we normalize the Haar measure. Therefore we conclude that $\tilde{S}_T^M(\mathrm{Frob})$ is the characteristic function of the double coset $(L\cap T)m_0(L\cap T)$. Recalling the definition of twisted $W_{\mathbf{G}}$-action, we see that the only elements in $W_{\mathbf{G}}$ that fix $\tilde{S}_T^M(\mathrm{Frob})$ are the permutations, i.e. elements of $W_{\mathbf{M}}$. This proves the claim.

Finally denote $\mathcal{H}_0(G,\mathbb{Q}) \subset \mathcal{H}(G,\mathbb{Q})$ and $\mathcal{H}_0(M,\mathbb{Q}) \subset \mathcal{H}(M,\mathbb{Q})$ the subalgebras of $\mathbb{Q}$-valued functions ϕ with support contained in the set of $\mathbb{Z}_p$-integral matrices (in the standard representation of $\mathbf{G}$). It is clear that they are stable under the untwisted involutions. Since the twisting in no way changes the support, both $\mathcal{H}_0(G,\mathbb{Q})$ and $\mathcal{H}_0(M,\mathbb{Q})$ are stable under the twisted involutions. Let $[p]$ denote the K-double-class of the diagonal matrix $p \cdot \mathrm{id}$ in $\mathcal{H}_0(G,\mathbb{Q})$. $\tilde{S}_M^G([p])$ is the L-double class of $p \cdot \mathrm{id}$ in $\mathcal{H}_0(M,\mathbb{Q})$ by an easy argument as above, using the fact that $p \cdot \mathrm{id}$ is in the center of G. We denote this element also by $[p]$ by abuse of notation. Since $p \cdot \mathrm{id}$ is in the center of G, the convolution of $[p]$ with the characteristic function of KxK (resp. LmL) is the characteristic function of $KpxK$ (resp. $LpmL$) for any element x in G (resp. m in M). Hence $\mathcal{H}(G,\mathbb{Q}) = \mathcal{H}_0(G,\mathbb{Q})[p]^{-1}$ and $\mathcal{H}(M,\mathbb{Q}) = \mathcal{H}_0(M,\mathbb{Q})[p]^{-1}$.

2. Algebraic and Cohomological Intersections

In this section we give a very brief summary of intersection theory for algebraic cycles and its l-adic realization, in order to explain the action of Hecke operators on the l-adic cohomology of A_g with good coefficient systems. The references are [Fu], [SGA 4], [SGA 4 1/2], [SGA 5].

Suppose k is a field. In [Fu] it is explained how algebraic cycles form a cohomology theory for k-schemes: If X is a smooth irreducible k-scheme, let $Z^d(X)$ denote the group of codimension d algebraic cycles on X. This is a covariant functor for proper morphisms. Dividing by rational equivalence we obtain the group $A^d(X)$. Up to a shift in degrees this is a covariant functor for proper morphisms of smooth irreducible k-schemes. One of the basic problems in algebraic geometry is the construction of intersections $Y \cdot Z \in A^{d+e}(X)$, for cycles $Y \in Z^d(X),\ Z \in Z^e(X)$. This is easiest for Chern classes. For example if $\mathcal{E}$ is a vector bundle on X, the i-th Chern class of $\mathcal{E}$ defines a morphism $c_i(\mathcal{E}) \cup (-) : A^d(X) \to A^{d+i}(X)$: For $Z \in Z^d(X)$ and $\mathcal{L}$ a line bundle $c_1(\mathcal{L}) \cup Z \in A^{d+1}(X)$ is represented by any Cartier divisor on Z which represents the restriction of $\mathcal{L}$ to Z. For $\mathcal{E}$'s of higher rank use projectivized bundles and the splitting principle. (Here "$\cup$" denotes the homological intersection product, corresponding to cohomological cup product.)

Finally if $Z \subset X$ has codimension d, and $Y \subset X$ a locally complete intersection of codimension e, we can define an intersection cycle $Y \cdot Z \in A^{d+e}(X)$, as follows: Let $\tilde{X}$ denote the blow-up of X along Y, and $\tilde{Z} \subset \tilde{X}$ the proper transform of Z. Furthermore let $E \subset \tilde{X}$ be the exceptional divisor, $F = E \cap \tilde{Z}$, and $\mathcal{E}$ the canonical bundle on E of rank $e-1$ which is the kernel of the projection onto $\mathcal{O}(1)$. Then $Y \cdot Z \in A^{d+e}(X)$ is the direct image under $\tilde{X} \to X$ of $c_{e-1}(\mathcal{E}^\vee) \cup F$. This construction actually gives a cycle contained in $Y \cap Z$ which represents $Y \cdot Z$, and is unique up to rational equivalence in $Y \cap Z$. In case Y and Z intersect properly, we recover the definition of $Y \cdot Z$ in terms of the "Tor formula", see [Se 1].

Intersections also can be defined via étale cohomology. Let k be a field, l be a prime invertible in k. If X is a smooth k-scheme, and $Z \subset X$ a subvariety of codimension e, we have natural morphisms $H^i(Z, \mathbb{Q}_l) \to H_Z^{i+2e}(X, \mathbb{Q}_l(e))$ and the image of $1 \in H^0(Z, \mathbb{Q}_l)$ is the class $c_Z \in H_Z^{2e}(X, \mathbb{Q}_l(e))$, the fundamental class of Z. Its main properties can be described as follows:

(i) if Z is a Cartier divisor, the image of c_Z in $H^2(X, \mathbb{Q}_l(1))$ is $c_1(\mathcal{O}(Z))$.

(ii) if $Z \subset X$ is locally a complete intersection of codimension e, $f : X^* \to X$ a morphism such that $f^{-1}(Z)$ is purely of codimension e in X^*, then the characteristic class of $f^{-1}(Z)$ is the pull-back of c_Z.

(iii) For general Z purely of dimension d and codimension e in X, recall that there is a trace map $\int_Z : H_!^{2d}(Z, \mathbb{Q}_l(d)) \to \mathbb{Q}_l$. If $Z \subset X$, for any $\alpha \in H^{2d}(X, \mathbb{Q}_l(e))$ such that "the restriction of α to Z has compact support", $\int_Z \alpha = \int_X c_Z \cup \alpha$.

Here the condition "the restriction of α to Z has compact support" means the following: there exists a closed subscheme $E \subset X$ such that $E \cap Z$ is complete, and α lies in the image of $H_E^{2d}(X, \mathbb{Q}_l(e))$ in $H^{2d}(X, \mathbb{Q}_l(e))$. An example of such a situation is the following: X is a product $X = X_1 \times X_2$, Z is proper over X_2, and α is the cup product of a class in $H^*(X_1)$ with a class in the image of $H_!^*(X_2)$.

We want to relate the two theories above. So again let k be a field, l invertible in k, X a k-scheme geometrically irreducible and smooth over k. Suppose $Z \subset X$ is a closed subvariety purely of codimension d, and $Y \subset X$ is a smooth subvariety of codimension e. The algebraic intersection $Y \cdot Z$ defines a class $c_{Y \cdot Z}$ in $H_{Y \cap Z}^{2(d+e)}(X, \mathbb{Q}_l(d+e))$. We claim that it is equal to the cup product of c_Y and c_Z: $c_E \cap c_{e-1}(\mathcal{E}^\vee)$ is the preimage of c_Y under $\tilde{X} \to X$, $\tilde{X}$ the blow-up of X in Y. It follows that the cup product $c_Y \cap c_Z$ is represented by the image of $c_F \cap c_{e-1}(\mathcal{E}^\vee)$ under $\tilde{X} \to X$ (notation as before). For such projections the cycle map is compatible with direct images, so the assertion follows.

We need to generalize this. Let k, l, X be as before. Suppose $\mathbb{V}_l$ is a locally constant $\mathbb{Q}_l$-adic sheaf on X. We we can construct intersection theories for pairs consisting of a cycle Z in X and a section of $\mathbb{V}_l$ over Z: We have direct images, and if Y is another cycle in X which happens to be a locally complete intersection, with a global section of $\mathbb{W}_l$ over Y, the intersection $Y \cdot Z$ is a cycle contained in $Y \cap Z$, with a section of $\mathbb{V}_l \otimes \mathbb{W}_l$, well determined up to rational equivalence. This corresponds to the cup product for the characteristic classes lying in $H_Z^{2d}(X, \mathbb{V})$ and $H_Y^{2e}(X, \mathbb{W})$ respectively.

Now let us explain algebraic correspondences. Suppose that X is smooth and geometrically irreducible over k, of dimension d, $\mathbb{V}_l$ a locally constant $\mathbb{Q}_l$-adic sheaf on X. Furthermore assume that $Z \subset X \times X$ is an irreducible closed subvariety of codimension d, such that both projections pr_i of Z to X are proper. Suppose furthermore that we have on Z a morphism $\phi^* : \mathrm{pr}_2^*(\mathbb{V}_l) \to \mathrm{pr}_1^*(\mathbb{V}_l)$. Then the pair (Z, ϕ^*) operates on $H^*(X, \mathbb{V}_l)$ and $H_!^*(X, \mathbb{V}_l)$, as follows: By Poincaré duality $H^n(X, \mathbb{V}_l)$ and $H_!^{2d-n}(X, \mathbb{V}_l^\vee(d))$ are dual, so it suffices to define a pairing between them. We use the pairing $(\alpha, \beta) \mapsto \int_Z \langle \phi(\mathrm{pr}_2^*(\alpha)) \times \mathrm{pr}_1^*(\beta) \rangle : H^n(X, \mathbb{V}_l) \times H_!^{2d-n}(X, \mathbb{V}_l^\vee(d)) \to H^n(Z, \mathbb{V}_l) \times H_!^{2d-n}(Z, \mathbb{V}_l^\vee(d)) \to$

$H_!^{2d}(Z, \mathbb{Q}_l(d)) \to \mathbb{Q}_l$. This defines an operation on $H^n(X, \mathbb{V}_l)$. We define an operation on $H_!^*(X, \mathbb{V}_l)$ similarly.

Another description can be given as follows: Suppose $j : X \subset \overline{X}$ is a smooth compactification. Then Z is a closed subset in $X \times \overline{X}$, as pr_1 is proper on Z. So we get a class c_{Z,ϕ^*} in $H_Z^{2d}(X \times \overline{X}, \mathrm{pr}_1^* \mathbb{V}_l \otimes \mathrm{pr}_2^* j_!(\mathbb{V}_l^\vee)(d))$. It maps to a class in $H^{2d}(X \times \overline{X}, \mathrm{pr}_1^* \mathbb{V}_l \otimes \mathrm{pr}_2^* j_!(\mathbb{V}_l^\vee)(d))$, and the pairing above is obtained by taking the cup product of $\mathrm{pr}_1^*(\beta) \otimes \mathrm{pr}_2^*(\alpha) \in H^{2d}(\overline{X} \times X, \mathrm{pr}_1^* j_!(\mathbb{V}_l^\vee) \otimes \mathrm{pr}_2^* \mathbb{V}_l)$ with this class and integrating over $\overline{X} \times \overline{X}$.

The product of two correspondences (Z_1, ϕ_1^*) and (Z_2, ϕ_2^*) is $\mathrm{pr}_{14*}((Z_1 \times Z_2) \cdot \Delta, \phi_2^* \circ \phi_1^*)$: Here $\Delta \subset X \times X \times X \times X$ is the locus where the second and the third coordinate coincide, the intersection takes place in $X \times X \times X \times X$, and pr_{14} means projection onto the two extreme factors. The product of correspondences induces the product operation on cohomology: That comes down to checking that in $H^{6d}(X \times \overline{X} \times X \times \overline{X}, \mathrm{pr}_1^* \mathbb{V}_l \otimes \mathrm{pr}_2^* j_!(\mathbb{V}_l^\vee) \otimes \mathrm{pr}_3^* \mathbb{V}_l \otimes \mathrm{pr}_4^* j_!(\mathbb{V}_l^\vee)(3d))$, the cup-product of $c_\Delta \in H_\Delta^{2d}(X \times \overline{X}, \mathrm{pr}_1^* \mathbb{V}_l \otimes \mathrm{pr}_2^* j_!(\mathbb{V}_l^\vee(d)))$ with the image of $c_{Z_1} \otimes c_{Z_2} \in H_{Z_1}^{2d}(X \times \overline{X}, \mathrm{pr}_1^* \mathbb{V}_l \otimes \mathrm{pr}_2^* j_!(\mathbb{V}_l^\vee)(d)) \otimes H_{Z_2}^{2d}(X \times \overline{X}, \mathrm{pr}_1^* \mathbb{V}_l \otimes \mathrm{pr}_2^* j_!(\mathbb{V}_l^\vee)(d))$ in $H^{4d}(X \times \overline{X} \times X \times \overline{X}, \mathrm{pr}_1^* \mathbb{V}_l \otimes \mathrm{pr}_2^* j_!(\mathbb{V}_l^\vee) \otimes \mathrm{pr}_3^* \mathbb{V}_l \otimes \mathrm{pr}_4^* j_!(\mathbb{V}_l^\vee)$, is equal to the direct image of the characteristic class of $((Z_1 \times Z_2) \cdot \Delta, \phi_2^* \circ \phi_1^*)$ (under $H^{6d}_{(Z_1 \times Z_2) \cdot \Delta}(X \times \overline{X} \times X \times \overline{X}, \mathrm{pr}_1^* \mathbb{V}_l \otimes \mathrm{pr}_2^* j_!(\mathbb{V}_l^\vee) \times \mathrm{pr}_3^* \mathbb{V}_l \otimes \mathrm{pr}_4^* j_!(\mathbb{V}_l^\vee)(3d) \to H^{6d}(X \times \overline{X} \times X \times \overline{X}, \mathrm{pr}_1^* \mathbb{V}_l \otimes \mathrm{pr}_2^* j_!(\mathbb{V}_l^\vee) \otimes \mathrm{pr}_3^* \mathbb{V}_l \otimes \mathrm{pr}_4^* j_!(\mathbb{V}_l^\vee)(3d)))$. But this follows from the comparison of algebraic and cohomological intersections, on the ambient scheme $X \times \overline{X} \times X \times \overline{X}$.

Finally we want to apply this to $X = \mathsf{A}_g \otimes k$ or $X = \mathsf{A}_{g,n} \otimes k$, k a prime field, l invertible in k. Of course this is a stack and not a scheme, but if we calculate with intersection numbers in $\mathbb{Q}$ we may introduce a level-m structure, $m \geq 3$, and may pass over to the corresponding covering, which is even quasi-projective. Our compactifications are smooth, so the previous discussion applies. The correspondences will come from generically finite proper maps $Z \to X \times X$, and isogenies $\mathrm{pr}_1^*(\mathsf{A}) \to \mathrm{pr}_2^*(\mathsf{A})$ on Z, A the universal abelian variety. In general $Z \to X \times X$ is not an embedding. To reduce to the previous case of a subvariety, one takes the direct image under $Z \to X \times X$ (as cycles).

3. Hecke Operators as Algebraic Correspondences Outside Characteristic p

3.1. Definition. A *p-isogeny* of two principally polarized abelian schemes $(A_1 \to S, \lambda_1)$ and $(A_2 \to S, \lambda_2)$ of relative dimension g is an isogeny $f : A_1 \to A_2$ over S, such that the degree of f is a power of p, and such that the pull-back $f^* \lambda_2$ of the polarization on A_2 is a multiple of the polarization on A_1. Clearly this multiple is a power of p: $f^* \lambda_2 = p^e \cdot \lambda_1$ for some $e \in \mathbb{N}$, and $\deg(f) = p^{eg}$.

In the situation above let S be the spectrum of an algebraically closed field of characteristic different from p. The principal polarizations λ_1, λ_2 define non-degenerate symplectic pairings on the p-adic Tate-modules $T_p(A_1)$, $T_p(A_2)$:

$$\langle\ ,\ \rangle_{T_p(A_i)} : T_p(A_i) \times T_p(A_i) \to \mathbb{Q}_p(1),\ i = 1, 2\ .$$

Note that $\mathbb{Q}_p(1)$ is isomorphic to $\mathbb{Q}_p$, each isomorphism corresponding to a choice of compatible system of p-th power roots of unity. The p-isogeny f induces a symplectic similitude on the Tate modules: $T_p(A_1) \to T_p(A_2)$, which after choice of (integral) symplectic basis is given by an element $\gamma \in G$. Observe that different choices of trivializations of $\mathbb{Q}_p(1)$ only change γ by (an element of $\mathbb{Z}_p^\times$) $\cdot\, I_{2g}$ and different choices of symplectic basis of $T_p(A_1)$, $T_p(A_2)$ only affect γ via left and right multiplication by elements of K. So the double-class $K\gamma K$ is well-defined by f and is called the *type* of the p-isogeny f. Note that in this way we only obtain double-classes $K\gamma K$ with an *integral* matrix.

If $K\gamma K$ is the the type of the p-isogeny f, the type of the dual isogeny f^t : $A_2 \to A_1$ is $K \cdot {}^t\gamma \cdot K$. Notice that transposition stabilizes K hence induces an involution on $\mathcal{H}(G, \mathbb{Q})$. We claim that this involution is trivial, or equivalently $K\gamma K = K \cdot {}^t\gamma \cdot K$ for any g in G. But this is obvious since every K-double-class contains an element of T (Cartan decomposition).

Let f be a p-isogeny over an arbitrary base S over $\mathbb{Z}[1/p]$. The types of the geometric fibres $f_{\overline{s}}$ of f are locally constant (w.r.t. étale, flat, or Zariski topology) on S, so over each connected component of S the isogeny has a well defined type.

Let Isog_g be the moduli stack of isogenies between principally polarized abelian schemes of relative dimension g, so that for an arbitrary base S, $\mathrm{Isog}_g(S)$ is the category in groupoids whose objects are the isogenies $\phi : A_1 \to A_2$ over S between principally polarized abelian schemes $(A_1 \to S, \lambda_1)$ and $(A_2 \to S, \lambda_2)$ of relative dimension g. Assigning to f its source $(A_1 \to S, \lambda_1)$ (resp. target $(A_2 \to S, \lambda_2)$) defines a morphism $\mathrm{pr}_1 : \mathrm{Isog}_g \to \mathsf{A}_g$ (resp. $\mathrm{pr}_2 : \mathrm{Isog}_g \to \mathsf{A}_g$). These two morphisms define a map $\mathrm{pr} = (\mathrm{pr}_1, \mathrm{pr}_2) : \mathrm{Isog}_g \to \mathsf{A}_g \times \mathsf{A}_g$.

The morphism pr allows us to understand Isog_g and reconstruct it, as follows: Consider the product $\mathsf{A}_g \times \mathsf{A}_g$ of A_g with itself over $\mathrm{Spec}(\mathbb{Z})$. Over it we have two abelian schemes $A_1 = \mathrm{pr}_1^*(A)$ and $A_2 = \mathrm{pr}_2^*(A)$ by pulling back the universal abelian scheme A via the two projections. Isogenies between A_1 and A_2 are given by their graphs, and so representable by a relative scheme $\mathrm{Isog}_g \to \mathsf{A}_g \times \mathsf{A}_g$ by the theory of Hilbert schemes.

The degree of the universal isogeny is locally constant on Isog_g. If we bound this degree the corresponding open-and-closed subset is unramified over $\mathsf{A}_g \times \mathsf{A}_g$ (ch. I, 1.7). Also the two projections of Z to A_g are proper by the valuative criterion: If an abelian variety A_1 is defined over a discrete valuation ring V, with fraction field K, any K-isogeny $A_1 \to A_2$ extends to a homomorphism over V (ch. I, prop. 2.7), which must be an isogeny by torsion point considerations. Notice that in general the projection maps are not finite, because in characteristic p we can have a non-trivial family of finite unipotent subgroup schemes inside an abelian variety. We can say more for separable isogenies, e.g.

for each component Z of Isog_g, the two projections to A_g are both finite étale over $\mathsf{A}_g[1/\deg(Z)]$.

The composition of isogenies defines a pairing

$$\mathrm{Isog}_g \times_{\mathrm{pr}_2, \mathsf{A}_g, \mathrm{pr}_1} \mathrm{Isog}_g \to \mathrm{Isog}_g \ ,$$

which sends a pair $\phi : A_1 \to A_2$ and $\psi : A_2 \to A_3$ of isogenies to their composition $\psi \circ \phi$. (Notice the order of composition here.) There is also an involution on Isog_g, by sending an isogeny $f : A_1 \to A_2$ to $f^\vee = \lambda_1^{-1} \circ f^t \circ \lambda_2 : A_2 \to A_1$, which is the dual of f after identifying A_1 and A_2 with their duals via the principal polarizations.

Over Isog_g we have the universal isogeny, and the p-isogenies live over an open and closed (relative) subscheme classifying p-isogenies, which we denote by p-Isog_g. Clearly the two projections of p-Isog_g to A_g are both étale over $\mathbb{Z}[1/p]$. It is equally clear that p-Isog_g is stable under composition and involution as defined above.

We will be interested in the action of Hecke algebra $\mathcal{H}(G, \mathbb{Q})$ on various cohomology groups of A_g over the base S, where S is the spectrum of $\mathbb{Z}[1/p]$, $\mathbb{F}_l$ ($l \neq p$), $\mathbb{Q}$ or $\mathbb{C}$, or more generally any scheme over $\mathbb{Z}[1/p]$. Let us first explain the general principle, which is simple (but a little vague):
Let $\mathbb{Z}[\mathrm{Isog}_g/S]$ be the free abelian group generated by the connected components of Isog_g, $\mathbb{Q}[\mathrm{Isog}_g/S] = \mathbb{Z}[\mathrm{Isog}_g/S] \otimes \mathbb{Q}$. The composition rule above gives $\mathbb{Q}[\mathrm{Isog}_g/S]$ and $\mathbb{Z}[\mathrm{Isog}_g/S]$ ring structures: use the product on the level of algebraic cycles. Explicitly, if Z, W are two connected components, let $\pi : Z \times_{\mathsf{A}_g} W \to \mathrm{Isog}_g$ be the composition morphism. Then the product $[Z] \cdot [W]$ in $\mathbb{Z}[\mathrm{Isog}_g/S]$ is the image of π as an algebraic cycle, taking multiplicities into account.

The free abelian group $\mathbb{Z}[p\text{-}\mathrm{Isog}_g/S]$ generated by the connected components of p-Isog_g is naturally contained in $\mathbb{Z}[\mathrm{Isog}_g/S]$ as a subring. Denote $\mathbb{Z}[p\text{-}\mathrm{Isog}_g/S] \otimes \mathbb{Q}$ by $\mathbb{Q}[p\text{-}\mathrm{Isog}_g/S]$, which is naturally contained in $\mathbb{Q}[p\text{-}\mathrm{Isog}_g/S]$.

We shall define a homomorphism $h : \mathcal{H}_0(G, \mathbb{Q}) \to \mathbb{Q}[p\text{-}\mathrm{Isog}_g/S]$ "in a natural way". As p is invertible in $\mathcal{O}_S$, h is defined by p-isogeny type considerations. We shall consider this in more detail later, but remark here that formally we should get an anti-homomorphism because of the ordering in the definition in the composition product of isogenies, but the fact that $\mathcal{H}(G, \mathbb{Q})$ is commutative saves us from this extra complication.

Let H^* be a "reasonable" $\mathbb{Q}$-valued cohomology theory, for instance l-adic cohomology, coherent cohomology, Betti cohomology, p-adic étale cohomology, de Rham cohomology, etc.. The algebra $\mathbb{Q}[p\text{-}\mathrm{Isog}_g/S]$ "acts by functoriality" on $H^*(\mathsf{A}_g, \mathbf{F})$, where $\mathbf{F}$ is a suitable sheaf of coefficients. Then $\mathcal{H}_0(G, \mathbb{Q})$ acts on $H^*(\mathsf{A}_g)$ via $h : \mathcal{H}(G, \mathbb{Q}) \to \mathbb{Q}[p\text{-}\mathrm{Isog}_g/S]$. This action can be extended to $\mathcal{H}(G, \mathbb{Q})$ by inverting $[p]$.

The definition of the action of $\mathbb{Q}[p\text{-}\mathrm{Isog}_g/S]$ on $H^*(\mathsf{A}_g, \mathbf{F})$ is given as follows: Let Z be a connected component of p-Isog_g/S, ϕ the universal isogeny over Z, and $\phi^* : \mathrm{pr}_2^* \mathbf{F} \to \mathrm{pr}_1^* \mathbf{F}$ is a morphism of sheaves naturally attached to ϕ. (The existence of ϕ^* is a property of $\mathbf{F}$ which we assume.) The following composition gives the action of (Z, ϕ^*) on $H^*(\mathsf{A}_g, \mathbf{F})$:

$$H^*(\mathsf{A}_g, \mathbf{F}) \xrightarrow{\mathrm{pr}_2^*} H^*(Z, \mathrm{pr}_2^* \mathbf{F}) \xrightarrow{\phi^*} H^*(Z, \mathrm{pr}_1^* \mathbf{F}) \xrightarrow{\mathrm{tr}} H^*(\mathsf{A}_g, \mathbf{F}),$$

where $\mathrm{tr} : H^*(Z, \mathrm{pr}_1^* \mathbf{F}) \longrightarrow H^*(\mathsf{A}_g, \mathbf{F})$ is the trace map attached to pr_1, whose existence is part of the assumption that the cohomology theory H^* is "reasonable".

Since ϕ^* is usually given naturally, we shall say by abuse of terminology that $\mathbb{Q}[p\text{-Isog}_g/S]$ operates on $H^*(A_g, \mathbf{F})$. The compatibility with composition is checked by writing down a suitable diagram, and using the property that the trace map for a composition is the composition of the traces. Details are left to the reader. A basic example is given by the sheaves $\mathbb{V}_l(\lambda)$ in l-adic étale cohomology, where ϕ^* is induced by the isomorphism $T_l(A_1) \cong T_l(A_2)$, and the trace map is the usual one.

We remark that since the above operators are given by algebraic cycles, they preserve all natural structures on H^*, e.g. mixed Hodge structure (for Betti cohomology), Galois action, weight filtration, crystalline structure, etc. We shall also see that these operators are self-adjoint w.r.t. duality (Serre duality for coherent cohomology, Poincaré duality for Betti, étale, crystalline cohomology), since the dual isogeny of ϕ lies in the same component of $\mathrm{Isog}/\mathbb{Z}[1/p]$ as ϕ. This corresponds to the fact that the involution on $\mathcal{H}_0(G, \mathbb{Q})$ induced by " $\vee$ " is trivial.

For the rest of this section, we will consider p-isogenies and assume that p is invertible in $\mathcal{O}_S$. We shall define $h : \mathcal{H}_0(G, \mathbb{Q}) \to \mathbb{Q}[p\text{-Isog}_g/S]$, compatibly with base change of schemes over $\mathbb{Z}[1/p]$. Therefore it suffices to define it for $S = \mathrm{Spec}\mathbb{Z}[1/p]$. One can define h for S a spectrum of a field k of characteristic $\neq p$ in a similar way, naturally compatible with specialization from $\mathrm{Spec}\mathbb{Z}[1/p]$.

Definition of $h : \mathcal{H}_0(G, \mathbb{Q}) \to \mathbb{Q}[p\text{-Isog}_g/S]$. To any double-class $K\gamma K$ with γ an integral matrix contained in G, we associated $Z(K\gamma K) :=$ the union of the connected components of $p\text{-Isog}_g$ where the isogeny is a p-isogeny of type $K\gamma K$. In fact we obtain only one connected component. To prove this claim (which will not be needed in the sequel) it suffices to check over the complex numbers since each connected component of $p\text{-Isog}_g/\mathrm{Spec}\mathbb{Z}[1/p]$ is finite étale over A_g. Over $\mathbb{C}$, we have only to check that $Z(K\gamma K)(\mathbb{C})$ is connected in the complex topology, which is easy: A_g is the quotient $\Gamma \backslash \mathbb{H}_g$ with $\Gamma = \mathrm{Sp}(2g, \mathbb{Z})$, while $Z(K\gamma K)$ is isomorphic to $(\Gamma \cap \gamma^{-1}\Gamma\gamma)\backslash \mathbb{H}_g$, where the two projections are induced by the identity on $\mathbb{H}_g$ the (left) action of g on $\mathbb{H}_g$ respectively. (The reader can consult the following section for more details concerning the complex uniformized situation, which is elementary, so there is no circle in our argument.) This shows that $Z(K\gamma K)$ is indeed connected. In fact this argument shows that its geometric fibres are all connected.

The above assignment defines a $\mathbb{Q}$-linear map $h : \mathcal{H}_0(G, \mathbb{Q}) \to \mathbb{Q}[p\text{-Isog}_g] =$ the $\mathbb{Q}$-vector-space spanned by the algebraic cycles of relative dimension $d = g(g+1)/2$ on $p\text{-Isog}_g$. For example $[p] \in \mathcal{H}_0(G, \mathbb{Q})$ maps to the isogeny "multiplication by p". "Set-theoretically" the action of a double-class $K\gamma K$ sends A_1

to the set of all A_2 such that there exists an isogeny $\phi : A_1 \to A_2$ of type $K\gamma K$. We remark that the previous argument shows that h is in fact an isomorphism for S the spectrum of $\mathbb{Z}[1/p]$, or any field of characteristic $\neq p$.

We claim that this mapping is compatible with products, where the ring structure was before. It suffices to check this over the complex numbers, where things are rather obvious (for example from the "set-theoretic" description). We call the reader's attention to the remark above that the commutativity of $\mathcal{H}_0(G, \mathbb{Q})$ saves us the nuisance of h being formally an anti-homomorphism.

Now $\mathcal{H}(G, \mathbb{Q})$ acts on the étale cohomology sheaves $Rf_*\mathbb{V}_l$ and $Rf_!\mathbb{V}_l$ attached to $\mathbf{G}$-representations V, as before via "pull-back and push-forward", where $f : \mathsf{A}_g[1/p\ell] \to \mathrm{Spec}(\mathbb{Z}[1/p\ell])$ is the structural morphism. In particular, the $\mathcal{H}(G, \mathbb{Q})$-action on the étale cohomology groups $H^*(\mathsf{A}_g \otimes \overline{\mathbb{Q}}, \mathbb{V}_l)$ and $H_!^*(\mathsf{A}_g \otimes \overline{\mathbb{Q}}, \mathbb{V}_l)$ commutes with the $\mathrm{Gal}(\overline{\mathbb{Q}}/\mathbb{Q})$-action. Moreover, $\mathcal{H}(G, \mathbb{Q})$ is self-adjoint for Poincaré duality. For the other cohomology theories one still has to work out some details: The problem is that we need some extension of the correspondences to the compactification $\overline{X}$. The easiest way is to normalize $\overline{X} \times \overline{X}$ in Isog_g. If we consider isogenies of degree m the structure at the boundary of this normalization can be checked via toroidal methods in characteristics not dividing m (for more details see the complex version below), and it is formally isomorphic to a suitable torus embedding (which may not be smooth). This way one can construct the necessary pull-backs and traces and obtain an action of $\mathcal{H}(G, \mathbb{Q})$ on the $\mathbb{Q}$-valued logarithmic de Rham cohomology $H^*_{\mathrm{DR}}(\overline{X}, \overline{\mathcal{V}})$ and on the coherent cohomology groups $H^*(\overline{X} \otimes \mathbb{Q}, \overline{\mathcal{E}})$, for vector bundles $\mathcal{E}$ given by representations E of $\mathbf{Q}$. This action preserves the relations between various theories: That is, the action on de Rham cohomology induces on the associated graded the action on coherent cohomology, which corresponds via Hodge-Tate to the action on p-adic étale cohomology. On the crystalline side (at the prime p) it is not quite obvious how to define the Hecke action, as we might not be able to find good compactifications of Isog_g in characteristic p. However we at least obtain an action of the Hecke algebras away from p, compatible with the action on p-adic étale cohomology.

Complex Uniformized Situation

In this subsection we take the base to be $\mathbb{C}$ and exploit the complex uniformization of $\mathsf{A}_g/\mathbb{C}$.

Let us first review some notation. As before $\mathbb{H}_g$ denotes the Siegel upper half spacc. A typical element will be denoted by Ω, so that Ω is a symmetric $g \times g$ matrix with coefficients in $\mathbb{C}$ with $\mathrm{Im}\Omega$ positive-definite. The universal family A of principally polarized abelian varieties with a symplectic basis of $H_1(A, \mathbb{Z})$ is defined over $\mathbb{H}_g$, and the fibre A_Ω over $\Omega \in \mathbb{H}_g$ is the quotient of $\mathbb{C}^g$ by the lattice $\Omega \cdot \mathbb{Z}^g + \mathbb{Z}^g$.

Let $\mathbf{G}(\mathbb{R})^+$ be the connected component of $\mathbf{G}(\mathbb{R})$, so that a typical element γ in $\mathbf{G}(\mathbb{R})^+$ is a real $g \times g$ matrix in block the form

$$x = \begin{pmatrix} A & B \\ C & D \end{pmatrix}$$

with $A\cdot{}^tB-B\cdot{}^tA=C\cdot{}^tD-D\cdot{}^tC=0,\ A\cdot{}^tD-B\cdot{}^tC=\eta(x)\cdot I_g,\ \eta(x)>0$ (equivalently, ${}^tD\cdot B-{}^tB\cdot D={}^tC\cdot A-{}^tA\cdot C=0,\ {}^tD\cdot A-{}^tB\cdot C=\eta(x)\cdot I_g$). $\mathbf{G}(\mathbb{R})^+$ acts on $\mathbb{H}_g$ via the familiar formula $x\cdot\Omega=(A\Omega+B)\cdot(C\Omega+D)^{-1}$, where x is in block form as above.

We define an automorphism σ of order 2 on $\mathbf{G}(\mathbb{R})$ by

$$x=\begin{pmatrix}A&B\\C&D\end{pmatrix}\mapsto x^\sigma=\begin{pmatrix}D&-C\\-B&A\end{pmatrix}=\eta(x)\cdot{}^t\gamma^{-1}=J\cdot x\cdot J^{-1}$$

$$\text{with } J=\begin{pmatrix}0&I_g\\-I_g&0\end{pmatrix},$$

which obviously stabilizes $\mathbf{G}(\mathbb{R})^+$. We also define a $\mathrm{GL}_g(\mathbb{C})$-valued 1-cocycle on $(\mathbf{G}(\mathbb{R})^+,\mathbb{H}_g)$:

$$J(x,\Omega)=\eta(x)\cdot{}^t(C\cdot\Omega+D)^{-1}\ \forall x\in\mathbf{G}(\mathbb{R})^+,\ \forall\Omega\in\mathbb{H}_g\ .$$

The cocycle rule $J(x\cdot y,\Omega)=J(x,y\cdot\Omega)\cdot J(y,\Omega)$ can be checked by straightforward computation, using the fact that transpose-inverse defines an automorphism of $\mathrm{GL}_g(\mathbb{C})$. Note that $J(c\cdot x,\Omega)=c\cdot J(x,\Omega)\ \forall c\in\mathbb{R}^\times$.

Now we introduce the useful formula

$$(*)\qquad J(x,\Omega)\cdot(\Omega,I_g)=(x\cdot\Omega,I_g)\cdot x^\sigma\ \forall x\in\mathbf{G}(\mathbb{R})^+,\ \forall\Omega\in\mathbb{H}_g\ .$$

Here (Ω,I_g) and $(x\cdot\Omega,I_g)$ are regarded as $g\times 2g$ matrices in block form. This formula is actually very easy to prove: It is equivalent to

$${}^t(C\cdot\Omega+D)^{-1}\cdot(\Omega,I_g)=(x\cdot\Omega,I_g)\cdot{}^tx^{-1}\ .$$

Taking transpose-inverse reduces this to a triviality, remembering that Ω and $x\cdot\Omega$ are symmetric.

We first explain the significance of the above formula. We will take x to be an element in $\mathbf{G}(\mathbb{Q})\cap\mathbf{G}(\mathbb{R})^{\neq}$ and write γ instead of x. Let n_1,n_2 be an arbitrary elements in $\mathbb{Q}^g$ and Ω any period matrix in $\mathbb{H}_g$. The formula (*) gives:

$$J(\gamma,\Omega)\cdot(\Omega,I_g)\cdot\begin{pmatrix}n_1\\n_2\end{pmatrix}=(\gamma\cdot\Omega,I_g)\cdot\gamma^\sigma\cdot\begin{pmatrix}n_1\\n_2\end{pmatrix}\ .$$

This tells us

(a) The linear transformation $J(\gamma,\Omega):\mathbb{C}^g\to\mathbb{C}^g$ induces an isogeny $f(\gamma): A_\Omega\to A_{\gamma\cdot\Omega}$ if γ is an integral matrix. *Assume this is the case.* Then the pull-back of the polarization of $A_{\gamma\cdot\Omega}$ is equal to $\eta(\gamma)\cdot$(the principal polarization of A_Ω, because the involution " $^\vee$ " defined in 4.1 corresponds to the dual isogeny and $\gamma\cdot\gamma^\vee=\gamma^\vee\cdot\gamma=\eta(\gamma)\cdot I_{2g}$. Notice that this simple construction defines a component $\Gamma\cap\gamma^{-1}\Gamma\gamma\backslash\mathbb{H}_g$ of $\mathrm{Isog}(\mathbb{C})$ of the universal family of abelian varieties over $\Gamma\backslash\mathbb{H}_g$ of degree $\lambda(\gamma)^g$. It is easy to see that every component of $\mathrm{Isog}(\mathbb{C})$

arises this way. Clearly $\eta(\gamma)$ is a p-isogeny iff the elementary divisors of γ (as an integral matrix) are all powers of p.

(b) The complex uniformization of A_Ω and $A_{\gamma\cdot\Omega}$ gives us canonical isomorphisms A_Ω [torsion] $\cong (\mathbb{Q}/\mathbb{Z})^{2g}$, $A_{\gamma\cdot\Omega}$ [torsion] $\cong (\mathbb{Q}/\mathbb{Z})^{2g}$ which preserve the symplectic pairings ($\mathbb{Q}^{2g}$ being endowed with the standard symplectic pairing). Under these isomorphisms, the map between torsion points of A_Ω and $A_{\gamma\cdot\Omega}$ induced by $f(\gamma)$ is given by γ^σ. (But not γ itself!)

We carry out the "general principle" outlined before for coherent (analytic) cohomology (with coefficients coming from homogeneous holomorphic vector bundles $\mathbb{H}_g$) and singular cohomology (with coefficients in locally constant sheaves on $\Gamma\backslash\mathbb{H}_g$ attached to rational representations of $\mathbf{G}$).

First we deal with coherent cohomology. We will be interested in coefficient sheaves coming from finite dimensional complex rational representations of $\mathbf{Q}/\mathbb{C}$ as in chapter VI. To such a representation E, there is an associated homogeneous holomorphic vector bundle (identified with the coherent sheaf of its local sections) on the compact dual of $\mathbb{H}_g$; its restriction on $\mathbb{H}_g$ will be denoted $\mathcal{E}$. Note that $\mathbf{G}(\mathbb{R})^+$ acts on $\mathcal{E}$. The assignment $E \mapsto \mathcal{E}$ from representations of $\mathbf{Q}$ to $\mathbf{G}(\mathbb{R})^+$ equivariant sheaves on $\mathbb{H}_g$ is compatible with direct sums, duals and tensor products. The (necessary) map ϕ^* exists because of the homogeneity of $\mathcal{E}$: take it to be the transpose of the map induced by γ^{-1} on $\mathcal{W}(\rho)$. The trace maps $H^*(Z, \mathrm{pr}_1^*\,\mathcal{E}) \longrightarrow H^*(\mathsf{A}_g, \mathcal{E})$ obviously exist, because the two projections of Isog/$\mathbb{C}$ to $\mathsf{A}_g/\mathbb{C}$ are finite étale on each connected component of Isog/$\mathbb{C}$). This finishes the definition of the action of $\mathcal{H}_0(G,\mathbb{Q})$ on $H^*(X,\mathcal{E})$. However we don't care much about $H^*(X,\mathcal{E})$. The coherent cohomology of the canonical extensions $\overline{\mathcal{E}}$ (resp. $\overline{\mathcal{E}}(-D)$, $D =$ divisor at ∞) are much more interesting in view of chap. VI thm. 5.5, and we use a variant of the "pull-back then push-forward" procedure to define the action.

Recall that $H^m(\overline{X},\overline{\mathcal{E}})$ is dual to $H^{d-m}(\overline{X},\overline{\mathcal{E}}^\vee \otimes \Omega^d_{\overline{X}}) = H^{d-m}(\overline{X},\mathcal{E}^\vee \otimes \overline{\mathcal{W}}(2\rho^M)(-D)$ by Serre duality, where $d = g\cdot(g+1)/2 = \dim(X)$, and D is the divisor at infinity. Thus to give an action of (Z,ϕ^*) on $H^m(\overline{X},\overline{\mathcal{E}})$, it suffices to give a pairing $H^m(\overline{X},\overline{\mathcal{E}}) \times H^{d-m}(\overline{X},\overline{\mathcal{E}}^\vee \otimes \Omega^d_{\overline{X}}) \to \mathbb{C}$. For this we form a toroidal compactification $\overline{Z}$ of Z, which exists since Z is a quotient of $\mathbb{D}$ by a congruence subgroup of $\mathrm{Sp}(2g,\mathbb{Z})$, and we have Mumford's theory of toroidal compactification. We can choose the cone decompositions used for $\overline{Z}$ such that the two projections extend to maps $\overline{Z} \to \overline{\mathsf{A}}_g$. Now the two pull-backs of $\overline{\mathcal{E}}$ become canonically isomorphic to the canonical extension of $\mathcal{E}(Z)$. We have maps $\mathrm{pr}_1^* : H^m(\overline{A}_g,\overline{\mathcal{E}}) \to H^m(\overline{Z},\overline{\mathcal{E}})$, $\mathrm{pr}_2^* : H^{d-m}(\overline{A}_g,\overline{\mathcal{E}}^\vee \otimes \Omega^d_{\overline{A}}) \to H^{d-m}(\overline{Z},\overline{\mathcal{E}}^\vee \otimes \Omega^d_{\overline{Z}})$, and a trace map $\int_{\overline{Z}} : H^d(\overline{Z},\Omega^d_{\overline{Z}}) \to \mathbb{C}$. Note that ϕ^* extends to $\overline{Z}$. Now we define the required pairing $H^m(\overline{A},\overline{\mathcal{E}}) \times H^{d-m}(\overline{X},\overline{\mathcal{E}}^\vee \otimes \Omega^d_{\overline{X}}) \to \mathbb{C}$ by sending (α,β) to $\int_{\overline{Z}} \phi^*\mathrm{pr}_2^*\alpha \cup \mathrm{pr}_1^*\beta$. The action of (Z,ϕ^*) on $H^m(\overline{A},\overline{\mathcal{E}}(-D))$, D the divisor at infinity, is defined similarly. One checks that these operations are compatible with compositions, hence define an action of $\mathcal{H}_0(G,\mathbb{Q})$ on $H^*(\overline{\mathsf{A}}_g,\overline{\mathcal{E}})$ (resp. $H^*(\overline{\mathsf{A}}_g,\overline{\mathcal{E}}(-D))$). If E is an irreducible representation of $\mathbf{Q}$ (hence an $\mathbf{M}$-representation), this operation maps $[p] \in \mathcal{H}_0(G,\mathbb{Q})$ into multiplication with some power of p (the same power by which the diagonal matrix $p \cdot I$ acts on ρ). By dévissage, $[p]$ acts as an automorphism of $H^*(\overline{\mathsf{A}}_g, \overline{\mathcal{E}})$ (resp. $H^*(\overline{\mathsf{A}}_g, \overline{\mathcal{E}}(-D))$ for general E. It follows that we can extend the action of $\mathcal{H}_0(G, \mathbb{Q})$ to an action of $\mathcal{H}(G, \mathbb{Q})$ on $H^*(\overline{\mathsf{A}}_g, \overline{\mathcal{E}})$ (resp. $H^*(\overline{\mathsf{A}}_g, \overline{\mathcal{E}}(-D))$). This action is natural with respect to maps between $\mathcal{E}$'s which are defined by $\mathbf{Q}$-linear maps between the corresponding E's, and even with respect to homogeneous differential operators. Finally the elements of $\mathcal{H}(G, \mathbb{Q})$ are self-adjoint with respect to Serre duality (which corresponds to passing to dual isogenies). These properties can all be checked using results of chapter VI.

Here are some familiar examples:

(a) When E is equal to the multiplier representation $\eta, \mathcal{E}(\eta)|_{\mathbb{H}_g}$ is trivial as a $\mathrm{Sp}(2g, \mathbb{R})$-homogeneous vector bundle, hence $\mathcal{E}(\eta)$ is trivial as an invertible sheaf on $\Gamma \backslash \mathbb{H}_g$. But as $\mathbf{G}(\mathbb{R})^+$-equivariant invertible sheaf, $\mathcal{E}(\eta)$ is certainly not trivial. In terms of the discussion of chap. IV,5, it is the F-filtration that distinguishes $\mathcal{E}(\eta)$ from $\mathcal{E}$ (trivial). We shall denote $\mathcal{E}(\eta)$ by $\mathbb{C}(1)$, since it corresponds to the Tate twist (either in the context of Galois representations or mixed Hodge structures), as we saw in chapter VI.

(b) When E is the restriction of the standard representation of $\mathbf{G}$ to $\mathbf{Q}$, we get $\mathcal{H}^1_{\mathrm{DR}}(A/\mathbb{H}_g)^\vee \cong \mathcal{H}^1_{\mathrm{DR}}(A/\mathbb{H}_g)(1)$, the sheaf of first relative de Rham cohomology of the universal family with the obvious action induced by the above uniformization.

(c) If E is the inflation of the standard representation of GL_g via

$$\mathbf{Q} \to \mathrm{GL}_g , \qquad \begin{pmatrix} A & B \\ 0 & D \end{pmatrix} \mapsto A ,$$

then $\mathcal{E} = \Omega^1_{A/\mathbb{H}_g}(1)$ is the sheaf of relative holomorphic 1-forms of the universal family (up to twist), which is (canonically isomorphic to) the sheaf of the dual of the relative Lie algebra $\mathbf{Lie}(A/\mathbb{H}_g)$. Therefore $\det(\mathcal{E})(-g)$ is the Hodge line bundle $\omega = \Lambda^g \mathcal{H}^0(\Omega^1_{A/\mathbb{H}_g})$.

We illustrate the action of G on $H^0(\omega^{\otimes k})$ = the space of all Siegel modular forms of weight k w.r.t. $\mathrm{Sp}(2g, \mathbb{Z})$. Let $\gamma \in G \cap M(2g, \mathbb{Z}_p)$ and write the double coset $K\gamma K$ as a (finite) disjoint union of left K-cosets: $K\gamma K = \coprod K\gamma_i$ with $\gamma_i \in G \cap M(2g, \mathbb{Z})$ represented in block form as (A_i, B_i, C_i, D_i). Let $s \in H^0(\omega^{\otimes k})$, which we write $f(\Omega) \cdot (\Lambda^d \, d\Omega)^{\otimes k}$ via the complex uniformization, where $\Lambda^d \, d\Omega = d\Omega_{11} \wedge d\Omega_{12} \wedge \ldots \wedge d\Omega_{1g} \wedge d\Omega_{22} \wedge \ldots \wedge d\Omega_{gg}$. It takes some patience and care to unravel all the definitions, and we find that $K\gamma K$ sends $f(\Omega) \cdot (\Lambda^d \, d\Omega)^{\otimes k}$ to

$$\begin{aligned} &\sum_i \det(J(\gamma_i, \Omega)^k \cdot f(\gamma_i \cdot \Omega) \cdot (\Lambda^d \, d\Omega)^{\otimes k} \\ &\qquad = \sum_i \lambda(\gamma)^{kg} \cdot \det(C_i \cdot \Omega + D_i)^{-k} \cdot f(\gamma_i \cdot \Omega) \cdot (\Lambda^d \, d\Omega)^{\otimes k} . \end{aligned}$$

This formula does not coincide with those given in the literature, for example by Hecke. For example according to Shimura's book, for $g = 1$ one should have

$$\sum_i \eta(\gamma)^{2k-1} \cdot \det(c_i \cdot \tau + d_i)^{-k} \cdot f(\gamma_i \cdot \tau)$$

instead of

$$\sum_i \eta(\gamma)^{2k} \cdot \det(c_i \cdot \tau + d_i)^{-k} \cdot f(\gamma_i \cdot \tau) .$$

More generally in the literature of Siegel modular forms people have been using

$$\sum_i \eta(\gamma)^{kg-d} \cdot \det(C_i \cdot \Omega + D_i)^{-k} \cdot f(\gamma_i \cdot \Omega), \text{ with } d = g \cdot (g+1)/2 = \dim_{\mathbb{C}} \mathbb{H}_g .$$

What is happening? The answer is the following: when $g = 1$, it is customary to think of a holomorphic modular form of weight k as a section of $\omega^{\otimes k-2} \otimes \Omega^1_{\mathbb{H}}$ and a holomorphic cusp form of weight k as a section of $\omega^{\otimes k-2} \otimes \Omega^1$ on the compactified modular curve. On the modular curve, $[n]^*$ acts on $\omega^{\otimes k}$ by n^k, but acts on $\omega^{\otimes k-2} \otimes \Omega^1_{\mathbb{H}}$ by n^{k-2}. You can also say that the two homogeneous line bundles ω and $\Omega^1_{\mathbb{H}}$ are isomorphic w.r.t. $\mathrm{SL}_2(\mathbb{R})$, but differ by the central character η if considered w.r.t. $\mathrm{GL}_2(\mathbb{R})^+$ under the action given explicitly before. In general, the answer is that in the classical formula, people were really thinking of holomorphic modular forms of weight k as sections of $\omega^{\otimes k-g-1} \otimes \Omega^d_{\mathbb{H}}$. The point is that the differential of the the action of $\mathbf{G}(\mathbb{R})^+$ is given by

$$\begin{aligned}
&d((A \cdot \Omega + B) \cdot (C \cdot \Omega + D)^{-1}) \\
&\quad = A \cdot d\Omega \cdot (C {\cdot} \Omega + D)^{-1} - (A {\cdot} \Omega + B) \cdot (C {\cdot} \Omega + D)^{-1} \cdot C \cdot d\Omega \cdot (C {\cdot} \Omega + D)^{-1} \\
&\quad = \{A - (A \cdot \Omega + B) \cdot (C \cdot \Omega + D)^{-1} \cdot C\} \cdot d\Omega \cdot (C \cdot \Omega + D)^{-1} \\
&\quad = \{A - (A \cdot \Omega + B) \cdot {}^tC \cdot {}^t(C \cdot \Omega + D)^{-1}\} \cdot d\Omega \cdot (C \cdot \Omega + D)^{-1} \\
&\qquad (\text{since } (C \cdot \Omega + D)^{-1} \cdot C = {}^tC \cdot {}^t(C \cdot \Omega + D)^{-1}, \text{ which is easy to check}) \\
&\quad = (A \cdot {}^tD - B \cdot {}^tC) \cdot (C \cdot \Omega + D)^{-1} \cdot C \cdot d\Omega \cdot (C \cdot \Omega + D)^{-1} \\
&\quad = \eta(\gamma) \cdot (C \cdot \Omega + D)^{-1} \cdot C \cdot d\Omega \cdot (C \cdot \Omega + D)^{-1} .
\end{aligned}$$

From this we deduce easily that as holomorphic homogeneous line bundles on $\mathbb{H}_g$ w.r.t. $\mathbf{G}(\mathbb{R})^+$, $S^2(\Omega_{A/\mathsf{A}_g})$ and $\Omega^d_{\mathbb{H}_g}$ differ by η, so that their determinants $\omega^{\otimes(g+1)}$ and $\Omega^d_{\mathbb{H}_g}$ differ by the central character η^d.

A similar theory holds for the locally constant sheaves of $\mathbb{Q}$ (or $\mathbb{R}$ or $\mathbb{C}$)-vector spaces $\mathbb{V}$ on $\mathsf{A}_g(\mathbb{C}) = \Gamma\backslash\mathbb{H}_g$ defined by (finite-dimensional) rational representations (ρ, V) of $\mathbf{G}$: Take ϕ^* to be the transpose of the map on $\mathbb{V}$ given by γ^{-1}, where $K\gamma K$ corresponds to ϕ. This way $\mathcal{H}(G, \mathbb{Q})$ acts on $H^*(\mathsf{A}_g(\mathbb{C}), \mathbb{V})$ and $H^*_!(\mathsf{A}_g(\mathbb{C}), \mathbb{V})$. Here we may define $H^*(\mathsf{A}_g(\mathbb{C}), \mathbb{V})$ either as the singular cohomology of a topological space (or better, an orbifold), or as algebraic de Rham cohomology. The action of Z on $H^m(A_g, \mathbb{V})$ is given by a pairing between $H^m(A_g, \mathbb{V})$ and its Poincaré dual $H^{2d-m}_!(A_g, \mathbb{V}^\vee(d))$, defined as the composition of $\mathrm{pr}_2^* \times \mathrm{pr}_1^* : H^m(A_g, \mathbb{V}) \times H^{2d-m}_!(A_g, \mathbb{V}^\vee(d)) \to H^m(Z, \mathbb{V}) \times H^{2d-m}_!(Z, \mathbb{V}^\vee(d))$, the cup product pairing, and the trace map $\int_Z : H^{2d}_!(Z, \mathbb{Q}(2d)) \to \mathbb{Q}$.

As before, the locally constant sheaf attached to the multiplier representation η is trivial, although η itself is not. One way to distinguish locally constant sheaves attached η is to use the F-filtration and the weight filtration as introduced in chap.VI. The sheaf attached to η will be denoted $\mathbb{Q}(1)$ (or $\mathbb{R}(1)$ or $\mathbb{C}(1)$).

It can be shown that the elements of $\mathcal{H}(G,\mathbb{Q})$ are self-adjoint w.r.t. Poincaré duality, using formal properties of global duality and trace as explained in SGA 3 XVIII. Also the action respects the various mixed and pure Hodge structures on $H^*(\mathsf{A}_g(\mathbb{C}),\mathbb{V})$, defined in the chapter VI. One only has to check that the various structures are functorial w.r.t. the trace maps.

There is a "level-n version", n prime to p: We consider $\mathsf{A}_{g,n}$ as a stack over $\mathbb{Z}[1/n]$, which classifies principally polarized abelian varieties A of dimension g, with an isomorphism $A[n] \cong (\mathbb{Z}/n\mathbb{Z})^{2g}$ which is a symplectic similitude. We also obtain $\mathrm{Isog}_{g,n}$, representing isogenies which respect the level-n structures. Repeating the arguments above we obtain actions of $\mathcal{H}(G,\mathbb{Q})$ on $H^*(\overline{\mathsf{A}}_{g,n},\overline{\mathcal{E}})$, $H^*(\mathsf{A}_{g,n},\mathbb{V})$, $H^*_!(\mathsf{A}_{g,n},\mathbb{V})$, $H^*(\mathsf{A}_{g,n}\otimes\overline{\mathbb{Q}},\mathbb{V}_l)$ and $H^*_!(\mathsf{A}_{g,n}\otimes\overline{\mathbb{Q}},\mathbb{V}_l)$. These actions are self-adjoint with respect to duality, compatible with various Hodge decompositions, and commute with the natural action of $\mathbf{G}(\mathbb{Z}/n\mathbb{Z})$ as well as with that of $\mathrm{Gal}(\overline{\mathbb{Q}}/\mathbb{Q})$. Some people will appreciate the fact that for $n \geq 3$ we obtain algebraic spaces, so that they do not have to define cohomology for stacks.

4. Isogenies in Characteristic p

We now intend to study Isog_g in characteristic p, so the base is $\mathrm{Spec}\,\mathbb{F}_p$ in this section unless otherwise specified. The overall strategy is "reduction to the ordinary locus".

Recall that an abelian variety A over a field of characteristic p is called *ordinary* if its p-rank over the algebraic closure is equal to $\dim(A)$, or equivalently if Frobenius is injective on $H^1(A,\mathcal{O}_A)$, or equivalently if the p-divisible group $A[p^\infty]$ is an extension of an étale p-divisible group by a multiplicative one. For example the Mumford construction applied to tori produces such abelian varieties when we look at the generic fibre. Isogenies preserve ordinariness, which can be seen easily from the last definition. (In fact isogenies preserve the p-rank.)

It is clear that for families of abelian varieties, being ordinary is a (Zariski) open condition on the base. When the base is $\mathbb{F}_p$ we denote by $\mathsf{A}_g^\circ \subset \mathsf{A}_g$ the open substack of A_g obtained by removed all abelian varieties in characteristic p which are not ordinary. (For a general base not necessarily of characteristics p, A_g° will denote the open substack of A_g whose fibres in characteristic different from p coincide with A_g, while in the fibre over $\mathbb{F}_p$ A_g° is the ordinary locus as defined above.) Since being ordinary is invariant under isogeny, we obtain finite and unramified maps $\mathrm{Isog}_g^\circ \to \mathsf{A}_g^\circ \times \mathsf{A}_g^\circ$, p-$\mathrm{Isog}_g^\circ \to \mathsf{A}_g^\circ \times \mathsf{A}_g^\circ$ by restricting to the ordinary locus.

Let us start with a local study of ordinary isogenies. Suppose k is an algebraically closed field of characteristic p, and $\phi : A_1 \to A_2$ is a p-isogeny of principally polarized g-dimensional ordinary abelian varieties over k, such that the pull-back of the polarization on A_2 is p^e times the polarization on A_1 (so that $\deg(\phi) = p^{eg}$). The associated Barsotti-Tate (p-divisible groups) $A_1[p^\infty]$ and $A_2[p^\infty]$ are both isomorphic to $\mu_{p^\infty}^g \oplus (\mathbb{Q}_p/\mathbb{Z}_p)^g$, and the map induced by ϕ between them respects these decompositions. If we choose $\mathbb{Z}_p$-bases for (the Serre dual of) $\mu_{p^\infty}^g$ and (the $\mathbb{Z}_p$-dual of) $(\mathbb{Q}_p/\mathbb{Z}_p)^g$ which are compatible with the symplectic pairings $\langle\ ,\ \rangle_i : A_i[p^\infty] \times A_i[p^\infty] \to \mu_{p^\infty}$, $i = 1, 2$, $T_p(\phi)$ acts on the first summand $\mu_{p^\infty}^g$ and second summand $(\mathbb{Q}_p/\mathbb{Z}_p)^g$ via $g \times g$-matrices a and d with coefficients in $\mathbb{Z}_p$ such that $a \cdot {}^t d = p^e \cdot I$. Notice that $m = \operatorname{diag}(a, d)$ is an element of M, whose L-double coset LmL is independent of all choices involved and will be called the *p-type of ϕ*. This is an extra structure in characteristic p, which one "cannot see" outside characteristic p. It is clear that the dual isogeny f^t has p-type $Lm^\vee L$, if f has p-type LmL. Hence ordinary isogenies in characteristic p have a well-defined type in $L\backslash M/L$, and one checks that for families of isogenies between ordinary abelian varieties in characteristic p the type is locally constant on the base.

We observe that the "p-type" as defined above is in some sense a refinement of the type defined for generic characteristic, because M is a subgroup of G. The essential reason is that an ordinary p-divisible group in characteristic p has a canonically defined filtration which does not exist in generic characteristic. This seems like a good time to explain the general theme of this section: In characteristic p, we have extra structures that don't have generic counterparts. That we have a "refined isogeny type" is just an example of this phenomenon. Another well-known and extremely important example is the existence of Frobenius. In our context, the Frobenius of (principally polarized) abelian varieties defines a component of p-Isog_g. Our major goal in this section is to relate these "extra p-isogenies" with the bigger ring of Hecke operators $\mathcal{H}(M, \mathbb{Q})$ via the extra structure of p-divisible groups in characteristic p.

Next let us investigate deformations. Suppose R is a local artinian ring with algebraically closed residue field k of characteristic p, A_1 is a principally polarized g-dimensional abelian scheme over R, and $\phi : A_1 \otimes_R k \to A_2 \otimes_R k$ a p-isogeny ($\phi^* \lambda_2 = p^e \cdot \lambda_1$ for some e) between ordinary abelian varieties of type LmL. We ask the following question: For fixed A_1 can we extend ϕ to an isogeny over R, and how many different extensions exist? Obviously the question can be reformulated in terms of the kernel $H \subset A_1[p^e] \otimes k$ of ϕ: Can we extend H to a totally isotropic finite flat subgroup of $A_1[p^e]$ over R? This question can be easily computed using Serre-Tate local moduli "mechanically" (cf. [Ka 3]). Here we do it directly.

Let $H_m \subset H$ denote the multiplicative part of H, $H_e = H/H_m$ the étale quotient. H_e is constant, so extends to a constant group scheme over R, which we denote $H_{e/R}$, or simply H_e. H_m lifts uniquely to a flat subgroup of $A_1[p^e]$ by the rigidity of groups of multiplicative type; the lift must be isotropic and of multiplicative type, and we denote it $H_{m/R}$ or H_m again. Thus we have to lift $H_e \subset A_1[p^e]/H_m$.

We remark that the extension $0 \to A_1[p^e]_m \to A_1[p^e] \to A_1[p^e]_e \to 0$ is locally trivial in the flat topology on $\operatorname{Spec}(R)$. In fact any such extension $0 \to H' \to H \to H'' \to 0$ is: base change to the inverse image of $\mathrm{id}_{H''}$ under $\mathbf{Hom}_{\mathbb{Z}/p^n}(H'', H) \to \mathbf{Hom}_{\mathbb{Z}/p^n}(H'', H'')$. Hence locally in the flat topology we can lift H to a finite flat subgroup $H_{/R}$ of $A_1[p^e]$, or equivalently H_e to a finite flat subgroup of $A_1[p^e]/H_m$. The different liftings $H_{e/R} \subset A_1[p^e]/H_m$ form a principally homogeneous space under the finite flat R-group scheme $\mathbf{Hom}_{\mathbb{Z}/p^n}(H_e, A_1[p^e]_m/H_m)$.

The next question is which liftings $H_{/R} \subset A_1[p^e]$ are totally isotropic. Consider the restriction to $H_{/R}$ of the alternating form on $A_1[p^e]$. Obviously $H_{m/R}$ is totally isotropic, but in fact it is in the radical of this restriction, i.e. $\langle H_{m/R}, H_{/R}\rangle = 1$, since the induced pairing $H_{m/R} \times (H_{/R}/H_{m/R}) \to \mu_{p^e}$ is trivial when restricted to $\operatorname{Spec} k$. Thus the restriction to $H_{/R}$ of the alternating form on $A_1[p^e]$ is induced from an alternating form on H_e. As $A_1[p^e]_m/H_m$ is the Cartier dual of H_e (with the duality induced by the alternating form on $A_1[p^e]$), we always may change the lifting to an isotropic one. Moreover the liftings $H_{/R} \subset A_1[p^e]$ such that $H_{/R}$ is totally isotropic form a principally homogeneous space under the group of symmetric homomorphisms $H_e \to A_1[p^e]_m/H_m$, namely the Cartier dual of the group $S^2(H_e)$. Let us remark that H_e is constant and isomorphic to $\mathbb{Z}_p^g/d(\mathbb{Z}_p^g)$, so the group $S^2(H_e)$ is determined by the type $L(a,d)L$ of the isogeny f. This settles the question of deformations. We summarize our results in

4.1. Proposition. *Notation as above. We have*

(i) The projections $\mathrm{pr}_i : \mathrm{Isog}_g^\circ \to \mathsf{A}_g^\circ$ *are finite and flat.*

(ii) Over the part of $\mathrm{Isog}_g^\circ \otimes \mathbb{F}_p$ *which classifies isogenies of type* LmL *with* $m = \mathrm{diag}(a,d)$, *the geometric fibres of* pr_1 *have pure multiplicity equal to the* $\mathrm{Card}(S^2(\mathbb{Z}_p^g/d(\mathbb{Z}_p^g)))$, *while the geometric fibres of* pr_2 *have pure multiplicity* $\mathrm{Card}(S^2(\mathbb{Z}_p^g/{}^t a(\mathbb{Z}_p^g)))$. *In fact, we have shown that the formal completion at a geometric point of* Isog_g° *of type* LmL *has the structure of a principally homogeneous space under* $\mathbf{Hom}_{\mathbb{Z}}(S^2(\mathbb{Z}_p^g/d(\mathbb{Z}_p^g)), G_m)$ *(resp.* $\mathbf{Hom}_{\mathbb{Z}}(S^2(\mathbb{Z}_p^g/{}^t a(\mathbb{Z}_p^g)), G_m)$) *via* pr_1 *(resp.* pr_2*).*

We will use prop. 4.1 to define a map $\hbar : \mathcal{H}_0(M,\mathbb{Q}) \to \mathbb{Q}[p\text{-}\mathrm{Isog}_g^\circ \otimes \mathbb{F}_p]$ from $\mathcal{H}_0(M,\mathbb{Q})$ into the $\mathbb{Q}$-vector space spanned by irreducible components of $p\text{-}\mathrm{Isog}_g^\circ \otimes \mathbb{F}_p$ as free generators.

Definition of $\hbar$. $\hbar$ sends the characteristic function of a double-class LmL ($m = (a,d)$) to $1/\mathrm{Card}(S^2(\mathbb{Z}_p^g/d(\mathbb{Z}_p^g)))\cdot$ (the algebraic cycle determined by the union of the connected components of $p\text{-}\mathrm{Isog}_g^\circ \otimes \mathbb{F}_p$ over which the universal isogeny has type LmL).

Set-theoretically (with multiplicity) the geometric fibres of this cycle under pr_1 may be described as follows: Choose a principally polarized g-dimensional ordinary abelian variety A_1 over an algebraically closed field k of characteristic p. The fibre consists of all isogenies $f : A_1 \to A_2$ of type LmL. It is clear that this description determines the cycle uniquely. Our work consisted in showing that there exists a cycle with these properties. For example the element $\mathrm{Frob} \in \mathcal{H}(M,\mathbb{Q})$ maps to the cycle which classifies isogenies $A_1 \to A_2$ between ordinary abelian varieties of dimension g whose kernel is the maximally isotropic subgroup of multiplicative type inside $A_1[p]$ (i.e. the multiplicative part $A_1[p]_m$ of $A_1[p]$). All these isogenies are necessarily isomorphic to the relative Frobenius $\mathrm{Frob} = \mathrm{Frob}_p(A_1/k) : A_1 \to A_1^{(p)} = \mathrm{Frob}_p^*(A_1)$. Hence, as its name already suggests, $\mathrm{Frob} \in \mathcal{H}(M, \mathbb{Q})$ maps to the Frobenius-isogeny.

Before we go on to explore the map $\hbar$, perhaps we should indicate why we put in the factor $1/\mathrm{Card}(S^2(\mathbb{Z}_p^g/d(\mathbb{Z}_p^g)))$ in the definition of $\hbar$. The answer is: $\mathrm{Card}(S^2(\mathbb{Z}_p^g/d(\mathbb{Z}_p^g)))$ is the multiplicity of $\mathrm{pr}_1 : \mathrm{Isog}_g^\circ \to A_g$ at various connected components; dividing by this factor makes $\hbar$ is compatible with the previous $h : \mathcal{H}_0(G, \mathbb{Q}) \to \mathbb{Q}[p\text{-}\mathrm{Isog}_g^\circ \otimes \mathbb{Z}[1/p]]$ and the reduction map (to characteristic p). Moreover, $\hbar$ commutes with involutions, where on $\mathcal{H}_0(M, \mathbb{Q})$ we use the *twisted* involution $\phi \mapsto (m \mapsto e^{2\rho_n}(m) \cdot \phi(m^\vee))$ on $\mathcal{H}(M, \mathbb{Q})$. (We shall see that the twisted Satake homomorphism $\tilde{S}_M^G$ not only has the advantage of being rational, but really comes up naturally when considering the specialization map. Hence the twisted involution is the truly natural one to use.)

Here are some properties. Again our mapping is compatible with multiplication: The composition $\mathrm{Isog}_g^\circ \times_{A_g^\circ} \mathrm{Isog}_g^\circ \to \mathrm{Isog}_g^\circ$ defines a product on algebraic cycles in characteristic p, as before: If Z_1 and Z_2 are two such cycles, the intersection of $Z_1 \times Z_2$ with the pre-image of the diagonal in $A_g^\circ \times A_g^\circ$ is a cycle in $(\mathrm{Isog}_g^\circ \times_{A_g^\circ} \mathrm{Isog}_g^\circ) \otimes \mathbb{F}_p$, as the diagonal is locally a complete intersection, so that intersection-multiplicities can be defined by the "Tor-formula", or by Koszul complexes, see [Se 1]. We take the image of this cycle in $\mathrm{Isog}_g^\circ \otimes \mathbb{F}_p$. In this context things could be further simplified, but we prefer to state them this way because we intend to generalize later.

With this composition of cycles the map above becomes a homomorphism of rings, compatible with involutions: If Lm_1L and Lm_2L are two integral double-classes, the product in $\mathcal{H}_0(M, \mathbb{Q})$ of their characteristic functions is obtained by decomposing the product ("with multiplicities") Lm_1Lm_2L into double classes LmL. On the level of cycles this translates into fixing an A_1, looking at all isogenies $f : A_1 \to A_2$ of type Lm_1L, and $g : A_2 \to A_3$ of type Lm_2L, and then considering the products $g \cdot f$. But this is just a description of the product-cycle. For the involutions: The characteristic function of LmL is sent to $e^{-2\rho_n}(m)\cdot$the characteristic function of $Lm^\vee L$. For cycles the involution sends the union of the connected components of $\mathrm{Isog}_g^\circ \otimes \mathbb{F}_p$ where the universal isogeny has type LmL, to the corresponding union for $Lm^\vee L$. So we only have to check coefficients: in other words, for $m = (a, d)$ we should have: $e^{-2\rho_n}(m) = \mathrm{order}(S^2(\mathbb{Z}_p^g/a^t(\mathbb{Z}_p^g)))/\mathrm{order}(S^2(\mathbb{Z}_p^g/d(\mathbb{Z}_p^g)))$. But this is easy: If $p^{e_1}, \dots, p^{e_g}$ are the elementary divisors of a, the elementary divisors of d are $p^{e-e_1}, \dots, p^{e-e_g}$ ($a \cdot d^t = p^e \cdot I$). The left side is the inverse of the modulus of the action $b \to a \cdot b \cdot d^{-1} = p^{-e} \cdot a \cdot b \cdot a^t$ on symmetric $g \times g$-matrices b. It is given by p to the power $\sum_{i \le j}(e_i + e_j) - e \cdot g \cdot (g+1)/2$. For the right side we obtain $\sum_{i \le j} \min(e_i, e_j) - \sum_{i \le j} \min(e - e_i, e - e_j) = \sum_{i \le j}(e_i + e_j - e)$, which coincides with the left. So we have shown that $\hbar : \mathcal{H}_0(M, \mathbb{Q}) \to \mathbb{Q}[p\text{-}\mathrm{Isog}_g^\circ \otimes \mathbb{F}_p]$ is a homomorphism of (commutative) algebras, commuting with involutions.

Finally there is a reduction map from cycles on $p\text{-}\mathrm{Isog}_g \otimes \mathbb{Q}$ to cycles on $p\text{-}\mathrm{Isog}_g^\circ \otimes \mathbb{F}_p$: Given $Z \subset p\text{-}\mathrm{Isog}_g \otimes \mathbb{Q}$, take the scheme-theoretic closure in

p-Isog_g°, and intersect it with the special fibre over $\mathbb{F}_p$. The reduction map preserves products, so we get a homomorphism $\mathrm{red}_p : \mathbb{Q}[p\text{-}\mathrm{Isog}_g^\circ \otimes \mathbb{Z}[1/p]] \to \mathbb{Q}[p\text{-}\mathrm{Isog}_g^\circ \otimes \mathbb{F}_p]$. Composing with $h : \mathcal{H}_0(G, \mathbb{Q}) \to \mathbb{Q}[p\text{-}\mathrm{Isog}_g^\circ \otimes \mathbb{Z}[1/p]]$, we get a homomorphism from $\mathcal{H}_0(G, \mathbb{Q})$ into the $\mathbb{Q}$-vector space spanned by cycles on p-$\mathrm{Isog}_g^\circ \otimes \mathbb{F}_p$. We claim that this homomorphism coincides with the composition of $S_M^G : \mathcal{H}_0(G, \mathbb{Q}) \to \mathcal{H}_0(M, \mathbb{Q})$, and the previously defined map $\hbar : \mathcal{H}_0(M, \mathbb{Q}) \to \mathbb{Q}[p\text{-}\mathrm{Isog}_g^\circ \otimes \mathbb{F}_p]$.

We check this by looking at geometric points: Suppose $A_{1/k}$ is a principally polarized ordinary g-dimensional abelian variety over an algebraically closed field k of characteristic p. We find a complete discrete valuation ring V of unequal characteristics, with residue field k and fraction-field F, such that $A_{1/k}$ extends to an abelian scheme $A_{1/V}$ over V. We may extend V so that the p^e-torsion points of $A_{1/F}$ are all F-rational, for some $e \geq 0$ to be chosen later. The Barsotti-Tate group $T_p(A_{1/V})$ of A_1 over V sits in an extension $0 \to T_p(A_{1/V})_m \to T_p(A_{1/V}) \to T_p(A_{1/V})_e \to 0$, of its étale by its multiplicative part. We obtain an extension structure on the Tate-module $T_p(A_{1/F})$ by restriction to $\mathrm{Spec}(F)$.

The reduction operation for cycles commutes with base change $\mathrm{Spec}(V) \to A_g^\circ$. It follows that for an $\mathbb{Z}_p$-integral matrix $\gamma \in G$ the reduction of the cycle associated to the characteristic function of $K\gamma K$ acts on $A_{1/K}$ by choosing all possible p-isogenies $A_{1/F} \to A_{2/F}$ of type $K\gamma K$; that is, the inclusion $T_p(A_{1/F}) \to T_p(A_{2/F})$ can be described by an element of $K\gamma K$. If e is big enough all these isogenies are defined over F (or V), and correspond after choice of a symplectic basis to the lattices $T_p(A_{2/F}) = \xi^{-1}(T_p(A_{1/F})) \supset T_p(A_{1/F})[1/p]$, where ξ runs through a system of representatives for $K \backslash K\gamma K$. The type of the induced isogeny over k is described by the effect on $T_p(A_{1/k})_m$ and $T_p(A_{1/k})_e$, so if we choose representatives $\xi \in Q$ the types are determined by the images of ξ in M. For a fixed $m \in M$ the number of $\xi \in K\backslash KgK = (K \cap Q)\backslash(KgK \cap Q)$ for which this image lies in Lm is equal to the volume (normalized so that K has volume 1) of $KgK \cap KUm$, which is equal to the volume (in U, normalized so that $U \cap K$ has volume 1) of $U \cap (KgKm^{-1})$. If ϕ is the characteristic function of $K\gamma K$, this number is just $\tilde{S}_M^G(\phi)(m) = \int_U \phi(um)\, du$. This is invariant under right translation with L, and it follows that the correspondence associated to $K\gamma K$ reduces to the sum of correspondences associated to LmL's, with coefficients as prescribed by $\tilde{S}_M^G$. We have proved the claim that $\hbar : \mathcal{H}_0(M, \mathbb{Q}) \to \mathbb{Q}[p\text{-}\mathrm{Isog}_g^\circ \otimes \mathbb{F}_p]$ is compatible with $h : \mathcal{H}_0(G, \mathbb{Q}) \to \mathbb{Q}[p\text{-}\mathrm{Isog}_g^\circ \otimes \mathbb{Z}[1/p]]$, with respect to the Satake homomorphism $\tilde{S}_M^G$ and the reduction map $\mathbb{Q}[p\text{-}\mathrm{Isog}_g^\circ \otimes \mathbb{Z}[1/p]] \to \mathbb{Q}[p\text{-}\mathrm{Isog}_g^\circ \otimes \mathbb{F}_p]$. In short: The action of $\mathcal{H}_0(M, \mathbb{Q})$ extends the $\mathcal{H}_0(G, \mathbb{Q})$-action.

Of course we are not satisfied with just the ordinary locus. However, we shall prove below that any isogeny between principally polarized abelian varieties can be deformed to an ordinary isogeny. It follows that all irreducible components of p-$\mathrm{Isog}_g \otimes \mathbb{F}_p$ meet p-$\mathrm{Isog}_g^\circ \otimes \mathbb{F}_p$, so that p-$\mathrm{Isog}_g^\circ \otimes \mathbb{F}_p$ is dense open in p-$\mathrm{Isog}_g \otimes \mathbb{F}_p$, and by restriction, the rings $\mathbb{Z}[p\text{-}\mathrm{Isog}_g \otimes \mathbb{F}_p]$ and $\mathbb{Z}[p\text{-}\mathrm{Isog}_g^\circ \otimes \mathbb{F}_p]$ are isomorphic. So the results above carry over to general isogenies.

Finally we have level-n versions, as usual.

For any rational l-adic representation V of $\mathbf{G}$ we have constructed l-adic sheaves $\mathbb{V}_l$ on $A_{g,n} \otimes \mathbb{F}_p$. The Hecke correspondences operate on them, with $[p]$ mapping to an isomorphism of $\mathbb{V}_l$, and we obtain the following result:

4.2. Theorem. *The actions of $\mathcal{H}(G, \mathbb{Q})$ on $H^*(A_{g,n} \otimes \overline{\mathbb{F}}_p, \mathbb{V}_l)$ and $H^*_!(A_{g,n} \otimes \overline{\mathbb{F}}_p, \mathbb{V}_l)$ extend to actions of $\mathcal{H}(M, \mathbb{Q})$, which contains $\mathcal{H}(G, \mathbb{Q})$ via $\tilde{S}^G_M$. These actions commute with* $\mathrm{Gal}(\overline{\mathbb{F}}_p/\mathbb{F}_p)$ *and* $\mathbf{G}(\mathbb{Z}/n\mathbb{Z})$, *respect involutions, and* Frob $\in \mathcal{H}(M, \mathbb{Q})$ *maps to the Frobenius* Frob_p^*.

Corollary. Frob_p *satisfies an equation of degree 2^g over the Hecke algebra $\mathcal{H}(G, \mathbb{Q})$.*

We still have to prove that the ordinary p-isogenies are dense in the space of all p-isogenies over $\mathbb{F}_p$. For the proof we are immediately reduced to a result about deformation of isogenies:

4.3. Proposition. *Suppose k is an algebraically closed field, $\phi : A_1 \to A_2$ an isogeny of principally polarized g-dimensional abelian varieties over k, such that ϕ^* (polarization on A_2) $= p^e$·polarization on A_1. If $p - \mathrm{rank}(A_1) = p - \mathrm{rank}(A_2) = r < g$, there exists a lifting of ϕ to an isogeny of principally polarized abelian varieties over $k[[t]]$ such that over $k((t))$ the p-rank is $> r$.*

Proof. We follow the paper of P. Norman and F. Oort [N-O]. They prove that any polarized abelian variety can be deformed into an ordinary polarized abelian variety in characteristic p. We shall adapt their method to deform isogenies. Note that it is not always possible to deform endomorphisms, since the Picard number may jump.

Deformations of abelian varieties are controlled by their associated Barsotti-Tate groups according to the Serre-Tate theorem. Barsotti-Tate groups and their deformations are more or less controlled by their Dieudonné modules, depending on the situation. There are three Dieudonné theories: (i) the classical (contravariant) Dieudonné theory, valid over a perfect field, reconsidered by Fontaine to include the case of a mixed characteristic complete valuation ring with absolute ramification index $\leq p-1$. (ii) crystalline (contravariant) Dieudonné theory, due to Grothendieck and Messing, giving the structure of a crystal on the nilpotent crystalline site. (iii) the (covariant) Cartier-Dieudonné theory. The covariant Dieudonné theory tends to produce fairly large rings over rings like $k[[t]]$, but has the advantage of being valid for any base ring of characteristic p, thus allowing one to examine deformations to *any* order without worrying about obstruction, which is exactly what we need. Standard references for the Cartier-Dieudonné theory are [Laz] and [Haz]. Let us describe the (covariant) Dieudonné modules.

Over a perfect field k, we consider free modules E over the Witt vectors $W(k)$, with semi-linear mappings F and V such that $F \cdot V = V \cdot F = p \cdot \mathrm{id}$, $F \cdot \alpha = \alpha^\sigma \cdot F$ and $\alpha \cdot V = V \cdot \alpha^\sigma$, for $\alpha \in W(k)$ and α^σ the transform of α under Frobenius. A *principal quasi-polarization* on E is an alternating form

$\langle\ ,\ \rangle : E \times E \to W(k)$, which is a perfect pairing over $W(k)$, such that V and F are adjoints: $\langle F(x), y\rangle = \langle x, V(y)\rangle^\sigma$. A principally polarized abelian variety A over k defines such a Dieudonné module E, of rank $2g$, with a principal quasi-polarization. (V is the module of all p-typical curves in the formal completion of A.) Furthermore the k-vector space $E/V(E)$ is isomorphic to the Lie algebra of A, and the p-rank of A is equal to $g - \dim(E/(F^g(E) + V(E)))$ = the stable rank of F on $E/V(E)$.

The isogeny $\phi : A_1 \to A_2$ induces a symplectic similitude $\phi_* : E_1 \to E_2$ on Dieudonné modules, with multiplier p^e. We have to deform these data (E_1, E_2, ϕ_*, alternating forms) to $k[[t]]$. The basic idea is to use an endomorphism N of E_2 which stabilizes E_1 (as a submodule of E_2), such that $N^2 = 0$, and N is skew-symmetric for $\langle\ ,\ \rangle : \langle N(x), y\rangle + \langle x, N(y)\rangle = 0$. Then we keep E_1, E_2, f_* and $\langle\ ,\ \rangle$ constant, but replace V by $V \cdot (1 + tN)$ and F by $(1 - tN) \cdot F$. However $k[[t]]$ is not perfect, and the Dieudonné ring $D(k[[t]])$ in question becomes much bigger: it consists of all formal sums of the form $\sum_{m,n\geq 0} V^m[a_{m,n}]F^n$, $a_{m,n} \in k[[t]]$, subject to the condition that for any m, $a_{m,n} = 0$ for all but finitely many n and subject to the usual commutation rules, and $W(k[[t]])$ is embedded in it via $(x_n)_{n\geq 0} \mapsto \sum_{n\geq 0} V^n[x_n]F^n$. To deal with Dieudonné modules over such a big ring, we have to replace the above description of Dieudonné modules by a different one, as in [N-O]. It amounts to introducing a sort of "normal forms" for Dieudonné modules for Barsotti-Tate groups: In this description Dieudonné modules over a Noetherian complete local ring R over a perfect field k (of characteristic p) can be given by an *invertible* $n \times n$-matrix $A = (\alpha_{ij})$ with entries in the Witt vectors $W(R)$. To such a matrix $A = (\alpha_{ij})$ we associate the Dieudonné module generated over the Dieudonné ring $D(R)$ by $e_1, \dots, e_n$, with relations $F(e_i) = \sum_j \alpha_{ij} \cdot e_j$, $1 \leq i \leq d$, $e_i = V(\sum_j \alpha_{ij} \cdot e_j)$, $d + 1 \leq i \leq n$. (Here d is the dimension of the corresponding Barsotti-Tate group.) It is in fact a free module over $W(R)$. There also exists a description of polarizations along these lines. If $R = k$ is a perfect field, we get such a description from the old one by choosing a $W(k)$-basis $e_1, \dots, e_n$ of E such that $e_1, \dots, e_d$ induces a k-basis of $E/V(E)$, and $e_{d+1}, \dots, e_n \in V(E)$. The deformation over $k[[t]]$ discussed above is obtained by multiplying the matrix A on the right with the matrix of $(1 - tN)$. However, the details are a little bit more complicated than this.

From now on we deal with quasi-polarized Dieudonné modules (not necessarily principal), forgetting for the moment that they come from abelian varieties. First we split off the étale and multiplicative parts: For $E = E_1$ or E_2 we have direct sum decompositions $E = E_e \oplus E_m \oplus E_{ii}$, such that V is bijective on the étale part E_e, F is bijective on the multiplicative part E_m, and F and V are topologically nilpotent on the bi-infinitesimal part E_{ii}. For the alternating form the first two summands are orthogonal to the third, and dual to each other. We shall only deform the third part, which ensures that the p-rank cannot decrease. So let us replace E_1 and E_2 by their bi-infinitesimal parts.

We use ϕ_* to identify E_1 with its image in E_2. Denote the alternating form on E_2 by $\langle\ ,\ \rangle$, so that $p^{-e} \cdot \langle\ ,\ \rangle$ defines the alternating form on E_1. We think of the underlying $W(R)$-modules and the alternating pairings as fixed, and deform the operation of F and V. Choose r maximal such that $E_1 \subset V^r(E_2)$, and replace E_2 by $V^r(E_2)$: Any deformation over $k[[t]]$ of $V^r(E_2)$ extends over $k[[t^{p^{-r}}]]$ to a deformation of E_2. So we may assume that E_1 is not contained in $V(E_2)$. Also if $E_1 = E_2$, the arguments in [N-O] give a good deformation of $E_1 = E_2$ (this case would be considered trivial in [N-O]), so we can assume that E_1 and E_2 are different. We shall find a skew-symmetric endomorphism N of (the underlying $W(k)$-module of) E_2 with $N^2 = 0$, such that N stabilizes E_1, and such that there exist elements $y, z \in E_2$ with $N(z) = y, z \in V(E_2)$, y not in $V(E_2)$, and $F(y) - z \in p \cdot E_2$. It follows that modulo $V(E_2)$, y is an eigenvector of $(1 - tN) \cdot F$, with eigenvalue t. So if we use N to deform E_1 and E_2 over $k[[t]]$, the resulting Dieudonné module over $k((t))$ has a non-trivial multiplicative part.

To construct y and z, start with an element $x \in E_1$ such that x is not in $V(E_2)$, but $F(x)$ is (F is nilpotent on $E_2/V(E_2)$). As $V(E_2)/p \cdot E_2$ is the kernel of F on $E_2/p \cdot E_2$, x and $F(x)$ are k-independant in $E_2/p \cdot E_2$. We want to find elements $y, z \in E_1$ with $\langle y, z \rangle = 0$, such that $x - y$ and $F(x) - z$ are in $p \cdot E_2$. (This is stronger than was required before.) We obtain them from x and $F(x)$ by adding elements of $E_1 \cap p \cdot E_2 \supset p \cdot E_1$. It suffices to find y and z satisfying the above conditions with $\langle y, z \rangle \in p^{e+1} \cdot W(k)$, since then y and z are k-independant in $E_1/p \cdot E_1$ (because x and $F(x)$ are k-independent in $E_2/p \cdot E_2$), so that the necessary further adjustments can be made by adding to them elements of $p \cdot E_1$.

Now consider the symplectic space $E_1/p \cdot E_1$ over k. If $M \subset E_1/p \cdot E_1$ denotes the kernel of $E_1/p \cdot E_1 \to E_2/p \cdot E_2$, M has k-dimension $\geq g = 1/2 \cdot \mathrm{rank}(E_1)$ (as the image is totally isotropic). Now there are only finitely many choices:

– If $\langle x, F(x) \rangle \in p^{e+1} \cdot W(k)$, we are done.

– If x is not perpendicular to M, we change $F(x)$ by an element of M.

– The same if $F(x)$ is not perpendicular to M.

– If x and $F(x)$ are both perpendicular to M, M cannot be isotropic, as otherwise its span with x is isotropic of dimension $> g$. Add elements of M to x and $F(x)$.

So we always can find y and z as above.

Consider the isotropic sublattice $W(k) \cdot y \oplus W(k) \cdot z$ of E_1 and E_2, with y, z satisfying the conditions above. We find elements $u, v \in E_2$ which span a complementary (for $\langle\, , \rangle$) sublattice in E_2 (i.e. quotient is torsion-free), and then necessarily $p^e \cdot u$, $p^e \cdot v \in E_1$ span a complementary sublattice in E_1. Define N by the rule that N annihilates everything perpendicular to y, z, u and v, and that on the span of y and z, N is given by $N(y) = 0$, $N(z) = y$, while on the span of u and v there is a unique way to define N such that it becomes skew-symmetric. This finishes the construction of a deformation, and the proofs of proposition 4.3 and theorem 4.2. □

These methods also allow us to describe the action of Hecke operators on q-expansions of modular forms (of weight k). They act on the space of all possible q-expansions, and this action is determined by the study of possible p-isogenies between degenerating abelian varieties of the form $A_1 = T/\imath(X)$ (T the torus with character group X, and $\imath$ the period-mapping). For these we always have $A_2 = \tilde{T}/\imath(p^{-e} \cdot \tilde{X})$, where $\tilde{X}$ is a sublattice of X containing $p^e \cdot X$ and $\tilde{T}$ the corresponding torus, which is a quotient of T. As these abelian varieties are all ordinary, the action of $\mathcal{H}(G, \mathbb{Q})$ extends to $\mathcal{H}(M, \mathbb{Q})$, with the M-type of an isogeny given by $\tilde{X} \subset X$. We easily derive the following result:

4.4. Proposition. *Suppose* $m = (a, d) \in M$ *is* $\mathbb{Z}_p$*-integral, with multiplier* $\eta(m) = p^e$. *Then* LmL *maps a formal* q*-expansion*

$$f = \sum_{\lambda \in S^2(X)} f(\lambda) \cdot \lambda \quad (f(\lambda) \in \det(X)^{\otimes k}) \text{ to}$$

$$\sum_{\tilde{X} \subset X} \sum_{\tilde{\lambda} \in S^2(\tilde{X}) \cap p^e \cdot S^2(X)} f(\tilde{\lambda}) \cdot (p^{-e} \cdot \tilde{\lambda}) \, .$$

The sum is over all lattices $\tilde{X} \subset X$ *of type* $\mathrm{GL}(g, \mathbb{Z}_p) \cdot a \cdot \mathrm{GL}(g, \mathbb{Z}_p)$, $f(\tilde{\lambda}) \in \det(\tilde{X})^{\otimes k} \subset \det(X)^{\otimes k}$, *and* $p^{-e} \cdot \tilde{\lambda}$ *is considered as an element of* $S^2(X)$.

We also may consider vector bundles associated to different representations of $\mathrm{GL}(g)$ (and q-expansions of vector-valued forms), and the result carries over.

Again everything extends to level structures.

5. Applications

The discussion of Hecke operators in this chapter serves only as background for further work on the arithmetic of Siegel modular forms. There are two difficulties: on the Lefschetz side we need a trace formula for Hecke operators, where "fixed points at the boundary" cause problems. On the other hand our knowledge of harmonic analysis for the symplectic group is very incomplete, and this aspect is best left to the trace-formula experts. We shall conclude this chapter by brief indications of some (potential) applications.

The various Hecke algebras $\mathcal{H}_p(G, \mathbb{Q})$, for different choices of p, operate on the cohomology of A_g or $\mathsf{A}_{g,n}$. These actions commute, so we can decompose the cohomology into generalized eigenspaces. These are parametrized by characters λ_p of $\mathcal{H}_p(G, \mathbb{Q})$. As the structure of this algebra is known, we can classify the characters, let us say with values in $\overline{\mathbb{Q}}$: They are given by invertible numbers $\alpha_1, \ldots, \alpha_g, \beta_1, \ldots, \beta_g \in \overline{\mathbb{Q}}^*$, modulo multiplying α_i and β_i by a common factor γ_i, with $\prod \gamma_i = 1$, and modulo the operation of the Weyl group $W_{\mathbf{G}} = \mathcal{S}_g \ltimes \{\pm 1\}^g$. Sometimes one calls the elements $\alpha_i \cdot \beta_i^{-1}$ *Satake parameters at* p. Together with the product $\alpha_1 \cdot \ldots \cdot \alpha_g$ they parametrize characters up to $W_{\mathbf{G}}$-action. The Ramanujan conjecture in representation theory predicts (not always correctly) that for characters λ_p related to cusp-forms (of weight $\geq g + 1$) the corresponding principal series representation is tempered, which for the layman translates into the fact that all conjugates of the Satake parameters should have absolute value 1. We have already mentioned the equation of degree 2^g for Frobenius over the Hecke algebra. Using the Weil conjectures (as shown by P. Deligne), one can show that for any archimedian valuation of $\overline{\mathbb{Q}}$, the product of the Satake parameters has absolute value 1 (for suitable normalization of the Satake parameters). For $g \geq 2$ this is weaker than the Ramanujan conjecture, which is fortunate as the latter is known to be false in general.

For any p, the action of $\mathcal{H}_p(G, \mathbb{Q})$ on l-adic cohomology extends to $\mathcal{H}_p(M, \mathbb{Q})$. It is an open problem how much this extended action depends on l. If a Lefschetz trace formula for Hecke operators is available, one can show that at least the alternating sum of the traces on $H^*(\mathsf{A}_g \otimes \overline{\mathbb{F}}_p, \mathbb{V}_l)$ is in $\mathbb{Q}$, and independent of l, for any element $\phi \in \mathcal{H}_p(M, \mathbb{Q})$, and any system of l-adic sheaves $\mathbb{V}_l$ defined by a $\mathbb{Q}$-representation of $\mathbf{G}$, as the trace is given by geometric data independant of l.

Suppose λ_p is a system of characters giving rise to a non-trivial eigenspace on cusp forms. Associated with it is a compatible system system of l-adic representations, on the corresponding eigenspaces in l-adic cohomology. It is conjectured that in certain "good" cases these representations have rank 2^g, and are compositions of maps from the $\mathrm{Gal}(\overline{\mathbb{Q}}/\mathbb{Q})$ into the general spin-group $\mathrm{GSpin}(2g+1)$ (the dual group to $\mathrm{GSp}(2g)$), and of the spin representation of $\mathrm{GSpin}(2g+1)$. Also in the Hodge decomposition of the associated complex representation each of the 2^g infinity-types (parametrized by $W^{\mathbf{M}}$) should have dimension one. If this could be shown, we would obtain a Hodge-Tate, even crystalline decomposition of the same type, which would sometimes allow us to make conclusions about the representations. However here we come into the realm of "thin air speculation", so we better stop and leave the subject to those who should know how to tackle the problems, namely, the trace formula experts.

Bibliography

[EGA] Grothendieck, A. and Dieudonné, J.: Eléments de Géométrie Algébrique. Springer-Verlag, Berlin, 1971 and Publ. Math. I.H.E.S. 4, 8, 11, 17, 20, 24, 28, 32 (1961-1967).

[SGA 1] Grothendieck, A.: Revêtements étales et groupe fondamentale. Lecture Notes in Math. 224, Springer-Verlag, 1971.

[SGA 2] Grothendieck, A.: Cohomologie locale des faisceaux cohérents et théorèmes de Lefschetz locaux et globaux. North Holland Pub. Co., 1968.

[SGA 3] Demazure, M., Grothendieck, A.: Schémas en groupes I,II,III. Lecture Notes in Math. 151, 152, 153, Springer-Verlag, 1971.

[SGA 4] Artin, M., Grothendieck, A., Verdier, J.-L.: Théorie des topos et cohomologie étale des schémas I, II, III. Lecture Notes in Math. 269, 270, 305, Springer-Verlag, 1971.

[SGA 4 1/2] Deligne, P.: Cohomologie étale. Lecture Notes in Math. 569, Springer-Verlag, 1977.

[SGA 5] Grothendieck, A.: Cohomologie l-adic et fonctions L. Lecture Notes in Math. 589, Springer-Verlag, 1977.

[SGA 6] Berthelot, P., Grothendieck, A., Illusie, L.: Théorie des intersections et théorème de Riemann-Roch. Lecture Notes in Math. 225, Springer-Verlag, 1971.

[SGA 7] Grothendieck, A., Deligne, P., Katz, N.: Groupes de monodromie en géométrie algébrique. Lecture Notes in Math. 288, 344, Springer-Verlag, 1971.

[GIT] Mumford, D., Fogarty, J.: Geometric invariant theory. Ergebnisse der Mathematik und ihrer Grenzgebiete, 2. Folge, Bd. 34, Springer-Verlag, Berlin, 1982, second edition.

[Ar 1] Artin, M.: Grothendieck topologies. Mimeographed lecture notes on a seminar by M. Artin, Spring 1962, Harvard University, Math. Dept., Cambridge, Mass., 1962.

[Ar 2] Artin, M.: The implicit function theorem in algebraic geometry. Colloq. in Algebraic Geometry, Bombay, 1969, 13-34.

[Ar 3] Artin, M.: Algebraic approximation of structures over complete local rings. Publ. Math. I.H.E.S. 36 (1969), 23-58.

[Ar 4] Artin, M.: Algebraization of formal moduli I. Global Analysis, Papers in Honor of K. Kodaira. 21-71, Princeton Univ. Press, Princeton, 1969.

[Ar 5] Artin, M.: Algebraization of formal moduli II. Existence of modifications. Ann. of Math. 91 (1970), 88-135.

[Ar 6] Artin, M.: Versal deformations and algebraic stacks. Invent. Math. 27 (1974), 165-189.

[Ar 7] Artin, M.: Théorème de représentabilité pour les espaces algébriques. Séminaire de Mathématiques Supérieures 44, Presses de l'Université de Montréal, 1973.

[AW] Artin, M. and Winters, G.: Degenerate fibers and stable reduction. Topology 1 (1971), 373-383.

[AMRT] Ash, A., Mumford, D., Rapoport, M., Tai, Y.S.: Smooth compactification of locally symmetric varieties. Math. Sci. Press, Brookline, 1975.

[B] Baily, W.L.: On the theory of theta functions, the moduli of abelian varieties and the moduli of curves. Ann. of Math. 75 (1962), 342-381.

[B-B] Baily, W.L. and Borel, A.: On the compactification of arithmetically defined quotients of bounded symmetric domains. Ann. of Math. 84 (1966), 442-528.

[BGG] Bernstein, I.N., Gelfand, I.M., Gelfand, S.I.: Differential operators on the base affine space and a study of $\mathfrak{g}$-modules. In: Lie groups and their representations (editor I.M. Gelfand). Adam Hilger, London, 1975.

[Be 1] Berthelot, P.: Cohomologie cristalline des schémas de caracteristique $p > 0$. Lecture Notes in Math. 407, Springer-Verlag, 1974.

[Be 2] Berthelot, P.: Géométrie rigide et cohomologie des variétés algébriques de caractéristique p. Soc. Math. France mémoire 23 (1986), 7-32.

[BO 1] Berthelot, P. and Ogus, A.: Notes on crystalline cohomology. Math. Notes, Princeton 21 (1978).

[BO 2] Berthelot, P. and Ogus, A.: F-isocrystals and de Rham cohomology, I.. Invent. Math. 72 (1983), 159-199.

[Br 1] Breen, L.: Fonctions thêta et théorème du cube. Lecture Notes in Math. 980, Springer-Verlag, 1983.

[Br 2] Breen, L.: The cube structure on the determinant bundle. To appear in Proc. Symp. Pure Math. 49, A.M.S., part 1, 663-673.

[Br-La] Brylinski, J.L. and Labresse, J.P.: Cohomologies d'intersection et fonctions L de certaines variétés de Shimura. Ann. Sci. Ec. Norm. Super. 17 (1984), 361-412.

[C-C-K] Carlson, J., Cattani, E.H., Kaplan, A.G.: Mixed Hodge structures and compactification of Siegel's space. In: Algebraic geometry, Proc. of Angiers Conference, Noordhof 1980, 77-105.

[Ch 1] Chai, C.-L.: Compactification of Siegel moduli schemes. London Math. Soc. Lecture Notes Series 107, Cambridge Univ. Press, 1985.

[Ch 2] Chai, C.-L.: Arithmetic compactification of the Siegel moduli space. Proc. Symp. Pure Math. 49, A.M.S., (1989), part 2, 19-44.

[ArG] Cornell, G. and Silverman, J. (ed.): Arithmetic geometry. Springer-Verlag, 1986.

[D 1] Deligne, P.: Equations différentiels à points singulier réguliers. Lecture Notes in Math. 163, Springer-Verlag, 1970.

[D 2] Deligne, P.: Théorie de Hodge II. Publ. Math. I.H.E.S. 40 (1971), 5-58.

[D 3] Deligne, P.: Théorie de Hodge III. Publ. Math. I.H.E.S. 44 (1975) 5-77.

[D 4] Deligne, P.: La conjecture de Weil, I. Publ. Math. I.H.E.S. 43 (1971), 273-307.

[D 5] Deligne, P.: La conjecture de Weil, II. Publ. Math. I.H.E.S. 52 (1980), 138-252.

[D 6] Deligne, P.: Preuve des conjectures de Tate et de Shafarevich, d'après G. Faltings. Séminaire Bourbaki 616 (1983/1984).

[D 7] Deligne, P.: Le lemme de Gabber. Astérisque 127 (1985), 131-150.

[De-Mum] Deligne, P., Mumford, D.: The irreducibility of the space of curves of given genus. Publ. Math. I.H.E.S. 36 (1969), 75-109.

[De-Ri] Deligne, P., Ribet, K.: Values of abelian L-functions at negative integers over totally real fields. Invent. Math. 59 (1980), 227-286.

[DMOS] Deligne, P., Milne, J.S., Ogus, A., Shih, K.: Hodge cycles, motives, and Shimura varieties. Lecture Notes in Math. 900, Springer-Verlag, 1982.

[DGR] Demazure, M., Giraud, J. and Raynaud, M.: Schémas Abéliennes, Séminaire de Géometrie Algébrique 1967/68, Faculté des Sciences de l'Université de Paris, Orsay (notes polycopiées).

[F 1] Faltings, G.: On the Cohomology of Locally Symmetric Hermitian Spaces. Lecture Notes in Math. 1029, 55-98.

[F 2] Faltings, G.: Endlichkeitssätze für abelsche Varietäten über Zahlkörpern. Invent. Math. 73 (1983), 349-366.

[F 3] Faltings, G.: Arithmetische Kompaktifizierung des Modulraums der abelschen Varietäten. Lecture Notes in Math. 1111, Springer-Verlag, 1985, 321-383.

[F 4] Faltings, G.: Hodge-Tate structures and modular forms. Math. Ann. 278 (1987), 133-149.

[F 5] Faltings, G.: p-adic Hodge theory. Jour. of the A.M.S. 1 (1988), 255-299.

[F 6] Faltings, G.: Crystalline cohomology and p-adic galois-representations. Proceedings 1st JAMI-conference. Johns-Hopkins Univ. Press (1990), 25-79.

[F 7] Faltings, G.: F-isocrystals on open varieties. Results and conjectures. (to appear in Grothendieck's 60th birthday proceedings, Birkhäuser, Boston).

[Fo 1] Fontaine, J.M.: Modules galoisiens, modules filtrés et anneaux de Barsotti-Tate. Astérisque 65 (1979), 3-80.

[Fo 2] Fontaine, J.M.: Sur certaines types de représentations p-adiques du groupe de Galois d'un corp local, construction d'un anneau de Barsotti-Tate. Ann. of Math. 115 (1982), 529-577.

[Fo 3] Fontaine, J.M.: Cohomologie de de Rham, cohomologie cristalline et représentations p-adiques. Lecture Notes in Math. 1016, Springer-Verlag, 1983, 86-108.

[Fo-La] Fontaine, J.M. and Lafaille, G.: Construction de représentations p-adiques. Ann. Sci. Ec. Norm. Super. 15 (1982), 547-608.

[Fo-Me] Fontaine, J.M. and Messing, W.: p-adic periods and p-adic étale cohomology. In: Current Trends in Arithmetic Algebraic Geometry, Contemporary Mathematics 67 (1987), 179-207.

[Fu] Fulton, W.: Intersection Theory. Ergebnisse der Mathematik und ihrer Grenzgebiete, 3. Folge, Bd. 2, Springer-Verlag, Berlin, 1984.

[FW] Faltings, G. and Wüstholz, G.: Rational Points. Vieweg, Braunschweig, 1984.

[Fr] Freitag, E.: Siegelsche Modulfunktionen. Springer-Verlag, Berlin, 1983.

[Gi] Giraud, J.: Cohomologie non-abélienne. Springer-Verlag, 1971.

[GHV] Greub, W., Halperin, S., and Vanstone, R.: Connections, curvature, and cohomology III. Academic Press, New York, 1976.

[Gre] Green, W.: Heights in families of abelian varieties. Duke Math. J. 58 (1989), 617-632.

[Gr 1] Grothendieck, A.: Techniques de construction et théorèmes d'existence en géométrie algébrique, IV: les schémas de Hilbert. Sém. Bourbaki 1960/61, exposé 221.

[Gr 2] Grothendieck, A.: Groupes de Barsotti-Tate et cristaux de Dieudonné. Séminaire de Mathématiques Supérieures 45, Presses de l'Université de Montréal, 1974.

[Gr 3] Grothendieck, A.: Un théorème sur les homomorphismes de schémas abéliennes. Invent. Math. 2 (1966), 59-78.

[G-M] Grothendieck, A. and Murre, J.-P.: The Tame Fundamental Group of a Formal Neighbourhood of a Divisor with Normal Crossings on a scheme. Lecture Notes in Math. 208, Springer-Verlag, 1971.

[Haz] Hazewinkel, M.: Formal groups and applications. Academic Press, 1978.

[Hak] Hakim, M.: Schémas relatifs sur les topos annelés. Ergebnisse der Mathematik und ihrer Grenzgebiete, Bd. 64. Springer-Verlag, 1972.

[He] Helgason, S.: Differential geometry, Lie groups and symmetric spaces. Academic Press, New York-San Francisco-London, 1978.

[Hi] Hironaka, H.: Resolution of singularities of an algebraic variety over a field of characteristic zero. Ann. of Math. 79 (1964), 109-326.

[Ho] Horrocks, G.: Vector bundles on the punctured spectrum of a local ring. Proc. Lond. Math. Soc. 14 (1964), 689-713.

[I 1] Igusa, J.-I.: On the algebraic theory of elliptic modular functions. J. Math. Soc. Japan 20 (1968), 96-106.

[I 2] Igusa, J.-I.: On the theory of compactifications, A.M.S. Summer Institute on Algebraic Geometry (Woods Hole, 1964), mimeographed lecture notes.

[I 3] Igusa, J.-I.: On the graded ring of theta constants I, II. Am. J. Math. 86 (1964), 219-246, ibid. 88 (1966), 221-236.

[I 4] Igusa, J.-I.: A desingularization problem in the theory of Siegel modular functions. Math. Ann. 168 (1967), 228-260.

[I 5] Igusa, J.-I.: Theta functions. Die Grundlehren der mathematischen Wissenschaften, Bd. 194, Springer-Verlag, 1972.

[I 6] Igusa, J.-I.: On the variety associated with the ring of thetanullverte. Am. J. Math. 103 (1981), 377-398.

[Ka 1] Katz, N.: Nilpotent connections and the monodromy theorem: Applications of a result of Turrittin. Publ. I.H.E.S. 39 (1970), 175-232.

[Ka 2] Katz, N.: Algebraic solutions of differential equations (p-curvature and the Hodge filtration). Invent. Math. 18 (1972), 1-118.

[Ka 3] Katz, N.: Serre-Tate local moduli. In: Surfaces Algébriques, exposé Vbis, Lecture Notes in Math. 868, Springer-Verlag, 1981, 138-202.

[Ka-La] Katz, N. and Lang, S.: Finiteness theorems in geometric classfield theory. Enseign. Math. 27, 1981, 285-319.

[KKMS] Kempf, G., Knudson, F., Mumford, D., Saint-Donat, B.: Toroidal embeddings I. Lecture Notes in Math. 339, Springer-Verlag, 1973.

[Kn] Knutson, D.: Algebraic spaces. Lecture Notes in Math. 203, Springer-Verlag, 1971.

[L 1] Lang, S.: Abelian varieties. Interscience, New York, 1959. Reprinted by Springer-Verlag, 1983.

[L 2] Lang, S.: Fundamentals of diophantine geometry. Springer-Verlag, 1983.

[L 3] Lang, S.: Introduction to modular Forms. Springer-Verlag, Berlin, 1976.

[Lar] Larsen, M. J.: Unitary groups and 1-adic representations. Princeton thesis, 1988.

[Laz] Lazard, M.: Commutative formal groups. Lecture Notes in Math. 443, Springer-Verlag, 1975.

[Li] Lipman, J.: Desingularization of two-dimensional schemes. Ann. of Math. 107 (1978), 151-207.

[M-D] Manin, Y. and Drinfel'd, V.G.: Periods of p-adic Schottky groups. J. Reine. Angew. Math. 262/263 (1973), 239-247.

[Ma-Me] Mazur, B., Messing, W.: Universal extensions and one-dimensional crystalline cohomology. Lecture Notes in Math. 370, Springer-Verlag, 1974.

[Me] Messing, W.: The crystals associated to Barsotti-Tate groups: with applications to abelian schemes. Lecture Notes in Math. 264, Springer-Verlag, 1972.

[MB 1] Moret-Bailly, L.: Pinceaux de variétés abéliennes. Astérisque 129 (1985).

[MB 2] Moret-Bailly, L.: Un théorème de pureté pour les familles de courbes lisses. C. R. Acad. Sci., Paris, t. 300, Série I, no. 14, (1985), 489-492.

[Mori] Morikawa, H.: On theta functions and abelian varieties over valuation fields of rank one. Nagoya Math. J. 20 (1962), 1-27, ibid. 21 (1962), 231-250.

[Mum 1] Mumford, D.: Abelian varieties. Tata Inst. Fund. Research Studies in Math. vol. 5, Oxford Univ. Press, Oxford, 1970.

[Mum 2] Mumford, D.: On the equations defining abelian varieties I, II, III. Invent. Math. 1 (1966), 287-354, ibid. 3 (1967), 71-135, 215-244.

[Mum 3] Mumford, D.: Bi-extensions of formal groups. Algebraic Geometry, Oxford Univ. Press, Oxford, 1969, pp. 307-322.

[Mum 4] Mumford, D.: Varieties defined by quadratic equations. In: Questioni sulle varieta algeriche, Corsi der C.I.M.E. Edizioni Cremonese, Roma, 1969.

[Mum 5] Mumford, D.: An analytic construction of degenerating abelian varieties over a complete local ring. Compos. Math. 24, Fasc. 3, (1972), 239-272.

[Mum 6] Mumford, D.: Hirzebruch's proportionality theorem in the non-compact case. Invent. Math. 42 (1977), 239-272.

[Mum 7] Mumford, D.: Tata Lectures on Theta, I. Progress in Math. vol. 28, Birkhäuser, Boston, 1983.

[Mur] Murre, J.P.: Representation of unramified functors. Applications. Séminaire Bourbaki 294, 1964/65.

[Nam 1] Namikawa, Y.: A new compactification of the Siegel space and degeneration of abelian varieties. I, II. Math. Ann. 221 (1976), 97-141, ibid. 201-241.

[Nam 2] Namikawa, Y.: Toroidal degeneration of abelian varieties. II. Math. Ann. 245 (1979), 117-150.

[Nam 3] Namikawa, Y.: Toroidal Compactification of Siegel Spaces. Lecture Notes in Math. 812, Springer-Verlag, 1980.

[N] Néron, P.: Modèles mimimaux des variétés abéliennes sur les corps locaux et globaux. Publ. Math. I.H.E.S. 21 (1964), 5-128.

[Nor 1] Norman, P.: An algorithm for computing local moduli of abelian varieties. Ann. of Math. 101 (1975), 499-509.

[Nor 2] Norman, P.: Lifting abelian varieties. Invent. Math. 64 (1981), 431-443.

[N-O] Norman, P. and Oort, F.: Moduli of abelian varieties. Ann. of Math. 112 (1980), 413-439.

[Oda 1] Oda, T.: The first de Rham cohomology groups and Dieudonné modules. Ann. Sci. Ec. Norm. Super. 2 (1959), 63-135.

[Oda 2] Oda, T.: Lectures on torus embeddings and applications. Tata Inst. of Fund. Research 58 (1978), Springer-Verlag.

[Oda 3] Oda, T.: Convex bodies and algebraic geometry. Ergebnisse der Math., 3. Folge, Band 5, Springer-Verlag, 1989.

[Og] Ogus, A.: F-isocrystals and de Rham cohomology II, convergent isocrystals. Duke Math. J. 51 (1984), 765-850.

[Oort] Oort, F.: Finite group schemes, local moduli for abelian varieties and lifting problems. In: Algebraic Geometry, Oslo, 1970, pp. 223-254, also in Compos. Math. 23 (1972), 265-296.

[Ra 1] Raynaud, M.: Modèle de Néron. C. R. Acad. Sci., Paris, t. 262, Sér. A. (1966), 345-347.

[Ra 2] Raynaud, M.: Un critère d'effectivité de descente. Dans "Les épimorphismes d'anneaux", Séminaire P. Samuel (Algèbre commutative) Anneé 1967/68,É cole Normale Supérieure de Jeunes Filles, no 5.

[Ra 3] Raynaud, M.: Faisceaux amples sur les schémas en groupes et les espaces homogènes. Lecture Notes in Math. 119, Springer-Verlag, 1970.

[Ra 4] Raynaud, M.: Variétés abéliennes et géométrie rigide. Actes du Congrès Intern. Math. 1970, tome 1, pp. 473-477.

[Ra 5] Raynaud, M.: Géométrie rigide-analytique, Table ronde d'analyse nonarchimédienne. Mém. Soc. Math. Fr. Nouv. Ser. 39-40 (1974), 319-327.

[Ra 6] Raynaud, M.: Spécialization du foncteur de Picard. Publ. Math. I.H.E.S. 38 (1971), 27-123.

[Ra′] Raynaud, M.: Théorème de Lefschetz en cohomologie cohérente et cohomologie étale. Bull. Soc. Math. Fr., Mémoire #41, (1975).

[Rap] Rapoport, M.: Compactifications de l'espace de modules de Hilbert-Blumenthal. Compos. Math. 36 (1978), 255-335.

[Ri] Ribet, K.: P-adic interpolation via Hilbert modular forms. Proc. Symp. Pure Math. 29 (1975), 581-592.

[Sa 1] Satake, I.: On the compactification of the Siegel space. J. Indian Math. Soc. 20 (1956), 259-281.

[Sa 2] Satake, I.: Theory of spherical functions on reductive algebraic groups over p-adic fields. Publ. Math. I.H.E.S. 18 (1963), 5-69.

[Sa 3] Satake, I.: Algebraic Structures on Symmetric Domains. Publ. of the Math. Soc. of Japan 14, Iwanami Shoten, Publishers and Princeton Univ. Press, 1980.

[Se 1] Serre, J.-P.: Algèbre Locale, Multiplicités. Lecture Notes in Math. 11, Springer-Verlag, 1975.

[Se 2] Serre, J.-P.: Rigidité du foncteur de Jacobi d'échelon $n \geq 3$, Appendice d'exposé 17, Séminaire Henri Cartan 13e année, 1960/61.

[Sch] Schlessinger, M.: Functors of Artin rings. Trans. Amer. Math. Soc. 130 (1968), 208-222.

[Sh 1] Shimura, G.: Moduli and Fibre systems of abelian varieties. Ann. of Math. 83 (1966), 294-338.

[Sh 2] Shimura, G.: On canonical models of arithmetic quotients of bounded symmetric domains I, II. Ann. of Math. 91 (1970), 144-222, ibid. 92 (1970), 528-549.

[Sh 3] Shimura, G.: On the Fourier coefficients of modular forms of several variables. Nachr. Akad. Wiss. Gött. (1975), 261-268.

[Sz 2] Szpiro, L. (ed.): Séminaire sur les pinceaux de courbes de genre au moins deux. Astérisque 86 (1981).

[Sz 2] Szpiro, L. (ed.): La conjecture de Mordell. Séminaire Bourbaki 619 (1983/1984).

[Sz 3] Szpiro, L.: Séminaire sur les pinceaux arithmétiques: La conjecture de Mordell. Astérisque 127 (1985).

[S-T] Serre, J.-P. and Tate, J.: Good reduction of abelian varieties. Ann. of Math. 88 (1968), 492-517.

[T] Tate, J.: *P*-divisible groups. Conference on local fields, Driebergen, Netherlands, 1966, edited by T.A. Springer, Springer-Verlag, 1967.

[Weyl] Weyl, H.: The classical groups. Princeton Univ. Press, Princeton, 1939.

[Weil] Weil, A.: Variétés abéliennes et courbes algébriques. Hermann, Paris, 1984.

[Za 1] Zarhin, J.G.: Abelian varieties in characteristic p. Math. Notes of the Academy of Sciences of the USSR 19 (1976), 240-244.

[Za 2] Zarhin, J.G.: Finiteness theorem for isogenies of abelian varieties over function fields of finite characteristic. Funct. Anal. Appl. 8 (1974), 301-303.

[Za 3] Zarhin, J.G.: Endomorphisms of abelian varieties over fields of finite characteristic. Math. USSR, Izvestija 9 (1975), ibid. 2, 255-260.

[Za 4] Zarhin, J.G.: Endomorphisms of abelian varieties and points of finite order in characteristic p. Math. Notes of the Academy of Sciences of the USSR 21 (1977), 734-744.

[Zari] Zariski, O.: The reduction of singularities of an algebraic surface. Ann. of Math. 40 (1939), 639-689.

[Z-S] Zariski, O. and Samuel, P., with cooperation of I.S. Cohen: Commutative Algebra II. Van Nostrand, Princeton, N.J., 1960.

Glossary of Notations

A_g the moduli stack classifying principally polarized abelian varieties of dimension g I.4.3, p.19; I.4.11, p.23

$\overline{A}_g$ a toroidal compactification of A_g IV.5.6, p.116

A_g^* the minimal compactification of A_g V.2.3, p.150

$A_{g,n}$ the moduli stack classifying principally polarized abelian varieties with (symplectic) level-n structure I.4.4, p.19; IV.6.1, p.121

$\overline{A}_{g,n}$ a toroidal compactification of $A_{g,n}$ IV.6, p.128; IV.6.7, p.130

$\mathsf{A}_{g,n}^*$ the minimal compactification of $A_{g,n}$ V.2.5, p.152

B the standard Borel subgroup of Sp($2g$) V.1.4, p.221

$\underline{B}(G)$ a bilinear map from $\underline{X}(G^t) \times \underline{X}(G)$ to Div(S) III.10.1, p.88

$B_{\mathrm{cris}}^+(\mathcal{O}_{\overline{K}})$, $B_{\mathrm{cris}}(\overline{K})$ Barsotti-Tate rings VI.6, p.239/240

B_{HT} Hodge-Tate ring VI.6, p.239

$B(X)$ $\mathbb{Z}$-valued symmetric bilinear forms on $X \cong \mathbb{Z}^g$, dual to $S^2(X)$ IV.2.1, p.96

$\tilde{B}_s(X)$ $B(X) \times (X^*)^s$ V.1.1, p.195

Corr correspondences, as a functor I.1.7, p.4

$\mathrm{CG}(S^\bullet(\mathcal{E}^\vee))$, $\mathrm{CG}(\Lambda^\bullet(\mathcal{E}))$ the categories of complexes of graded $S^\bullet(\mathcal{E}^\vee)$-(resp. $\Lambda^\bullet(\mathcal{E})$-) modules VI.3, p.211

$\mathrm{CF}(\Omega^\bullet)$, $\mathrm{CF}(\mathcal{D})$ the categories of complexes of filtered $\Omega^\bullet$-(resp. $\mathcal{D}$-) modules VI.3, p.215

$\mathrm{CF}(\overline{\Omega}^\bullet)$, $\mathrm{CF}(\overline{\mathcal{D}})$ variants of $\mathrm{CF}(\Omega^\bullet)$, $\mathrm{CF}(\mathcal{D})$ VI.3, p.217

$C(X)$ the rational positive semi-definite cone in $B(X) \otimes_{\mathbb{Z}} \mathbb{R}$ IV.2.1, p.96

$\tilde{C}_s(X)$ a cone in $\tilde{B}_s(X) \otimes_{\mathbb{Z}} \mathbb{R}$ VI.1, p.195

$C_{\lambda\bullet}$ the BGG-resolution of $V(\lambda)$ VI.5, p.231

$\mathrm{DG}(S'(\mathcal{E}^\vee))$, $\mathrm{DG}(\Lambda^\bullet(\mathcal{E}))$ derived categories of $\mathrm{CG}(S'(\mathcal{E}^\vee))$, $\mathrm{CG}(\Lambda^\bullet(\mathcal{E}))$ VI.3, p.211

$\mathrm{DF}(\Omega^\bullet)$, $\mathrm{DF}(\mathcal{D})$, $\mathrm{DF}(\overline{\Omega}^\bullet)$, $\mathrm{DF}(\overline{\mathcal{D}})$ derived categories of $\mathrm{CF}(\Omega^\bullet)$, $\mathrm{CF}(\mathcal{D})$, $\mathrm{CF}(\overline{\Omega}^\bullet)$, $\mathrm{CF}(\overline{\mathcal{D}})$ VI.3, p.217

DD, $\mathrm{DD}_{\mathrm{ample}}$, $\mathrm{DD}_{\mathrm{pol}}$, $\mathrm{DD}_{\mathrm{IS}}$ semiabelian degeneration data (with extra structure) II.0, p.33; III.2, p.57/58

DEG, $\mathrm{DEG}_{\mathrm{ample}}$, $\mathrm{DEG}_{\mathrm{pol}}$, $\mathrm{DEG}_{\mathrm{IS}}$ semiabelian degeneration (with extra structure) II.0, p.33; III.2, p.56/60

D_{HT} VI.6, p.239

$\overline{\mathcal{E}}$ canonical extension of $\mathcal{E}$ to $\overline{A}_g$ VI.4.2, p.225

ét étale

fppf faithfully flat of finite presentation

fpqc faithfully flat and quasicompact

G GSp($2g$) = Chevalley group scheme of symplectic similitudes VI.4, p.219

$\mathbf{G}^0$ Sp($2g$) = Chevalley group scheme of symplectic transformations VI.4, p.220

$\mathcal{G}(\mathcal{L})$ theta group attached to $\mathcal{L}$ I.5, p.25; II.4, p.39; II.5, p.46

H VI.5, p.230

H_w VI.5, p.232

$\mathbb{H}_g$ Siegel upper half space of genus g I.6, p.29

$\mathcal{H}(G, \mathbb{Q})$, $\mathcal{H}_0(G, \mathbb{Q})$ Hecke algebras of G VII.1.1, p.244/248

$\mathcal{H}(M,\mathbb{Q})$, $\mathcal{H}_0(M,\mathbb{Q})$ Hecke algebras of M VII.1.1, p.244/248
$h : \mathcal{H}_0(G,\mathbb{Q}) \to \mathbb{Q}[p\text{-Isog}_g]$ Hecke algebra as algebraic correspondences VII.3, p.253
$\bar{h} : \mathcal{H}_0(M,\mathbb{Q}) \to \mathbb{Q}[p\text{-Isog}_g \otimes \mathbb{F}_p]$ Hecke algebra as algebraic correspondences in characteristic $p > 0$ VII.4, p.261
Isog, p-Isog isogenies VII.3, p.251/252
Isog^0, $p\text{-Isog}^0$ ordinary isogenies VII.4, p.259
$K(\mathcal{L})$ kernel of $\lambda(\mathcal{L})$ I.5, p.25; II.4, p.39; II.5, p.46
Kos Koszul complex VI.3, p.213
$\mathcal{K}^\bullet_\lambda$, $\overline{\mathcal{K}}^\bullet_\lambda$ dual BGG-complex VI.5, p.232
$l(w)$ length of an element w in the Weyl group VI.5, p.229
M $GL_g \times G_m$, embedded in GSp($2g$) VI.4, p.220
$\mathbf{M}^0$ $M \cap G^0$ VI.4, p.220
$\mathcal{MF}^\nabla(X)$ VI.6, p.241
N the unipotent radical of B VI.4, p.221
$N_{g,r}$ the unipotent radical of $P_{g,r}$ IV.6.3, p.122
N_g, $\overline{N}_g$, N_g^+, N_g^-, $\mathcal{N}_g^{\text{ord}}$, $\mathcal{N}_g^{\text{ord},1}$, $\mathcal{N}_g^{\text{ord},2}$ moduli stacks with theta structures IV.7, p.132–135
$P_{g,r}$ a standard parabolic subgroup in Sp($2g$) IV.6.3, p.122/123
Q upper triangular blocks in GSp($2g$) VI.4, p.221
$\mathbf{Q}^0$ $Q \cap G^0$ VI.4, p.221
$Q_\rho(X)$ IV.7, p.132
S_T^G, S_M^G, S_T^M Satake homomorphisms VII.1, p.245
$\tilde{S}_T^G$, $\tilde{S}_M^G$, $\tilde{S}_T^M$ twisted Satake homomorphisms VII.1, p.246
Sp($2g,\mathbb{Z}$) the group of symplectic integral transformations I.6, p.29
Sp($2g,\mathbb{Z}$)(n) the principal congruence subgroup of Sp($2g,\mathbb{Z}$) of level n IV.6.6, p.128
S_ξ, S_σ local charts at the boundary of $\overline{A}_g$ IV.3, p.106
$S_{\xi,n}$ local charts at the boundary of $\overline{A}_{g,n}$ IV.6, p.127
$S_{\xi,n,d}$ IV.6, p.127
$U(\mathfrak{g})$, $U(\mathfrak{q})$ enveloping algebra of $\mathfrak{g},\mathfrak{q}$ VI.5, p.227
$V(\lambda)$ highest weight module for Sp($2g$) or GSp($2g$) VI.5, p.228
$\mathbb{V}(\lambda)$, $\mathcal{V}(\lambda)$ locally constant system and vector bundle attached to $V(\lambda)$ VI.5, p.232
W_G, W_M Weyl groups VI.5, p.229
W^M a system of representatives for $W_M \setminus W_G$ VI.5, p.229
$W(\mu)$ highest weight module for M VI.5, p.228
$\mathcal{W}(\mu)$ vector bundle attached to $W(\mu)$ VI.5, p.232
w_0 the longest element in W_G VI.5, p.229
w_1 the longest element in W_M VI.5, p.229
$\underline{X}(T)$ the étale sheaf of characters of a torus I.2.1, p.7
$\underline{X}(G)$ the étale constructible sheaf of characters of the toric parts of a semi-abelian scheme G I.2.10, p.12
$X \twoheadrightarrow X_\xi$, $X \twoheadrightarrow X_\sigma$ quotients of free abelian groups IV.3, p.104/105
$\Gamma_{\sigma,n}$ the stabilizer of σ in $GL(X_\sigma)(n)$ IV.6.7, p.131
$\gamma^\vee$ involution of GSp($2g$) VI.4, p.220
η the multiplier representation of GSp($2g$) VI.4, p.219
$\Theta(\mathcal{L})$ I.1.3, p.1
$\tilde{\theta}_{\text{null}}$ the thetanull map I.5, p.28
$\theta(\Omega, z)$ theta series I.6, p.29
$\theta(\Omega, a, b)$ theta constant V.3, p.160
$\lambda(\mathcal{L})$ homomorphisms from an abelian variety to its dual attached to $\mathcal{L}$ I.1.6, p.3
$\Lambda(\mathcal{L})$ I.1, p.2
$\Lambda_\mathbb{Z}$, $\Lambda_{\mathbb{Z}/n\mathbb{Z}}$ lattices with symplectic pairings IV.6, p.128/129
Ξ_ξ, $\overline{\Xi}$ IV.3, p.105
$\Xi_{\xi,n}$, $\overline{\Xi}_{\xi,n}$ IV.6, p.127
ρ_G, ρ_M half a sum of positive roots VI.5, p.228
$\{\sigma_\alpha\}$ a cone decomposition of $C(X)$ IV.1, p.93, IV.2.2, p.96
Σ_n a conical complex IV.6, p.129
$\Sigma_\mathbb{Z}$ a sublattice in $\Lambda_\mathbb{Z}$ with $\Sigma_\mathbb{Z}^\perp \subseteq \Sigma_\mathbb{Z}$ IV.6, p.128
$\Sigma_{g,r}$, $\Psi_{g,r}$ IV.6.3, p.122
σ an automorphism of GSp($2g$) VI.4, p.220; VII.3, p.255
$\{\tau_\beta\}$ a cone decomposition of $\tilde{C}_s(X)$ VI.1.3, p.196/197
Φ, Ψ functors between CF($\Omega^\bullet$) and CF($\mathcal{D}$) and their variants VI.3, p.212/215
Φ_G, Φ_M root systems VI.5, p.228

Φ^M $\Phi_G - \Phi_M$ VI.5, p.228
Ω_A, Ω_G sheaf of invariant 1-forms on A, G I.4, p.24
ω_A, ω_G top exterior products of Ω_A, Ω_G I.4, p.24
$\overline{\Omega}^\bullet_X$, $\overline{\Omega}^\bullet_{Y/X}$ (relative) differentials with logarithmic poles VI.1, p.195

Index

Appendix

An Analytic Construction of Degenerating Abelian Varieties over Complete Rings

David Mumford

This paper is reproduced from Compositio Mathematica, Vol. 24, Fasc. 3, 1972, pp. 239-272 with kind permission of Kluwer Academic Publishers, Dordrecht.

AN ANALYTIC CONSTRUCTION OF DEGENERATING ABELIAN VARIETIES OVER COMPLETE RINGS

by

David Mumford

This paper is a sequel to the earlier paper on the construction of degenerating curves. The basic ideas of the 2 papers are very similar and we refer the reader to the introduction of that paper for discussion and motivation. We limit ourselves here to an outline of the contents of the present paper.

Throughout this paper, A will stand for a fixed excellent integrally closed noetherian ring with quotient field K, $I \subset A$ will be an ideal such that $I = \sqrt{I}$ and A is complete with respect to the I-adic topology. We will be interested in group schemes over the base scheme:

$$S = \mathrm{Spec}(A)$$

which have particular properties over

(1) the generic point $\eta \in S$

(2) the closed subscheme $S_0 = \mathrm{Spec}\,(A/I)$.

If X is a scheme over S, X_η and X_0 will stand for the induced schemes over $\{\eta\}$ and S_0. A smooth commutative group scheme G of finite type over any base Z will be called *semi-abelian* if all its fibres G_z are connected algebraic groups without unipotent radical, i.e., each G_z is an extension of an abelian variety by a torus ($=$ a form of $\boldsymbol{G}_m^r$ over $k(z)$). The *rank function* will be the map

$$z \mapsto r(z)$$

associating to each point z the dimension of the torus part of G_z; r is easily checked to be upper $\frac{1}{2}$-continuous (e.g., by looking at the cardinality of the fibres of the subscheme $\mathrm{Ker}(n_G) \subset G$, étale over Z where n is prime to the residue characteristics of Z). The semi-abelian group schemes G of constant rank are *globally over* Z extensions of an abelian scheme over Z by a torus over Z. We are interested in constructing semi-abelian group schemes G/S such that

(1) G_η is an abelian variety,

(2) G_0 has constant rank.

According to an idea of John Tate, if $r = \mathrm{rank}\,(G_0)$, then G should be canonically represented as a 'quotient' of a semi-abelian group scheme $\tilde{G}$ of constant rank r *over the whole of* S by a discrete subgroup Y of L-valued points of $\tilde{G}$ with $Y \cong \mathbf{Z}^r$. (L a finite algebraic extension of K). To simplify matters, we will consider only the case where G_0 is a split torus, i.e., $G_0 \cong \mathbf{G}_m^r \times S_0$, in which case $\tilde{G} \cong \mathbf{G}_m^r \times S$ and the points of Y are K-valued. Our plan is the following:

i) start with a set of periods $Y \subset \tilde{G}(K)$ where $\tilde{G} \cong \mathbf{G}_m^r \times S$, satisfying suitable conditions;

ii) construct a kind of compactification:

$$\begin{array}{ccc} \tilde{G} & \subset & \tilde{P} \\ & {\scriptstyle \text{open}} & \\ \searrow & & \swarrow \\ & S & \end{array}$$

such that the action of Y by translation extends to $\tilde{P}$, and Y acts freely and discontinuously (in the *Zariski* topology) on $\tilde{P}_0$. Unlike the case of curves, $\tilde{P}$ is neither unique nor canonical!

iii) Take the I-adic completion $\tilde{\mathfrak{P}}$ of $\tilde{P}$, construct $\mathfrak{P} = \tilde{\mathfrak{P}}/Y$, algebrize $\mathfrak{P}$ to a scheme P projective over S, and take a suitable open subset $G \subset P$:

$$\begin{array}{ccccc} \tilde{G} & \underset{\text{open}}{\subset} & \tilde{P} & \underset{\text{compl.}}{\overset{I\text{-adic}}{\dashleftarrow}} & \tilde{\mathfrak{P}} \\ & & & & \Big\downarrow {\scriptstyle \text{formal morphism}} \\ G & \underset{\text{open}}{\subset} & P & \underset{\text{compl.}}{\overset{I\text{-adic}}{\dashleftarrow}} & \mathfrak{P} \end{array}$$

iv) prove that G is a semi-abelian group scheme over S independent of the choice of $\tilde{P}$, G_η is abelian, and $G_0 \cong \tilde{G}_0 \cong \mathbf{G}_m^r \times S_0$.

We will also show that this uniformization $\tilde{G} \to G$ is uniquely determined by G, i.e. if $\tilde{G}/Y_1 \cong \tilde{G}/Y_2$, then $Y_1 = Y_2$. However we will not discuss at all whether all semi-abelian G as in (iv) admit such a uniformization. This seems to be a fairly difficult question. In case $\dim A = 1$, and A is local, Raynaud and I have proven (independently) that all G's as in (iv) do admit uniformizations. Raynaud's method is highly analytic and mine is an application of my theory of 2-adic theta functions*. We both have reason to believe that our methods will extend to the general case, but this has not yet been done. Instead I conclude the paper with many examples. For me, one of the most enjoyable features of this research was the beauty of the examples which one works out without a great deal of extra effort. In fact, the non-uniqueness of $\tilde{P}$ gives one freedom to seek for the most elegant solutions in any particular case.

1. Periods

To begin our program, let $\tilde{G}$ be a given split torus over S. Then if X is the character group of $\tilde{G}$,for all $\alpha \in X$, the character is a canonical element

$$\mathscr{X}^\alpha \in \Gamma(\tilde{G}, \mathcal{O}_{\tilde{G}}).$$

Moreover $\tilde{G}$ is affine and can be described explicitly as:

$$\tilde{G} = \operatorname{Spec} A[\cdots, \mathscr{X}^\alpha, \cdots]_{\alpha \in X} \Big/ \left(\begin{matrix} \mathscr{X}^\alpha \cdot \mathscr{X}^\beta = \mathscr{X}^{\alpha+\beta} \\ \mathscr{X}^0 = 1 \end{matrix} \right)$$

Let:

$$\tilde{G}(K) = \text{group of } K\text{-valued points of } \tilde{G} \cong (K^*)^r.$$

Note that if $y \in G(K)$ and $\alpha \in X$, then the character $\mathscr{X}^\alpha$ takes a value on y which is an element $\mathscr{X}^\alpha(y) \in K^*$.

(1.1) Definition: A set of *periods* is a subgroup $Y \subset \tilde{G}(K)$ isomorphic to $\boldsymbol{Z}^r$.

The only assumption that we will make about Y is that they admit a polarization, in the following sense:

(1.2) Definition: A *polarization* for the periods Y is a homomorphism

$$\phi \colon Y \to X$$

such that:

i) $\mathscr{X}^{\phi(y)}(z) = \mathscr{X}^{\phi(z)}(y)$, all $y, z \in Y$,

ii) $\mathscr{X}^{\phi(y)}(y) \in I$ for all $y \in Y$, $y \neq 0$.

Note that by (ii) ϕ must be injective, hence also $[X \colon \phi Y] < +\infty$. Before going further, we stop to prove a basic lemma of a technical nature concerning periods and polarizations which will be very useful:

[1] On the equations defining abelian varieties, Inv. Math., Vol 1 and 3.

(1.3) Basic Lemma: *Suppose that for every $y \in Y$, $y \neq 0$, a positive integer n_y is given. Then there exists a finite set of elements $y_1, \cdots, y_k \in Y$, $y_i \neq 0$ and a finite subset $S \subset Y$ such that for all $z \in Y - S$,*

$$\mathscr{X}^{\phi(z)}(y_i) \in [\mathscr{X}^{\phi(y_i)}(y_i)]^{n_{y_i}} \cdot A \text{ for some } i.$$

Proof: Since Y is finitely generated, there is only a finite set of minimal prime ideals $\mathfrak{p} \subset A$ such that

$$\operatorname{ord}_{\mathfrak{p}} \mathscr{X}^{\phi(y)}(z) \not\equiv 0, \text{ all } y, z \in Y.$$

Let these prime ideals be $\mathfrak{p}_1, \cdots, \mathfrak{p}_n$, and let v_i be the valuation $\operatorname{ord}_{\mathfrak{p}_i}$. The axioms for a polarization tell us that

$$Q_i(y, z) = v_i[\mathscr{X}^{\phi(y)}(z)]$$

is a positive semi-definite quadratic form on Y. We can extend Q_i uniquely to a $\boldsymbol{R}$-valued quadratic form on $Y \otimes \boldsymbol{R}$. Since A is integrally closed, the assertion to be proved is equivalent to:

$$Q_j(z, y_i) \geqq n_{y_i} Q_j(y_i, y_i), \text{ for all } j.$$

For all $y \in Y$, $y \neq 0$, let

$$C_y = \left\{ z \in Y \otimes \boldsymbol{R} \;\middle|\; \begin{array}{l} Q_j(z, y) > n_y Q_j(y, y) \text{ for} \\ \text{all } j \text{ such that } y \notin \text{Null-space } (Q_j) \end{array} \right\}$$

Note that C_y is a convex open subset of $Y \otimes \boldsymbol{R}$ such that $\lambda C_y \subset C_y$ if $\lambda \in \boldsymbol{R}$, $\lambda \geqq 1$. To prove the lemma, it suffices to check that:

(*) $$C_{y_1} \cup \cdots \cup C_{y_k} \supset Y \otimes \boldsymbol{R} - (\text{a compact set})$$

for some $y_1, \cdots, y_k$. But first I claim that

(**) $$\bigcup_{N=1}^{\infty} \bigcup_{\substack{y \in Y \\ y \neq 0}} \frac{1}{N} C_y = Y \otimes \boldsymbol{R} - (0).$$

In fact, suppose $z \in Y \otimes \boldsymbol{R}$ and $z \neq 0$. If we approximate z by an element of $Y \otimes \boldsymbol{Q}$ which lies in the same Null-spaces of the Q_j that z does, and multiply this element by a large positive integer, we find a $y \in Y$ such that

$$z \in \text{Null-space } (Q_j) \Rightarrow y \in \text{Null-space } (Q_j)$$
$$z \notin \text{Null-space } (Q_j) \Rightarrow Q_j(y, z) > 0.$$

Therefore $N \cdot z \in C_y$ if $N \gg 0$, hence $z \in 1/N\, C_y$. This proves (**). By the compactness of the unit sphere in $Y \otimes \boldsymbol{R}$, it follows that there are $y_1, \cdots, y_k \in Y$, $y_i \neq 0$, and positive integers N_i such that

$$\bigcup_{i=1}^{k} \frac{1}{N_i} C_{y_i} \supset (\text{unit sphere}).$$

Then $\bigcup_{i=1}^{k} C_{y_i}$ contains all spheres of radius $\geqq \max(N_i)$ so (*) is proven.

QED

Another basic fact which is very useful is:

(1.4) LEMMA: *For any $\alpha \in X$, there exists an $n \geqq 1$ such that $\mathscr{X}^{n\phi(y)+\alpha}(y) \in A$ for all $y \in Y$.*

PROOF. Let $\mathfrak{p}_1, \cdots, \mathfrak{p}_n$ and $Q_1, \cdots, Q_n$ be as before. Note that using $\phi: Y \hookrightarrow X$, we can identify X with a subgroup of $Y \otimes \boldsymbol{R}$ so that

$$Q_i(\alpha, y) = \mathrm{ord}_{\mathfrak{p}_i} \mathscr{X}^{\alpha}(y), \qquad \alpha \in X, y \in Y.$$

We must show that for $n \gg 0$,

$$Q_i(ny+\alpha, y) \geqq 0, \text{ all } i, \text{ all } y.$$

But

$$nQ_i(ny+\alpha, y) = Q_i\left(ny+\frac{\alpha}{2}, ny+\frac{\alpha}{2}\right) - \tfrac{1}{4}Q_i(\alpha, \alpha).$$

Since Y projects into a lattice in $Y \otimes \boldsymbol{R}/\text{Null-space } Q_i$, it follows that if n is large enough, then for all y, either

a) $y \in \text{Null-space } (Q_i)$

or

b) $ny+\alpha/2$ is arbitrarily 'big' in $Y \otimes \boldsymbol{R}/\text{Null-space } Q_i$. In either case, $Q_i(ny+\alpha, y) \geqq 0$. *QED*

2. Relatively complete models

We now return to our basic problem: given a set of periods Y for which a polarization exists, construct *canonically* a 'quotient' of $\tilde{G}$ by Y. The main tool will be the following:

(2.1) DEFINITION. A *relatively complete model* of $\tilde{G}$ with respect to periods Y and polarization ϕ will be a collection of 5 pieces of data:

a) an integral scheme $\tilde{P}$ locally of finite type over A,

b) an open immersion $i: \tilde{G} \hookrightarrow \tilde{P}$ (we will henceforth identify $\tilde{G}$ with its image in $\tilde{P}$),

c) an invertible sheaf $\tilde{L}$ on $\tilde{P}$,

d) an action of the torus $\tilde{G}$ on $\tilde{P}$ and $\tilde{L}$ (we denote the action of an S'-valued point a of $\tilde{G}$ by

$$T_a : \tilde{P} \to \tilde{P}$$

and

$$T_a^* : \tilde{L} \to \tilde{L}),$$

e) and action of Y on $\tilde{P}$ and $\tilde{L}$ (we denote the action of $y \in Y$ by

$$S_y : \tilde{P} \to \tilde{P}$$

and

$$S_y^* : \tilde{L} \to \tilde{L}),$$

such that

(i) there exists an open $\tilde{G}$-invariant subset $U \subset \tilde{P}$ of finite type over S such that

$$\tilde{P} = \bigcup_{y \in Y} S_y(U),$$

(ii) for all valuations v on $\boldsymbol{R}(\tilde{G})$, the field of rational functions on $\tilde{G}$, for which $v \geqq 0$ on A, v has a center on $\tilde{P}$ if and only if

$$* \ [\forall\, \alpha \in X, \exists y \in Y \text{ such that } v(\mathscr{X}^\alpha(y) \cdot \mathscr{X}^\alpha) \geqq 0].$$

(These valuations are allowed to have rank > 1!)

(iii) The action of $\tilde{G}$ and Y on $\tilde{P}$ extends the 'translation" action of $\tilde{G}$ and Y on $\tilde{G}$ given by the group law on $\tilde{G}$.

(iv) The actions of Y and $\tilde{G}$ on $\tilde{L}$ satisfy the identity:

$$T_a^* S_y^* = \mathscr{X}^{\phi(y)}(a) \cdot S_y^* T_a^*$$

all $y \in Y$ all S'-valued points a of $\tilde{G}$.

(v) $\tilde{L}$ is ample on $\tilde{P}$.[2]

We turn first to the construction of relatively complete models. We need one more definition:

(2.2) Definition. A *star* Σ is a finite subset of X such that $0 \in \Sigma$, $-\Sigma = \Sigma$ and Σ contains a basis of X.
Let θ be an indeterminate and consider the big graded ring:

$$\mathscr{R} = \sum_{k=0}^{\infty} \left(K[\cdots, \mathscr{X}^\alpha, \cdots] \Big/ \left\{ \begin{matrix} \mathscr{X}^\alpha \cdot \mathscr{X}^\beta - \mathscr{X}^{\alpha+\beta} \\ \mathscr{X}^0 - 1 \end{matrix} \right\} \right) \cdot \theta^k.$$

Let Y act on $\mathscr{R}$ via operators S_y^*:

$$S_y^*(c) = c, \text{ all } c \in K,$$
$$S_y^*(\mathscr{X}^\alpha) = \mathscr{X}^\alpha(y) \cdot \mathscr{X}^\alpha, \text{ all } \alpha \in X,$$
$$S_y^*(\theta) = \mathscr{X}^{\phi(y)}(y) \cdot \mathscr{X}^{2\phi(y)} \cdot \theta.$$

[2] By definition, we take this to mean that the sections of L^n, $n \geqq 1$, form a basis of the topology of $\tilde{P}$. In EGA, II. 4.5., Grothendieck only defines ample on quasi-compact schemes, but this seems to be the best property among his equivalent defining properties for our purposes.

Let

(2.3) DEFINITION. $R_{\phi,\Sigma}$ = subring of $\mathscr{R}$ generated over A by the elements $S_y(\mathscr{X}^\alpha\theta)$, $y \in Y$, $\alpha \in \Sigma$,
i.e.,

$$R_{\phi,\Sigma} = A[\cdots, \mathscr{X}^{\phi(y)+\alpha}(y) \cdot \mathscr{X}^{2\phi(y)+\alpha} \cdot \theta, \cdots].$$

This ring is only what we want in case $\mathscr{X}^{\phi(y)+\alpha}(y) \in A$ for all $y \in Y$, $\alpha \in \Sigma$, i.e., in case

$$R_{\phi,\Sigma} \subset A[\cdots, \mathscr{X}^\alpha \cdot \theta, \cdots].$$

However, we can always make sure that this extra condition is satisfied if we replace the polarization ϕ by $n\phi$ for a large enough n. This follows from lemma (1.4). From now on, we assume that ϕ has been so chosen that $\mathscr{X}^{\phi(y)+\alpha}(y) \in A$, all $y \in Y$, $\alpha \in \Sigma$.

Now consider Proj$(R_{\phi,\Sigma})$. I claim it is a relatively complete model of $\tilde{G}$! Since it is Proj of a graded ring generated by elements of degree 1, it carries a canonical ample invertible sheaf $\mathcal{O}(1)$. Moreover, the automorphisms S_y^* of $R_{\phi,\Sigma}$ induce automorphisms S_y of Proj$(R_{\phi,\Sigma})$ and a compatible automorphism S_y^* of $\mathcal{O}(1)$, i.e., an action of Y on Proj$(R_{\phi,\Sigma})$ and on$\mathcal{O}(1)$. To get an action of $\tilde{G}$ on Proj$(R_{\phi,\Sigma})$ and on $\mathcal{O}(1)$, it suffices to define, for every A-algebra B and every B-valued point a of $\tilde{G}$, an automorphism T_a^* of $B \otimes_A R_{\phi,\Sigma}$, in a way which is functorial and compatible with compositions. In fact, let

$$\begin{aligned} T_a^*(c) &= c, \text{ all } c \in A, \\ T_a^*(\mathscr{X}^\alpha) &= \mathscr{X}^\alpha(a) \cdot \mathscr{X}^\alpha, \text{ all } \alpha \in X, \\ T_a^*(\theta) &= \theta. \end{aligned}$$

This clearly has all these properties and gives us our action. Since:

$$T_a^* \cdot S_y^* = R \cdot S_y^* \cdot T_a^*$$

where

$$\begin{aligned} R(c) &= c, \text{ all } c \in A \\ R(\mathscr{X}^\alpha) &= \mathscr{X}^\alpha, \text{ all } \alpha \in X \\ R(\theta) &= \mathscr{X}^{2\phi(y)}(a) \cdot \theta, \end{aligned}$$

it follws that the T_a's and S_y's skew-commute as required for the polarization 2ϕ.

Next Proj$(R_{\phi,\Sigma})$ is covered by the affine open sets:

$$U_{\alpha,y} = \operatorname{Spec} A\left[\cdots, \frac{\mathscr{X}^{\phi(z)+\beta}(z)}{\mathscr{X}^{\phi(y)+\alpha}(y)} \cdot \mathscr{X}^{2\phi(z-y)+\beta-\alpha}, \cdots\right]_{\substack{\beta\in\Sigma\\ z\in Y}}.$$

Although we have here an infinite number of open sets, the action S_{y_0} of $y_0 \in Y$ carries $U_{\alpha,y}$ to $U_{\alpha,y+y_0}$, so there are only a finite number of orbits mod Y in this collection of open sets. All the affine rings are integral domains contained in $K(\cdots, \mathscr{X}^{\alpha}, \cdots)$. Moreover,

$$\begin{aligned} U_{0,0} &= \operatorname{Spec} A[\cdots, \mathscr{X}^{\phi(z)+\beta}(z) \cdot \mathscr{X}^{2\phi(z)+\beta}, \cdots]_{\substack{z\in Y\\ \beta\in\Sigma}} \\ &= \operatorname{Spec} A[\cdots, \mathscr{X}^{\beta}, \cdots]_{\beta\in\Sigma} \\ &= \operatorname{Spec} A[\cdots. \mathscr{X}^{\beta}, \cdots]_{\beta\in X} = \tilde{G}. \end{aligned}$$

Thus $\operatorname{Proj}(R_{\phi,\Sigma})$ is an integral scheme over A containing $\tilde{G}$ as a dense open subset. Next we prove:

(2.4) PROPOSITION. *$U_{\alpha,y}$ is of finite type over A.*

PROOF. Since $U_{\alpha,y}$ is isomorphic to $U_{\alpha,z}$ for any $y, z \in Y$, is suffices to check this for $U_{\alpha,0}$. The affine ring of $U_{\alpha,0}$ is generated by the infinite set of monomials:

$$M_{\beta,y} = \mathscr{X}^{\phi(y)+\beta}(y) \cdot \mathscr{X}^{2\phi(y)+\beta-\alpha}.$$

It is easy to check that:

$$(*)\ M_{\beta,y} = [\mathscr{X}^{2\phi(y-z)-\alpha+\beta}(z)] \cdot M_{\alpha,z} \cdot M_{\beta,y-z}.$$

Now for all z, choose a positive integer n_z such that

$$\mathscr{X}^{\phi(z)}(z)^{2n_z-3} \cdot \mathscr{X}^{-\alpha+\beta}(z) \in A, \text{ all } \beta \in \Sigma.$$

Then by the basic lemma (1.3), there is a finite set $z_1, \cdots, z_k$ such that for all $y \in Y$ except for a finite set $S \subset Y$,

$$\mathscr{X}^{\phi(y)}(z_i) \in [\mathscr{X}^{\phi(z_i)}(z_i)]^{n_{z_i}} \cdot A, \text{ some } i.$$

Therefore

$$\begin{aligned} \mathscr{X}^{2\phi(y-z_i)-\alpha+\beta}(z_i) &= [\text{elt. of } A \cdot \mathscr{X}^{\phi(z_i)}(z_i)^{n_{z_i}}]^2 \cdot \mathscr{X}^{-2\phi(z_i)}(z_i) \cdot \mathscr{X}^{-\alpha+\beta}(z_i) \\ &= (\text{elt. of } A) \cdot \mathscr{X}^{\phi(z_i)}(z_i) \in I, \end{aligned}$$

so that

$$M_{\beta,y} = (\text{elt. of } I) \cdot M_{\alpha,z_i} \cdot M_{\beta,y-z_i}.$$

Thus $M_{\alpha_1 z}, \cdots, M_{\alpha,z_k}$ plus the $M_{\beta,y}$ with $y \in S$ generate the whole ring.

QED

It follows that if $U = \bigcup_{\alpha\in\Sigma} U_{\alpha,0}$, then U is an open subset of $\operatorname{Proj}(R_{\phi,\Sigma})$ of finite type over A such that

$$\bigcup_{y\in Y} S_y(U) = \operatorname{Proj}(R_{\phi,\Sigma}).$$

It remains to check the following completeness property for $\mathrm{Proj}(R_{\phi,\Sigma})$: if v is a valuation of $\boldsymbol{R}(\tilde{G})$, $v \geqq 0$ on A, then

$$\begin{bmatrix} v \text{ has a center} \\ \text{on Proj } (R_{\phi,\Sigma}) \end{bmatrix} \Leftrightarrow \begin{bmatrix} \forall \alpha \in x, \exists y \in Y \text{ such that} \\ v(\mathscr{X}^{\alpha}(y) \cdot \mathscr{X}^{\alpha}) \geqq 0 \end{bmatrix}.$$

Clearly v has a center on $\mathrm{Proj}(R_{\phi,\Sigma})$ if and only if

$$\min_{\substack{\alpha \in \Sigma \\ y \in Y}} v(\mathscr{X}^{\phi(y)+\alpha}(y) \cdot \mathscr{X}^{2\phi(y)+\alpha}) \text{ exists.}$$

As for the other side, note that it is equivalent to any of the statements:

$$(*)'[\forall z \in Y, \exists y \in Y \text{ such that } v(\mathscr{X}^{\phi(z)}(y) \cdot \mathscr{X}^{\phi(z)}) \geqq 0]$$
$$(*)''[\forall z \in Y, \exists \alpha \in X \text{ such that } v(\mathscr{X}^{\alpha}(z) \cdot \mathscr{X}^{\phi(z)}) \geqq 0]$$
$$(*)'''[\forall z \in Y, \exists n \geqq 1 \text{ such that } v(\mathscr{X}^{\phi(z)}(z)^{n} \cdot \mathscr{X}^{\phi(z)}) \geqq 0].$$

Now to check the implication '$\Rightarrow$', suppose that:

$$\min_{\substack{\alpha \in \Sigma \\ y \in Y}} v(\mathscr{X}^{\phi(y)+\alpha}(y) \cdot \mathscr{X}^{2\phi(y)+\alpha}) = v(\mathscr{X}^{\phi(y_0)+\alpha_0}(y_0) \cdot \mathscr{X}^{2\phi(y_0)+\alpha_0}).$$

If $(*)''$ is false, then take the $z \in Y$ for which $v(\mathscr{X}^{\alpha}(z) \cdot \mathscr{X}^{\phi(z)}) < 0$, all $\alpha \in X$, and simply note that

$$\begin{aligned} &v(\mathscr{X}^{\phi(y_0+z)+\alpha_0}(y_0+z) \cdot \mathscr{X}^{2\phi(y_0+z)+\alpha_0}) \\ &\quad = v(\mathscr{X}^{\phi(y_0)+\alpha_0}(y_0) \cdot \mathscr{X}^{2\phi(y_0)+\alpha_0}) + v(\mathscr{X}^{\phi(z)}) + v(\mathscr{X}^{2\phi(y_0)+\phi(z)+\alpha_0}(z) \cdot \mathscr{X}^{\phi(z)}) \\ &\quad < v(\mathscr{X}^{\phi(y_0)+\alpha_0}(y_0) \cdot \mathscr{X}^{2\phi(y_0)+\alpha_0}). \end{aligned}$$

Conversely, assume that $(*)'''$ holds. For all $y \in Y$, choose n_y large enough so that

$$v(\mathscr{X}^{\phi(y)}(y)^{n_y} \cdot \mathscr{X}^{2\phi(y)}) \geqq 0$$

and

$$\mathscr{X}^{\phi(y)}(y)^{n_y - 1} \cdot \mathscr{X}^{\alpha}(y) \in A, \text{ all } \alpha \in \Sigma.$$

By the basic lemma (1.3), there exist $y_1, \cdots, y_k \in Y$ such that for all $z \notin S$, a finite set, $\mathscr{X}^{\phi(z)}(y_i) \in [\mathscr{X}^{\phi(y_i)}(y_i)]^{n_{y_i}} \cdot A$ for some i. But then

$$\begin{aligned} &v(\mathscr{X}^{\phi(z)+\alpha}(z) \cdot \mathscr{X}^{2\phi(z)+\alpha}) \\ &\quad = v(\mathscr{X}^{\phi(z-y_i)+\alpha}(z-y_i) \cdot \mathscr{X}^{2\phi(z-y_i)+\alpha}) \\ &\qquad + v(\mathscr{X}^{2\phi(z)}(y_i) \cdot \mathscr{X}^{-\phi(y_i)+\alpha}(y_i) \cdot \mathscr{X}^{2\phi(y_i)}) \\ &\quad \geqq v(\mathscr{X}^{\phi(z-y_i)+\alpha}(z-y_i) \cdot \mathscr{X}^{2\phi(z-y_i)+\alpha}) \\ &\qquad + v(\mathscr{X}^{\phi(y_i)}(y_i)^{n_{y_i}} \cdot \mathscr{X}^{2\phi(y_i)}) \\ &\qquad + v(\mathscr{X}^{\phi(y_i)}(y_i)^{n_{y_i}-1} \cdot \mathscr{X}^{\alpha}(y_i)) \\ &\quad \geqq v(\mathscr{X}^{\phi(z-y_i)+\alpha}(z-y_i) \cdot \mathscr{X}^{2\phi(z-y_i)+\alpha}) \end{aligned}$$

so that the minimum in question exists for some $z \in S$. This completes the proof of:

(2.5) THEOREM. *Let $\tilde{G}$ be a split torus over S, let $Y \subset \tilde{G}(K)$ be a set of periods and $\phi\colon Y \to X$ a polarization. Then if ϕ is replaced by $n\phi$ for $n \in \mathbf{Z}$ sufficiently large, $\mathrm{Proj}(R_{\phi,\Sigma})$ is a relatively complete model of $\tilde{G}$ over S relative to Y and 2ϕ.*

3. The Construction of the quotient

We can now forget about $\mathrm{Proj}(R_{\phi,\Sigma})$ and deal with an arbitrary relatively complete model $\tilde{P}$. The first thing I want to prove is that $\tilde{P}$ is not 'too much' bigger than $\tilde{G}$:

(3.1) PROPOSITION. *Let $y \in Y$ and let $f = \mathscr{X}^{\phi(y)}(y)$. Then in the open set $\tilde{P}_f \subset \tilde{P}$, $\mathscr{X}^{\phi(y)}$ is a unit, i.e., $\in \Gamma(\tilde{P}_f, \mathcal{O}^*_{\tilde{P}})$.*

PROOF. By the axioms for a polarization, it follows easily that for all $\alpha \in X$,

$$\mathscr{X}^{n\phi(y)+\alpha}(y) \in A \text{ if } n \gg 0.$$

Therefore $\mathscr{X}^{\alpha}(y) \in A_f$. In other words, the section y of $\tilde{G}$ over $\{\eta\}$ extends to a section over S_f too. For clarity, call this y'. Then we have automorphisms S_y and $T_{y'}$ of $\tilde{P}_f$ and S^*_y and $T^*_{y'}$, of $\tilde{L}_f = \tilde{L}|_{\tilde{P}_f}$. Since $S_y = T_{y'}$, on G_f, S_y equals $T_{y'}$, everywhere, so

$$S^*_y = \begin{pmatrix}\text{mult. by a unit } \lambda \\ \text{in } \tilde{P}_f\end{pmatrix} \cdot T^*_{y'}.$$

The law of skew-commutativity of S^*_y with the operators T^*_a shows that

$$\lambda(a+b) = \mathscr{X}^{\phi(y)}(b) \cdot \lambda(a)$$

for all points a, b of $\tilde{G}_\eta$ in some K-algebra B. Therefore, on $\tilde{G}_\eta$, the function $\lambda/\mathscr{X}^{\phi(y)}$ is constant along the fibres, i.e.,

$$\lambda = \zeta \cdot \mathscr{X}^{\phi(y)}, \ \zeta \in K.$$

But λ and $\mathscr{X}^{\phi(y)}$ are units on the bigger open set $\tilde{G}_f$, hence ζ must be a unit in A_f. Then λ and ζ^{-1} are units in $\tilde{P}_f$ so $\mathscr{X}^{\phi(y)}$ is also. *QED*

(3.2) COROLLARY. $\tilde{G}_\eta = \tilde{P}_\eta$.

Next, let's look at the closed subscheme $\tilde{P}_0$ of $\tilde{P}$.

(3.3) PROPOSITION. *Every irreducible component of $\tilde{P}_0$ is proper over A/I.*

PROOF. First apply the completeness condition (ii) to prove that if Z

is any component of $\tilde{P}_0$ and if v is any valuation of its quotient field $\boldsymbol{R}(Z)$, with $v \geqq 0$ on A/I, then v has a center on Z. In fact, let v_1 be a valuation of $\boldsymbol{R}(\tilde{G})$, $v \geqq 0$ on A, whose center is Z, and let v_2 be the composite of the valuations v and v_1. Since for all $z \in Y$, if $n \gg 0$, $\mathscr{X}^{\phi(z)}(z)^n \cdot \mathscr{X}^{\phi(z)}$ is regular and zero at the generic point of Z by (3.1), it follows that $v_1(\mathscr{X}^{\phi(z)}(z)^n \cdot \mathscr{X}^{\phi(z)}) > 0$, hence $v_2(\mathscr{X}^{\phi(z)}(z)^n \cdot \mathscr{X}^{\phi(z)}) > 0$. So by the completeness condition, v_2 has a center on $\tilde{P}$, hence v has a center on Z.

The Proposition now follows from the rather droll:

(3.4) LEMMA. *Let f: $X \to Y$ be a morphism locally of finite type, with X an irreducible scheme but Y arbitrary. If f satisfies the valuative criterion for properness for all valuations, then f is proper.*

PROOF. The usual valuative criterion (cf. EGA. II-7-3) would hold if we know that f was of finite type. It suffices to prove that f is quasi-compact. To prove this, we may as well replace X by X_{red}, Y by Y_{red}; then looking locally on the base, we can assume Y is affine, say Spec(A); and finally we can assume f is dominating. Then A is a subring of the function field $\boldsymbol{R}(X)$ of X: let $\mathscr{X}$ be Zariski's Riemann Surface of $\boldsymbol{R}(X)/A$ (cf. Zariski-Samuel, vol. II, p. 110). Set-theoretically, $\mathscr{X}$ is the set of valuations v of $\boldsymbol{R}(X)$, $v \geqq 0$ on A. By the valuation criterion, every v has a center on X, so there is a natural map π: $\mathscr{X} \to X$ taking v to its center. Now $\mathscr{X}$ is a quasi-compact topological space and π is continuous and surjective. Therefore X is quasi-compact. *QED*

(3.5) COROLLARY. *The closure $\bar{U}_0$ of U_0 in the scheme $\tilde{P}_0$ is proper over S_0.*

PROOF. Since U_0 is of finite type over S_0, $\bar{U}_0$ has only a finite number of irreducible components, and by the Proposition, each is of finite type over S_0. *QED*

The next thing I want to prove is that Y acts freely and discontinuously on $\tilde{P}_0$ in the Zariski-topology:

(3.6) PROPOSITION. *Let $U \subset \tilde{P}$ be the open set given by the definition of a relatively complete model. Let $\bar{U}_0$ be the closure of U_0 in $\tilde{P}_0$. There is a finite subset $S \subset Y$ such that*

$$S_y(\bar{U}_0) \cap S_z(\bar{U}_0) = \phi$$

if $y - z \notin S$.

PROOF. Let $F \subset \tilde{P}$ be the closed subset which is the locus of geometric points left fixed by the action of $\tilde{G}$. The action of $\tilde{G}$ on the invertible sheaf $\tilde{L}|_F$ is a 1-dimensional representation of $\tilde{G}$ over the base scheme

F. Therefore for every *connected* subset $F' \subset F$, there is a character $\alpha \in X$ such that $\tilde{G}$ acts on $\tilde{L}|_{F'}$ via the character α. Moreover, if $y \in Y$, then $S_y(F')$ will be another connected subset of F, and by the skew-commutativity of the actions of Y and $\tilde{G}$, $\tilde{G}$ will act on $\tilde{L}|_{S_y(F')}$ through the character $\alpha+\phi(y)$. Now $F \cap \bar{U}_0$ has only a finite number of connected components: let $\alpha_1, \cdots, \alpha_n$ be the characters $\tilde{G}$ associated to the action of $\tilde{G}$ on $\tilde{L}$ on these sets. Then $\tilde{G}$ acts on $\tilde{L}$ at the points of $F \cap S_y(\bar{U}_0)$ through the characters $\alpha_1+y, \cdots, \alpha_n+y$. Now suppose $S_y(\bar{U}_0) \cap S_z(\bar{U}_0) \neq \phi$. Since this intersection is proper over Spec(A/I), by the Borel fixed point theorem (cf. A. Borel-Linear algebraic groups; Benjamin, 1959; pag. 242, th. 10.4):

$$F \cap S_y(\bar{U}_0) \cap S_z(\bar{U}_0) \neq \phi.$$

Looking at the action of $\tilde{G}$ on $\tilde{L}$ here, it follows that one of the characters α_i+y must equal one of the characters α_j+z. Let $S = \{\cdots, \alpha_i-\alpha_j, \cdots\}$. Thus $y-z \in S$.

QED

(3.7) Corollary. *Y acts freely on* $\tilde{P}_0$.

Proof. If some $y \in Y$, $y \neq 0$, had a fixed point x in $\tilde{P}_0$, then since y has infinite order, x would be left fixed by an infinite subgroup of Y and this would contradict the Proposition. *QED*

(3.8) Theorem. $\tilde{P}_0$ *is connected.*

Proof. Note that since A is complete in the I-adic topology, and A has no idempotents, A/I has no idempotents either, i.e., S_0 is connected. Therefore $\tilde{G}_0$ is a connected open subset of $\tilde{P}_0$ and it determines a canonical connected component of $\tilde{P}_0$. Now suppose there is a 2nd connected component. Choose a point $x \in \tilde{P}_0$ in this 2nd component and let v be a discrete rank 1 valuation of $\boldsymbol{R}(\tilde{G})$, $v \geqq 0$ on A, with center x. Let $A' = \{x \in K | v(x) \geqq 0\}$. Let $S' = \text{Spec}(A')$ and $\tilde{P}' = \tilde{P} \times_S S'$. Now A' is a discrete, rank 1, valuation ring with quotient field K, so S and S' have the 'same' generic point, and $\tilde{P}$, $\tilde{P}'$ have the 'same' generic fibre. Let $\tilde{P}''$ be the closure in $\tilde{P}'$ of its generic fibre. $\tilde{P}''$ is an integral scheme with the same quotient field as $\tilde{P}$, namely $\boldsymbol{R}(\tilde{G})$, and it is locally of finite type over the valuation ring A'. Now S' has only 2 points – its generic point and its closed point. Let $\tilde{P}_0''$ be the fibre of $\tilde{P}''$ over the closed point. There is a natural morphism

$$\tilde{P}_0'' \to \tilde{P}_0.$$

I claim that (a) $x \in$ Image, and (b) $\tilde{G}_0 \subset$ Image, hence the image meets $\geqq 2$ connected components, hence $\tilde{P}_0''$ is disconnected too. The reason for (a) is that if R_v = valuation ring of v, then we are given morphisms:

$$\begin{array}{ccc} \mathrm{Spec}\,(R_v) & \longrightarrow & \tilde{P} \\ \downarrow & & \downarrow \\ S' & \longrightarrow & S \end{array}$$

hence we get a morphism $\mathrm{Spec}(R_v) \to \tilde{P}'$, which takes the generic point of $\mathrm{Spec}(R_v)$ to the generic fibre of $\tilde{P}'$. Therefore this morphism factors through $\tilde{P}''$. The image $x' \in \tilde{P}_0''$ of the closed point lies over x. The reason for (b) is that $\tilde{G} \subset \tilde{P}$, and $\tilde{G}$ is smooth of S; so $\tilde{G} \times_s S' \subset \tilde{P}'$, and it is smooth over S'; so $\tilde{G} \times_s S' \subset \tilde{P}''$.

Now because of the completeness property (ii) of $\tilde{P}$ we know exactly which valuations of $\boldsymbol{R}(\tilde{G})$ have centers in $\tilde{P}''$ too. We have reduced the Theorem to:

(3.9) Lemma. *Let A be a discrete rank 1 valuation ring with maximal ideal (π), let G be a split torus over $S = \mathrm{Spec}(A)$, let P be an integral scheme, locally of finite type over S containing G as a dense open subset. Assume*:

1) *the generic fibres G_η, P_η are equal,*

2) *for all valuations v of $\boldsymbol{R}(G)$, $v \geqq 0$ on A and $v(\pi) > 0$, v has a center on P if and only if for all $\alpha \in X$, (the character group of G), $-nv(\pi) \leqq v(\mathscr{X}^\alpha) \leqq nv(\pi)$, if $n \gg 0$.*

Then the closed fibre P_0 of P is connected.

Proof. Introduce a basis in X, and let $\mathscr{X}_1, \cdots, \mathscr{X}_r$ be the corresponding characters. For all n, let

$$\begin{aligned} P^{(n)} &= \mathrm{Spec}\, A[\pi^n \mathscr{X}_1 \cdot \pi^n \mathscr{X}_1^{-1}, \cdots, \pi^n \mathscr{X}_r, \pi^n \cdot \mathscr{X}_r^{-1}] \\ &\cong \mathrm{Spec}\, A[U_1, V_1, \cdots, U_r, V_r]/(U_1 V_1 - \pi^{2n}, \cdots, U_r V_r - \pi^{2n}). \end{aligned}$$

This scheme is a relative complete intersection in A_s^{2r} over S and is smooth over S, hence regular, outside a subset of codimension 2. Therefore $P^{(n)}$ is a normal scheme. $P^{(n)}$ and P have the same function field, so let $Z^{(n)} \subset P^{(n)} \times_A P$ be the join of this birational correspondence. By (1) and (2), all valuations v of $\boldsymbol{R}(G)$ with a center on $P^{(n)}$ also have a center on P, so $Z^{(n)} \to P^{(n)}$ satisfies the valuative criterion for properness. Since $Z^{(n)}$ is at least locally of finite type over $P^{(n)}$, by lemma (3.4), $Z^{(n)}$ is proper over $P^{(n)}$. Therefore by Zariski's connectedness theorem, all fibres of $Z^{(n)}$ over $P^{(n)}$ are connected. Now the closed fibre of $P^{(n)}$ is isomorphic to:

$$\mathrm{Spec}\,(k[U, V]/(U \cdot V)) \underset{\mathrm{Spec}\, k}{\times} \cdots \underset{\mathrm{Spec}\,(k)}{\times} \mathrm{Spec}\,(k[U, V]/(U \cdot V))$$

where $k = A/(\pi)$, which is certainly connected. Therefore the closed fibre of $Z^{(n)}$ is connected. Therefore if

$$W_n = \overline{p_2(Z_0^{(n)})} \subset P,$$

W_n is connected too. But I claim that $P_0 = \bigcup_n W_n$. In fact, every valuation v with a center x on P_0 has a center on *some* $P^{(n)}$ because of (2). Therefore x lifts to a point of $Z^{(n)}$ for some n, hence $x \in W_n$. This shows that P_0 is connected. *QED*

We are now ready to begin the construction of G. The first step is:

(3.10) THEOREM. *For every $n \geqq 1$, there exists a scheme P_n, projective over A/I^n, an ample sheaf $\mathcal{O}(1)$ on P_n, and an étale surjective morphism:*

$$\pi: \tilde{P} \times_A A/I^n \to P_n$$

such that set-theoretically, $\pi(x) = \pi(y)$ if and only if x and y are in the same Y-orbit, and such that $\mathcal{O}(1)$ on $\tilde{P} \times_A A/I^n$ is the pull-back of $\mathcal{O}(1)$on P_n.

PROOF. First, let $k \geqq 1$ be an integer such that under the action of the subgroup $kY \subset Y$, no 2 points of any open set

$$S_y(U) \times_A A/I^n$$

are identified. Then we can form a quotient:

$$\pi' : \tilde{P} \times_A A/I^n \to P'_n$$

by the subgroup kY by the simple device of gluing these basic open sets together on bigger overlaps. Observe that since Y acts on $\mathcal{O}(1)$ we get a 'descended' form of $\mathcal{O}(1)$ on P'_n.

Choose coset representatives $y_1, \cdots, y_l \in Y$ for the cosets of kY in Y. Now notice that the restriction of π':

$$\operatorname{Res}(\pi') : \bigcup_{y \in \{y_1, \cdots, y_l\}} S_y(U) \times_A A/I \to P'_n$$

is surjective, hence so is the restriction:

$$\operatorname{Res}(\pi') : \bigcup_{y \in \{y_1, \cdots, y_l\}} S_y(\overline{U_0}) \to P'_n.$$

But the scheme on the left is a finite union of schemes proper over A/I. Therefore P'_n is proper over A/I^n. Moreover $\mathcal{O}(1)$ on P'_n pulls back to $\mathcal{O}(1)$ on the left, and $\mathcal{O}(1)$ here is ample. Therefore $\mathcal{O}(1)$ on P'_n is ample too (for instance, by Nakai's criterion, cf. S. Kleiman – A note on the Nakai-Moisezon test for ampleness of a divisor; Amer. J. Math. *87* (1965), 221–226).

Finally, the finite group Y/kY acts freely on the projective scheme P'_n and on the ample sheaf $\mathcal{O}(1)$, so a quotient $P_n = P'_n/(Y/kY)$ exists. Moreover by descent, P_n carries an ample $\mathcal{O}(1)$ too, so P_n has all the required properties. *QED*

These schemes P_n obviously fit together to form a formal scheme $\mathfrak{P}$ proper over A. Moreover the sheaves $\mathcal{O}(1)$ fit together into an ample sheaf $\mathcal{O}(1)$ on $\mathfrak{P}$. We now apply the fundamental *formal existence* theorem (EGA Ch. 3, (5.4.5)) of Grothendieck: this shows that $\mathfrak{P}$ is the formal completion of a unique scheme P, proper over A, and that $\mathcal{O}(1)$ on $\mathfrak{P}$ comes by completion from an $\mathcal{O}(1)$ on P relatively ample over A.

Now inside all of our schemes, we want to pick out a big open set:

(*I*.) $\bigcup_{y \in Y} S_y(\tilde{G}) \subset \tilde{P}$

(*II*.) $G_n =^{def} \bigcup_{y \in Y} (S_y(\tilde{G}) \times_A A/I^n)/Y \subset P_n$

(*III*.) $\varprojlim_n G_n = \mathfrak{G} \subset \mathfrak{P}$.

Note that $G_n \cong \tilde{G} \times_A A/I^n$, so that $\mathfrak{G} \cong$ I-adic completion of $\tilde{G}$. To pick out an open subset of P whose completion is $\mathfrak{G}$, proceed as follows:

(*I*.) Let $\tilde{B} = \tilde{P} - \bigcup_{y \in Y} S_y(\tilde{G})$ and make $\tilde{B}$ into a reduced closed subscheme of $\tilde{P}$.

(*II*.) Let $B_n = (\tilde{B} \times_A A/I^n)/Y \subset P_n$.

(*III*.) Let $\varprojlim_n B_n = \mathfrak{B} \subset \mathfrak{P}$.

Then $\mathfrak{B}$ is the formal completion of a reduced closed subscheme $B \subset P$. Finally let $G = P - B$. Then, by construction the I-adic completion of G is $\mathfrak{G}$, i.e., the I-adic completions of G and of $\tilde{G}$ are canonically isomorphic.

This G is our final goal. We will eventually prove that G is a semi-abelian group scheme.

4. G is semi-abelian

We begin by proving that G is smooth over S. This is best proved in a more general context:

(4.1) Proposition. Given

(1) 2 *schemes, locally of finite type over* S:

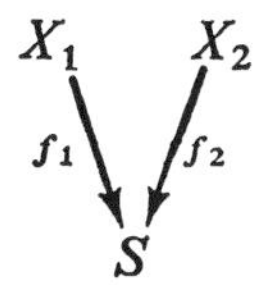

with X_2 *proper over* S.

(2) *an étale surjective morphism of their I-adic completions*:

$$\pi\colon \mathscr{X}_1 \to \mathscr{X}_2$$

(3) *closed reduced subschemes* $B_1 \subset X_1$, $B_2 \subset X_2$, *assume that the following holds*:

(4) $X_1 - B_1$ *is smooth over S, of relative dimension r*,

(5) *if* $\mathfrak{B}_i$ = *I-adic completion of* B_i, *then we have the inclusion of formal subschemes (not just subsets)*

$$\mathfrak{B}_1 \subset \pi^{-1}(\mathfrak{B}_2),$$

then we conclude that X_2–B_2 *is smooth over S, of relative dimension r*.

PROOF. First, let's check that X_2–B_2 is flat over S. Let $\mathscr{M} \subset \mathscr{N}$ be 2 A-modules. Consider the 2 kernels:

$$0 \to \mathscr{K}_1 \to \mathscr{M} \otimes \mathcal{O}_{X_1} \to \mathscr{N} \otimes \mathcal{O}_{X_1}$$
$$0 \to \mathscr{K}_2 \to \mathscr{M} \otimes \mathcal{O}_{X_2} \to \mathscr{N} \otimes \mathcal{O}_{X_2}.$$

Since X_1–B_1 is flat over S, $\operatorname{Supp}(\mathscr{K}_1) \subset B_1$, hence for all $x \in X_1$, $(\mathscr{K}_1)_x \cdot (I_{B_1})_x^n = (0)$ for some n. Taking I-adic completions, we get exact sequences:

$$0 \to \hat{\mathscr{K}}_1 \to \mathscr{M} \otimes \mathcal{O}_{\mathscr{X}_1} \to \mathscr{N} \otimes \mathcal{O}_{\mathscr{X}_1}$$
$$0 \to \hat{\mathscr{K}}_2 \to \mathscr{M} \otimes \mathcal{O}_{\mathscr{X}_2} \to \mathscr{N} \otimes \mathcal{O}_{\mathscr{X}_2}.$$

and since π is flat, it follows that $\hat{\mathscr{K}}_1 \cong \pi^*\hat{\mathscr{K}}_2$. Since for all $x \in \mathscr{X}_1$, $(\hat{\mathscr{K}}_1)_x \cdot (I_{\mathfrak{B}_1})_x^n = (0)$ for some n, and since by (5), $I_{B_1} \supset \pi^*(I_{B_2})$, it follows that $(\hat{\mathscr{K}}_2)_{\pi(x)} \cdot (I_{\mathfrak{B}_2})_{\pi(x)}^n = (0)$. This means that in an open neighborhood of $f_2^{-1}(S_0)$, $\mathscr{K}_2$ is killed by I_{B_2}, hence $\operatorname{Supp}(\mathscr{K}_2) \subset B_2$. Since X_2 is proper over S, all closed points of X_2 lie over S_0 so that $\operatorname{Supp}(\mathscr{K}_2) \subset B_2$ everywhere.

To show that X_2–B_2 is actually smooth over S, it suffices to show that in addition to being flat, it is differentiably smooth: i.e., $\Omega^1_{X_2/S}$ is locally free of rank r outside B_2, and $S^n(\Omega^1_{X_2/S}) \to \mathfrak{P}^n_{X_2/S}$ is an isomorphism outside B_2 (cf. EGA, IV[4]. 16). Since we know these are true for X_1/S, we deduce, in particular, that at all points x of $f_1^{-1}(S_0)$,

a) for all $g \in I(B_1)_x$, $(\Omega^1_{X_1/S})_x \otimes_{\mathcal{O}_{x,X_1}} \mathcal{O}_{x,X_1}[1/g]$ is locally free of rank r over $\mathcal{O}[1/g]$,

b) Ker and Coker of $S^n(\Omega^1_{X_1/S})_x \to \mathfrak{P}^n_{X_1/S,x}$ are killed by powers of $I(B_1)_x$.

These 2 facts imply the corresponding facts for the formal scheme $\mathscr{X}_1$. Since π is étale, $\pi^*(\Omega^1_{\mathscr{X}_2/S}) \cong \Omega^1_{\mathscr{X}_1/S}$ and $\pi^*(\mathfrak{P}^n_{\mathscr{X}_2/S}) \cong \mathfrak{P}^n_{\mathscr{X}_1/S}$, so by assumption (5), we get the corresponding facts for $\mathscr{X}_2$. Finally, these imply that (a) and (b) hold for X_2/S at points $x \in f_2^{-1}(S_0)$. Since X_2 is proper over S, they hold everywhere on X_2. Thus $X_2 - B_2$ is differentiably smooth over S. *QED*

(4.2) COROLLARY. *G is smooth over S.*

PROOF. Take $X_1 = \tilde{P}$, $X_2 = P$, $B_1 = \tilde{B}$, $B_2 = B$, and use the fact that $\tilde{G}$ is smooth over S. *QED*

Our next step is to show that P and hence G is irreducible:

(4.3) PROPOSITION. *P is irreducible.*

PROOF. To prove this, we may assume that P is normal. In fact, we can replace $\tilde{P}$ by its normalization and then, since A is excellent, $\tilde{\mathfrak{P}}$ is normal, hence $\mathfrak{P}$ and P are normal. Then to show that P is irreducible, it suffices to show that P is connected. But we know that $\tilde{P}_0$ is connected. Therefore $P_0 = \tilde{P}_0/Y$ is connected. But P is proper over S and so therefore P is connected too. *QED*

Next, we take up the key problem of proving that G is independent of the choice of the polarization ϕ and the model $\tilde{P}$.

(4,4) DEFINITION. A subtorus $\tilde{H} \subset \tilde{G}$ is *integrable* if

$$\operatorname{rank}(Y \cap \tilde{H}(K)) = \dim \tilde{H}.$$

The key step in our proof of independence is that an integrable subtorus $\tilde{H} \subset \tilde{G}$ defines a closed subscheme $H \subset G$ in the following way:

a) let W_1 be the closure of $\tilde{H}$ in $\tilde{P}$ (considered as a reduced closed subscheme of $\tilde{P}$). If $Y^* = Y \cap \tilde{H}(K)$, then W_1 is Y^*-invariant.

b) let $\mathfrak{W}_1$ be the I-adic completion of W_1. It is also Y^*-invariant. Then

$$\mathfrak{W}_2 = \bigcup_{y \in Y/Y^*} S_y(\mathfrak{W}_1)$$

turns out to be a locally finite union, so this defines $\mathfrak{W}_2$ as a reduced closed subscheme of $\tilde{\mathfrak{P}}$.

c) Let $\mathfrak{W}_3 = \mathfrak{W}_2/Y \subset \mathfrak{P}$.

d) Let $W_3 \subset \mathrm{P}$ be the reduced closed subscheme whose I-adic completion is $\mathfrak{W}_3$.

e) $H = W_3 \cap G$.

The finiteness assertion in (b) is the only non-trivial step where integrability has to be used. It results from:

(4.5) PROPOSITION. *Let $\tilde{H} \subset \tilde{G}$ be an integrable subtorus and let $Y^* = Y \cap \tilde{H}(K)$. Let W_1 be the closure of $\tilde{H}$ in a relatively complete model $\tilde{P}$ of $\tilde{G}$. Then there is a finite set $S \subset Y$ such that*

$$W_1 \cap S_y(U_0) = \phi \text{ if } y \notin S+Y^*.$$

PROOF. Let $X^* \subset X$ be the group of characters that are identically 1 on $\tilde{H}$. Then $\mathscr{X}^\alpha(y) = 1$ if $\alpha \in X^*$, $y \in Y^*$. In particular, $Y^* \cap \phi^{-1}(X^*) = (0)$ since if $y \in Y^* \cap \phi^{-1}(X^*)$, $y \neq 0$, then $\mathscr{X}^{\phi(y)}(y)$ is in I and is 1. If $r = \dim \tilde{G}$, $s = \dim \tilde{H}$, then rank $Y^* = s$ and rank $\phi^{-1}(X^*) =$ rank $X^* = r-s$. Therefore $Y^*+\phi^{-1}(X^*)$ has finite index in Y. Let $k \geqq 1$ be an integer such that

$$kY \subseteq Y^*+\phi^{-1}(X^*) \subseteq Y.$$

Let $Y^{**} = \{y \in Y | ny \in Y^*, \text{ some } n \geqq 1\}$. Then Y^{**}/Y^* is a finite group killed by k and Y/Y^{**} is torsion-free. Consider the quotient torus $\tilde{G}_1 = \tilde{G}/\tilde{H} \cdot Y^{**}$. Its character group is the subgroup $X^{**} \subset X^*$ of characters which are 1 on all of Y^{**}. Note that $kX^* \subset X^{**}$. In $\tilde{G}_1$, consider the group of periods $Y_1 = Y/Y^{**}$. Define a polarization $\psi\colon Y_1 \to X^{**}$ as follows:

let $\bar{y} \in Y_1$,
lift $\bar{y}$ to an element $y \in Y$,
write $ky = y^*+w$, where $y^* \in Y^*$, $\phi(w) \in X^*$;
set $\psi(\bar{y}) = k \cdot \phi(w)$.

If $\bar{y}_0$ is a second element of Y_1, and $ky_0 = y_0^*+w_0$ as above, then

$$\begin{aligned}\mathscr{X}^{\psi(\bar{y})}(\overline{y_0}) &= \mathscr{X}^{k\phi(w)}(y_0)\\ &= \mathscr{X}^{\phi(w)}(ky_0)\\ &= \mathscr{X}^{\phi(w)}(y_0^*+w_0)\\ &= \mathscr{X}^{\phi(w)}(w_0)\\ &= \mathscr{X}^{\phi(w_0)}(w) = \cdots = \mathscr{X}^{\psi(\overline{y_0})}(\bar{y})\end{aligned}$$

so that ψ is a bona fide polarization. We shall apply the basic lemma (1.3) to this torus and this polarization.

To find the appropriate n_y's, recall that for all $y \in Y$, $\mathscr{X}^{\phi(y)}$ is a regular function on $\tilde{P}$ outside of the locus $\mathscr{X}^{\phi(y)}(y) = 0$. In particular, since U is of finite type over A,

$$[\mathscr{X}^{\phi(y)}(y)]^n \cdot \mathscr{X}^{\phi(y)}$$

is a regular function on all of U if n is large. Increasing n by one, we can even make it a function that vanishes on U_0. Now if $\bar{y} \in Y_1$ and $y \in Y$

lies over $\bar{y}$ with $ky = y^* + w$ as above, then

$$\mathscr{X}^{\psi(\bar{y})}(\bar{y}) = \mathscr{X}^{\phi(w)}(w)$$

and

$$\mathscr{X}^{\psi(\bar{y})} = [\mathscr{X}^{\phi(w)}]^k.$$

So choose an integer $n_{\bar{y}}$ such that

$$(*) \qquad [\mathscr{X}^{\psi(\bar{y})}(\bar{y})]^{n_{\bar{y}}} \cdot \mathscr{X}^{\psi(\bar{y})} \text{ is regular on } U, \text{ zero on } U_0.$$

Applying the lemma, we find a finite set $y_1, \cdots, y_k \in Y$ and a finite subset $S \subset Y$ such that

$$(**) \qquad \begin{cases} \text{for all } z \in Y,\ z \notin S + Y^{**}, \\ \mathscr{X}^{\psi(\bar{z})}(\bar{y}_i) = (\text{elt. of } A) \cdot \mathscr{X}^{\psi(\bar{y}_i)}(\bar{y}_i)^{n_{y_i}}, \text{ for some } i. \end{cases}$$

Now consider the function $\mathscr{X}^{\psi(\bar{y}_i)}$ on $S_z(U)$. Via the isomorphism S_z of $S_z(U)$ and U, it corresponds to the function

$$\mathscr{X}^{\psi(\bar{y}_i)}(z) \cdot \mathscr{X}^{\psi(\bar{y}_i)}$$

on U. Combining (*) and (**), it follows that this function is regular on U and zero on U_0. Therefore $\mathscr{X}^{\psi(\bar{y}_i)}$ is regular on $S_z(U)$ and zero on $S_z(U_0)$. But $\mathscr{X}^{\psi(\bar{y}_i)} \equiv 1$ on W_1 so this shows that $W_1 \cap S_z(U_0) = \phi$. Since z was an arbitrary element of Y outside $S + Y^{**}$, this proves the Proposition. *QED*

We are now ready to prove:

(4.6) THEOREM. *Let $(\tilde{G}_i, Y_i, \phi_i, \tilde{P}_i)$, $i = 1, 2$ be 2 tori plus periods, polarizations, and relatively complete models. Let G_i, $i = 1, 2$ be the 2 schemes constructed as above. Then for all S-homomorphisms $\tilde{\alpha}: \tilde{G}_1 \to \tilde{G}_2$ such that $\tilde{\alpha}(Y_1) \subset Y_2$, there is a unique S-morphism $\alpha: G_1 \to G_2$ such that under the canonical isomorphisms of the I-adic completions of G_i and $\tilde{G}_i$, α and $\tilde{\alpha}$ are formally identical.*

PROOF OF THEOREM. Consider the torus $\tilde{G}_1 \times_s \tilde{G}_2$; $Y_1 \times Y_2$ is a set of periods for this torus, $X_1 \times X_2$ is its character group and $\phi_1 \times \phi_2$: $Y_1 \times Y_2 \to X_1 \times X_2$ is a polarization. Moreover $P_1 \times_S P_2$ is easily seen to be a relatively complete model for $\tilde{G}_1 \times \tilde{G}_2$ relative to these periods and this polarization. Now suppose $\tilde{\alpha}: \tilde{G}_1 \to \tilde{G}_2$ is an S-homomorphism such that $\tilde{\alpha}(Y_1) \subset Y_2$. Look at the graph

$$\tilde{H} = \text{Image of } (1, \tilde{\alpha}) : \tilde{G}_1 \to G_1 \times_S \tilde{G}_2.$$

It is a subtorus of $G_1 \times_S G_2$, and because $\tilde{\alpha}(Y_1) \subset Y_2$, it is integrable. As in the last Proposition, it induces a closed subscheme of $G_1 \times_S G_2$ as follows:

$$W_1 = \text{closure of } \tilde{H} \text{ in } \tilde{P}_1 \times_S \tilde{P}_2$$
$$\mathfrak{W}_2 = \bigcup_{y \in Y_1 \times Y_2} S_y(\mathfrak{W}_1)$$
$$\mathfrak{W}_3 = \mathfrak{W}_2/Y_1 \times Y_2 \subset \mathfrak{P}_1 \times_S \mathfrak{P}_2$$
$$H = W_3 \cap (G_1 \times_S G_2).$$

I claim that H is the graph of a morphism from G_1 to G_2. First of all, we prove that the projection

$$p_1 : W_3 \to P_1$$

is smooth of relative dimension 0 outside $B_1 = P_1 - G_1$. This follows by essentially the same argument used in the proof of Proposition (4.1). In fact, p_1: $W_1 \to \tilde{\mathfrak{P}}_1$ is smooth of rel. dim. 0 outside $\tilde{B} = \tilde{\mathfrak{P}}_1 - \tilde{G}_1$. Now locally at every point $\mathfrak{W}_2$ is the formal completion of a finite union $S_{y_1}(W_1) \cup \cdots \cup S_{y_k}(W_1)$, $y_i \in Y_1 \times Y_2$, and since this is also smooth of rel. dim. 0 outside $\tilde{B}_1$, so is p_1: $\mathfrak{W}_2 \to \tilde{\mathfrak{P}}_1$. Here by 'smooth outside $\tilde{B}_1$', we do *not* mean merely smooth at points of $\tilde{\mathfrak{P}}_1$–$\tilde{B}_1$: instead we mean smoothness in the sense of properties (a) and (b) in the proof of the proposition, viz. smoothness after localizing by the ideal $I(\tilde{B}_1)$. This property descends to smoothness for $p_1 : \mathfrak{W}_3 \to \mathfrak{P}_1$, hence for $p_1 : W_3 \to P_1$. Secondly, we prove that $W_3 \cap (P_1 \times_S B_2) \subset B_1 \times_S B_2$. This follows by the same method, by descending a stronger ideal-theoretic property on the $\tilde{}$-schemes. In fact, since for every finite set of y_i's:

$$[S_{y_1}(W_1) \cup \cdots \cup S_{y_k}(W_1)] \cap (\tilde{P}_1 \times_S \tilde{B}_2) \subset \tilde{B}_1 \times_S \tilde{B}_2,$$

it follows that on $\tilde{\mathfrak{P}}_1 \times \tilde{\mathfrak{P}}_2$,

$$I(\mathfrak{W}_2) + I(\tilde{\mathfrak{P}}_1 \times_S \tilde{\mathfrak{B}}_2) \supset I(\tilde{\mathfrak{B}}_1 \times_S \tilde{\mathfrak{B}}_2)^N$$

for some N. This property descends, and on algebraizing, proves that $W_3 \cap (P_1 \times_S B_2) \subset B_1 \times_S B_2$. Since $p_1 : W_3 \to P_1$ is a proper morphism, this proves that the restriction $p_1 : H \to G_1$ is also proper. Combining these 2 halves, it follows that $p_1 : H \to G_1$ is finite and étale. But finally the formal I-adic completion of $\mathfrak{H}$ is obviously the graph of the formal morphism from G_1 to G_2 defined by $\tilde{H}$. Therefore $p_1 : H \to G_1$ has degree 1 over S_0, hence because G_1 is irreducible, it has degree 1 everywhere. This proves that there is an S-morphism α: $G_1 \to G_2$ extending the formal morphism defined by $\tilde{\alpha}$. Finally, since G_1 is irreducible, such an α is clearly determined by its restriction to $\mathfrak{G}_1$. *QED*

(4.7) COROLLARY. *The scheme G depends only on the torus $\tilde{G}$ and the periods Y, and is independent of the polarization ϕ and the relatively complete model $\tilde{P}$.*

PROOF. Apply the theorem to 2 4-tuples $(\tilde{G}, Y, \phi_1, \tilde{P}_1)$ and $(\tilde{G}, Y, \phi_2, \tilde{P}_2)$ and to the identity $1_{\tilde{G}} : \tilde{G} \to \tilde{G}$.

(4.8) COROLLARY. *G is a group scheme over S.*

PROOF. Apply the theorem to $(\tilde{G} \times_S \tilde{G},\ Y \times Y,\ \phi \times \phi,\ \tilde{P} \times_S \tilde{P})$ and $(\tilde{G}, Y, \phi, \tilde{P})$ and the multiplication map from $\tilde{G} \times_S \tilde{G}$ to $\tilde{G}$. This yields a map $\mu \colon G \times_S G \to G$. Apply the theorem to the inverse from $\tilde{G}$ to $\tilde{G}$. This yields a map $i \colon G \to G$. Since μ and i make the formal completion $\mathfrak{G}$ of G into a formal group scheme and since G is irreducible, μ and i satisfy the same identities on G on as $\mathfrak{G}$ and therefore G is a group scheme. *QED*

(4.9) COROLLARY. *G_η is an abelian variety.*

PROOF. Since $\tilde{G}_\eta = \tilde{P}_\eta$, the generic fibre $\tilde{B}_\eta$ of $\tilde{B}$ is empty. It follows that the structure sheaf $\mathcal{O}_{\tilde{B}}$ is killed by some non-zero $\tau \in A$. Therefore the structure sheaf $\mathcal{O}_B$ is also killed by τ. Therefore $B_\eta = \phi$, hence $G_\eta = P_\eta$ is proper over K. Since G is irreducible, G_η is also irreducible, hence an abelian variety. *QED*

Before we can prove that G is, in fact, a semi-abelian group scheme, we have to prove that *every* fibre of G over S is connected. This follows from a description of the points of G of finite order:Start with $\tilde{G}, Y, \phi, \tilde{P}$ again.
Let $G^* = \bigcup_{y \in Y} S_y(\tilde{G}) \subset \tilde{P}$.
Let $\sigma_y \colon S \to G^*$ be the section such that $\sigma_y(\eta) = y$.
Let $Z_y^{(n)}$ be the subscheme of points z such that $nz = y$, i.e., the fibre product:

$$\begin{array}{ccc} Z_y^{(n)} & \hookrightarrow & G^* \\ \downarrow & & \downarrow{\scriptstyle n} \\ S & \underset{\sigma_y}{\hookrightarrow} & G^* \end{array}$$

Note that S_z induces an isomorphism of $Z_y^{(n)}$ with $Z_{y+nz}^{(n)}$.

(4.10) THEOREM. *The kernel $G^{(n)} \subset G$ of multiplication by n is isomorphic over S to the disjoint union of the schemes $Z_y^{(n)}$ as y runs over a set of cosets of Y/nY.*

PROOF. First consider the closure $\bar{Z}_y^{(n)}$ of $Z_y^{(n)}$ in $\tilde{P}$. By the valuation property of the relatively complete model $\tilde{P}$, it follows that all valuations of $\bar{Z}_y^{(n)}$ have centers on $\bar{Z}_y^{(n)}$. By Lemma (3.4), it follows that $\bar{Z}_y^{(n)}$ is proper over S. Let $\overline{\ \ }_y^{(n)}$ be its I-adic completion. We have seen above that if

$$W_1^{(n)} = \text{closure in } \tilde{P} \times_S \tilde{P} \text{ of the graph of } x \mapsto nx$$

then

$$\mathfrak{W}_2^{(n)} = \bigcup_{y \in Y} S_{(0,u)}(\mathfrak{W}_1^{(n)}) \subset \tilde{\mathfrak{P}} \times_S \tilde{\mathfrak{P}}$$

is a locally finite union. Since $\bar{Z}_y^{(n)} \times_S \sigma_y(S) \subset W_1^{(n)}$, hence $\bar{Z}_y^{(n)} \times_S \sigma_0(S) \subset S_{(0,-y)}(W_1^{(n)})$, it follows that

$$\bigcup_{y \in Y} \bar{\mathfrak{Z}}_y^{(n)} \subset \tilde{\mathfrak{P}}$$

is a locally finite union, and that

$$\bigcup_{y \in Y} \bar{\mathfrak{Z}}_y^{(n)} = \mathfrak{W}_2^{(n)} \cap (\tilde{\mathfrak{P}} \times_S \sigma_0(S)).$$

Taking quotients by Y and $Y \times Y$, we obtain a formal closed subscheme $\bigcup_{y \in Y} \bar{\mathfrak{Z}}_y^{(n)}/Y \subset \mathfrak{P}$ such that

$$\bar{\mathfrak{Z}}^{(n)} = \bigcup_{y \in Y} \bar{\mathfrak{Z}}_y^{(n)}/Y = \mathfrak{W}_3^{(n)} \cap (\mathfrak{P} \times_S \sigma_0(S))$$

where $\mathfrak{W}_3^{(n)} = \mathfrak{W}_2^{(n)}/Y \times Y$. It follows that $\bar{\mathfrak{Z}}^{(n)}$ algebrizes to a subscheme $\bar{Z}^{(n)} \subset P$ such that

$$\bar{Z}^{(n)} = W_3^{(n)} \cap (P \times_S \sigma_0(S)),$$

hence $Z^{(n)} = \bar{Z}^{(n)} \cap G = H^{(n)} \cap (G \times_S \sigma_0(s))$ where $H^{(n)} \subset G \times_S G$ is the graph of the morphism $x \mapsto nx$. Thus $Z^{(n)}$ is the kernel in G of multiplication by n. Now for every *finite* subset $Y_0 \subset Y$, we have a formal morphism as follows:

$$\alpha : \bigcup_{y \in Y_0} \bar{\mathfrak{Z}}_y^{(n)} \to \bar{\mathfrak{Z}}^{(n)}.$$

Since these formal schemes are the completions of the (algebraic) schemes $\bigcup_{y \in Y_0} \bar{Z}_y^{(n)}$ and $\bar{Z}^{(n)}$, which are *proper* over S, this formal morphism extends an (algebraic) morphism:

$$\alpha : \bigcup_{y \in Y_0} \bar{Z}_y^{(n)} \to \bar{Z}^{(n)}.$$

If $\tilde{B} = \tilde{P}\text{–}\tilde{G}$ and $B = P\text{–}G$, recall that $\tilde{\mathfrak{B}}$ is the inverse image of $\mathfrak{B}$ in the étale map $\tilde{\mathfrak{P}} \to \mathfrak{P}$. Therefore $\tilde{\mathfrak{B}} \cap \bigcup_y \bar{\mathfrak{Z}}_y^{(n)}$ is the inverse image of $\mathfrak{B} \cap \bar{\mathfrak{Z}}^{(n)}$, hence $\tilde{B} \cap \bigcup_{y \in Y_0} \bar{Z}_y^{(n)}$ is the inverse image of $B \cap \bar{Z}^{(n)}$. Therefore α restricts to a proper morphism:

$$\operatorname{res}(\alpha) : \bigcup_{y \in Y_0} Z_y^{(n)} \to Z^{(n)}.$$

Next, note that

$$\alpha : \bigcup_{y \in Y} \bar{\mathfrak{Z}}_y^{(n)} \to \bar{\mathfrak{Z}}^{(n)}$$

is étale and surjective. It follows that for each fixed $y_0 \in Y$, there is a finite set $Y_0 \subset Y$ such that α: $\bigcup_{y \in Y_0} \bar{\mathfrak{Z}}_y^{(n)} \to \bar{\mathfrak{Z}}^{(n)}$ is étale at all points of

$\overline{\mathfrak{Z}}_{y_0}^{(n)}$, and is surjective. Therefore its algebraization has the same properties. On the other hand, when we intersect with $\tilde{G}$, the union $\bigcup_{y \in Y_0} Z_y^{(n)}$ is disjoint, so it follows that

$$\operatorname{res}(\alpha) : \bigcup_{y \in Y_0} Z_Y^{(n)} \to Z^{(n)}$$

is étale for every Y_0, and surjective for Y_0 big enough.

Now it is clear that the diagram:

$$\begin{array}{ccc} Z_y^{(n)} & \xrightarrow{S_z} & Z_{y+nz}^{(n)} \\ & {}_{\alpha}\searrow \quad \swarrow_{\alpha} & \\ & Z^{(n)} & \end{array}$$

commutes, so, in fact, if Y_0 is a set of coset representatives of Y/nY

$$\operatorname{res}(\alpha) : \bigcup_{y \in Y_0} Z_Y^{(n)} \to Z^{(n)}$$

is already surjective. More than that: identifying formally

$$\bigcup_{y \in Y_0} Z_y^{(n)} \text{ with } \bigcup_{y \in Y} Z_Y^{(n)}/Y$$

we see that $\bigcup_{y \in Y_0} Z_y^{(n)}$ has a natural group scheme structure and it is clear that $\operatorname{res}(\alpha)$ is a homomorphism. Therefore $\operatorname{res}(\alpha)$ has degree 1 everywhere if it has degree 1 over the points $\sigma_0(S) \subset G$, and this will follow if $\operatorname{res}(\alpha)$ has degree 1 over the points $\sigma_0(S_0) \subset G_0$. But it is easy to see that *over* S_0, $Z_y^{(n)} = \phi$ unless $y \in nY$, $Z_0^{(n)}$ is the kernel of n in the part of the torus $\tilde{G}$ over S_0, and $\operatorname{res}(\alpha)$ is just the restriction to the kernel of n of the canonical isomorphism of $\tilde{G}$ and G over S_0, Thus $\operatorname{res}(\alpha)$ has degree 1. *QED*

(4.11) Corollary. *Let $s \in S$. Let*

$$Y_1 = \{y \in Y \mid \mathscr{X}^{\phi(y)}(y) \notin m_{s,S}\}.$$

Then Y_1 is a subgroup of Y (in fact, it is a direct summand) and the kernel of n in G_s fits into an exact sequence:

$$0 \to (\text{ker of } n \text{ in } \tilde{G}_s) \to (\text{ker of } n \text{ in } G_s) \to \frac{1}{n} Y_1/Y_1 \to 0.$$

As n increases, we obtain in the limit an exact sequence

$$0 \to (\text{torsion in } \tilde{G}_s) \to (\text{torsion in } G_s) \to Y_1 \otimes \boldsymbol{Q}/\boldsymbol{Z} \to 0.$$

Proof. This follows immediately from the Theorem and the remark that $Z_y^{(n)}$ has a non-empty fibre over y if and only if $y \in Y_1 + nY$.

(4.12) COROLLARY. *G_s is connected, hence G is semi-abelian.*

PROOF. By Corollary 1, its torsion is p-divisible for every prime p, hence G_s is connected and without unipotent radical.

5. Examples-dim $G/S = 1$

Let's look first at the 1-dimensional case $\tilde{G} = G_m \times S$ where $r = 1$.

Then $X \cong \mathbf{Z}$. Choose such an isomorphism. Then we have a distinguished generator $1 \in X$, and we denote the character $\mathscr{X}^1$ imply by $\mathscr{X}$, so that

$$\tilde{G} = \operatorname{Spec} A[\mathscr{X}, \mathscr{X}^{-1}].$$

All possible ϕ's will be positive multiples of one basic ϕ which is an isomorphism of Y and X. We assume this isomorphism chosen as our ϕ. All periods are multiples of the basic period $y_0 = \phi^{-1}(1)$: let $\tau = \mathscr{X}(y_0)$.

Note that $\tau \in I$. Identifying $\tilde{G}(K)$ with K^*, Y becomes the set of powers $\{\tau^m\}_{m \in \mathbf{Z}}$. The simplest possible Σ consists in the 3 elements $\{1, 0, -1\}$. Then

$$\begin{aligned} R_{\phi,\Sigma} &= A[\cdots, \mathscr{X}^{\phi(ky_0)+\{\varepsilon\}}(ky_0) \cdot \mathscr{X}^{2\phi(ky_0)+\{\varepsilon\}} \cdot \theta, \cdots]_{\varepsilon=0,\pm1} \\ &= A[\cdots, \tau^{k^2+k} \cdot \mathscr{X}^{2k+1} \cdot \theta, \tau^{k^2} \cdot \mathscr{X}^{2k} \cdot \theta, \tau^{k^2-k} \cdot \mathscr{X}^{2k-1} \cdot \theta, \cdots] \end{aligned}$$

with basic automorphism:

$$\begin{aligned} S(\mathscr{X}) &= \tau \cdot \mathscr{X} \\ S(\theta) &= \tau \cdot \mathscr{X}^2 \cdot \theta \end{aligned}$$

One checks easily that

$$\begin{aligned} U_{1,0} &= \operatorname{Spec} A[\tau\mathscr{X}, \mathscr{X}^{-1}] \\ U_{0,0} &= \operatorname{Spec} A[\mathscr{X}, \mathscr{X}^{-1}] \\ U_{-1,0} &= \operatorname{Spec} A[\mathscr{X}, \tau\mathscr{X}^{-1}] \end{aligned}$$

Using the automorphism S, it follows that

$$\begin{aligned} U_{1,ky_0} &= \operatorname{Spec} A[\tau^{k+1}\mathscr{X}, \tau^{-k}\mathscr{X}^{-1}] \\ U_{0,ky_0} &= \operatorname{Spec} A[\tau^{k}\mathscr{X}, \tau^{-k}\mathscr{X}^{-1}] \\ U_{-1,ky_0} &= \operatorname{Spec} A[\tau^{k}\mathscr{X}, \tau^{-k+1}\mathscr{X}^{-1}] \end{aligned}$$

In particular $U_{1,ky_0} = U_{-1,(k-1)y_0}$. One checks immediately that the the closed fibre of $\operatorname{Proj}(R_{\phi,\Sigma})$ is an infinite union of non-singular rational curves, connected in a chain and crossing each other transversely.

We may visualize $\operatorname{Proj}(R_{\phi,\Sigma})$ like this (when A is a discrete valuation ring):

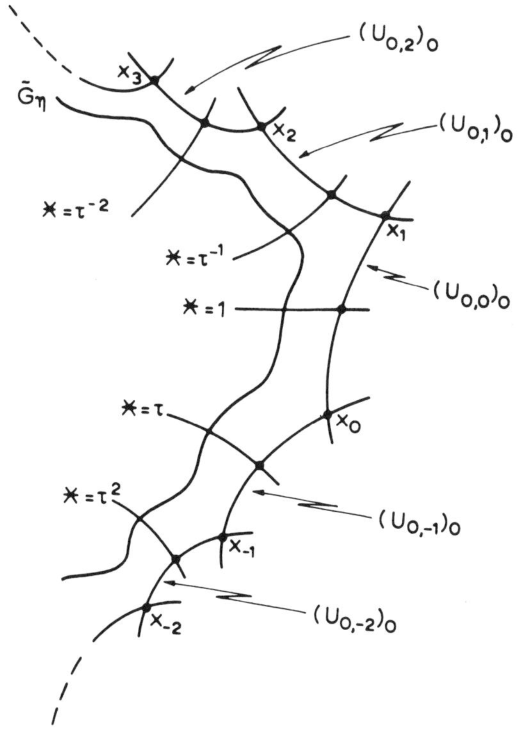

Dividing by Y, we obtain a formal scheme $\mathfrak{P}$ whose closed fibre is an irreducible rational curve with one ordinary double point. P has the same property and looks like this:

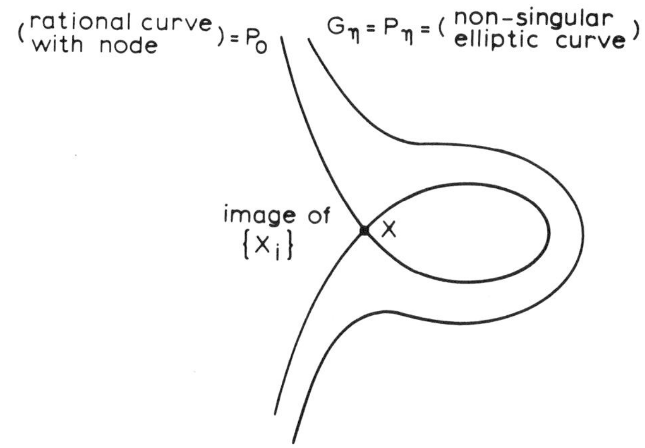

The part S which is to be thrown out will be just the single point x if A is a discrete valuation ring; in general, it will be a section of P through x over the subscheme $(\tau = 0)$ of Spec (A).

6. Examples-dim S = 1

Next, I want to look at higher dimensional tori $\tilde{G}$, but in the case where dim $A = 1$, i.e. A is a complete discrete valuation ring. Let r be the dimension of $\tilde{G}$ over S. In this case, we can describe the models $\tilde{P}$ very intuitively by a polyhedral decomposition of r-dimensional Euclidean space and we can see in this way the degree of choice in $\tilde{P}$ very clearly. We shall limit ourselves to *normal* relative complete models $\tilde{P}$ which admit an affine open covering $\{U_i\}$ such that each U_i is $\tilde{G}$-invariant. (It seems to me likely that every $\tilde{P}$ has such a covering; at least the Proj$(R_{\Phi,\Sigma})$'s do). Our method is this: for every finite algebraic extension $L \supset K$, let A_L be the integral closure of A in L. A_L is again a complete discrete valuation ring. By the completeness property of $\tilde{P}$, every L-valued point ξ of $\tilde{G}$ extends uniquely to an A_L-valued point of $\tilde{P}$. We shall describe various subsets Σ of $\tilde{P}_0$ by describing those ξ whose closure meets Σ.

To begin with, the only thing that will really matter about a ξ is the *rate of growth of the various characters of* $\tilde{G}$ *on* ξ. This can be conveniently measured as follows:

(a) let $E = Hom(X, \mathbf{R})$, $X =$ char. gp. of $\tilde{G}$ (this is a covariant functor of $\tilde{G}$),

(b) $\forall \alpha \in X$, let $l_\alpha : E \to \mathbf{R}$ be the linear functional given by evaluation at α,

(c) fix, once and for all, an embedding of the value group of A in $\mathbf{R}$. This naturally extends to an embedding of the value groups of each A_L in $\mathbf{R}$,

(d) $\forall \xi \in \tilde{G}(L)$, define $||\xi|| \in E$ by $||\xi||(\alpha) = \text{ord}\, \mathscr{X}^\alpha(\xi)$.

Note that for each L, $\{||\xi|| \mid \xi \in \tilde{G}(L)\} = N_L$ is a lattice in E; as L becomes more and more ramified over K, N_L gets denser and denser. Inside N_K, we have the smaller lattice $\{||\xi|| \mid \xi \in Y\}$ induced by the periods. We shall identify this with Y:

$$Y \subset N_K \subset \cdots \subset N_L \subset \cdots \subset E.$$

Now consider the finitely generated normal affine schemes U/S ($S = \text{Spec}\, A$), such that $U_\eta = \tilde{G}_\eta$ and such that $\tilde{G}$ acts on U extending the translation action of $\tilde{G}_\eta$ on U_η. It is easy to check that all such U are all of the type:

$$U = \text{Spec}\, A[\cdots, \pi^{r_i}\mathscr{X}^{\alpha_i}, \cdots]_{1 \leq i \leq N}$$

$$\text{where } (\pi) = \text{maximal ideal of } A$$

$$r_i \in \mathbf{Z}$$

$$\alpha_i \in X.$$

Concerning such U, we have the following:

(6.1) PROPOSITION. *There is a* 1–1 *correspondence between*:

(I) *normal affine schemes* U, *with* $U_\eta = \tilde{G}_\eta$ *and invariant under the action of* $\tilde{G}$, *and*

(II) *closed, bounded polyhedra* $\Delta \subset E$ *with vertices in* $\mathbf{Q} \cdot N_K$.

This is set up by the relation: $\forall L \supset K$, $\forall \xi \in \tilde{G}(L)$,

$$\left\{\begin{array}{l} \xi \text{ extends to an} \\ A_L\text{-valued point of } U \end{array}\right\} \Leftrightarrow \{\|\xi\| \in \Delta\}.$$

PROOF. In fact, if $\Gamma(U, \mathcal{O}_U)$ is generated by $\pi^{r_i}\mathcal{X}^{\alpha_i}$, then define Δ by

$$\Delta = \{x \in E \mid l_{\alpha_i}(x) \geqq -r_i \operatorname{ord}(\pi), 1 \leqq i \leqq N\}.$$

Conversely, given Δ, define

$$U = \operatorname{Spec} A[\cdots, Y_{r,\alpha}, \cdots],$$

with $Y_{r,\alpha} = \pi^r\mathcal{X}^\alpha$ for all (r, α) such that $l_\alpha + r \cdot \operatorname{ord}(\pi)$ is non-negative on Δ. We leave it to the reader to check that these set up inverse maps between the sets (I) and (II) of the Proposition.

(6.2) DEFINITION. Let $U_\Delta = \operatorname{Spec}(R_\Delta)$ be the affine scheme associated to Δ by the Proposition.

By a *closed face* of a polyhedron Δ, we mean Δ itself or a polyhedron $\Delta_1 = H \cap \Delta$, where H is a hyperplane, and Δ lies completely on side of H. By an *open face*, we mean a closed face minus all properly smaller subfaces. It is clear that every polyhedron Δ is a finite disjoint union of its open faces.

(6.3) PROPOSITION. *There is a* 1–1 *correspondence between*:

(I) *orbits* Z *of* $\tilde{G}_0$ *on the closed fibre* $(U_\Delta)_0$ *of* U_Δ, *and*

(II) *open faces* $\sigma \subset \Delta$.

This is set up by the relation: $\forall L \supset K$, $\forall \xi \in \tilde{G}(L)$,

$$(*) \qquad \left\{\begin{array}{l} \xi \text{ extends to a} \\ A_L\text{-valued point meeting } Z \end{array}\right\} \Leftrightarrow \{\|\xi\| \in \sigma\}.$$

Moreover $\dim Z + \dim\sigma = r$.

PROOF. Since the union of the open faces is Δ, and the union of the orbits is $U_\Delta \times \operatorname{Spec}(\bar{k})$ it will suffice to show that for all σ there is an orbit Z satisfying (*). Let $R_\Delta = A[\cdots, \pi^{r_i}\mathcal{X}^{\alpha_i}, \cdots]_{1 \leqq i \leqq N}$. Then

$\Delta = \{x \in E | l_{\alpha_i}(x) + r_i \cdot \text{ord } \pi \geqq 0\}$. Let σ be defined by:

$$l_{\alpha_i}(x) + r_i \cdot \text{ord } \pi = 0, \; i \in I_1$$
$$l_{\alpha_i}(x) + r_i \cdot \text{ord } \pi > 0, \; i \in I_2$$

where $I_1 \cup I_2 = \{1, \cdots, N\}$. Write $\pi^{r_i}\mathscr{X}^{\alpha_i} = Y_i$ for simplicity. Then for all $L \supset K$, and A_L-valued points ξ such that $||\xi|| \in \Delta$, we see that

$$||\xi|| \in \sigma \Leftrightarrow \begin{cases} Y_i\ (\xi) \text{ a unit for } i \in I_1 \\ Y_i\ (\xi) \in \text{max. ideal of } A_L,\ i \in I_2 \end{cases}$$
$$\Leftrightarrow \xi \text{ meets } Z$$

where $Z \subset (U_\Delta)_0$ is defined by $Y_i \neq 0$, $i \in I_1$ (resp. $= 0$, $i \in I_2$). Now Z is clearly $\tilde{G}_0$-invariant. Moreover Z is easily checked to be the locus in the affine Y_i-space ($i \in I_1$) defined by:

i) $Y_i \neq 0$, all $i \in I_1$

ii) $\Pi Y_i^{s_i} = 1$ whenever $s_i \in \mathbf{Z}$, $s_i \geqq 0$ satisfy $\sum s_i \alpha_i = 0$ (Since $\sigma \neq \emptyset$, $\sum s_i \alpha_i = 0 \Rightarrow \Sigma s_i r_i = 0$!).

If $\tilde{H} \subset \tilde{G}$ is the subtorus defined by $\mathscr{X}^{\alpha_i} = 1$, $i \in I_1$, then it follows immediately that Z is isomorphic to $\tilde{G}_0/\tilde{H}_0$ if a point of Z with coordinates $Y_i = a_i$ is associated by a point of G_0/H_0 with coordinate $\mathscr{X}^{\alpha_i} = a_i$.

(6.4) Corollary. *$(U_\Delta)_0$ is a finite union of orbits, hence each component contains a unique open orbit, and the set of components is in* 1–1 *correspondence with the vertices of Δ.*

(6.5) Corollary. *If Δ_1, Δ_2 are polyhedra, then U_{Δ_1} is an open subset of U_{Δ_2} if and only if Δ_1 is a closed face of Δ_2.*

Proof. By the Prop., it is certainly necessary that Δ_1 be a closed face of Δ_2. Conversely, if Δ_1 is a closed face of Δ_2, then $\Delta_1 = \Delta_2 \cap H$ where H is a hyperplane defined by $l_\alpha(X) + r \cdot \text{ord } \pi = 0$ and $l_\alpha(x) + r \cdot \text{ord } \pi \geqq 0$ on Δ_2. Then $\pi^r \cdot \mathscr{X}^\alpha \in \Gamma(\mathcal{O}_{\Delta_2})$ and U_{Δ_1} is the affine open subset of U_{Δ_2} defined by $\pi^r \cdot \mathscr{X}^\alpha \neq 0$.

(6.6) Corollary. *There is a* 1–1 *correspondence between:*

(I) *normal schemes $\tilde{P}$ locally of finite type over S such that*

(a) $\tilde{P}_\eta = \tilde{G}_\eta$,

(b) *the translation action of $\tilde{G}$ extends to $\tilde{P}$ and P is covered by $\tilde{G}$-invariant affine open sets, and*

(c) *for all valuations v on $\mathbf{R}(\tilde{G})$ if $v \geqq 0$ on A and if* $[\forall \alpha \in X, \exists n$ *such that* $n \cdot v(\pi) \geqq v(\mathscr{X}^\alpha) \geqq -nv(\pi)]$ *hold, then v has a centre on $\tilde{P}$, and*

(II) *Polyhedral decompositions of E, i.e. a set of polyhedron* $\Delta_\alpha \subset E$ *such that* $\bigcup \Delta_\alpha = E$, *every closed face of a* Δ_α *is a* Δ_β, $\Delta_\alpha \cap \Delta_\beta \neq \phi \Rightarrow \Delta_\alpha \cap \Delta_\beta = \Delta_\gamma$ *(some* γ*), all of whose vertices are in* $\boldsymbol{Q} \cdot \boldsymbol{N}_K$.

PROOF. Clear

In the 1–1 correspondence of this last corollary, certain properties of $\tilde{P}$ carry over nicely to properties of $\{\Delta_\alpha\}$:

(A) $\tilde{P} \supset \tilde{G}$ if and only one of the Δ_α's is the origin $\{0\}$.

(B) If $Y \subset \tilde{G}(K)$ is a set of periods, the action of Y on $\tilde{G}_\eta$ extends to $\tilde{P}$ if and only if the polyhedral decomposition is Y-invariant, i.e. $\forall \alpha$, y, $\Delta_\alpha + y =$ some Δ_β.

(C) $\tilde{P}$ is smooth over A at all generic points of $\tilde{P}_0$ if and only if the vertices of all Δ_α are points of N_K.

(D) $\tilde{P}$ is regular everywhere and smooth A at all generic points of $\tilde{P}_0$ if and only if the vertices of all Δ_α are in N_K, each Δ_α of dimension r is an r-simplex with volume $1/r!$ in a coordinate system in E making N_K into the integral points.

Here (A), (B) and (C) are almost immediate, and we omit the proof of (D), which is harder, because we do not need it. So far, there appears to be tremendous choice in constructing a relatively complete model $\tilde{P}$. However, the existence of an ample $\tilde{L}$ corresponding to a polarization $\phi\colon Y \to X$ puts very strong restrictions on the polyhedral decomposition $\{\Delta_\alpha\}$. The exact conditions on $\{\Delta_\alpha\}$ for the existence of $\tilde{L}$ are very messy, so we will only state, without proof, one partial result. First we must recall some constructions used in sphere packing problems. For all the following, cf. Rogers, *Packing and Covering*, Camb. Univ. Press, Ch. 7, § 1 and Ch. 8, § 1. Assume given a Euclidean metric on E. Start with a discrete set of points $\Sigma \subset E$ such that for some r, $E =$ union of balls around points of Σ with radius r. We may now construct 2 canonical polyhedral decompositions E:

(A) the *Voronoi decomposition*: $\forall \sigma \in \Sigma$, let

$$\Delta_\sigma = \{y \in E \mid \|y - \sigma\| \leqq \|y - \tau\|, \text{ all } \tau \in \Sigma\}.$$

These are the top dimensional polyhedra. We get their faces by looking, for all $\sigma_1, \cdots, \sigma_k \in \Sigma$, at:

$$\Delta_{\sigma_1, \cdots, \sigma_k} = \left\{ y \in E \;\middle|\; \begin{array}{c} \|y-\sigma_1\| = \cdots = \|y-\sigma_k\| \leqq \|y-\tau\| \\ \text{all } \tau \in \Sigma \end{array} \right\}.$$

(B) *the Delaunay decomposition*: For all $\sigma_1, \cdots, \sigma_k$ such that

$\Delta_{\sigma_1,\cdots,\sigma_k} \neq \phi$ and such that for y in its interior,

$$||y-\sigma_1|| = \cdots = ||y-\sigma_k|| < ||y-\tau|| \ (\tau \in \Sigma, \tau \neq \sigma_i),$$

let $\Delta^* =$ convex hull of $\sigma_1, \cdots, \sigma_k$. These polyhedra Δ^* form a decomposition of E whose vertices are exactly the points of Σ.

Now returning to our tori, let $Y \subset \tilde{G}(K)$ be a set of periods and let $\phi: Y \to X$ be a polarization. The quadratic form

$$||y||^2 = \text{ord } \mathscr{X}^{\phi(y)}(y), \ y \in Y$$

extends rationally to a Euclidean norm $|| \ ||$ on E. We find:

(6.7) PROPOSITION. *Let $\tilde{G}$, Y and ϕ be given as above. Let $\tilde{G} \subset \tilde{P}$ be given where $\tilde{P}$ is a normal scheme locally of finite type over S such that*

i) $\tilde{P}_\eta = \tilde{G}_\eta$,

ii) *$\tilde{G}$ and Y act on $\tilde{P}$,*

iii) *P has the valuative completeness property of Corollary* (6.6),

iv) *$\tilde{P}$ is covered by $\tilde{G}$-invariant affine open sets.*

Let $\tilde{P}$ correspond to the polydedral decomposition $\{\Delta_\alpha\}$ and let Σ be the 0*-simplices in $\{\Delta_\alpha\}$. If $\{\Delta_\alpha\}$ is the Delaunay decomposition of E associated to Σ, with the Euclidean norm $|| \ ||$ defined by ϕ, then there exists an ample invertible sheaf $\tilde{L}$ on $\tilde{P}$, plus an action of $\tilde{G}$ and Y on $\tilde{L}$, making $\tilde{P}$ into a relatively complete model corresponding to the polarization $n \cdot \phi$ (some $n \geqq 1$).*

(Proof omitted)

7. A final example

I want to give one final example which

(i) illustrates the situation where both the base and the fibre are more than one dimensional, and

(ii) is literally the keystone in the compactification of the moduli space of 2-dimensional abelian varieties.

This example was proposed by Deligne before we saw the general theory. Take as the ground ring:

$$A = k[[a, b, c]]$$

and take $\tilde{G}$ to be 2-dimensional. To express the symmetry of the situation, it is convenient not to introduce a basis of the character group X of $\tilde{G}$, but rather 3 generators $\mathscr{X}$, $\mathscr{Y}$, $\mathscr{Z}$ for X, related by one identity:

$$\mathscr{X} \cdot \mathscr{Y} \cdot \mathscr{Z} \equiv 1$$

The group of periods Y will similarly be generated by the 3 periods r, s, t with $r+s+t = 0$, given by

$$\begin{array}{lll} \mathscr{X}(r) = bc, & \mathscr{Y}(r) = c^{-1}, & \mathscr{Z}(r) = b^{-1}, \\ \mathscr{X}(s) = c^{-1}, & \mathscr{Y}(s) = ac, & \mathscr{Z}(s) = a^{-1}, \\ \mathscr{X}(t) = b^{-1}, & \mathscr{Y}(t) = a^{-1}, & \mathscr{Z}(t) = ab. \end{array}$$

The polarization ϕ will be defined by

$$\begin{aligned} \phi(r) &= \mathscr{X} \\ \phi(s) &= \mathscr{Y} \\ \phi(t) &= \mathscr{Z} \end{aligned}$$

and Σ will consist in the seven characters:

$$\Sigma = \{1, \mathscr{X}, \mathscr{X}^{-1}, \mathscr{Y}, \mathscr{Y}^{-1}, \mathscr{Z}, \mathscr{Z}^{-1}\}.$$

The resulting ring $R_{\phi,\Sigma}$ will have some automorphisms, in addition to those given by translation with respect to Y:

a) an automorphism σ such that $\sigma^3 = \mathrm{id}$,

$$\begin{aligned} &\sigma(a) = b, \sigma(b) = c, \sigma(c) = a \\ &\sigma(\mathscr{X}) = \mathscr{Y}, \sigma(\mathscr{Y}) = \mathscr{Z}, \sigma(\mathscr{Z}) = \mathscr{X} \\ &\sigma(\theta) = \theta \end{aligned}$$

b) an involution $\tau(\tau^2 = \mathrm{id})$,

$$\begin{aligned} &\tau = \mathrm{id} \text{ on } A \\ &\tau(\mathscr{X}) = \mathscr{X}^{-1}, \tau(\mathscr{Y}) = \mathscr{Y}^{-1}, \tau(\mathscr{Z}) = \mathscr{Z}^{-1} \\ &\tau(\theta) = \theta. \end{aligned}$$

Now Proj $(R_{\phi,\Sigma})$ is covered by the $U_{\alpha,y}$'s with $\alpha \neq 0$ and these are all isomorphic under these various automorphisms. We must calculate the ring of one of them. Look at $U_{\mathscr{X},0}$; its ring is generated by:

$$\mathscr{X}^{\phi(y)+\alpha}(y) \cdot \mathscr{X}^{2\phi(y)+\alpha} \cdot \mathscr{X}^{-1}, \text{ all } y \in Y, \alpha \in \Sigma.$$

One then checks easily that:

$$\begin{aligned} U_{\mathscr{X},0} &= \operatorname{Spec} A[\mathscr{Y}, c\mathscr{Y}^{-1}, \mathscr{Z}, b\mathscr{Z}^{-1}] \\ &\cong \operatorname{Spec} A[X_1, X_2, X_3, X_4]/(X_1 X_2 - c, X_3 X_4 - b). \end{aligned}$$

In particular, this is regular, so $\mathrm{Proj}(R_{\phi,\Sigma})$, $\mathfrak{P}$ and P are all regular 5-dimensional schemes. Moreover, $(U_{\mathscr{X},0})_0$ has 4 components which are the orbits of $\tilde{G}_0$ containing the 4 sections:

$$\begin{aligned} (\mathscr{X}, \mathscr{Y}, \mathscr{Z}) &= (1, 1, 1) \\ &= (c^{-1}, c, 1) \end{aligned}$$

$$= (b^{-1}, 1, b)$$
$$= (b^{-1}c^{-1}, c, b).$$

Applying the automorphisms σ, τ and translation periods, it is not hard to construct the whole closed fibre of $\mathrm{Proj}(R_{\phi,\Sigma})$. We give the result in the following diagram (strangely isometric to the root diagram of G_2). Each triple stands for the component which is the $\tilde{G}_0$-orbit of the closed point on the section defined by that triple; 2 components meet along a curve if they are joined by a line. For each diamond, there is one point at which the 4 components corresponding to its 4 vertices meet. The components indicated by triples in square brackets are projective planes. each of them meets 3 other components along the three lines of a triangle, The components indicated by triples in round brackets are projective planes with 3 non-collinear points blown up; the 3 exceptional lines, plus the 3 lines joining pairs of blown-up points, form a hexagon along which these components meet 6 neighbouring components.

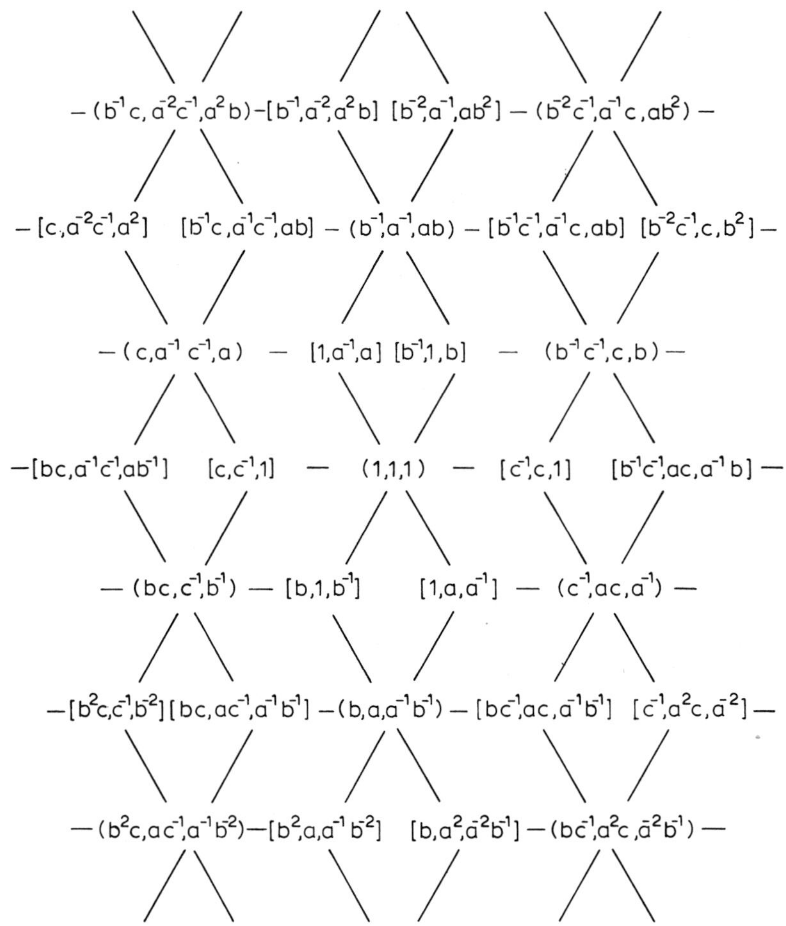

To find the structure of the closed fibre of P, we must divide by the action of Y. All the components of $\mathrm{Proj}(R_{\phi,\Sigma})_0$ collapse to 3 components: one isomorphic to the three times blown-up projective plane, which is the image of $(1, 1, 1)$; two isomorphic to projective planes which are the images, for example, of $[1, a^{-1}, a]$ and $[1, a, a^{-1}]$ respectively. They are glued as indicated below (I).

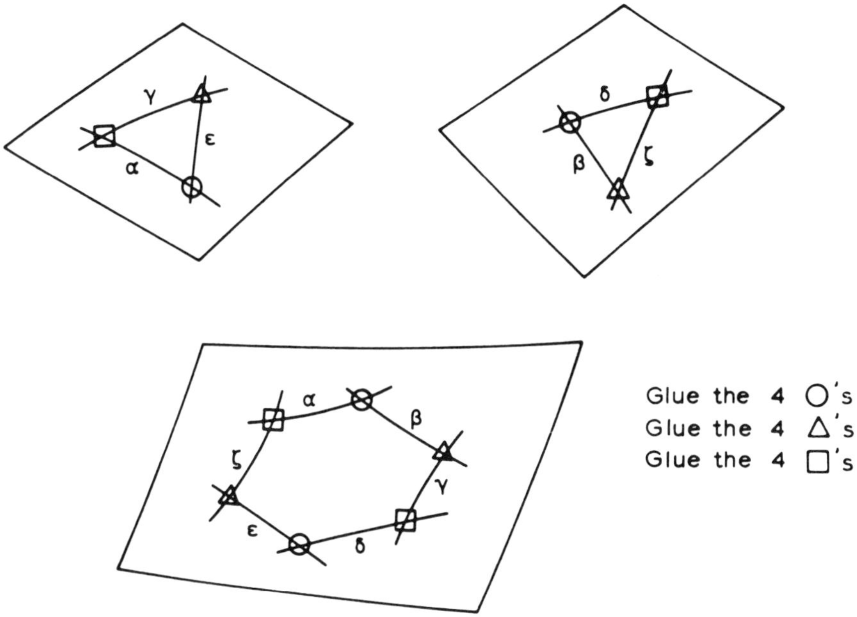

The other fibres of P over Spec (A) have the following structure: over any point x where $a \cdot b \cdot c \neq 0$, the fibre is a 2-dimensional abelian variety. Over a point where *only one* of the 3 coordinates a, b, c is 0, the fibre is a $\boldsymbol{P}^1$-bundle over an elliptic curve glued to itself along 2 disjoint sections of this fibration but with a shift (cf. illustration III). Over a point where exactly *two* of the coordinates a, b, c are 0, the fibre is $\boldsymbol{P}^1 \times \boldsymbol{P}^1$ glued to itself along 2 pairs of lines (cf. illustration II).

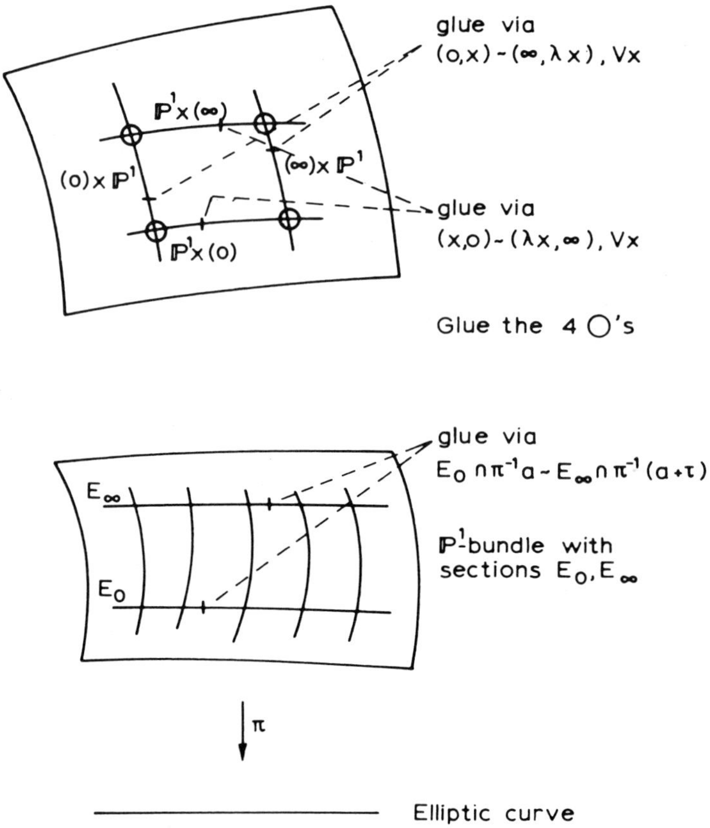

(Oblatum 16-III-1971)

Department of Mathematics,
Harvard University,
Cambridge,
Mass.,
U.S.A.

Volume 15: **T. Oda**

Convex Bodies and Algebraic Geometry

An Introduction to the Theory of Toric Varieties

1987. VIII, 212 pp. 42 figs. ISBN 3-540-17600-4

Volume 16: **G. van der Geer**

Hilbert Modular Surfaces

1988. IX, 291 pp. 39 figs. ISBN 3-540-17601-2

Volume 17: **G. A. Margulis**

Discrete Subgroups of Semisimple Lie Groups

1990. IX, 388 pp. ISBN 3-540-12179-X

Volume 18: **A. E. Brouwer, A. M. Cohen, A. Neumaier**

Distance-Regular Graphs

1989. XVII, 495 pp. ISBN 3-540-50619-5

Volume 19: **I. Ekeland**

Convexity Methods in Hamiltonian Mechanics

1990. X, 247 pp. 4 figs. ISBN 3-540-50613-6

Contents: Introduction. - Linear Hamiltonian Systems. - Convex Hamiltonian Systems. - Fixed-Period Problems: The Sublinear Case. - Fixed-Period Problems: The Superlinear Case. - Fixed-Energy Problems. - Open Problems. - Bibliography. - Index.

Volume 20: **A. I. Kostrikin**

Around Burnside

1990. XII, 255 pp. ISBN 3-540-50602-0

Contents: Preface. - Introduction. - The Descent to Sandwiches. - Local Analysis to thin Sandwiches. - Proof of the Main Theorem. - Evolution of the Method of Sandwiches. - The Problem of Global Nilpotency. - Finite p-Groups and Lie Algebras. - Appendix I. - Appendix II. - Epilogue. - References. - Author Index. - Subject Index. - Notation.

Volume 21: **S. Bosch, W. Lütkebohmert, M. Raynaud**

Néron Models

1990. X, 325 pp. 4 figs. ISBN 3-540-50587-3

Contents: Introduction. - What Is a Néron Model? - Some Background Material from Algebraic Geometry. - The Smoothening Process. - Construction of Birational Group Laws. - From Birational Group Laws to Group Schemes. - Descent. - Properties of Néron Models. - The Picard Functor. - Jacobians of Relative Curves. - Néron Models of Not Necessarily Proper Algebraic Groups. - Bibliography. - Subject Index.

Volume 22: **G. Faltings, C.-L. Chai**

Degeneration of Abelian Varieties

1990. XII, 316 pp. ISBN 3-540-52015-5

Volume 23: **M. Ledoux, M. Talagrand**

Probability in Banach Spaces

Isoperimetry and Processes

1991. Approx. 500 pp. ISBN 3-540-52013-9

Springer-Verlag
Berlin
Heidelberg
New York
London
Paris
Tokyo
Hong Kong
Barcelona

Ergebnisse der Mathematik und ihrer Grenzgebiete, 3. Folge

A Series of Modern Surveys in Mathematics

Volume 1: **A. Fröhlich**
Galois Module Structure of Algebraic Integers
1983. X, 262 pp. ISBN 3-540-11920-5

Volume 2: **W. Fulton**
Intersection Theory
1984. XI, 470 pp. ISBN 3-540-12176-5

Volume 3: **J. C. Jantzen**
Einhüllende Algebren halbeinfacher Lie-Algebren
1983. V, 298 S. ISBN 3-540-12178-1

Volume 4: **W. Barth, C. Peters, A. van de Ven**
Compact Complex Surfaces
1984. X, 304 pp. ISBN 3-540-12172-2

Volume 5: **K. Strebel**
Quadratic Differentials
1984. XII, 184 pp. 74 figs. ISBN 3-540-13035-7

Volume 6: **M. J. Beeson**
Foundations of Constructive Mathematics
Metamathematical Studies
1985. XXIII, 466 pp. ISBN 3-540-12173-0

Volume 7: **A. Pinkus**
n-Widths in Approximation Theory
1985. X, 291 pp. ISBN 3-540-13638-X

Volume 8: **R. Mañé**
Ergodic Theory and Differentiable Dynamics
Translated from the Portuguese by Silvio Levy
1987. XII, 317 pp. 32 figs. ISBN 3-540-15278-4

Volume 9: **M. Gromov**
Partial Differential Relations
1986. IX, 363 pp. ISBN 3-540-12177-3

Volume 10: **A. L. Besse**
Einstein Manifolds
1986. XII, 510 pp. 22 figs. ISBN 3-540-15279-2

Volume 11: **M. D. Fried, M. Jarden**
Field Arithmetic
1986. XVII, 458 pp. ISBN 3-540-16640-8

Volume 12: **J. Bochnak, M. Coste, M.-F. Roy**
Géométrie algébrique réelle
1987. X, 373 pp. 44 figs. ISBN 3-540-16951-2

Volume 13: **E. Freitag, R. Kiehl**
Etale Cohomology and the Weil Conjecture
With an Historical Introduction by J. A. Dieudonné
Translated from the German Manuscript
1987. XVIII, 317 pp. ISBN 3-540-12175-7

Volume 14: **M. R. Goresky, R. D. MacPherson**
Stratified Morse Theory
1988. XIV, 272 pp. 84 figs.
ISBN 3-540-17300-5

Springer-Verlag
Berlin
Heidelberg
New York
London
Paris
Tokyo
Hong Kong
Barcelona